T0305617

Forecasting and Analytics with the Augmented Dynamic Adaptive Model (ADAM)

Forecasting and Analytics with the Augmented Dynamic Adaptive Model (ADAM) focuses on a time series model in Single Source of Error state space form, called "ADAM" (Augmented Dynamic Adaptive Model). The book demonstrates a holistic view to forecasting and time series analysis using dynamic models, explaining how a variety of instruments can be used to solve real life problems. At the moment, there is no other tool in R or Python that would be able to model both intermittent and regular demand, would support both ETS and ARIMA, work with explanatory variables, be able to deal with multiple seasonalities (e.g. for hourly demand data) and have a support for automatic selection of orders, components and variables and provide tools for diagnostics and further improvement of the estimated model. ADAM can do all of that in one and the same framework. Given the rising interest in forecasting, ADAM, being able to do all those things, is a useful tool for data scientists, business analysts and machine learning experts who work with time series, as well as any researchers working in the area of dynamic models.

Key Features:

- It covers basics of forecasting,
- It discusses ETS and ARIMA models,
- It has chapters on extensions of ETS and ARIMA, including how to use explanatory variables and how to capture multiple frequencies,
- It discusses intermittent demand and scale models for ETS, ARIMA and regression,
- It covers diagnostics tools for ADAM and how to produce forecasts with it,
- It does all of that with examples in R.

Ivan Svetunkov is a Lecturer of Marketing Analytics at Lancaster University, UK and a Marketing Director of Centre for Marketing Analytics and Forecasting. He has PhD in Management Science from Lancaster University and a candidate degree in economics from Saint Petersburg State University of Economics and Finance, Russia. His areas of interests includes statistical methods of analytics and forecasting, focusing on demand forecasting in healthcare, supply chain and retail. He is a creator and a maintainer of several forecasting and analytics-related R packages, such as greybox, smooth and legion.

Forecasting and Analytics with the Augmented Dynamic Adaptive Model (ADAM)

Ivan Svetunkov

CRC Press is an imprint of the
Taylor & Francis Group, an **informa** business
A CHAPMAN & HALL BOOK

Designed cover image: © Shutterstock ID 1306244494, Vector Graphics Contributor PayPau

First edition published 2024
by CRC Press
2385 NW Executive Center Drive, Suite 320, Boca Raton FL 33431

and by CRC Press
4 Park Square, Milton Park, Abingdon, Oxon, OX14 4RN

CRC Press is an imprint of Taylor & Francis Group, LLC

Library of Congress Cataloging-in-Publication Data

Names: Svetunkov, Ivan, author.
Title: Forecasting and analytics with the augmented dynamic adaptive model (ADAM) / Ivan Svetunkov.
Description: First edition. | Boca Raton, FL : CRC Press, 2024. | Includes bibliographical references.
Identifiers: LCCN 2023024208 (print) | LCCN 2023024209 (ebook) | ISBN 9781032590370 (hardback) | ISBN 9781032589206 (paperback) | ISBN 9781003452652 (ebook)
Subjects: LCSH: Box-Jenkins forecasting. | Trend surface analysis. | Error analysis (Mathematics) | Adaptive sampling (Statistics) | Seasonal variations (Economics)
Classification: LCC QA280 .S94 2024 (print) | LCC QA280 (ebook) | DDC 511/.43--dc23/eng/20230821
LC record available at https://lccn.loc.gov/2023024208
LC ebook record available at https://lccn.loc.gov/2023024209

ISBN: 978-1-032-59037-0 (hbk)
ISBN: 978-1-032-58920-6 (pbk)
ISBN: 978-1-003-45265-2 (ebk)

DOI: 10.1201/9781003452652

Typeset in Latin Modern font
by KnowledgeWorks Global Ltd.

Publisher's note: This book has been prepared from camera-ready copy provided by the authors.

To the almighty Flying Spaghetti Monster,

without whom this book would have been much more boring. Ramen!

Contents

List of Tables

List of Figures

Preface

Who is this book for?

This monograph assumes that the reader has a good understanding of statistics and a good knowledge of regression analysis. Some chapters rely on more advanced topics, including likelihood and information criteria. These are explained, for example, in Svetunkov (2022). Some parts of the book also use linear algebra to show how various properties of state space models are derived. However, these can be skipped by a reader who is not comfortable with higher maths without losing in the understanding of the main idea of models. Furthermore, while the monograph is self-sufficient to a large extent, it does not contain a detailed discussion of special cases of state space models (e.g. the local level model is only discussed in passing in Section 4.3, while the local trend is covered in Section 4.4). The reader interested in knowing more about some special cases is referred to the Hyndman, Koehler, Ord, and Snyder (2008) monograph. Finally, while the monograph starts with the basics of forecasting and explains some of the simple forecasting techniques, it does not cover these in detail. A good book covering the fundamentals of forecasting is Ord, Fildes, and Kourentzes (2017).

In a more wide sense, this book is written for data scientists, computer scientists, and data, business, and marketing analysts who work in forecasting or want to learn more about univariate forecasting using statistical models.

What is ADAM?

ADAM stands for "Augmented Dynamic Adaptive Model". The term "adaptive model" means that the parameters of the model change over time according to some assumed process. The word "dynamic" reflects the idea that the model has time series related components (ETS, ARIMA). Finally, the word "augmented" is included because ADAM is the model that supports additional features not included in the conventional ETS/ARIMA. ADAM is a unified framework for constructing ETS/ARIMA/regression, based on more advanced statistical instruments. For example, classical ARIMA is built on the assumption of normality of the error term, but ADAM lifts this assumption and allows for using other distributions as well (e.g. Generalised Normal, Inverse Gaussian, etc). Another example: typically the conventional models are estimated either via the maximisation of the likelihood function or using basic losses like MSE or MAE (see Section 2.1), but ADAM includes a broader spectrum of losses and allows using custom ones. There is much more, and we will discuss different aspects of ADAM in detail later in this monograph. Here is a brief list of ADAM features:

1. ETS;
2. ARIMA;
3. Regression;
4. TVP regression;
5. Combination of (1), (2), and either (3) or (4);
6. Automatic selection/combination of states for ETS;
7. Automatic orders selection for ARIMA;
8. Variables selection for regression;
9. Normal and non-normal distributions;
10. Automatic selection of most suitable distributions;
11. Multiple seasonality;
12. Occurrence part of the model to handle zeroes in data (intermittent demand);
13. Modelling scale of distribution (GARCH and beyond);
14. Handling uncertainty of estimates of parameters.

All these extensions are needed to solve specific real life problems, so we will include examples and case studies later in the book to see how all of this can be used. The `adam()` function from the `smooth` package for R implements ADAM and supports the following features:

1. Model diagnostics using `plot()` and other methods;
2. Confidence intervals for parameters of models;
3. Automatic outliers detection;
4. Handling missing data;
5. Fine-tuning of persistence vector (smoothing parameters);
6. Fine-tuning of initial values of the state vector (e.g. level / trend / seasonality);
7. Two initialisation options (optimal / backcasting);
8. Advanced and custom loss functions;
9. Manual parameters for ETS, ARMA, and regression;
10. Fine-tuning of the optimiser (selection of optimisation algorithm and convergence criteria);
11. Scale model via the `sm()` method.

This monograph uses two packages from R, namely `greybox`, which focuses on forecasting using regression models, and `smooth`, which implements Single Source of Error (SSOE) state space models for time series analysis and forecasting. The monograph focuses on explaining how ADAM, one of the `smooth` functions (introduced in v3.0.0), works, also showing how it can be used in practice with examples from R.

If you want to run examples from the monograph, two R packages are needed (Svetunkov, 2023a, 2023b):

```
install.packages("greybox")
install.packages("smooth")
```

Some explanations of functions from the packages are given in my blog: Package greybox for R[1], Package smooth for R[2].

An important thing to note is that this monograph **does not use `tidyverse` packages**. I like base R, and, to be honest, I am sure that `tidyverse` packages are great, but I have never needed them in my research. So, I will not use pipeline operators, `tibble` or `tsibble` objects and `ggplot2`. I assume throughout the monograph that you can do all those nice tricks on your own if you want to.

How to cite ADAM

You can use the following to cite the monograph:

Svetunkov, I. (2023). Forecasting and Analytics with the Augmented Dynamic Adaptive Model (ADAM) (1st ed.). Chapman and Hall/CRC. `https://doi.org/10.1201/9781003452652`

If you use LaTeX, the following can be used instead:

```
@book{SvetunkovAdam,
    author = {Ivan Svetunkov},
    isbn = {978-1-003-45265-2},
    publisher = {Chapman and Hall/CRC},
    title = {Forecasting and Analytics with
        the Augmented Dynamic Adaptive Model (ADAM)},
    doi = {10.1201/9781003452652},
    url = {https://openforecast.org/adam/},
    year = {2023}
}
```

License

This monograph is licensed under Creative Common License by-nc-sa 4.0[3], which means that you can share, copy, redistribute, and remix the content of the monograph for non-commercial purposes as long as you give appropriate

---

[1] `https://forecasting.svetunkov.ru/en/category/r-en/greybox/`

[2] `https://forecasting.svetunkov.ru/en/category/r-en/smooth/`

[3] `https://creativecommons.org/licenses/by-nc-sa/4.0/`

Acknowledgments

I would like to thank Robert Fildes for his comments about this book and helping in shaping it into a monograph and J. Keith Ord for his invaluable suggestions for improvement of the monograph. I also would like to thank Tobias Schmidt and Benedikt Sonnleitner for correcting grammatical mistakes.

[4]https://creativecommons.org/licenses/by-nc-sa/4.0/

About the author

Ivan Svetunkov is a Lecturer of Marketing Analytics at Lancaster University[5], UK and a Marketing Director of the Centre for Marketing Analytics and Forecasting[6]. He has a PhD in Management Science from Lancaster University and a candidate degree in economics from Saint Petersburg State University of Economics and Finance, Russia. His areas of interests include statistical methods of analytics and forecasting, focusing on demand forecasting in healthcare, supply chains and retail. He is a creator and a maintainer of several forecasting- and analytics-related R packages, such as `greybox` (Svetunkov, 2023a), `smooth` (Svetunkov, 2023b), and `legion` (Svetunkov & Pritularga, 2023b).

[5] https://www.lancaster.ac.uk/lums/people/ivan-svetunkov
[6] https://www.lancaster.ac.uk/cmaf/

1

Introduction

I started writing this monograph in 2020 during the COVID-19 pandemic, having figured out that I was bored to death in isolation and needed to do something useful. By that time, I had done a substantial amount of work in the area of dynamic models and tried to publish several papers on statistical models developed in the Single Source of Error (SSOE) framework. I had even developed several R functions based on my own ideas, which were all available in the `smooth` package and attracted substantial interest in the forecasting and data science communities, but had not been published anywhere. Furthermore, a friend of mine, Nikos Kourentzes, had been telling me that when he used my functions, he could not reference them properly because of the lack of publications from my side. So, it became apparent that I needed to either publish lots of papers, covering different small aspects of what I had done, or write a monograph that would summarise everything in one place. Feeling lonely and depressed because of the lockdown, I chose the second option.

At this stage, I should mention that all my ideas rely on the framework from the monograph of Hyndman et al. (2008), which I have modified and upgraded. Their original book discussed the ETS (Error-Trend-Seasonality) model in the SSOE form, but I have decided to expand it, introducing more features that are required in day-by-day demand forecasting. So, for example, while the original ETS works very well on monthly, quarterly, and yearly data, my modifications support high frequency and/or intermittent data, work with explanatory variables, and overall represent a holistic view on demand forecasting in practice.

However, before we move to the discussion of the framework, I should point out that many parts of this monograph rely on such topics as scales of information, model uncertainty, likelihood, information criteria, and model building. All these topics are discussed in detail in the online lecture notes of Svetunkov (2022). It is recommended that you familiarise yourself with them before moving to ADAM's more advanced topics.

In this chapter, I explain what forecasting is, how it is different from planning and analytics, and what the main forecasting principles are. All these aspects will help you not to fail in trying to predict the future.

1.1 Forecasting, planning, and analytics

While there are many definitions of what a forecast is, I like the following one, proposed by Sergey Svetunkov (translated by me from Russian into English from Svetunkov & Svetunkov, 2014): **A forecast is a scientifically justified assertion about possible states of an object in the future.** This definition has several important elements. First, it does not have the word "probability" in it, because in some cases, forecasts do not rely on rigorous statistical methods and theory of probabilities. For example, the Delphi method allows obtaining judgmental forecasts, typically focusing on what to expect, not on the probability side. Second, an essential word in the definition is "**scientific**". If a prediction is made based on coffee grounds, then it is not a forecast. Judgmental predictions, on the other hand, can be considered forecasts if a person has a reason behind them. If they do not, this exercise should be called "foretelling", not forecasting. Finally, the word "future" is important as well, as it shows the focus of the discipline: without the future, there is no forecasting, only overfitting. As for the definition of **forecasting**, it is a process of producing forecasts – as simple as that.

Forecasting is a vital activity carried out by many organisations, some of which do it unconsciously or label it as "demand planning" or "predictive analytics". However, there is a difference between the terms "forecasting" and "planning". The latter relies on the former and implies the company's actions to adjust its decisions. For example, if we forecast that the sales will go down, a company may make some marketing decisions to increase the demand on the product. The first part relates to forecasting, while the second relates to planning. If a company does not like a forecast, it should change something in its activities, not in the forecast itself. It is important not to confuse these terms in practice.

Another crucial thing to keep in mind is that any forecasting activity should be done to inform decisions. Forecasting for the sake of forecasting is pointless. Yes, we can forecast the overall number of hospitalisations due to SARS-CoV-2 virus in the world for the next decade, but what decisions can be made based on that? If there are some decisions, then this exercise is worthwhile. If not, then this is just a waste of time.

Example 1.1. Retailers typically need to order some amount of milk that they will sell over the next week. They do not know how much they will sell, so they usually order, hoping to satisfy, let us say, 95% of demand. This situation tells us that the forecasts need to be made a week ahead, they should be cumulative (considering the overall demand during a week before the following order), and that they should focus on an upper bound of a 95% prediction interval. Producing only point forecasts would not be helpful in this situation.

Related to this is the question of forecast accuracy. In reality, accurate forecasts do not always translate to good decisions. This is because many different aspects of reality need to be taken into account, and forecasting focuses only on one of them. Capturing the variability of demand correctly is sometimes more useful than producing very accurate point forecasts – this is because many decisions are based on distributions of possible values rather than on point forecasts. The classical example of this situation is inventory management, where the ordering decisions are made based on quantiles of distribution to form safety stock. Furthermore, the orders are typically done in pallets, so it is not important whether the expected demand is 99 or 95 units if a pallet includes 100 units of a product. This means that whenever we produce forecasts, we need to consider how they will be used.

In some cases, accurate forecasts might be wasted if people make decisions differently and/or do not trust what they see. For example, a demand planner might decide that a straight line is not an appropriate forecast for their data and would start changing the values, introducing noise. This might happen due to a lack of experience, expertise, or trust in models (Spavound & Kourentzes, 2022), and this means that it is crucial to understand who will use the forecasts and how.

Finally, in practice, not all issues can be resolved with forecasting. In some cases, companies can make decisions based on other reasons. For example, promotional decisions can be dictated by the existing stocks of the product that need to be moved out. In another case, if the holding costs for a product are low, then there is no need to spend time forecasting the demand for it – a company can implement a simple replenishment policy, ordering, when the stock reaches some threshold. And in times of crisis, some decisions are dictated by the company's financial situation, not by forecasts: arguably, you do not need to predict demand for products that are sold out of prestige if they are not profitable, and a company needs to cut costs.

Summarising all the above, it makes sense to determine what decisions will be made based on forecasts, by whom and how. There is no need to waste time and effort on improving the forecasting accuracy if the process in the company is flawed and forecasts are then ignored, not needed, or amended inappropriately.

As for **analytics**, this is a relatively new term, meaning the systematic process of data analysis to support informed decisions. The term is broad and relies on many research areas, including forecasting, simulations, optimisation, etc. In this monograph, we will focus on the forecasting aspect, occasionally discussing how to analyse the existing processes (thus touching on the analytics part) and how various models could help make good practical decisions.

1.2 Forecasting principles

If you have decided that you need to forecast something, it makes sense to keep several important forecasting principles in mind.

First, as discussed earlier, you need to understand why the forecast is required, how it will be used, and by whom. Answers to these questions will guide you in deciding what technique to use, how specifically to forecast, and what should be reported. For example, if a client does not know machine learning, it might be unwise to use Neural Networks for forecasting – the client will potentially not trust the technique and thus will not trust the forecasts, switching to simpler methods. If the final decision is to order a number of units, it would be more reasonable to produce cumulative forecasts over the lead time (time between the order and product delivery) and form safety stock based on the model and assumed distribution.

When you understand what to forecast and how, the second principle comes into play: select the relevant error measure. You need to decide how to measure the accuracy of forecasting methods, keeping in mind that accuracy needs to be as close to the final decision as possible. For example, if you need to decide the number of nurses for a specific day in the A&E department based on the patients' attendance, then it would be more reasonable to compare models in terms of their quantile performance (see Section 2.2) rather than expectation or median. Thus, it would be more appropriate to calculate pinball loss instead of MAE or RMSE (see details in Chapter 2).

Third, you should always test your models on a sample of data not seen by them. Train your model on one part of a sample (called *train set* or *in-sample*) and test it on another one (called *test set* or *holdout sample*). This way, you can have some guarantees that the model will not overfit the data and that it will be reasonable when you need to produce a final forecast. Yes, there are cases when you do not have enough data to do that. All you can do in these situations is use simpler, robust models (for example, damped trend Exponential Smoothing by Roberts, 1982; and Gardner & McKenzie, 1985; or Theta by Assimakopoulos & Nikolopoulos, 2000) and to use judgment in deciding whether the final forecasts are reasonable or not. But in all the other cases, you should test the model on the data it is unaware of. The recommended approach, in this case, is rolling origin, discussed in more detail in Section 2.4.

Fourth, the forecast horizon should be aligned with specific decisions in practice. If you need predictions for a week ahead, there is no need to produce forecasts for the next 52 weeks. If you do that then on the one hand, this will be costly and excessive, and on the other hand, the accuracy measurement will not align with the company's needs. The related issue is the test set (or holdout) size selection. There is no unique guideline for this, but it should not be shorter than the forecasting horizon and preferrably it should align with the specific horizon coming from managerial decisions.

Fifth, the time series aggregation level should be as close to the specific decisions as possible. There is no need to produce forecasts on an hourly level for the next week (168 hours ahead) if the decision is based on the order of a product for the whole week. We would not need such a granularity of data for the decision; aggregating the actual values to the weekly level and then applying models will do the trick. Otherwise, we would be wasting a lot of time making complicated models work on an hourly level.

Sixth, you need to have benchmark models. Always compare forecasts from your favourite approach with those from Naïve, global average, and/or regression (they are discussed in Section 3.3) – depending on what you deal with specifically. If your fancy Neural Network performs worse than Naïve, it does not bring value and should not be used in practice. Comparing one Neural Network with another is also not a good idea because Simple Exponential Smoothing (see Section 3.4), being a much simpler model, might beat both networks, and you would never find out about that. If possible, also compare forecasts from the proposed approach with forecasts of other well-established benchmarks, such as ETS (Hyndman et al., 2008), ARIMA (Box & Jenkins, 1976) and Theta (Assimakopoulos & Nikolopoulos, 2000).

Finally, when comparing forecasts from different models, you might end up with several very similar performing approaches. If the difference between them is not significant, then the general recommendation is to select the faster and simpler one. This is because simpler models are more difficult to break, and those that work faster are more attractive in practice due to reduced energy consumption (save the planet and stop global warming! Dhar, 1999).

These principles do not guarantee that you will end up with the most accurate forecasts, but at least you will not end up with unreasonable ones.

1.3 Types of forecasts

Depending on circumstances, we might require different types of forecasts with different characteristics. It is essential to understand what your model produces to measure its performance correctly (see Section 2.1) and make correct decisions in practice. Several things are typically produced for forecasting purposes. We start with the most common one.

1.3.1 Point forecasts

Point forecast is the most often produced type of forecast. It corresponds to a trajectory from a model. This, however, might align with different types of statistics depending on the model and its assumptions. In the case of a pure additive model (such as linear regression), the point forecasts correspond to the conditional expectation (**mean**) from the model. The conventional interpretation of this value is that it shows what to expect on average if the situation would repeat itself many times (e.g. if we have the day with similar conditions, then the average temperature will be 10 degrees Celsius). In the case of time series, this interpretation is difficult to digest, given that time does not repeat itself, but this is the best we have. We will come back to the technicalities of producing conditional expectations from ADAM in Section 18.2.

Another type of point forecast is the (conditional) geometric expectation (**geometric mean**). It typically arises, when the model is applied to the data in logarithms and the final forecast is then exponentiated. This becomes apparent from the following definition of geometric mean:

$$\check{\mu} = \sqrt[T]{\prod_{t=1}^{T} y_t} = \exp \left(\frac{1}{T} \sum_{t=1}^{T} \log(y_t) \right), \tag{1.1}$$

where y_t is the actual value on observation t, and T is the sample size. To use the geometric mean, we need to assume that the actual values can only be positive. Otherwise, the root in (1.1) might produce imaginary units (because of taking a square root of a negative number) or be equal to zero (if one of the values is zero). In general, the arithmetic and geometric means are related via the following inequality:

$$\check{\mu} \leq \mu, \tag{1.2}$$

where $\check{\mu}$ is the geometric mean and μ is the arithmetic one. Although geometric mean makes sense in many contexts, it is more difficult to explain than the arithmetic one to decision makers.

Finally, sometimes **medians** are used as point forecasts. In this case, the point forecast splits the sample into two halves and shows the level below which 50% of observations will lie in the future.

Remark. The specific type of point forecast will differ depending on the model used in construction. For example, in the case of the pure additive model, assuming some symmetric distribution (e.g. Normal one), the arithmetic mean, geometric mean, and median will coincide. On the other hand, a model constructed in logarithms will assume an asymmetric distribution for the original data, leading to the following relation between the means and the median (in case of positively skewed distribution):

$$\begin{aligned} \check{\mu} &\leq \mu \\ \tilde{\mu} &\leq \mu' \end{aligned} \tag{1.3}$$

where $\tilde{\mu}$ is the median of distribution. The relation between geometric mean and median is more complicated and will differ from one distribution to another. In case of symmetric distributions or distributions becoming symmetric in logarithms, the two measures should coincide (at least in theory).

1.3.2 Quantiles and prediction intervals

As some forecasters say, all point forecasts are wrong. They will never correspond to the actual values because they only capture the model's mean (or median) performance, as discussed in the previous subsection. Everything that is not included in the point forecast can be considered as the uncertainty of demand. For example, we never will be able to say precisely how many cups of coffee a cafe will sell on the forthcoming Monday, but we can at least capture the main tendencies and the uncertainty around our point forecast.

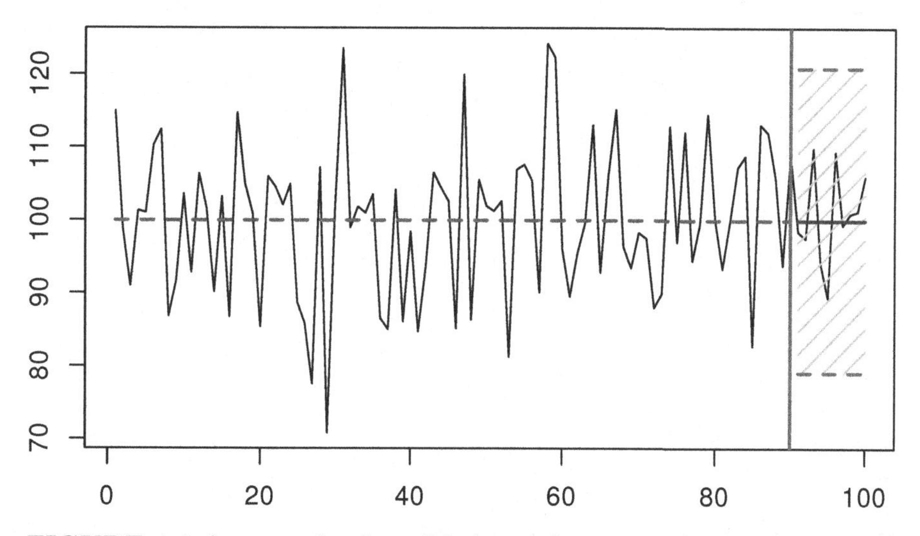

FIGURE 1.1 An example of a well-behaved data, point forecast, and a 95% prediction interval.

Figure 1.1 shows an example with a well-behaved demand, for which the best point forecast is the straight line. To capture the uncertainty of demand, we can construct the prediction interval, which will tell roughly where the demand will lie in $1 - \alpha$ percent of cases. The interval in Figure 1.1 has the width of 95% (significance level $\alpha = 0.05$) and shows that if the situation is repeated many times, the actual demand will be between 79.05 and 120.81. Capturing the uncertainty correctly is important because real-life decisions need to be made based on the full information, not only on the point forecasts.

We will discuss how to produce prediction intervals in more detail in Section 18.3. For a more detailed discussion on the concepts of prediction and confidence intervals, see Chapter 6 of Svetunkov (2022).

Another way to capture the uncertainty (related to the prediction interval) is via specific quantiles of distribution. The prediction interval typically has two sides, leaving $\frac{\alpha}{2}$ values outside of each side of distribution. Instead of producing the interval, in some cases, we might need just a specific quantile, essentially creating the one-sided prediction interval (see Section 18.4.2 for technicalities). The bound in this case will show the particular value below which the pre-selected percentage of cases would lie. This becomes especially useful in such contexts as the safety stock calculation (because we are not interested in knowing the lower bound, we want products in inventory to satisfy some proportion of demand).

1.3.3 Forecast horizon

Finally, an important aspect in forecasting is the horizon, for which we need to produce forecasts. Depending on the context, we might need:

1. Only a specific value h steps ahead, e.g., the temperature following Monday.
2. All values from 1 to h steps ahead, e.g. how many patients we will have each day next week.
3. Cumulative values for the period from 1 to h steps ahead, e.g. what the cumulative demand over the lead time (the time between the order and product delivery) will be (see discussion in Section 18.4.3).

It is essential to understand how decisions are made in practice and align them with the forecast horizon. In combination with the point forecasts and prediction intervals discussed above, this will give us an understanding of what to produce from the model and how. For example, in the case of safety stock calculation, it would be more reasonable to produce quantiles of the cumulative over the lead time demand than to produce point forecasts from the model (see, for example, discussion on safety stock calculation in Silver, Pyke, & Thomas, 2016).

1.4 Models, methods, and typical assumptions

While we do not aim to fully cover the topic of models, methods, and typical assumptions of statistical models, we need to make several important definitions to clarify what we will discuss later in this monograph. For a more detailed discussion, see Chapters 1 and 15 of Svetunkov (2022).

Chatfield, Koehler, Ord, and Snyder (2001) was the first to discuss the distinction between forecasting model and method, although the two are not thoroughly defined in their paper: "method, meaning a computational procedure for producing forecasts", and "a model, meaning a mathematical representation of reality". I think it is important to make a proper distinction between the two.

The Cambridge dictionary (Dictionary, 2021) defines **method** as a particular way of doing something. So, the method does not necessarily explain the structure or how some components or variables interact with each other; it only describes how a value (for example, point forecast) is produced. In our context, the forecasting method would be a formula that generates point forecasts based on some parameters and available data. It would not explain what structure underlies the data.

A statistical model, on the other hand, is a "mathematical representation of a real phenomenon with a complete specification of distribution and parameters" (Svetunkov & Boylan, 2023). It explains what happens inside the data, reveals the structure, and shows how the random variables interact with the structure.

While discussing statistical models, we should also define **true model**. It is "the idealistic statistical model that is correctly specified (has all the necessary components in the correct form), and applied to the data in population" (Svetunkov, 2022). Some statisticians also use the term **Data Generating Process** (DGP) when discussing the true model as a synonym. However, we need to distinguish between the two terms, as DGP implies that the data is somehow generated using a mathematical formula. In real life, the data is never generated from any function; it comes from a measurement of a complex process, influenced by many factors (e.g. behaviour of a group of customers based on their individual preferences and mental states). The DGP is useful when we want to conduct experiments on simulated data in a controlled environment, but it is not helpful when applying models to the data. Finally, the true model is an abstract notion because it is never known or reachable (e.g. we do not always have all the necessary variables). But it is still a useful one, as it allows us to see what would happen in therory if we knew the model and, more importantly, what would happen if the model we used was wrong (which is always the case in real life).

Related to this definition is the **estimated** or **applied model**, which is the statistical model that is applied to the available sample of data. This model will almost always be wrong because even if we know the specification of the true model for some mysterious reason, we would still need to estimate it on our data. In this case, the estimates of parameters would differ from those in the population, and thus the model will still be wrong.

Mathematically, in the simplest case the true model can be written as:

$$y_t = \mu_{y,t} + \epsilon_t, \tag{1.4}$$

where y_t is the actual observed value, $\mu_{y,t}$ is the structure, and ϵ_t is the true noise. If we manage to capture the structure correctly, the model applied to the sample of data would be written as:

$$y_t = \hat{\mu}_{y,t} + e_t, \tag{1.5}$$

where $\hat{\mu}_{y,t}$ is the estimate of the structure $\mu_{y,t}$ and e_t is the estimate of the noise ϵ_t (also known as "**residuals**"). If the structure is captured correctly, there would still be a difference between (1.4) and (1.5) because the latter is estimated on a sample of data. However, if the sample size increases and we use an adequate estimation procedure, then due to Central Limit Theorem (see Chapter 6 of Svetunkov, 2022), the distance between the two models will decrease, and asymptotically (with the increase of sample size) e_t would converge to ϵ_t. This does not happen automatically, and some assumptions must hold for this to happen.

1.4.1 Assumptions of statistical models

Very roughly, the typical assumptions of statistical models can be split into the following categories (Svetunkov, 2022):

1. Model is correctly specified:
   a. We have not omitted important variables in the model (underfitting the data);
   b. We do not have redundant variables in the model (overfitting the data);
   c. The necessary transformations of the variables are applied;
   d. We do not have outliers in the model;
2. Errors are independent and identically distributed (i.i.d.):
   a. There is no autocorrelation in the residuals;
   b. The residuals are homoscedastic (i.e. have constant variance);
   c. The expectation of residuals is zero, no matter what;
   d. The variable follows the assumed distribution;
   e. More generally speaking, the distribution of residuals does not change over time;
3. The explanatory variables are not correlated with anything but the response variable:

a. No multicollinearity;
b. No endogeneity.

Remark. The third group above relates more to the assumptions of model estimation rather than the model itself, but it is useful to have it in mind during the model-building process.

Many of these assumptions come from the idea that we have correctly captured the structure, meaning that we have not omitted any essential variables, we have not included redundant ones, and we transformed all the variables correctly (e.g. took logarithms, where needed). If all these assumptions hold, then we would expect the applied model to converge to the true one with the increase of the sample size. If some of them do not hold, then the point forecasts from our model might be biased, or we might end up producing wider (or narrower) prediction intervals than expected.

These assumptions with their implications on an example of multiple regression are discussed in detail in Chapter 15 of Svetunkov (2022). The diagnostics of dynamic models based on these assumptions are discussed in Chapter 14 of this monograph.

2

Forecasts evaluation

As discussed in Section 1.1, forecasts should serve a specific purpose. They should not be made "just because" but to help make decisions. The decision then dictates the kind of forecast that should be made – its form and its time horizon(s). It also dictates how the forecast should be evaluated – a forecast only being as good as the quality of the decisions it enables.

When you understand how your system works and what sort of forecasts you should produce, you can start an evaluation process, measuring the performance of different forecasting models/methods and selecting the most appropriate for your data.

This chapter discusses the most common approaches, focusing on evaluating point forecasts, then moving towards prediction intervals and quantile forecasts. After that, we discuss how to choose the appropriate error measure and, finally, ensure that the model performs consistently on the available data via rolling origin evaluation and statistical tests.

2.1 Measuring accuracy of point forecasts

We start with a setting in which we are interested in point forecasts only. In this case, we typically begin by splitting the available data into training and test sets, applying the models under consideration to the former, and producing forecasts on the latter, hiding it from the models. This is called the "fixed origin" approach: we fix the point in time from which to produce forecasts, produce them, calculate some appropriate error measure, and compare the models.

Different error measures can be used in this case. Which one to use depends on the specific need. Here I briefly discuss the most important measures and refer readers to Davydenko and Fildes (2013), Svetunkov (2019) and Svetunkov (2017) for the gory details.

The majority of point forecast measures rely on the following two popular metrics:

Root Mean Squared Error (RMSE):

$$\mathrm{RMSE} = \sqrt{\frac{1}{h}\sum_{j=1}^{h} \left(y_{t+j} - \hat{y}_{t+j}\right)^2}, \tag{2.1}$$

and **Mean Absolute Error** (MAE):

$$\mathrm{MAE} = \frac{1}{h}\sum_{j=1}^{h} \left|y_{t+j} - \hat{y}_{t+j}\right|, \tag{2.2}$$

where y_{t+j} is the actual value j steps ahead (values in the test set), $\hat{y}_{t+j}$ is the j steps ahead point forecast, and h is the forecast horizon. As you see, these error measures aggregate the performance of competing forecasting methods across the forecasting horizon, averaging out the specific performances on each j. If this information needs to be retained, the summation can be dropped to obtain "SE" and "AE" values.

In a variety of cases RMSE and MAE might recommend different models, and a logical question would be which of the two to prefer. It is well-known (see, for example, Kolassa, 2016) that the **mean value of distribution minimises RMSE**, and **the median value minimises MAE**. So, when selecting between the two, you should consider this property. It also implies, for example, that MAE-based error measures should not be used for the evaluation of models on intermittent demand (see Chapter 13 for the discussion of this topic) because zero forecast will minimise MAE, when the sample contains more than 50% of zeroes (see, for example, Wallström & Segerstedt, 2010).

Another error measure that has been used in some cases is Root Mean Squared Logarithmic Error (RMSLE, see discussion in Tofallis, 2015):

$$\mathrm{RMSLE} = \exp\left(\sqrt{\frac{1}{h}\sum_{j=1}^{h} \left(\log y_{t+j} - \log \hat{y}_{t+j}\right)^2}\right). \tag{2.3}$$

It assumes that the actual values and the forecasts are positive, and it is **minimised by the geometric mean**. I have added the exponentiation in the formula (2.3), which is sometimes omitted, bringing the metric to the original scale to have the same units as the actual values y_t.

The main difference in the three measures arises when the data we deal with is not symmetric – in that case, the arithmetic, geometric means, and median will all be different. Thus, the error measures might recommend different methods depending on what specifically is produced as a point forecast from the model (see discussion in Section 1.3.1).

2.1.1 An example in R

In order to see how the error measures work, we consider the following example based on a couple of forecasting functions from the `smooth` package for R (Hyndman et al., 2008; Svetunkov, Kourentzes, & Ord, 2022) and measures from `greybox`:

```
# Generate the data
y <- rnorm(100,100,10)
# Apply two models to the data
model1 <- es(y,h=10,holdout=TRUE)
model2 <- ces(y,h=10,holdout=TRUE)
# Calculate RMSE
setNames(sqrt(c(MSE(model1$holdout, model1$forecast),
                MSE(model2$holdout, model2$forecast))),
         c("ETS","CES"))
# Calculate MAE
setNames(c(MAE(model1$holdout, model1$forecast),
           MAE(model2$holdout, model2$forecast)),
         c("ETS","CES"))
# Calculate RMSLE
setNames(exp(sqrt(c(MSE(log(model1$holdout),
                        log(model1$forecast)),
                    MSE(log(model2$holdout),
                        log(model2$forecast))))),
         c("ETS","CES"))
```

```
##      ETS      CES
## 9.492744 9.494683

##      ETS      CES
## 7.678865 7.678846

##      ETS      CES
## 1.095623 1.095626
```

Remark. The point forecasts produced by ETS and CES correspond to conditional means, so ideally we should focus the evaluation on RMSE-based measures.

Given that the distribution of the original data is symmetric, all three error measures should generally recommend the same model. But also, given that the data we generated for the example are stationary, the two models will produce very similar forecasts. The values above demonstrate the latter point – the accuracy between the two models is roughly the same. Note that we have evaluated the same point forecasts from the models using different error measures, which would be wrong if the distribution of the data was skewed. In our case, the model relies on Normal distribution so that the point forecast would coincide with arithmetic mean, geometric mean, and median.

2.1.2 Aggregating error measures

The main advantage of the error measures discussed in the previous subsection is that they are straightforward and have a clear interpretation: they reflect the "average" distances between the point forecasts and the observed values. They are perfect for the work with only one time series. However, they are not suitable when we consider a set of time series, and a forecasting method needs to be selected across all of them. This is because they are scale-dependent and contain specific units: if you measure sales of apples in units, then MAE, RMSE, and RMSLE will show the errors in units as well. And, as we know, you should not add apples to oranges – the result might not make sense.

To tackle this issue, different error scaling techniques have been proposed, resulting in a zoo of error measures:

1. MAPE – Mean Absolute Percentage Error:
$$\mathrm{MAPE} = \frac{1}{h} \sum_{j=1}^{h} \frac{|y_{t+j} - \hat{y}_{t+j}|}{y_{t+j}}, \tag{2.4}$$
2. MASE – Mean Absolute Scaled Error (Hyndman & Koehler, 2006):
$$\mathrm{MASE} = \frac{1}{h} \sum_{j=1}^{h} \frac{|y_{t+j} - \hat{y}_{t+j}|}{\bar{\Delta}_y}, \tag{2.5}$$
where $\bar{\Delta}_y = \frac{1}{t-1}\sum_{j=2}^{t} |\Delta y_j|$ is the mean absolute value of the first differences $\Delta y_j = y_j - y_{j-1}$ of the in-sample data;
3. rMAE – Relative Mean Absolute Error (Davydenko & Fildes, 2013):
$$\mathrm{rMAE} = \frac{\mathrm{MAE}_a}{\mathrm{MAE}_b}, \tag{2.6}$$
where MAE_a is the mean absolute error of the model under consideration and MAE_b is the MAE of the benchmark model;
4. sMAE – scaled Mean Absolute Error (Petropoulos & Kourentzes, 2015):
$$\mathrm{sMAE} = \frac{\mathrm{MAE}}{\bar{y}}, \tag{2.7}$$
where $\bar{y}$ is the mean of the in-sample data;
5. and others.

Remark. MAPE and sMAE are typically multiplied by 100% to get the percentages, which are easier to work with.

There is no "best" error measure. All have advantages and disadvantages, but some are more suitable in some circumstances than others. For example:

1. MAPE is scale sensitive (if the actual values are measured in thousands of units, the resulting error will be much lower than in the case of hundreds of units) and cannot be estimated on data with zeroes. Furthermore, this error measure is biased, preferring when models underforecast the data (see, for example, Makridakis, 1993) and is not minimised by either mean or median, but by an unknown quantity. Accidentally, in the case of Log-Normal distribution, it is minimised by the mode (see discussion in Kolassa, 2016). Despite all the limitations, MAPE has a simple interpretation as it shows the percentage error (as the name suggests);
2. MASE avoids the disadvantages of MAPE but does so at the cost of losing a simple interpretation. This is because of the division by the first differences of the data (some interpret this as an in-sample one-step-ahead Naïve forecast, which does not simplify the interpretation);
3. rMAE avoids the disadvantages of MAPE, has a simple interpretation (it shows by how much one model is better than the other), but fails, when either MAE_a or MAE_b for a specific time series is equal to zero. In practice, this happens more often than desired and can be considered a severe error measure limitation. Furthermore, the increase of rMAE (for example, with the increase of sample size) might mean that either the method A is performing better than before or that the method B is performing worse than before – it is not possible to tell the difference unless the denominator in the formula (2.6) is fixed;
4. sMAE avoids the disadvantages of MAPE, has an interpretation close to it but breaks down, when the data is non-stationary (e.g. has a trend).

When comparing different forecasting methods, it might make sense to calculate several error measures for comparison. The choice of metric might depend on the specific needs of the forecaster. Here are a few rules of thumb:

- You should typically avoid MAPE and other percentage error measures because the actual values highly influence them in the holdout;
- If you want a robust measure that works consistently, but you do not care about the interpretation, then go with MASE;
- If you want an interpretation, go with rMAE or sMAE. Just keep in mind that if you decide to use rMAE or any other relative measure for research purposes, you might get in an unnecessary dispute with its creator, who might blame you of stealing his creation (even if you reference his work);

- If the data does not exhibit trends (stationary), you can use sMAE.

Furthermore, similarly to the measures above, there have been proposed RMSE-based scaled and relative error metrics, which measure the performance of methods in terms of means rather than medians. Here is a brief list of some of them:

1. RMSSE – Root Mean Squared Scaled Error (Makridakis, Spiliotis, & Assimakopoulos, 2022):
$$\mathrm{RMSSE} = \sqrt{\frac{1}{h} \sum_{j=1}^{h} \frac{(y_{t+j} - \hat{y}_{t+j})^2}{\bar{\Delta}_y^2}}; \tag{2.8}$$
2. rRMSE – Relative Root Mean Squared Error (Stock & Watson, 2004):
$$\mathrm{rRMSE} = \frac{\mathrm{RMSE}_a}{\mathrm{RMSE}_b}; \tag{2.9}$$
3. sRMSE – scaled Root Mean Squared Error (Petropoulos & Kourentzes, 2015):
$$\mathrm{sRMSE} = \frac{\mathrm{RMSE}}{\bar{y}}. \tag{2.10}$$

Similarly, RMSSLE, rRMSLE, and sRMSLE can be proposed, using the same principles as in (2.8), (2.9), and (2.10) to assess performance of models in terms of geometric means across time series.

Finally, when aggregating the performance of forecasting methods across several time series, sometimes it makes sense to look at the distribution of errors – this way, you will know which of the methods fails seriously and which does a consistently good job. If only an aggregate measure is needed then I recommend using **both the mean and median of the chosen metric**. The mean might be non-finite for some error measures, especially when a method performs exceptionally poorly on a time series (an outlier). Still, it will give you information about the average performance of the method and might flag extreme cases. The median at the same time is robust to outliers and is always calculable, no matter what the distribution of the error term is. Furthermore, comparing mean and median might provide additional information about the tail of distribution without reverting to histograms or the calculation of quantiles. Davydenko and Fildes (2013) argue for the use of geometric mean for relative and scaled measures. Still, as discussed earlier, it might become equal to zero or infinity if the data contains outliers (e.g. two cases, when one of the methods produced a perfect forecast, or the benchmark in rMAE produced a perfect forecast). At the same time, if the distribution of errors in logarithms is symmetric (which is the main argument of Davydenko & Fildes, 2013), then the geometric mean will be similar to the median, so there is no point in calculating the geometric mean at all.

2.1.3 Demonstration in R

In R, there is a variety of functions that calculate the error measures discussed above, including the `accuracy()` function from the `forecast` package and `measures()` from `greybox`. Here is an example of how the measures can be calculated based on a couple of forecasting functions from the `smooth` package for R and a set of generated time series:

```
# Apply a model to a test data to get names of error measures
y <- rnorm(100,100,10)
test <- es(y,h=10,holdout=TRUE)
# Define number of iterations
nsim <- 100
# Create an array for nsim time series,
# 2 models and a set of error measures
errorMeasures <- array(NA, c(nsim,2,length(test$accuracy)),
                       dimnames=list(NULL,c("ETS","CES"),
                                     names(test$accuracy)))
# Start a loop for nsim iterations
for(i in 1:nsim){
  # Generate a time series
  y <- rnorm(100,100,10)
  # Apply ETS
  testModel1 <- es(y,"ANN",h=10,holdout=TRUE)
  errorMeasures[i,1,] <- measures(testModel1$holdout,
                                  testModel1$forecast,
                                  actuals(testModel1))
  # Apply CES
  testModel2 <- ces(y,h=10,holdout=TRUE)
  errorMeasures[i,2,] <- measures(testModel2$holdout,
                                  testModel2$forecast,
                                  actuals(testModel2))
}
```

The default benchmark method for the relative measures above is Naïve. To see what the distribution of error measures would look like, we can produce violinplots via the `vioplot()` function from the `vioplot` package. We will focus on the rRMSE measure (see Figure 2.2).

```
vioplot::vioplot(errorMeasures[,,"rRMSE"])
```

The distributions in Figure 2.2 look similar, and it is hard to tell which one performs better. Besides, they do not look symmetric, so we will take logarithms to see if this fixes the issue with the skewness (Figure 2.2).

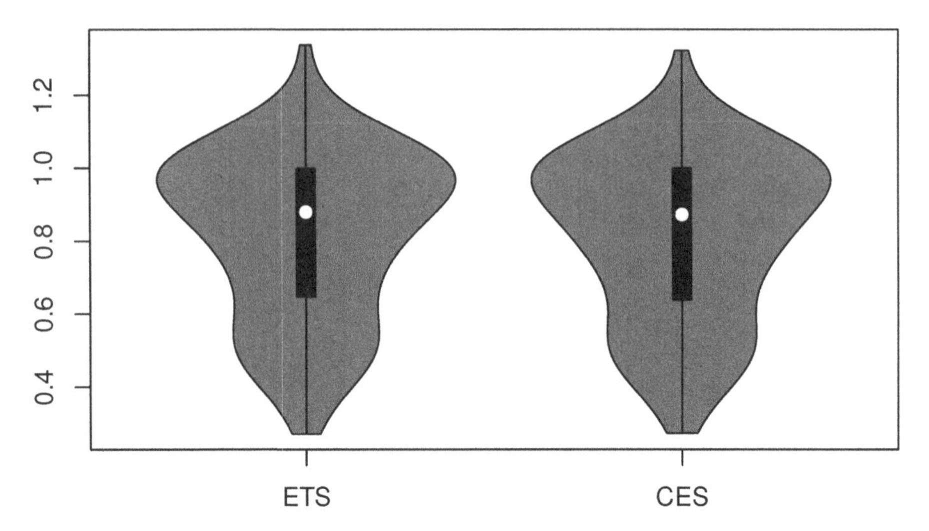

FIGURE 2.1 Distribution of rRMSE on the original scale.

```
vioplot::vioplot(log(errorMeasures[,,"rRMSE"]))
```

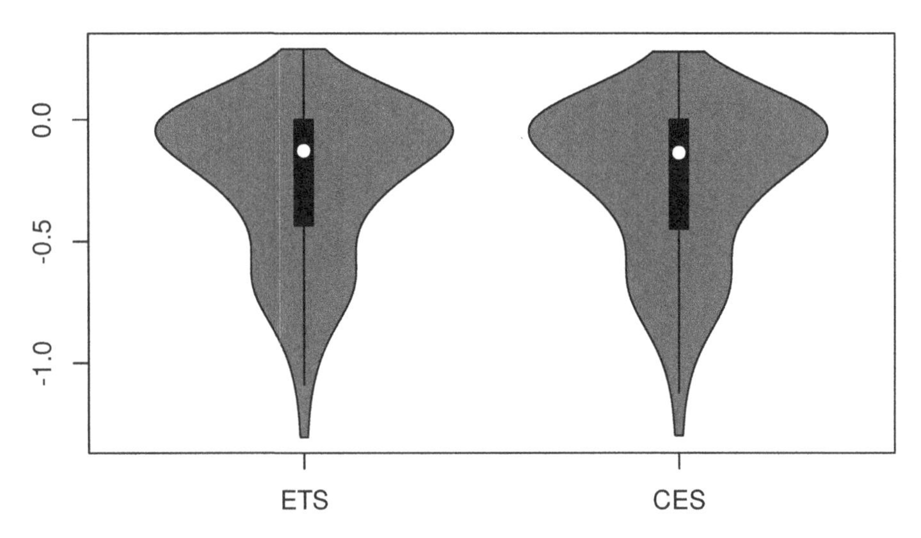

FIGURE 2.2 Distribution of rRMSE on the log scale.

Figure 2.2 demonstrates that the distribution in logarithms is still skewed, so the geometric mean would not be suitable and might provide a misleading information (being influenced by the tail of distribution). So, we calculate the mean and median rRMSE to check the overall performance of the two models:

```
# Calculate mean rRMSE
apply(errorMeasures[,,"rRMSE"],2,mean)
```

```
##       ETS       CES
## 0.8163452 0.8135303
```

```
# Calculate median rRMSE
apply(errorMeasures[,,"rRMSE"],2,median)
```

```
##       ETS       CES
## 0.8796325 0.8725286
```

Based on the values above, we cannot make any solid conclusion about the performance of the two models; in terms of both mean and median rRMSE, CES is doing slightly better, but the difference between the two models is not substantial, so we can probably choose the one that is easier to work with.

2.2 Measuring uncertainty

As discussed in Section 1.3.2, point forecasts are not sufficient for adequate decision making – prediction intervals and quantiles are needed to capture the uncertainty of demand. As with point forecasts, multiple measures can be used to evaluate them. There are several useful measures for the evaluation of intervals. We start with the simplest of them, coverage.

1. **Coverage** shows the percentage of observations lying inside the interval:

$$\text{Coverage} = \frac{1}{h} \sum_{j=1}^{h} \left(\mathbb{1} \left(y_{t+j} > l_{t+j} \right) \times \mathbb{1} \left(y_{t+j} < u_{t+j} \right) \right), \tag{2.11}$$

where l_{t+j} is the lower bound and u_{t+j} is the upper bound of the interval and $\mathbb{1}(\cdot)$ is the indicator function, returning one, when the condition is true and zero otherwise. Ideally, the coverage should be equal to the confidence level of the interval, but in reality, this can only be observed asymptotically (with the increase of the sample size) due to the inherited randomness of any sample estimates of parameters;
2. **Range** shows the width of the prediction interval:

$$\text{Range} = \frac{1}{h} \sum_{j=1}^{h} (u_{t+j} - l_{t+j}), \tag{2.12}$$

If the range of intervals from one model is lower than the range of the other one, then the uncertainty about the future values is lower for the first one. However, the narrower interval might not include as many actual values in the holdout sample, leading to lower coverage. So, there is a natural trade-off between the two measures;

3. **Mean Interval Score** (Gneiting & Raftery, 2007) combines the properties of the previous two measures:

$$\begin{aligned} \mathrm{MIS} = & \frac{1}{h} \sum_{j=1}^{h} \left((u_{t+j} - l_{t+j}) + \frac{2}{\alpha} (l_{t+j} - y_{t+j}) \mathbb{1}(y_{t+j} < l_{t+j}) + \right. \\ & \left. \frac{2}{\alpha} (y_{t+j} - u_{t+j}) \mathbb{1}(y_{t+j} > u_{t+j}) \right) , \end{aligned} \tag{2.13}$$

where α is the significance level. If the actual values lie outside of the interval, they get penalised with a ratio of $\frac{2}{\alpha}$, proportional to the distance from the interval bound. At the same time, the width of the interval positively influences the value of the measure: the wider the interval, the higher the score. The ideal model with MIS $= 0$ should have all the actual values in the holdout lying on the bounds of the interval and $u_{t+j} = l_{t+j}$, implying that the bounds coincide with each other and that there is no uncertainty about the future (which is not possible in real life);

4. **Pinball Score** (Koenker & Bassett, 1978) measures the accuracy of models in terms of specific quantiles (this is usually applied to different quantiles produced from the model, not just to the lower and upper bounds of 95% interval):

$$\mathrm{PS} = (1-\alpha) \sum_{y_{t+j} < q_{t+j}, j=1,\dots,h} |y_{t+j} - q_{t+j}| + \alpha \sum_{y_{t+j} \geq q_{t+j}, j=1,\dots,h} |y_{t+j} - q_{t+j}|, \tag{2.14}$$

where q_{t+j} is the value of the specific quantile of the distribution. PS shows how well we capture the specific quantile in the data. The lower the value of the pinball is, the closer the bound is to the specific quantile of the holdout distribution. If the PS is equal to zero, then we have done the perfect job in hitting that specific quantile. The main issue with PS is that it is very difficult to assess the quantiles correctly on small samples. For example, in order to get a better idea of how the 0.975 quantile performs, we would need to have at least 40 observations, so that 39 of them would be expected to lie below this bound $\left(\frac{39}{40} = 0.975\right)$. In fact, quantiles are not always uniquely defined (see, for example, Taylor, 2020), which makes the measurement challenging.

Similar to the pinball function, it is possible to propose the expectile-based score, but while expectiles have good statistical properties (Taylor, 2020), they are more difficult to interpret.

Range, MIS, and PS are unit-dependent. To aggregate them over several time series, they need to be scaled either via division by either the in-sample mean or in-sample mean absolute differences to obtain the scaled counterparts of the measures or via division by the values from the benchmark model to get the relative one. The idea here would be similar to what we discussed for MAE and RMSE in Section 2.1.

If you are interested in the model's overall performance, then MIS provides this information. However, it does not show what happens specifically (is there an issue in the distance from the bound or the width of the interval?) and it is difficult to interpret. Coverage and range are easier to interpret, but they only give information about the specific prediction interval. They typically must be traded off against each other (i.e. one can either cover more or have a narrower interval). Academics prefer pinball for uncertainty assessment, as it shows more detailed information about the predictive distribution from each model. However, while it is easier to interpret than MIS, it is still not as straightforward as coverage and range. So, the selection of the measure depends on the specific situation and the understanding of statistics by decision-makers.

2.2.1 Example in R

Continuing the example from Section 2.1, we could produce prediction intervals from the two models and compare them using MIS and pinball:

```
model1Forecast <- forecast(model1,h=10,interval="p",level=0.95)
model2Forecast <- forecast(model2,h=10,interval="p",level=0.95)

# Mean Interval Score
setNames(c(MIS(model1$holdout, model1Forecast$lower,
               model1Forecast$upper, 0.95),
           MIS(model2$holdout, model2Forecast$lower,
               model2Forecast$upper, 0.95)),
         c("Model 1", "Model 2"))
```

```
##  Model 1  Model 2
## 39.68038 46.91441
```

```
# Pinball for the upper bound
setNames(c(pinball(model1$holdout, model1Forecast$upper, 0.975),
           pinball(model2$holdout, model2Forecast$upper, 0.975)),
         c("Model 1", "Model 2"))
```

```
##  Model 1  Model 2
## 5.402511 6.703046
```

```
# Pinball for the lower bound
setNames(c(pinball(model1$holdout, model1Forecast$lower, 0.025),
           pinball(model2$holdout, model2Forecast$lower, 0.025)),
         c("Model 1", "Model 2"))
```

```
##  Model 1  Model 2
## 4.517584 5.025557
```

```
# Coverage
setNames(c(mean(model1$holdout > model1Forecast$lower &
                model1$holdout < model1Forecast$upper),
           mean(model2$holdout > model2Forecast$lower &
                model2$holdout < model2Forecast$upper)),
         c("Model 1", "Model 2"))
```

```
## Model 1 Model 2
##     0.9     0.9
```

The values above imply that the first model (ETS) performed better than the second one in terms of MIS and pinball loss (the interval was narrower). However, these measures do not tell much in terms of the performance of models when only applied to one time series. To see more solid results, we need to apply models to a set of time series, produce prediction intervals, calculate measures, and then look at their aggregate performance, e.g. via mean/median or quantiles. The loop and the analysis would be similar to the one discussed in Section 2.1.3, so we do not repeat it here.

2.3 How to choose appropriate error measure

While, in general, the selection of error measures should be dictated by the specific problem at hand, some guidelines might be helpful in the process. I have summarised them in the flowchart in Figure 2.3.

The flowchart does not provide excessive options and simplifies the possible process. It does not discuss the quantile and interval measures in detail, as there are many options for them. The idea of the flowchart is to list the most important measures, and its aim is to provide a guideline for selection based on:

1. Number of time series under consideration. If there are several of them and you need to aggregate the error measure, you need to use either

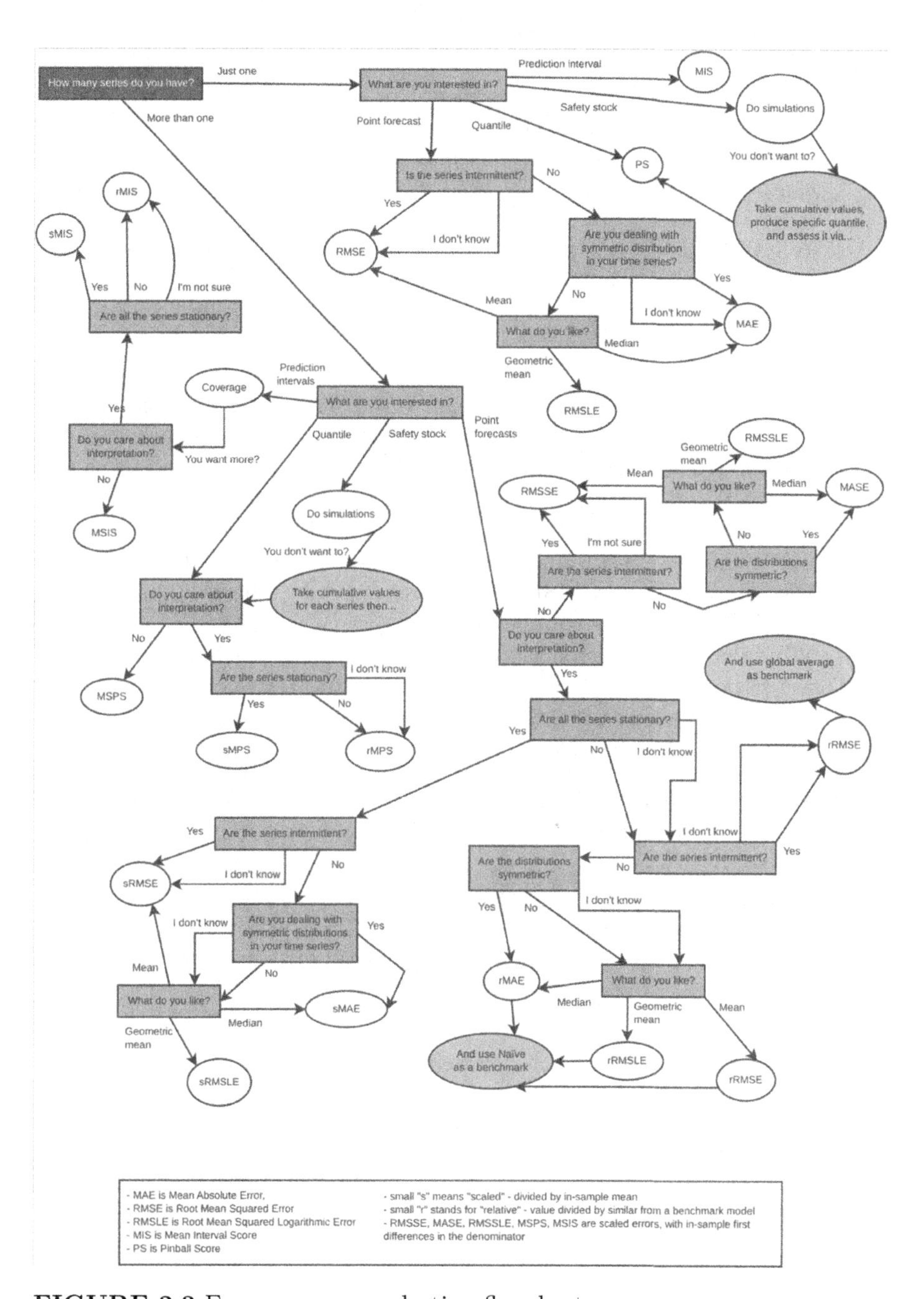

FIGURE 2.3 Error measures selection flowchart.

scaled or relative ones. In the case of just one time series, you do not need to scale the error measure;
2. What specifically you want to measure: point forecasts, quantiles, prediction interval, or something else;
3. Whether the interpretability of the error measure is essential or not. If not, then scaled measures similar to Hyndman and Koehler (2006) can be used. If yes, then the choice is between relative and scaled using mean measures;
4. Whether the data is stationary or not. If it is, then it is safe to use scaled measures similar to Petropoulos and Kourentzes (2015) because the division by the in-sample mean would be meaningful. Otherwise, you should either use Hyndman and Koehler (2006) scaling or relative measures;
5. Whether the data is intermittent or not. If it is and you are interested in point forecasts, then you should use RMSE based measures – other measures might recommend zero forecast as the best one;
6. Symmetry of distribution of demand. If it is symmetric (which does not happen very often), then the median will coincide with the mean and geometric mean, and it would not be important whether to use RMSE-, MAE-, or RMSLE- based measure. In that case, just use an MAE-based one (for simplicity reasons);
7. What you need (denoted as "What do you like?" in the flowchart). If you are interested in mean performance, then use RMSE based measures. The median minimises MAE, and the geometric mean minimises RMSLE. This relates to the discussion in Section 1.3.

The point forecast related error measures have been discussed in Section 2.1, while the interval and quantile ones – in Section 2.2.

Remark. I personally do not recommend using MAPE and SMAPE (symmetric MAPE) for the reasons discussed by Goodwin and Lawton (1999) and Hyndman and Koehler (2006). In fact, any percentage-based error measure has severe limitations and should be avoided if possible.

You can also download this flowchart in PDF format from the https://www.openforecast.org/adam/[1] website.

[1] https://www.openforecast.org/adam/images/errorMeasuresFlowChart-v2.pdf

2.4 Rolling origin

Remark. The text in this section is based on the vignette for the greybox package[2], written by the author of this monograph.

When there is a need to select the most appropriate forecasting model or method for the data, the forecaster usually splits the sample into two parts: in-sample (aka "training set") and holdout sample (aka out-sample or "test set"). The model is estimated on the in-sample, and its forecasting performance is evaluated using some error measure on the holdout sample.

Using this procedure only once is known as "fixed origin" evaluation. However, this might give a misleading impression of the accuracy of forecasting methods. If, for example, the time series contains outliers or level shifts, a poor model might perform better in fixed origin evaluation than a more appropriate one just by chance. So it makes sense to have a more robust evaluation technique, where the model's performance is evaluated several times, not just once. An alternative procedure known as "rolling origin" evaluation is one such technique.

In rolling origin evaluation, the forecasting origin is repeatedly moved forward by a fixed number of observations, and forecasts are produced from each origin (Tashman, 2000). This technique allows obtaining several forecast errors for time series, which gives a better understanding of how the models perform. This can be considered a time series analogue to cross-validation techniques (see Chapter 5 of James, Witen, Hastie, & Tibshirani, 2017). Here is a simple graphical representation, courtesy of Nikos Kourentzes[3].

The plot in Figure 2.4 shows how the origin moves forward and the point and interval forecasts of the model change. As a result, this procedure gives information about the performance of the model over a set of observations, not on a random one. There are different options of how this can be done, and here we discuss the main principles behind it.

2.4.1 Principles of rolling origin

Figure 2.5 (Svetunkov & Petropoulos, 2018) illustrates the basic idea behind rolling origin. White cells correspond to the in-sample data, while the light grey cells correspond to the three steps ahead forecasts. The time series in the figure has 25 observations, and forecasts are produced for eight origins starting from observation 15. In the first step, the model is estimated on the first in-sample set, and forecasts are created for the holdout. Next, another observation is added to the end of the in-sample set, the test set is advanced,

[2]https://cran.r-project.org/package=greybox
[3]https://kourentzes.com/forecasting/

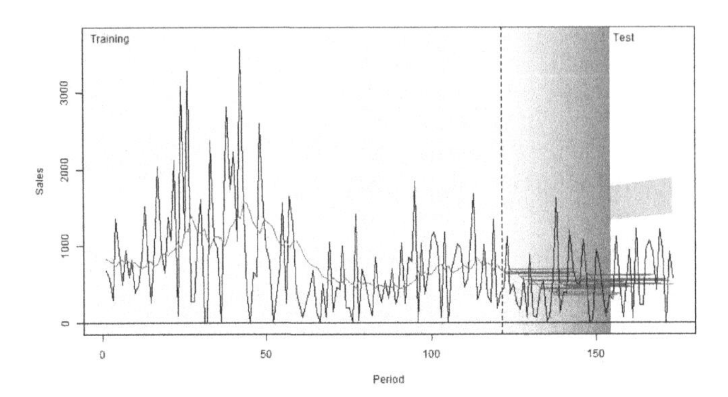

FIGURE 2.4 Visualisation of rolling origin by Nikos Kourentzes

and the procedure is repeated. The process stops when there is no more data left. This is a rolling origin with a **constant holdout** sample size. As a result of this procedure, eight one to three steps ahead forecasts are produced. Based on them, we can calculate the preferred error measures and choose the best performing model (see Section 2.1.2).

	1	2	3	4	5	6	7	8	9	10	11	12	13	14	15	16	17	18	19	20	21	22	23	24	25
Origin = 15																									
Origin = 16																									
Origin = 17																									
Origin = 18																									
Origin = 19																									
Origin = 20																									
Origin = 21																									
Origin = 22																									

FIGURE 2.5 Rolling origin with constant holdout size from Svetunkov and Petropoulos, (2018).

Another option for producing forecasts via rolling origin would be to continue with rolling origin even when the test sample is smaller than the forecast horizon, as shown in Figure 2.6. In this case, the procedure continues until origin 22, when the last complete set of three steps ahead forecasts can be produced, and then continues with a decreasing forecasting horizon. So the two steps ahead forecast is produced from origin 23, and only a one-step-ahead forecast is produced from origin 24. As a result, we obtain ten one-step-ahead forecasts, nine two steps ahead forecasts and eight three steps ahead forecasts. This is a rolling origin with a **non-constant holdout** sample size, which can be helpful with small samples when not enough observations are available.

Finally, in both cases above, we had the **increasing in-sample** size. However, we might need a **constant in-sample** for some research purposes. Figure 2.7 demonstrates such a setup. In this case, in each iteration, we add an observation

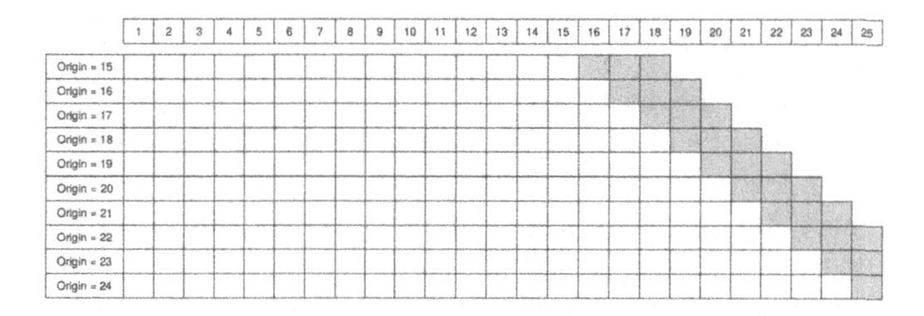

FIGURE 2.6 Rolling origin with non-constant holdout size.

to the end of the in-sample series and remove one from the beginning (dark grey cells).

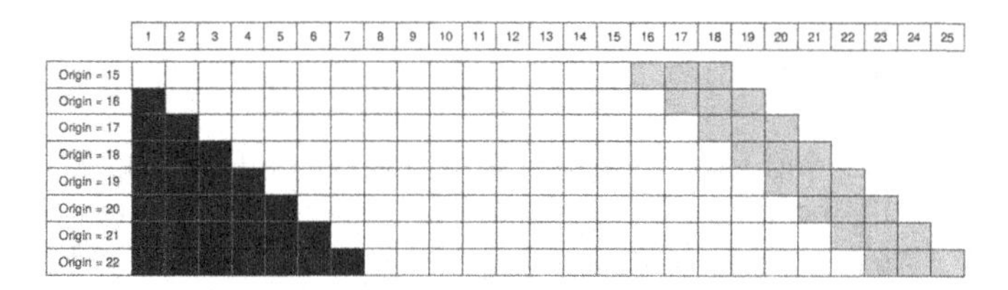

FIGURE 2.7 Rolling origin with constant in-sample size.

2.4.2 Implementing rolling origin in R

Now that we discussed the main idea of rolling origin, we can see how it can be implemented in R. In this section, we will implement rolling origin with a fixed holdout sample size and a changing in-sample. This aligns with what is typically done in practice when new data arrives: the model is re-estimated, and forecasts are produced for the next h steps ahead. For this example, we will use artificially created data and apply a Simple Moving Average (discussed in Subsection 3.3.3) implemented in the `smooth` package.

```
# Set sample to 100 observations
obs <- 100
# Generate the data
y <- rnorm(obs,100,10)
```

We will produce forecasts for the horizon of 10 steps ahead, $h = 10$ from 5 origins.

```
h <- 10
origins <- 5
```

We will create a list containing several objects of interest:

- `actuals` will contain all the actual values;
- `holdout` will be a matrix containing the actual values for the holdout. It will have `h` rows and `origins` columns;
- `mean` will contain point forecasts from our model. This will also be a matrix with the same dimensions as the `holdout` one.

```
returnedValues1 <- setNames(vector("list",3),
                            c("actuals","holdout","mean"))
returnedValues1$actuals <- y
returnedValues1$holdout <-
  returnedValues1$mean <-
  matrix(NA,h,origins,
         dimnames=list(paste0("h",1:h),
                       paste0("origin",1:origins)))
```

Finally, we write a simple loop that repeats the model fit and forecasting for the horizon `h` several times. The trickiest part is understanding how to define the train and test samples. In our example, the former should have `obs+1-origins-h` observations in the first step and `obs-h` in the last one so that we can have `h` observations in the test set throughout all origins, and we can repeat this `origins` times. One way of doing this is via the following loop:

```
for(i in 1:origins){
    # Fit the model
    testModel <- sma(y[1:(obs+i-origins-h)])
    # Drop the in-sample observations
    # and extract the first h observations from the rest
    returnedValues1$holdout[,i] <- head(y[-c(1:(obs-origins+i-h))], h)
    # Produce forecasts and write down the mean values
    returnedValues1$mean[,i] <- forecast(testModel, h=h)$mean
}
```

This basic loop can be amended to include anything else we want from the function or by changing the parameters of the rolling origin. After filling in the object `returnedValues1`, we can analyse the residuals of the model over the horizon and several origins in various ways. For example, Figure 2.8 shows boxplots across the horizon of 10 for different origins.

```
boxplot(returnedValues1$holdout-returnedValues1$mean)
```

In the ideal situation, the boxplots in Figure 2.8 should be similar, meaning that the model performs consistently over different origins. We do not see this

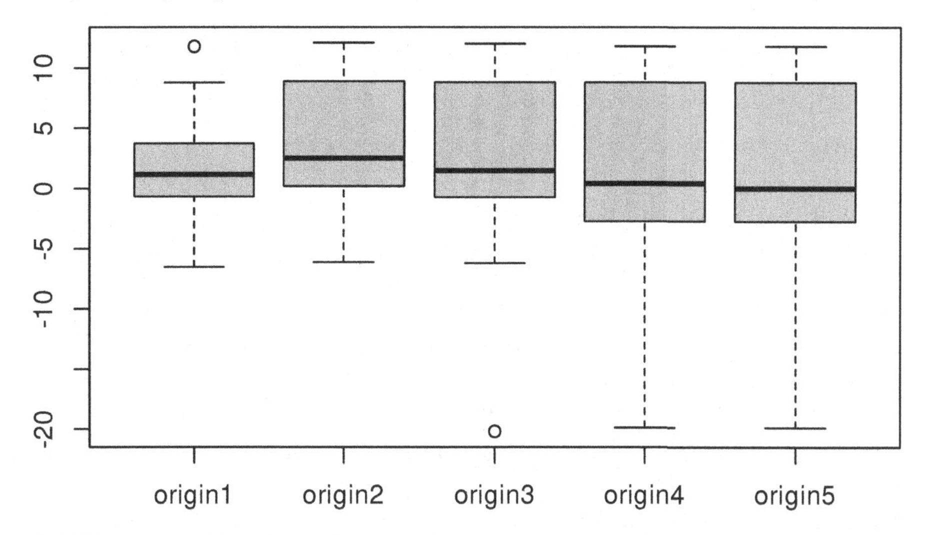

FIGURE 2.8 Boxplots of forecast errors over several origins.

in our case, observing that the distribution of errors changes from one origin to another.

While the example above already gives some information about the performance of a model, more useful information could be obtained if the performance of one model is compared to the others in the rolling origin experiment. This can be done manually for several models using the code above or it can be done using the function `ro()` from the `greybox` package.

2.4.3 Rolling origin function in R

In R, there are several packages and functions that implement rolling origin. One of those is the function `ro()` from the `greybox` package (written by Yves Sagaert and Ivan Svetunkov in 2016 on their way to the International Symposium on Forecasting in Riverside, US). It implements the rolling origin evaluation for any function you like with a predefined `call` and returns the desired `value`. It heavily relies on the two variables: `call` and `value` – so it is pretty important to understand how to formulate them to get the desired results. `ro()` is a very flexible function, but as a result, it is not very simple. In this subsection, we will see how it works in a couple of examples.

We start with a simple example, generating a series from Normal distribution:

```
y <- rnorm(100,100,10)
```

We use an ARIMA(0,1,1) model implemented in the `stats` package (this model is discussed in Section 8). Given that we are interested in forecasts from the model, we need to use the `predict()` function to get the desired values:

```
ourCall <- "predict(arima(x=data,order=c(0,1,1)),n.ahead=h)"
```

The call that we specify includes two important elements: `data` and `h`. `data` specifies where the in-sample values are located in the function that we want to use, and **it needs to be called "data"** in the call; `h` will tell our function, where the forecasting horizon is specified in the provided line of code. Note that in this example we use `arima(x=data,order=c(0,1,1))`, which produces a desired ARIMA(0,1,1) model and then we use `predict(..., n.ahead=h)`, which produces an h steps ahead forecast from that model.

Having the call, we also need to specify what the function should return. This can be the conditional mean (point forecasts), prediction intervals, the parameters of a model, or, in fact, anything that the model returns (e.g. name of the fitted model and its likelihood). However, there are some differences in what `ro()` returns depending on what the function returns. If it is a vector, then `ro()` will produce a matrix (with values for each origin in columns). If it is a matrix, then an array is returned. Finally, if it is a list, then a list of lists is returned.

In order not to overcomplicate things, we start with collecting the conditional mean from the `predict()` function:

```
ourValue <- c("pred")
```

Remark. If you do not specify the value to return, the function will try to return everything, but it might fail, especially if many values are returned. So, to be on the safe side, **always provide the `value` when possible.**

Now that we have specified `ourCall` and `ourValue`, we can produce forecasts from the model using rolling origin. Let's say that we want three steps ahead forecasts and eight origins with the default values of all the other parameters:

```
returnedValues1 <- ro(y, h=3, origins=8,
                      call=ourCall, value=ourValue)
```

The function returns a list with all the values that we asked for plus the actual values and the holdout sample. We can calculate some basic error measure based on those values, for example, scaled Absolute Error (Petropoulos & Kourentzes, 2015):

```
apply(abs(returnedValues1$holdout - returnedValues1$pred),
      1, mean, na.rm=TRUE) /
  mean(returnedValues1$actuals)
```

```
##         h1         h2         h3
## 0.07936963 0.07179073 0.05690950
```

In this example, we use the `apply()` function to distinguish between the different forecasting horizons and have an idea of how the model performs for each of them. These numbers do not tell us much on their own, but if we compare the performance of this model with an alternative one, we could infer if one model is more appropriate for the data than the other one. For example, applying ARIMA(0,2,2) to the same data, we will get:

```
ourCall <- "predict(arima(x=data,order=c(0,2,2)),n.ahead=h)"
returnedValues2 <- ro(y, h=3, origins=8,
                      call=ourCall, value=ourValue)
apply(abs(returnedValues2$holdout - returnedValues2$pred),
      1, mean, na.rm=TRUE) /
  mean(returnedValues2$actuals)
```

```
##         h1         h2         h3
## 0.08201084 0.07927290 0.06367389
```

Comparing these errors with the ones from the previous model, we can conclude which of the approaches is more suitable for the data.

We can also plot the forecasts from the rolling origin, which shows how the models behave:

```
par(mfcol=c(2,1), mar=c(4,4,3,1))
plot(returnedValues1, main="ARIMA(0,1,1)")
plot(returnedValues2, main="ARIMA(0,2,2)")
```

In Figure 2.9, the forecasts from different origins are close to each other. This is because the data is stationary, and both models produce flat lines as forecasts. The second model, however, has a slightly higher variability because it has more parameters than the first one (bias-variance trade-off in action).

The rolling origin function from the `greybox` package also allows working with explanatory variables and returning prediction intervals if needed. Some further examples are discussed in the vignette of the package. Just run the command `vignette("ro","greybox")` in R to see it.

Practically speaking, if we have a set of forecasts from different models we can

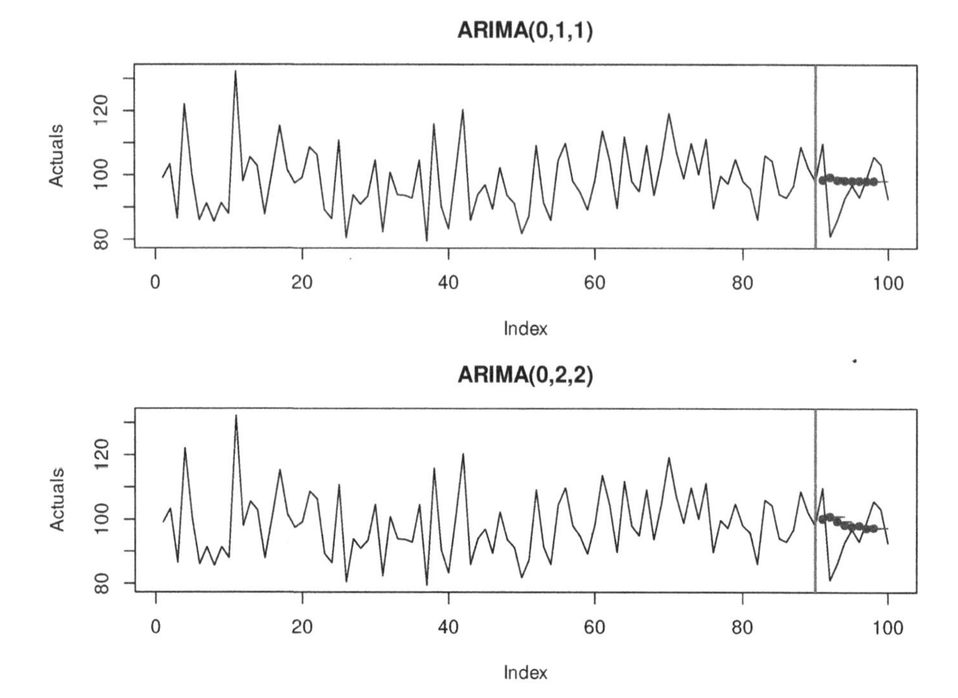

FIGURE 2.9 Rolling origin performance of two forecasting methods.

analyse the distribution of error measures and come to conclusions about the performance of models. Here is an example with an analysis of performance for $h = 1$ based on absolute errors:

```
aeValuesh1 <- cbind(abs(returnedValues1$holdout -
                        returnedValues1$pred)[1,],
                    abs(returnedValues1$holdout -
                        returnedValues2$pred)[1,])
colnames(aeValuesh1) <- c("ARIMA(0,1,1)","ARIMA(0,2,2)")
boxplot(aeValuesh1)
points(apply(aeValuesh1,2,mean),pch=16,col="red")
```

The boxplots in Figure 2.10 can be interpreted as any other boxplot applied to random variables (see, for example, discussion in Section 5.2 of Svetunkov, 2022).

Remark.

When it comes to applying `ro()` to models with explanatory variables, one can use the internal parameters `counti`, `counto`, and `countf`, which define the size

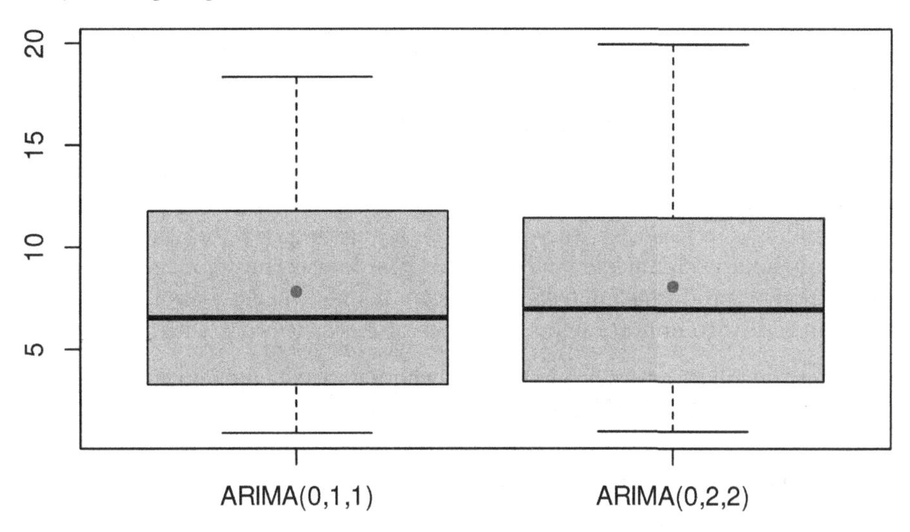

FIGURE 2.10 Boxplots of error measures of two methods.

of the in-sample, the holdout and the full sample, respectively. An example of the code in this situation is shown below with a function `alm()` from the `greybox` package being used for fitting a simple linear regression model.

```
# Generate the data
x <- rnorm(100, 100, 10)
xreg <- cbind(y=100+1.5*x+rnorm(100, 0, 10), x=x)
# Predict values from the model.
# counti and counto determine sizes for the in-sample and the holdout
ourCall <- "predict(alm(y~x, data=xreg[counti,,drop=FALSE]),
                        newdata=xreg[counto,,drop=FALSE])"
# Extract the mean only
ourValue <- "mean"
# Run rolling origin
testRO <- ro(xreg[,"y"],h=5,origins=5,ourCall,ourValue)
# plot the result
plot(testRO)
```

2.5 Statistical comparison of forecasts

After applying several competing models to the data and obtaining a distribution of forecast errors, we might find that some approaches performed very similarly. In this case, there might be a question, whether the difference between them is significant and which of the forecasting models we should select. If they produce similar forecasts then it might make sense to select the one that is less computationally expensive or easier to work with.

Consider the following artificial example, where we have four competing models and measure their performance in terms of RMSSE:

```
smallCompetition <- matrix(NA, 100, 4,
                           dimnames=list(NULL,
                                         paste0("Method",c(1:4))))
smallCompetition[,1] <- rnorm(100,1,0.35)
smallCompetition[,2] <- rnorm(100,1.2,0.2)
smallCompetition[,3] <- runif(100,0.5,1.5)
smallCompetition[,4] <- rlnorm(100,0,0.3)
```

We can check the mean and median error measures in this example in order to see, how the methods perform overall:

```
overalResults <-
  matrix(c(colMeans(smallCompetition),
           apply(smallCompetition, 2, median)),
         4, 2, dimnames=list(colnames(smallCompetition),
                             c("Mean","Median")))
round(overalResults,5)
```

```
##            Mean  Median
## Method1 1.04148 0.97523
## Method2 1.23750 1.23572
## Method3 1.00213 1.02843
## Method4 0.97789 0.92591
```

In this artificial example, it looks like the most accurate method in terms of mean and median RMSSE is Method 4, and the least accurate one is Method 2. However, the difference in terms of accuracy between Methods 1, 3, and 4 does not look substantial. So, should we conclude that Method 4 is the best? Let's first look at the distribution of errors using `vioplot()` function from `vioplot` package (Figure 2.11).

```
vioplot::vioplot(smallCompetition)
points(colMeans(smallCompetition), col="red", pch=16)
```

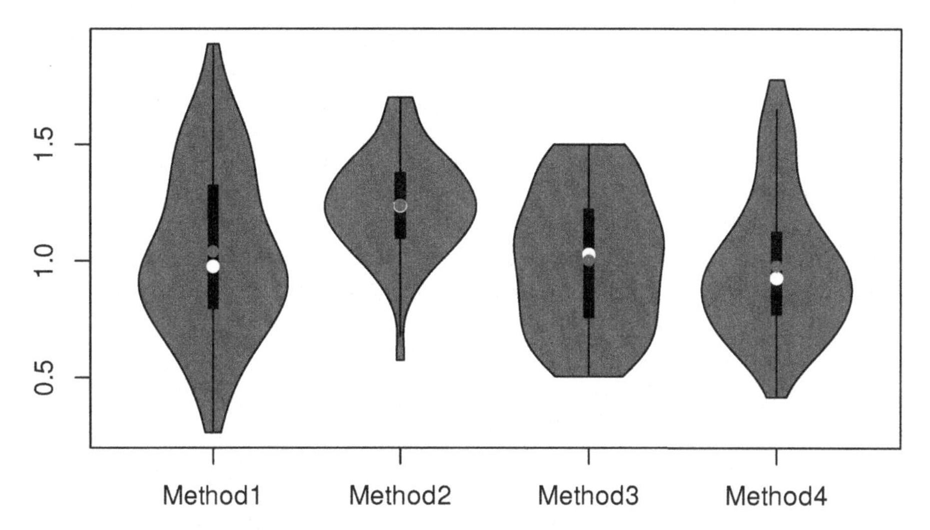

FIGURE 2.11 Boxplot of RMSE for the artificial example.

The violin plots in Figure 2.11 show that the distribution of errors for Method 2 is shifted higher than the distributions of other methods. It also looks like Method 2 is working more consistently, meaning that the variability of the errors is lower (the size of the box on the graph). It is difficult to tell whether Method 1 is better than Methods 3 and 4 or not – their boxes intersect and roughly look similar, with Method 4 having a slightly shorter box and Method 3 having the box positioned slightly lower.

This is all the basics of descriptive statistics, which allows concluding that in general, Methods 1, 3 and 4 do a better job than Method 2. This is also reflected in the mean and median error measures discussed above. So, what should we conclude?

Well, we should not make hasty decisions, and we should remember that we are dealing with a sample of data (100 time series), so inevitably, the performance of methods will change if we try them on different data sets. If we had a population of all the time series in the world, we could run our methods and make a more solid conclusion about their performances. But here, we deal with a sample. So it might make sense to see whether the difference in performance of the methods is significant. How can we do that?

We can **compare means** of distributions of errors using a parametric statistical test. We can try the F-test (see, for example Section 10.4 of Newbold, Carlson, & Thorne, 2020), which will tell us whether the mean performance of methods is similar or not. Unfortunately, this will not tell us how the methods compare. But the t-test (see Chapter 10 of Newbold et al., 2020) could be used to do that instead for pairwise comparison. One could also use a regression model with dummy variables for methods, giving us parameters and their confidence intervals (based on t-statistics), telling us how the means of methods compare. However, F-test, t-test, and t-statistics from regression rely on strong assumptions related to the distribution of the means of error measures (that it is symmetric, so that the Central Limit Theorem (CLT) works, see discussion in Section 6.2 of Svetunkov, 2022). If we had a large sample (e.g. a thousand series) and well-behaved distribution, we could try it, hoping that the CLT would work and might get something relatively meaningful. However, on 100 observations, this still could be an issue, especially given that the distribution of error measures is typically asymmetric (the estimate of the mean might be biased, which leads to many issues).

Alternatively, we can **compare medians** of distributions of errors. They are robust to outliers, so their estimates should not be too biased in case of skewed distributions on smaller samples. To have a general understanding of performance (is everything the same or is there at least one method that performs differently), we could try the Friedman test (Friedman, 1937), which could be considered a nonparametric alternative of the F-test. This should work in our case but will not tell us how specifically the methods compare. We could try the Wilcoxon signed-ranks test (Wilcoxon, 1945), which could be considered a nonparametric counterpart of the t-test. However, it only applies to two variables, while we want to compare four.

Luckily, there is the Nemenyi/MCB test (Demšar, 2006; Koning, Franses, Hibon, & Stekler, 2005; Kourentzes, 2012). What the test does, is it ranks the performance of methods for each time series and then takes the mean of those ranks and produces confidence bounds for those means. The means of ranks correspond to the medians, so by using this test, we compare the medians of errors of different methods. If the confidence bounds for different methods intersect, we can conclude that the medians are not different from a statistical point of view. Otherwise, we can see which of the methods has a higher rank and which has the lower one. There are different ways to present the test results, and there are several R functions that implement it, including `nemenyi()` from the `tsutils` package. However, we will use the function `rmcb()` from the `greybox`, which has more flexible plotting capabilities, supporting all the default parameters for the `plot()` method.

```
rmcb(smallCompetition, outplot="none") |>
    plot(outplot="mcb", main="")
```

Figure 2.12 shows that Methods 1, 3, and 4 are not statistically different on the 5% level – their intervals intersect, so we cannot tell the difference between them, even though the mean rank of Method 4 is lower than for the other

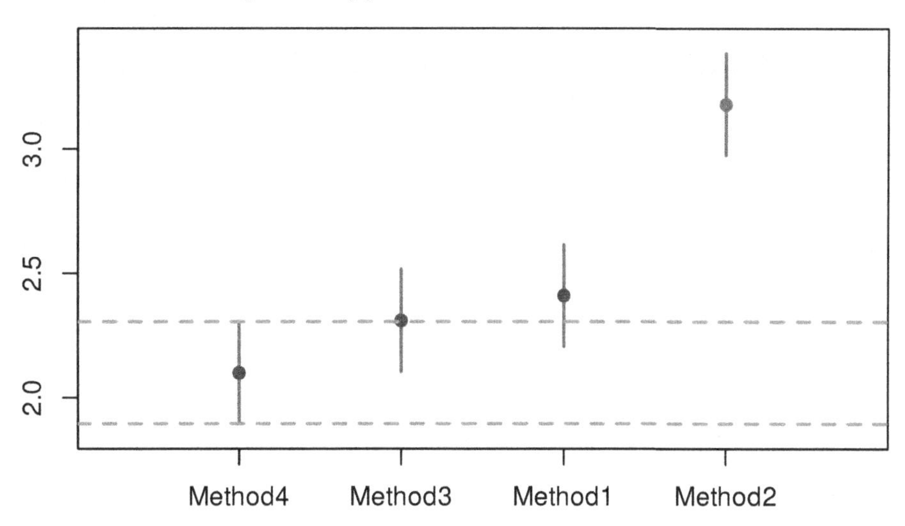

FIGURE 2.12 MCB test results for small competition.

methods. Method 2, on the other hand, is significantly worse than the other methods on the 5% level: it has the highest mean rank of all, and its interval does not intersect with the intervals of other methods.

Note that while this is a good way of presenting the results, all the MCB test does is a comparison of mean ranks. It does not tell much about the distribution of errors and neglects the distances between values (i.e. 0.1 is lower than 0.11, so the first method has a lower rank, which is precisely the same result as with comparing 0.1 and 100). This happens because by doing the test, we move from a numerical scale to the ordinal one (see Section 1.2 of Svetunkov, 2022). Finally, like any other statistical test, it will become more powerful when the sample increases. We know that the null hypothesis "variables are equal to each other" in reality is always wrong (see Section 7.1 of Svetunkov, 2022), so the increase of sample size will lead at some point to the correct conclusion: methods are statistically different. Here is a demonstration of this assertion:

```
largeCompetition <-
  matrix(NA, 100000, 4,
         dimnames=list(NULL, paste0("Method",c(1:4))))
# Generate data
largeCompetition[,1] <- rnorm(100000,1,0.35)
largeCompetition[,2] <- rnorm(100000,1.2,0.2)
largeCompetition[,3] <- runif(100000,0.5,1.5)
largeCompetition[,4] <- rlnorm(100000,0,0.3)
# Run the test
```

```
rmcb(largeCompetition, outplot="none") |>
    plot(outplot="mcb", main="")
```

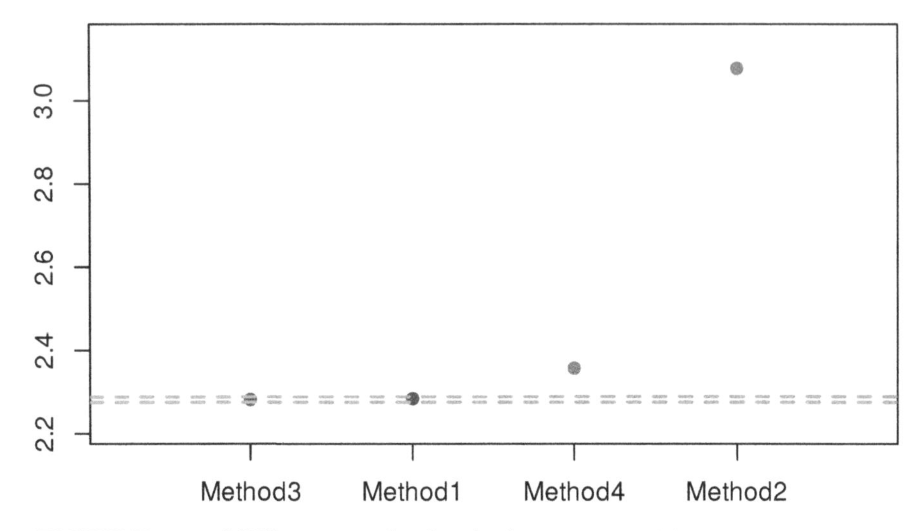

FIGURE 2.13 MCB test results for the large competition.

In the plot in Figure 2.13, Method 4 has become significantly worse than Methods 1 and 3 in terms of the mean ranks on the 5% level (note that it was winning in the small competition). The difference between Methods 1 and 3 is still not significant at 5%, but it would become so if we continued increasing the sample size. This example tells us that we need to be careful when selecting the best method, as this might change under different circumstances. At least we knew from the start that Method 2 was not suitable.

3

Time series components and simple forecasting methods

Before we turn to the state space framework, ETS, ARIMA, and other models, we need to discuss time series decomposition and the ETS taxonomy. These topics lie at the heart of ETS models and are essential for understanding further material.

In this chapter, we start with a discussion of time series components, then move to the idea of decomposing time series into distinct components and applying simple forecasting methods, including Naïve, Global Average, Simple Moving Average, and Simple Exponential Smoothing.

3.1 Time series components

The main idea behind many forecasting techniques is that any time series can include several unobservable components, such as:

1. **Level** of the series – the average value for a specific time period;
2. **Growth** of the series – the average increase or decrease of the value over a period of time;
3. **Seasonality** – a pattern that repeats itself with a fixed periodicity;
4. **Error** – unexplainable white noise.

Remark. Sometimes, the researchers also include **Cycle** component, referring to aperiodic long term changes of time series. We do not discuss it here because it is not useful for what follows.

The level is the fundamental component that is present in any time series. In the simplest form (without variability), when plotted on its own without other components, it will look like a straight line, shown, for example, in Figure 3.1.

```
level <- rep(100,40)
```

```
plot(ts(level, frequency=4),
    type="l", xlab="Time", ylab="Sales", ylim=c(80,160))
```

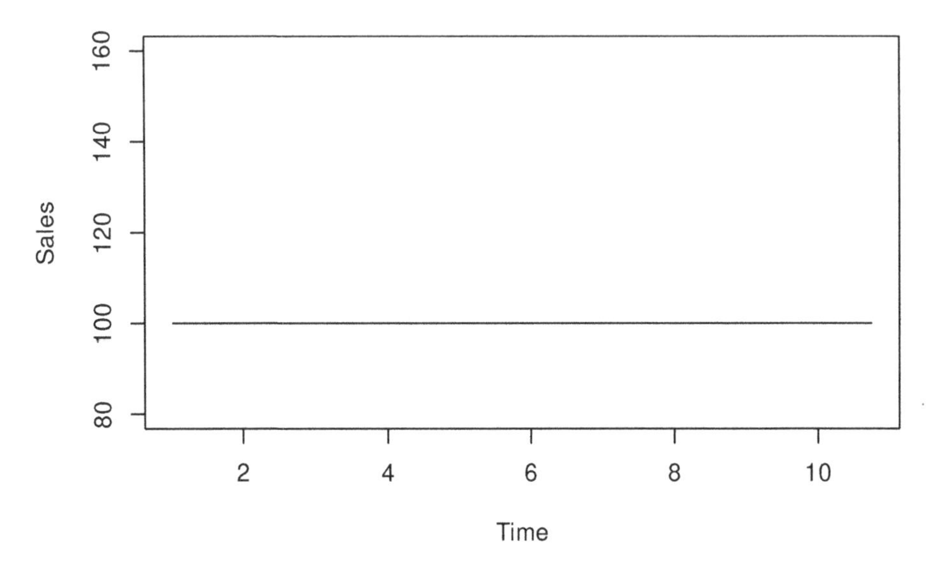

FIGURE 3.1 Level of time series without any variability.

If the time series exhibits growth, the level will change depending on the observation. For example, if the growth is positive and constant, we can update the level in Figure 3.1 to have a straight line with a non-zero slope as shown in Figure 3.2.

```
growth <- c(1:40)
plot(ts(level+growth, frequency=4),
    type="l", xlab="Time", ylab="Sales", ylim=c(80,160))
```

The seasonal pattern will introduce some similarities from one period to another. This pattern does not have to literally be seasonal, like beer sales being higher in summer than in winter (seasons of the year). Any pattern with a fixed periodicity works: the number of hospital visitors is higher on Mondays than on Saturdays or Sundays because people tend to stay at home over the weekend. This can be considered as the day of week seasonality. Furthermore, if we deal with hourly data, sales are higher during the daytime than at night (hour of the day seasonality). Adding a deterministic seasonal component to the example above will result in fluctuations around the straight line, as shown in Figure 3.3.

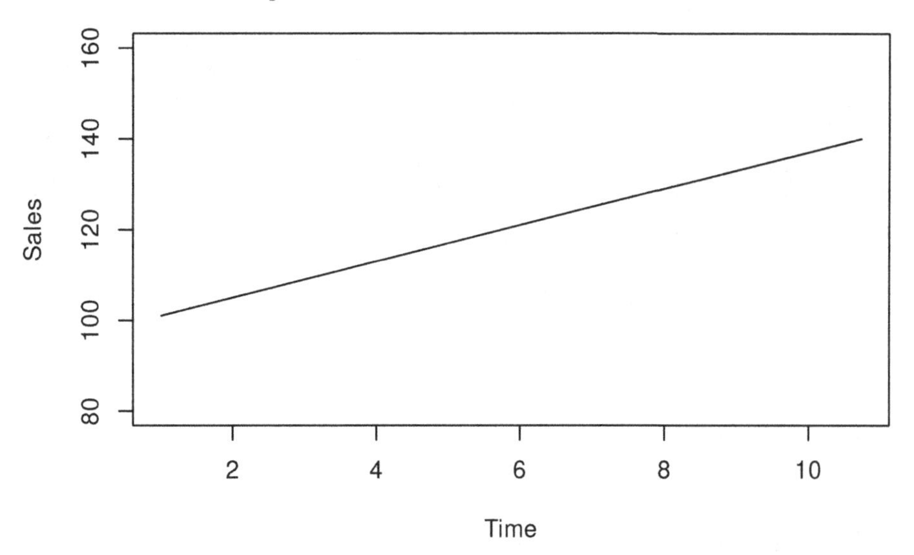

FIGURE 3.2 Time series with a positive trend and no variability.

```
seasonal <- rep(c(10,15,-20,-5),10)
plot(ts(level+growth+seasonal, frequency=4),
     type="l", xlab="Time", ylab="Sales", ylim=c(80,160))
```

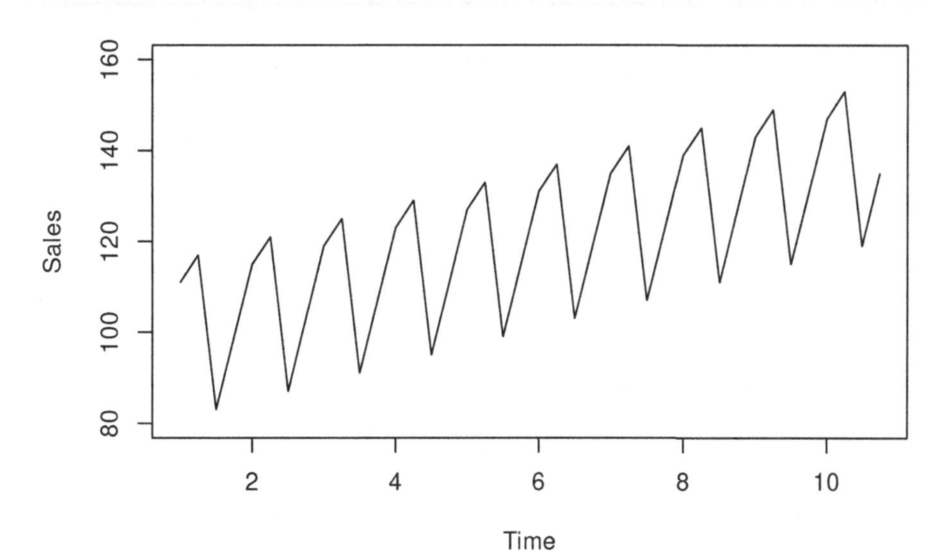

FIGURE 3.3 Time series with a positive trend, seasonal pattern, and no variability.

Remark. When it comes to giving names to different types of seasonality, you can meet terms like "monthly" and "weekly" or "intra-monthly" and "intra-weekly". In some cases these names are self explanatory (e.g. when we have monthly data and use the term "monthly" seasonality), but in general this might be misleading. This is why I prefer "frequency 1 of frequency 2" naming, e.g. "month of year" or "week of year", which is more precise and less ambiguous than the names mentioned above.

Finally, we can introduce the random error to the plots above to have a more realistic time series as shown in Figure 3.4.

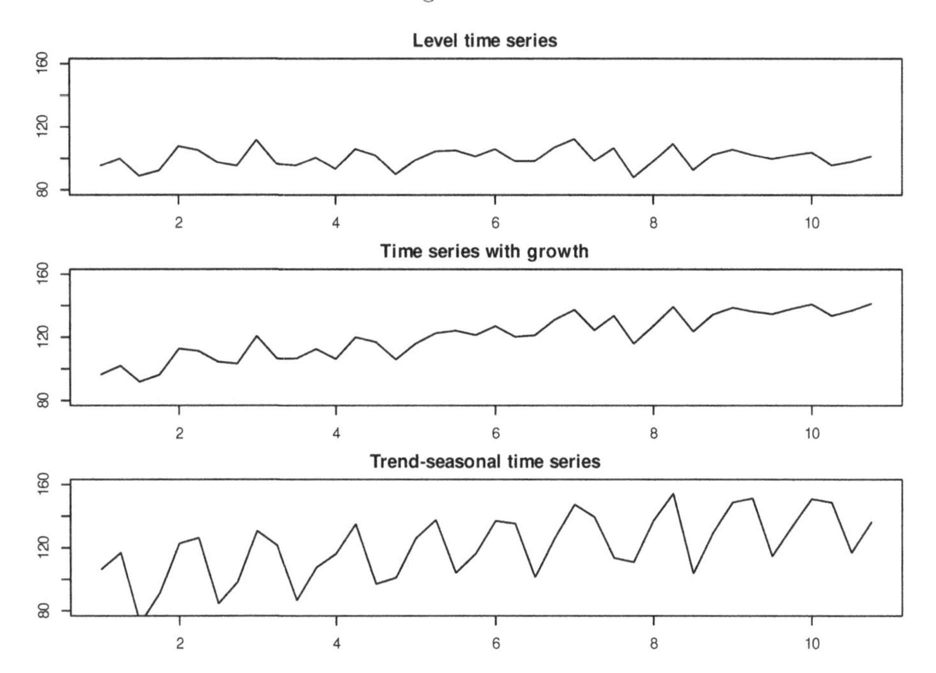

FIGURE 3.4 Time series with random errors.

Figure 3.4 shows artificial time series with the above components. The level, growth, and seasonal components in those plots are **deterministic**, they are fixed and do not evolve over time (growth is positive and equal to 1 from year to year). However, in real life, typically, these components will have more complex dynamics, changing over time and thus demonstrating their **stochastic** nature. For example, in the case of stochastic seasonality, the seasonal shape might change, and instead of having peaks in sales in December, the data would exhibit peaks in February due to the change in consumers' behaviour.

Remark. Each textbook and paper might use slightly different names to refer to the aforementioned components. For example, in classical decomposition (Warren M. Persons, 1919), it is assumed that (1) and (2) jointly represent a "trend" component so that a model will contain error, trend, and seasonality.

When it comes to ETS, the growth component (2) is called "trend", so the model consists of the four components: level, trend, seasonal, and the error term. We will use the ETS formulation in this monograph. According to it, the components can interact in one of two ways: additively or multiplicatively. The pure additive model, in this case, can be written as:

$$y_t = l_{t-1} + b_{t-1} + s_{t-m} + \epsilon_t, \tag{3.1}$$

where l_{t-1} is the level, b_{t-1} is the trend, s_{t-m} is the seasonal component with periodicity m (e.g. 12 for months of year data, implying that something is repeated every 12 months) – all these components are produced on the previous observations and are used on the current one. Finally, ϵ_t is the error term, which follows some distribution and has zero mean. The pure additive models were plotted in Figure 3.4. Similarly, the pure multiplicative model is:

$$y_t = l_{t-1} b_{t-1} s_{t-m} \varepsilon_t, \tag{3.2}$$

where ε_t is the error term with a mean of one. The interpretation of the model (3.1) is that the different components add up to each other, so, for example, the sales increase over time by the value b_{t-1}, and each January they typically change by the amount s_{t-m}, and in addition there is some randomness in the model. The pure additive models can be applied to data with positive, negative, and zero values. In the case of the multiplicative model (3.2), the interpretation is different, showing by how many times the sales change over time and from one season to another. The sales, in this case, will change every January by $(s_{t-m} - 1)\%$ from the baseline. The model (3.2) only works on strictly positive data (data with purely negative values are also possible but rare in practice).

It is also possible to define mixed models in which, for example, the trend is additive but the other components are multiplicative:

$$y_t = (l_{t-1} + b_{t-1}) s_{t-m} \varepsilon_t \tag{3.3}$$

These models work well in practice when the data has large values, far from zero. In other cases, however, they might break and produce strange results (e.g. negative values on positive data), so the conventional decomposition techniques only consider the pure models.

Remark. Sometimes the model with time series components is compared with the regression model. Just to remind the reader, the latter can be formulated as:

$$y_t = a_0 + a_1 x_{1,t} + a_2 x_{2,t} + \dots + a_n x_{n,t} + \epsilon_t,$$

where a_j is a j-th parameter for an explanatory variable $x_{j,t}$. One of the mistakes that is made in the forecasting context in this case, is to assume that the components resemble explanatory variables in the regression context. This is incorrect. The components correspond to the parameters of a regression model if we allow them to vary over time. We will show in Section 10.5 an example of how the seasonal component s_t can also be modelled via the parameters of a regression model.

3.2 Classical Seasonal Decomposition

3.2.1 How to do?

One of the classical textbook methods for decomposing the time series into unobservable components is "Classical Seasonal Decomposition" (Warren M. Persons, 1919). It assumes either a pure additive or pure multiplicative model, is done using Centred Moving Averages (CMA), and is focused on splitting the data into components, not on forecasting. The idea of the method can be summarised in the following steps:

1. Decide which of the models to use based on the type of seasonality in the data: additive (3.1) or multiplicative (3.2);
2. Smooth the data using a CMA of order equal to the periodicity of the data m. If m is an odd number then the formula is:

$$d_t = \frac{1}{m} \sum_{i=-(m-1)/2}^{(m-1)/2} y_{t+i}, \tag{3.4}$$

   which means that, for example, the value of d_t on Thursday is the average of all the actual values from Monday to Sunday. If m is an even number then a different weighting scheme is typically used, involving the inclusion of an additional value:

$$d_t = \frac{1}{m} \left(\frac{1}{2} \left(y_{t+(m-1)/2} + y_{t-(m-1)/2} \right) + \sum_{i=-(m-2)/2}^{(m-2)/2} y_{t+i} \right), \tag{3.5}$$

   which means that we use half of December of the previous year and half of December of the current year to calculate the CMA in June.

The values d_t are placed in the middle of the window going through the series (e.g. as above, on Thursday, the average will contain values from Monday to Sunday).

The resulting series is deseasonalised. This is because when we average (e.g. sales in a year) we automatically remove the potential seasonality, which can be observed each period individually. A drawback of using CMA is that we inevitably lose $\frac{m}{2}$ observations at the beginning and the end of the series.

In R, the `ma()` function from the `forecast` package implements CMA.

3. De-trend the data:

- For the additive decomposition, this is done using: ${y'}_t = y_t - d_t$;
- For the multiplicative decomposition, it is: ${y'}_t = \frac{y_t}{d_t}$;

After that the series will only contain the residuals together with the seasonal component (if the data is seasonal).

4. If the data is seasonal, the average value for each period is calculated based on the de-trended series, e.g. we produce average seasonal indices for each January, February, etc. This will give us the set of seasonal indices s_t. If the data is non-seasonal, we skip this step;

5. Calculate the residuals based on what you assume in the model:

- additive seasonality: $e_t = y_t - d_t - s_t$;
- multiplicative seasonality: $e_t = \frac{y_t}{d_t s_t}$;
- no seasonality: $e_t = {y'}_t$.

After doing steps (1) – (5), we will obtain several variables containing the components of a time series: s_t, d_t, and e_t.

Remark. The functions in R typically allow selecting between additive and multiplicative seasonality. There is no option for "none", and so even if the data is not seasonal, you will nonetheless get values for s_t in the output. Also, notice that the classical decomposition assumes that there is a deseasonalised series d_t but does not make any further split of this variable into level l_t and trend b_t.

3.2.2 A couple of examples

An example of the classical decomposition in R is the `decompose()` function from the `stats` package. Here is an example with a pure multiplicative model and `AirPassengers` data (Figure 3.5).

```
decompose(AirPassengers,
          type="multiplicative") |>
  plot()
```

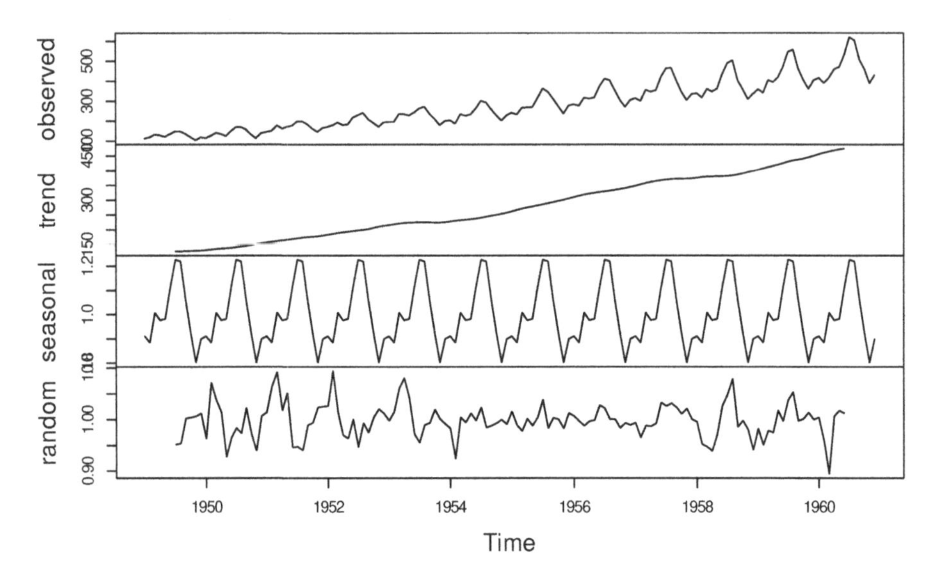

FIGURE 3.5 Decomposition of AirPassengers time series according to multiplicative model.

We can see from Figure 3.5 that the function has smoothed the original series and produced the seasonal indices. Note that the trend component has gaps at the beginning and the end. This is because the method relies on CMA (see above). Note also that the error term still contains some seasonal elements, which is a downside of such a simple decomposition procedure. However, the lack of precision in this method is compensated for by the simplicity and speed of calculation. Note again that the trend component in the `decompose()` function is in fact $d_t = l_t + b_t$.

Figure 3.6 shows an example of decomposition of the **non-seasonal data** (we assume the pure additive model in this example).

```
ts(c(1:100)+rnorm(100,0,10),frequency=12) |>
  decompose(type="additive") |>
  plot()
```

As we can see from Figure 3.6, the original data is not seasonal, but the

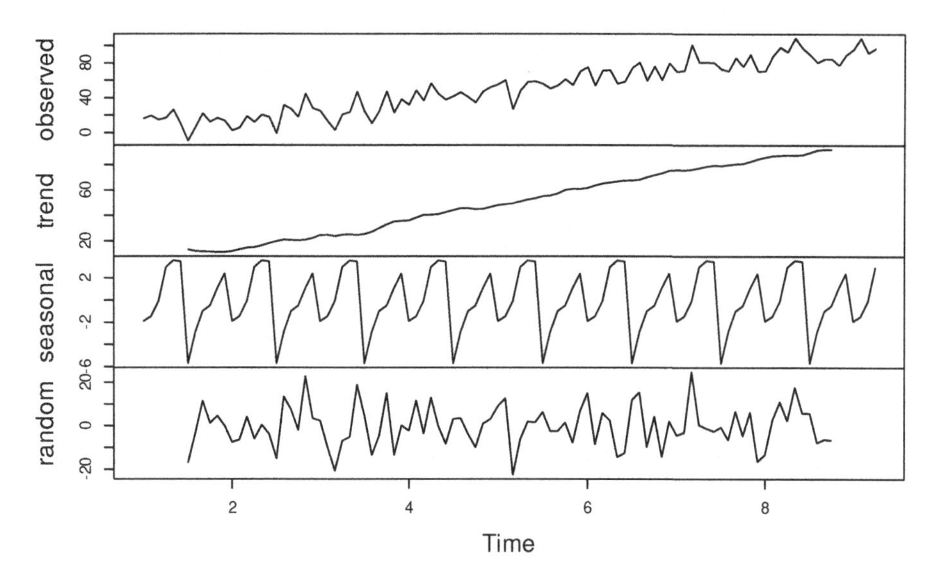

FIGURE 3.6 Decomposition of AirPassengers time series according to the additive model.

decomposition assumes that it is and proceeds with the default approach, returning a seasonal component. You get what you ask for.

3.2.3 Multiple seasonal decomposition

A simple extension to the classical decomposition explained in Subsection 3.2.1 is the decomposition of data with multiple seasonal frequencies. In that case we need to use several CMAs and apply them sequentially to get rid of seasonaliities starting from the lower values and moving to the higher ones. This is better explained with an example of half-hourly data with two seasonal patterns: 48 half-hours per day and 336 (48×7) half-hours per week. Assuming the pure multiplicative model, the decomposition procedure can be summarised as:

1. Smooth the data:
    a. Use with CMA(48). This way we get rid of the lower level frequency and obtain the smooth pattern for the higher seasonal frequency of 336, $d_{1,t}$;
    b. Smooth the same actual values with CMA(336). This way we get rid of both seasonal patterns and extract the trend, $d_{2,t}$;
2. Extract seasonal patterns:

   a. Divide the actual values by $d_{1,t}$ to get seasonal indices for half-hours of day: $y'_{1,t} = \frac{y_t}{d_{1,t}}$;
   b. Divide $d_{1,t}$ by $d_{2,t}$ to get seasonal indices for half-hours of week: $y'_{2,t} = \frac{d_{1,t}}{d_{2,t}}$;
3. Get seasonal indices:
   a. Average out the values of $y'_{1,t}$ for each period to get seasonal indices for half-hours of day $s_{1,t}$;
   b. Do the same as (3.a) for each period of $y'_{2,t}$ to get half-hours of week $s_{2,t}$;
4. Finally, extract the residuals via: $e_t = \frac{y_t}{d_{2,t}s_{1,t}s_{2,t}}$ and obtain the components of the decomposed series.

This procedure can be automated for data with more than two frequencies. The logic would be the same, we would just need to introduce more sub-steps (c, d, e) for each of the steps above (1, 2, 3). The procedure described above is implemented in the `msdecompose()` function from the `smooth` package. Here is how it works on the half-hourly data (from Taylor, 2003a) provided in the `taylor` object from the `forecast` package (Figure 3.7):

```
msdecompose(forecast::taylor, lags=c(48,336),
            type="multiplicative") |>
    plot(which=12, main="")
```

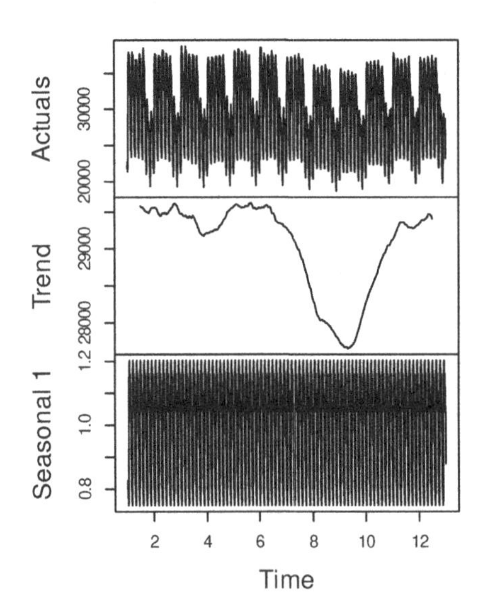

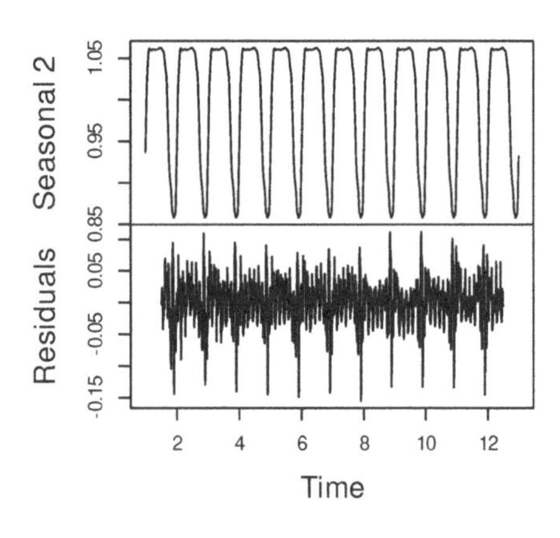

FIGURE 3.7 Decomposition of half-hourly electricity demand series according to the multiple seasonal classical decomposition.

The main limitation of this approach (similar to the conventional decomposition) is that it assumes that the seasonality does not evolve over time. This is a serious limitation if the seasonality needs to be used in forecasting. However, the procedure is simple and fast, and can be used as a starting point for the estimation of more complicated models (see, for example, Section 12.1).

3.2.4 Other decomposition techniques

There are other techniques that decompose series into error, trend, and seasonal components but make different assumptions about each component. The general procedure, however, always remains the same:

1. Smooth the original series;
2. Extract the seasonal components;
3. Smooth them out.

The methods differ in the smoother they use (e.g. LOESS uses a bisquare function instead of CMA), and in some cases, multiple rounds of smoothing are performed to make sure that the components are split correctly.

There are many functions in R that implement seasonal decomposition. Here is a small selection:

- `decomp()` from the `tsutils` package does classical decomposition and fills in the tail and head of the smoothed trend with forecasts from Exponential Smoothing;
- `stl()` from the `stats` package uses a different approach – seasonal decomposition via LOESS. It is an iterative algorithm that smooths the states and allows them to evolve over time. So, for example, the seasonal component in STL can change;
- `mstl()` from the `forecast` package does the STL for data with several seasonalities.

3.2.5 "Why bother?"

"Why decompose?" you may wonder at this point. Understanding the idea behind decomposition and how to perform it helps with the understanding of ETS, which relies on it. From a practical point of view, it can be helpful if you want to see if there is a trend in the data and whether the residuals contain outliers or not. It will *not* show you whether the data is seasonal or not as the seasonality is *assumed* in the decomposition (I stress this because many students think otherwise). Additionally, when seasonality cannot be added to a particular model under consideration, decomposing the series, predicting the trend, and then reseasonalising can be a viable solution. Finally, the values from the decomposition can be used as starting points for the estimation of components in ETS or other dynamic models relying on the error-trend-seasonality.

3.3 Simple forecasting methods

Now that we understand that time series might contain different components and that there are approaches for their decomposition, we can introduce several simple forecasting methods that can be used in practice, at least as benchmarks. Their usage aligns with the idea of forecasting principles discussed in Section 1.2. We discuss the most popular methods for the following specific types of time series:

1. Level time series: Naïve (Subsection 3.3.1), Global Average (Subsection 3.3.2), Simple Moving Average (Subsection 3.3.3), Simple Exponential Smoothing (Section 3.4);
2. Trend time series: Random Walk with drift (Subsection 3.3.4), Global Trend (Subsection 3.3.5);
3. Seasonal time series: Seasonal Naïve (Subsection 3.3.6).

3.3.1 Naïve

Naïve is one of the simplest forecasting methods. According to it, the one-step-ahead forecast is equal to the most recent actual value:

$$\hat{y}_t = y_{t-1}. \tag{3.6}$$

Using this approach might sound naïve indeed, but there are cases where it is very hard to outperform. Consider an example with temperature forecasting. If we want to know what the temperature outside will be in five minutes, then Naïve would be typically very accurate: the temperature in five minutes will be the same as it is right now.

The statistical model underlying Naïve is called the "Random Walk" and is written as:

$$y_t = y_{t-1} + \epsilon_t. \tag{3.7}$$

The variability of ϵ_t will impact the speed of change of the data: the higher it is, the more rapid the values will change. Random Walk and Naïve can be represented in Figure 3.8. In the example of R code below, we use the Simple Moving Average (discussed later in Subsection 3.3.3) of order 1 to generate the data from Random Walk and then produce forecasts using Naïve.

```
sim.sma(1, 120) |>
    sma(order=1,
        h=10, holdout=TRUE) |>
    plot(which=7, main="")
```

As shown from the plot in Figure 3.8, Naïve lags behind the actual time series

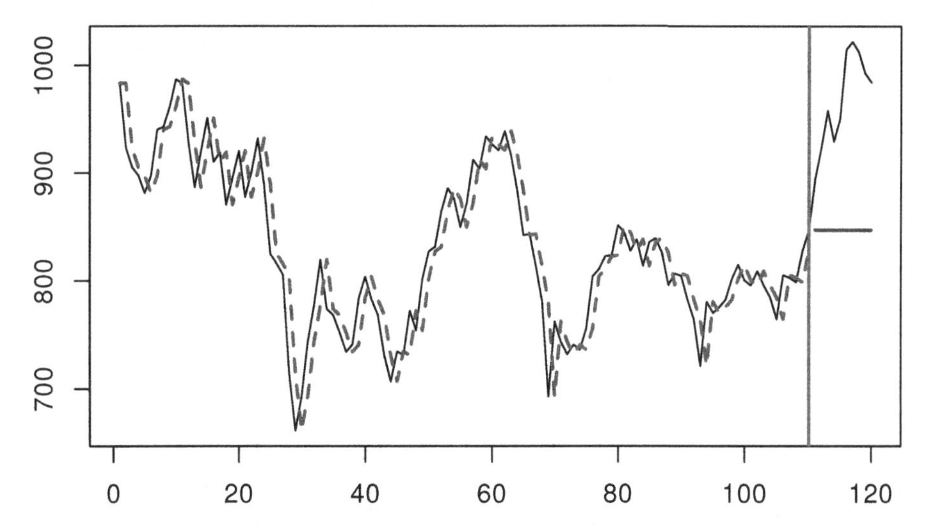

FIGURE 3.8 A Random Walk example.

by one observation because of how it is constructed via equation (3.6). The point forecast corresponds to the straight line parallel to the x-axis. Given that the data was generated from Random Walk, the point forecast shown in Figure 3.8 is the best possible forecast for the time series, even though it exhibits rapid changes in the level.

Note that if the time series exhibits level shifts or other types of unexpected changes in dynamics, Naïve will update rapidly and reach the new level instantaneously. However, because it only has a memory of one (last) observation, it will not filter out the noise in the data but rather copy it into the future. So, it has limited usefulness in demand forecasting (although it has applications in financial analysis). However, being the simplest possible forecasting method, it is considered one of the basic forecasting benchmarks. If your model cannot beat it, it is not worth using.

3.3.2 Global Mean

While Naïve considers only one observation (the most recent one), the Global Mean (aka "global average") relies on all the observations in the data:

$$\hat{y}_t = \bar{y} = \frac{1}{T} \sum_{t=1}^{T} y_t, \tag{3.8}$$

where T is the sample size. The model underlying this forecasting method is called "global level" and is written as:

$$y_t = \mu + \epsilon_t, \tag{3.9}$$

so that the $\bar{y}$ is an estimate of the fixed expectation μ. Graphically, this is represented with a straight line going through the series as shown in Figure 3.9.

```
rnorm(120, 100, 10) |>
  es("ANN", persistence=0,
     h=10, holdout=TRUE) |>
  plot(which=7, main="")
```

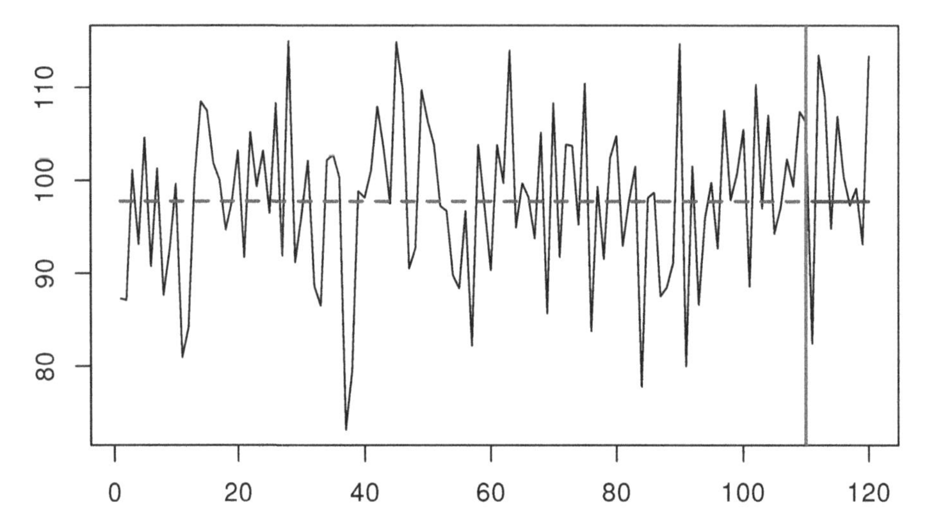

FIGURE 3.9 A global level example.

The series shown in Figure 3.9 is generated from the global level model, and the point forecast corresponds to the forecast from the Global Mean method. Note that the method assumes that the weights of the in-sample observations are equal, i.e. the first observation has precisely the exact weight of $\frac{1}{T}$ as the last one (being as important as the last one). Suppose the series exhibits some changes in level over time. In that case, the Global Mean will not be suitable because it will produce the averaged out forecast, considering values for parts before and after the change. However, Global Mean works well in demand forecasting context and is a decent benchmark on intermittent data (discussed in Chapter 13).

3.3.3 Simple Moving Average

Naïve and Global Mean can be considered as opposite points in the spectrum of possible level time series (although there are series beyond Naïve, see for example ARIMA(0,1,1) with $\theta_1 > 0$, discussed in Chapter 8). The series between them exhibits slow changes in level and can be modelled using different forecasting approaches. One of those is the Simple Moving Average (SMA), which uses the mechanism of the mean for a small part of the time series. It relies on the formula:

$$\hat{y}_t = \frac{1}{m} \sum_{j=1}^{m} y_{t-j}, \tag{3.10}$$

which implies going through time series with something like a "window" of m observations and using their average for forecasting. The order m determines the length of the memory of the method: if it is equal to 1, then (3.10) turns into Naïve, while in the case of $m = T$ it transforms into Global Mean. The order m is typically decided by a forecaster, keeping in mind that the lower m corresponds to the shorter memory method, while the higher one corresponds to the longer one.

Svetunkov and Petropoulos (2018) have shown that SMA has an underlying non-stationary AR(m) model with $\phi_j = \frac{1}{m}$ for all $j = 1, \dots, m$. While the conventional approach to forecasting from SMA is to produce the straight line, equal to the last fitted value, Svetunkov and Petropoulos (2018) demonstrate that, in general, the point forecast of SMA does not correspond to the straight line. This is because to calculate several steps ahead point forecasts, the actual values in (3.10) are substituted iteratively by the predicted ones on the observations for the holdout.

```
y <- sim.sma(10,120)
par(mfcol=c(2,2), mar=c(2,2,2,1))
# SMA(1)
sma(y$data, order=1,
    h=10, holdout=TRUE) |>
    plot(which=7)
# SMA(10)
sma(y$data, order=10,
    h=10, holdout=TRUE) |>
    plot(which=7)
# SMA(20)
sma(y$data, order=20,
    h=10, holdout=TRUE) |>
    plot(which=7)
# SMA(110)
sma(y$data, order=110,
    h=10, holdout=TRUE) |>
    plot(which=7)
```

Figure 3.10 demonstrates the time series generated from SMA(10) and several SMA models applied to it. We can see that the higher orders of SMA lead to smoother fitted lines and calmer point forecasts. On the other hand, the SMA

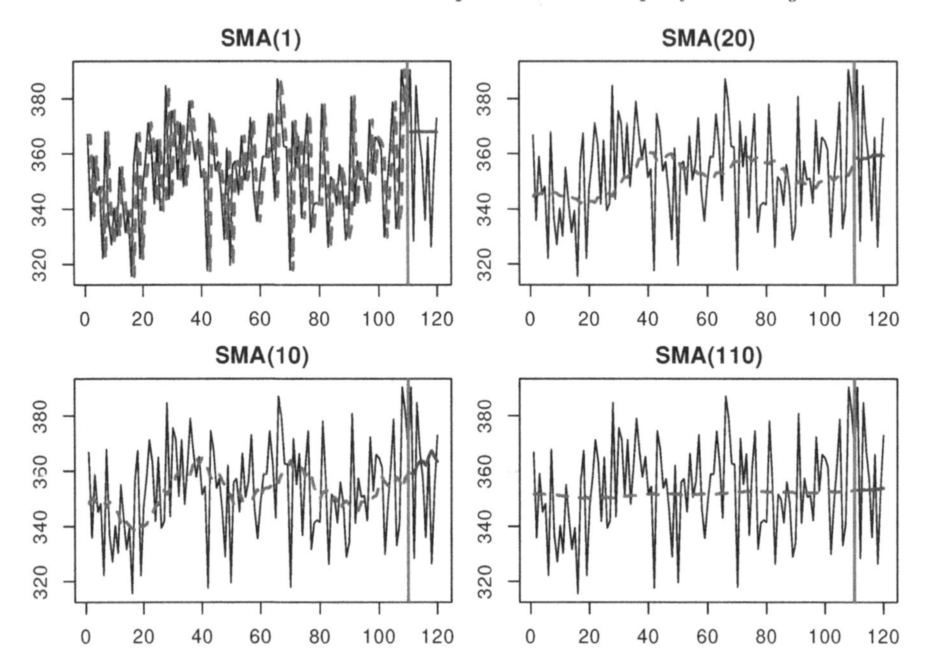

FIGURE 3.10 Examples of SMA time series and several SMA models of different orders applied to it.

of a very high order, such as SMA(110), does not follow the changes in time series efficiently, ignoring the potential changes in level. Given the difficulty with selecting the order m, Svetunkov and Petropoulos (2018) proposed using information criteria for the order selection of SMA in practice.

Finally, an attentive reader has already spotted that the formula for SMA corresponds to the procedure of CMA of an odd order from equation (3.4). They are similar, but they have a different purpose: CMA is needed to smooth out the series, and the calculated values are inserted in the middle of the average, while SMA is used for forecasting, and the point forecasts are inserted at the last period of the average.

3.3.4 Random Walk with drift

So far, we have discussed the methods used for level time series. But as mentioned in Section 3.1, there are other components in the time series. In the case of the series with a trend, Naïve, Global Mean, and SMA will be inappropriate because they would be missing the essential component. The simplest model that can be used in this case is called "Random Walk with drift", which is formulated as:

$$y_t = y_{t-1} + a_1 + \epsilon_t, \tag{3.11}$$

where a_1 is a constant term, the introduction of which leads to increasing or decreasing trajectories, depending on the value of a_1. The point forecast from this model is calculated as:

$$\hat{y}_{t+h} = y_t + a_1 h, \tag{3.12}$$

implying that the forecast from the model is a straight line with the slope parameter a_1. Figure 3.11 shows what the data generated from Random Walk with drift and $a_1 = 10$ looks like. This model is discussed in Subsection 8.1.5.

```
sim.ssarima(orders=list(i=1), lags=1, obs=120,
            constant=10) |>
    msarima(orders=list(i=1), constant=TRUE,
            h=10, holdout=TRUE) |>
    plot(which=7, main="")
```

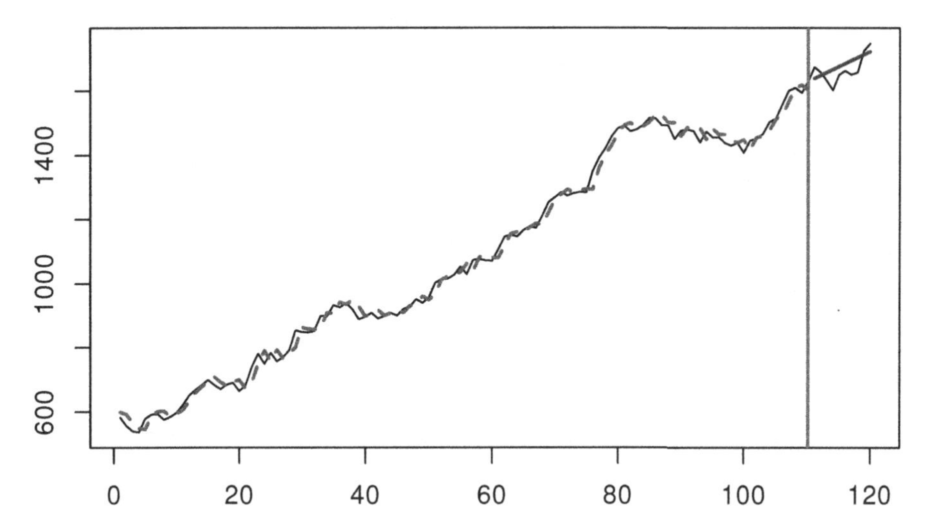

FIGURE 3.11 Random Walk with drift data and the model applied to that data.

The data in Figure 3.11 demonstrates a positive trend (because $a_1 > 0$) and randomness from one observation to another. The model is helpful as a benchmark and a special case for several other models because it is simple and requires the estimation of only one parameter.

3.3.5 Global Trend

Continuing the discussion of the trend time series, there is another simple model that is sometimes used in forecasting. The Global Trend model is formulated as:

$$y_t = a_0 + a_1 t + \epsilon_t, \tag{3.13}$$

where a_0 is the intercept and a_1 is the slope of the trend. The positive value of a_1 implies that the data exhibits growth, while the negative means decline. The size of a_1 characterises the steepness of the slope. The point forecast from this model is:

$$\hat{y}_{t+h} = a_0 + a_1(t+h), \tag{3.14}$$

implying that the forecast from the model is a straight line with the slope parameter a_1. Figure 3.12 shows how the data generated from a Global Trend with $a_0 = 10$ and $a_1 = 5$ looks like.

```
xreg <- data.frame(y=rnorm(120, 10+5*c(1:120), 10))
alm(y~trend, xreg, subset=c(1:100)) |>
    forecast(tail(xreg, 20), h=20) |>
    plot(main="")
```

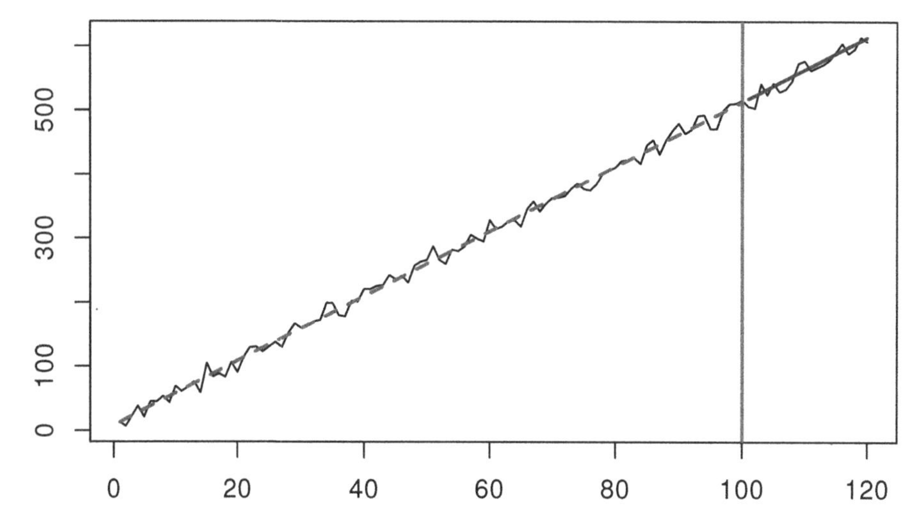

FIGURE 3.12 Global Trend data and the model applied to it.

The data in Figure 3.12 demonstrates a linear trend with some randomness around it. In this situation, the slope of the trend is fixed and does not change over time in contrast to what we observed in Figure 3.11 for a Random Walk with drift model. In some cases, the Global Trend model is the most suitable for the data. Some of the models discussed later in this monograph have the Global Trend as a special case when some restrictions on parameters are imposed (see, for example, Subsection 4.4.1).

3.3.6 Seasonal Naïve

Finally, in the case of seasonal data, there is a simple forecasting method that can be considered as a good benchmark in many situations. Similar to

Naïve, Seasonal Naïve relies only on one observation, but instead of taking the most recent value, it uses the value from the same period a season ago. For example, for producing a forecast for January 1984, we would use January 1983. Mathematically this is written as:

$$\hat{y}_t = y_{t-m}, \tag{3.15}$$

where m is the seasonal frequency. This method has an underlying model, Seasonal Random Walk:

$$y_t = y_{t-m} + \epsilon_t. \tag{3.16}$$

Similar to Naïve, the higher variability of the error term ϵ_t in (3.16) is, the faster the data exhibits changes. Seasonal Naïve does not require estimation of any parameters and thus is considered one of the popular benchmarks to use with seasonal data. Figure 3.13 demonstrates how the data generated from Seasonal Random Walk looks and how the point forecast from the Seasonal Naïve applied to this data performs.

```
y <- sim.ssarima(orders=list(i=1), lags=4,
                 obs=120, mean=0, sd=50)
msarima(y$data, orders=list(i=1), lags=4,
        h=10, holdout=TRUE) |>
   plot(which=7, main="")
```

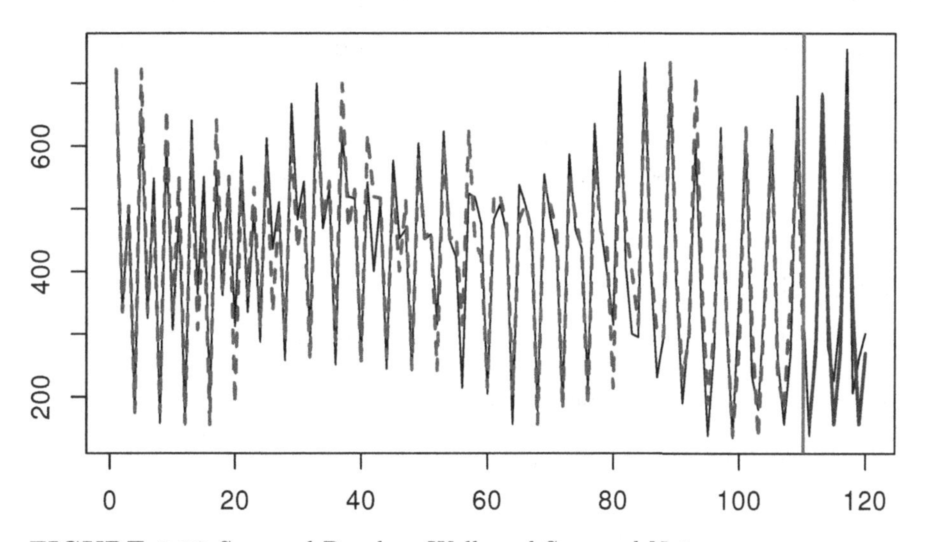

FIGURE 3.13 Seasonal Random Walk and Seasonal Naïve.

Similarly to the previous methods, if other approaches cannot outperform Seasonal Naïve, it is not worth spending time on those approaches.

3.4 Simple Exponential Smoothing

One of the most powerful and efficient forecasting methods for level time series (which is also very popular in practice according to Weller & Crone, 2012) is Simple Exponential Smoothing (sometimes also called "Single Exponential Smoothing"). It was first formulated by Brown (1956) and can be written as:

$$\hat{y}_{t+1} = \hat{\alpha} y_t + (1 - \hat{\alpha})\hat{y}_t, \tag{3.17}$$

where $\hat{\alpha}$ is the smoothing parameter, which is typically restricted within the (0, 1) region (this region is arbitrary, and we will see in Section 4.7 what is the correct one). This is one of the simplest forecasting methods. The smoothing parameter is typically interpreted as a weight between the latest actual value and the one-step-ahead predicted one. If the smoothing parameter is close to zero, then more weight is given to the previous fitted value $\hat{y}_t$ and the new information is neglected. If $\hat{\alpha} = 0$, then the method becomes equivalent to the Global Mean method, discussed in Subsection 3.3.2. When it is close to one, then most of the weight is assigned to the actual value y_t. If $\hat{\alpha} = 1$, then the method transforms into Naïve, discussed in Subsection 3.3.1. By changing the smoothing parameter value, the forecaster can decide how to approximate the data and filter out the noise.

Also, notice that this is a recursive method, meaning that there needs to be some starting point $\hat{y}_1$ to apply (3.17) to the existing data. Different initialisation and estimation methods for SES have been discussed in the literature. Still, the state of the art one is to estimate $\hat{\alpha}$ and $\hat{y}_1$ together by minimising some loss function (Hyndman, Koehler, Snyder, & Grose, 2002). Typically MSE (see Section 2.1) is used, minimising the squares of one step ahead in-sample forecast error.

3.4.1 Examples of application

Here is an example of how this method works on different time series. We start with generating a stationary series and using the `es()` function from the `smooth` package. Although it implements the ETS model, we will see in Section 4.3 the connection between SES and ETS(A,N,N). We start with the stationary time series and $\hat{\alpha} = 0$:

```
rnorm(100,100,10) |>
    es(model="ANN", h=10, persistence=0) |>
    plot(which=7, main="")
```

As we see from Figure 3.14, the SES works well in this case, capturing the deterministic level of the series and filtering out the noise. In this case, it works

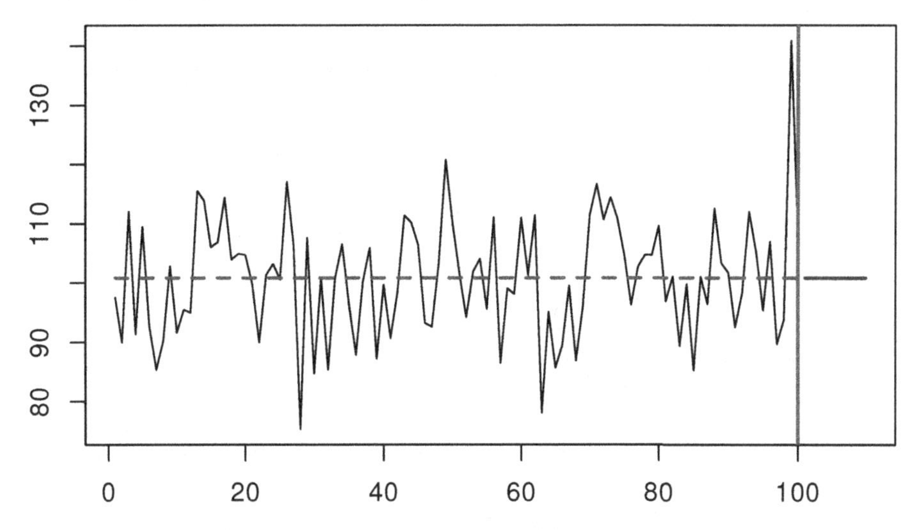

FIGURE 3.14 An example with a time series and SES forecast. $\hat{\alpha} = 0$.

like a global average applied to the data. As mentioned before, the method is flexible, so if we have a level shift in the data and increase the smoothing parameter, it will adapt and get to the new level. Figure 3.15 shows an example with a level shift in the data.

```
y <- c(rnorm(50,100,10),rnorm(50,130,10))
es(y, model="ANN", h=10, persistence=0.1) |>
    plot(7, main="")
```

With $\hat{\alpha} = 0.1$, SES manages to get to the new level, but now the method starts adapting to noise a little bit – it follows the peaks and troughs and repeats them with a lag, but with a much smaller magnitude (see Figure 3.15). Increasing the smoothing parameter, the model will react to the changes much faster, at the cost of responding more to noise. This is shown in Figure 3.16 with different smoothing parameter values.

If we set $\hat{\alpha} = 1$, we will end up with the Naïve forecasting method (see Section 3.3.1), which is not appropriate for our example (see Figure 3.17).

So, when working with SES, we need to make sure that the reasonable smoothing parameter is selected. This can be done automatically via minimising the in-sample MSE (see Figure 3.18):

```
ourModel <- es(y, model="ANN", h=10, loss="MSE")
plot(ourModel, 7, main=paste0("SES with alpha=",
                               round(ourModel$persistence,3)))
```

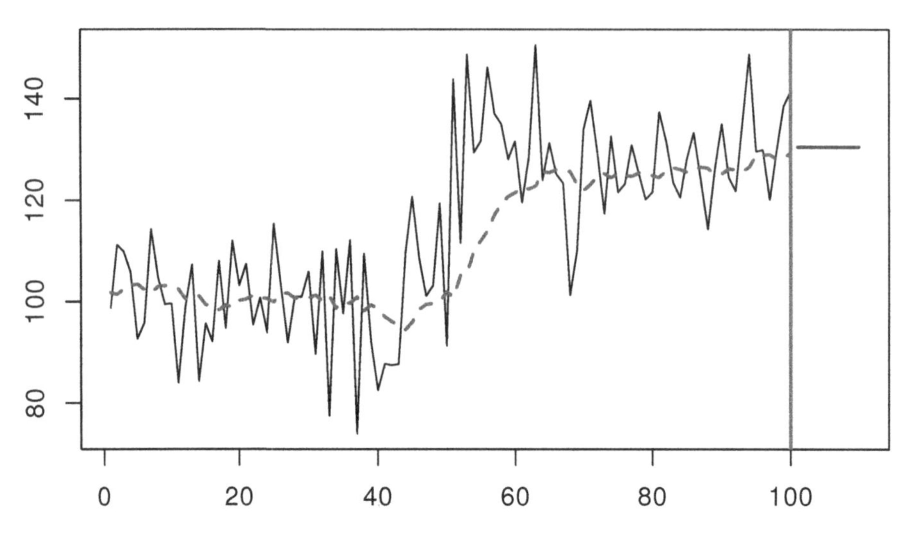

FIGURE 3.15 An example with a time series and SES forecast. $\hat{\alpha} = 0.1$.

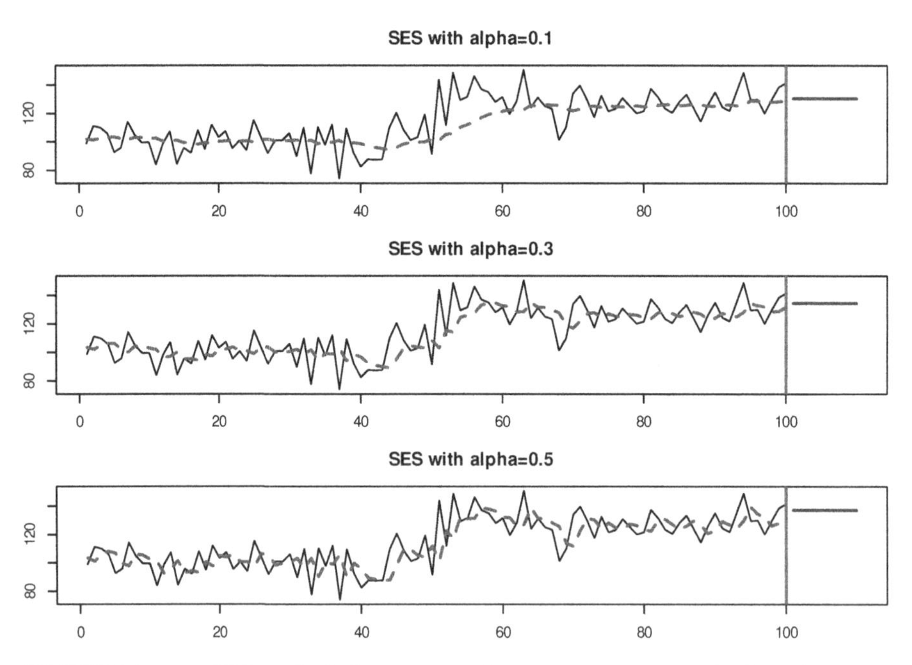

FIGURE 3.16 SES with different smoothing parameters applied to the same data.

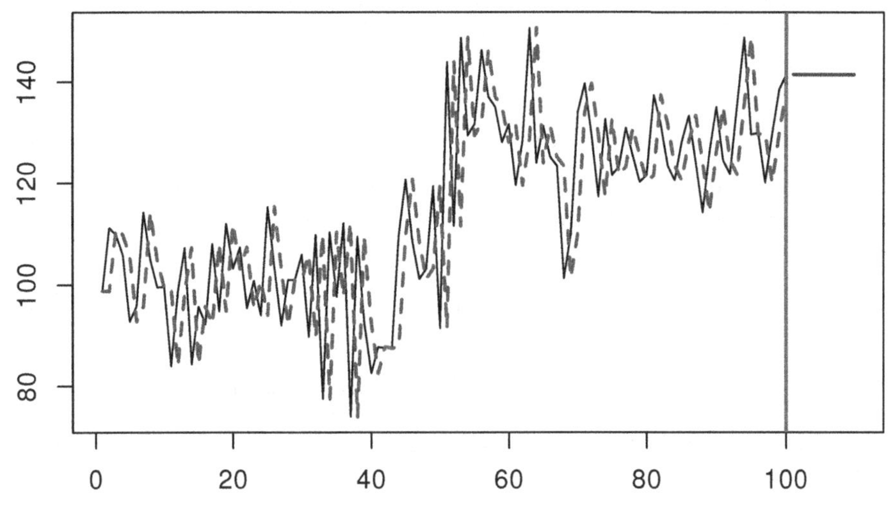

FIGURE 3.17 SES with $\hat{\alpha} = 1$.

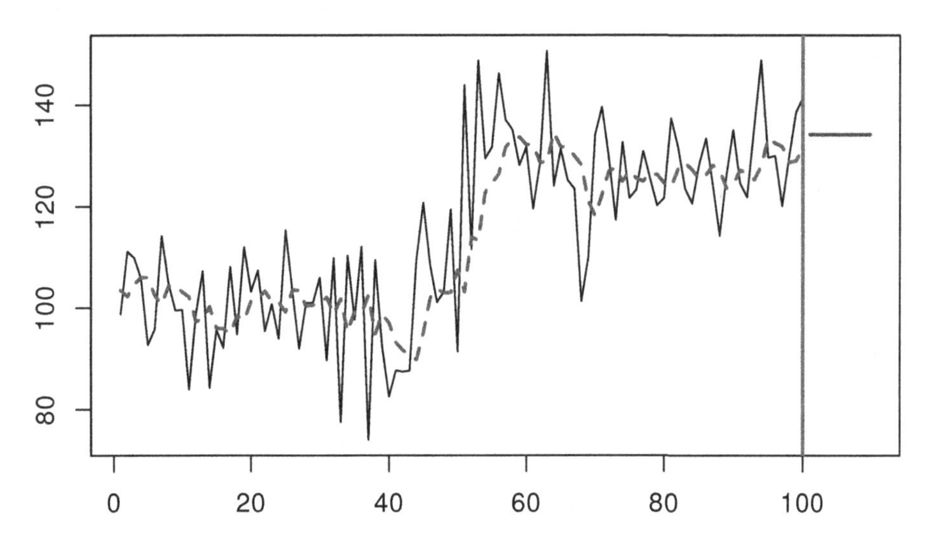

FIGURE 3.18 SES with optimal smoothing parameter.

This approach won't guarantee that we will get the most appropriate $\hat{\alpha}$. Still, it has been shown in the literature that the optimisation of smoothing parameters on average leads to improvements in terms of forecasting accuracy (see, for example, Gardner, 1985).

3.4.2 Why "exponential"?

Now, **why is it called "exponential"**? Because the same method can be represented in a different form, if we substitute $\hat{y}_t$ in right-hand side of (3.17) by the formula for the previous step:

$$\begin{aligned} \hat{y}_t =& \hat{\alpha} y_{t-1} + (1 - \hat{\alpha}) \hat{y}_{t-1}, \\ \hat{y}_{t+1} =& \hat{\alpha} y_t + (1 - \hat{\alpha}) \hat{y}_t = \\ & \hat{\alpha} y_t + (1 - \hat{\alpha}) \left(\hat{\alpha} y_{t-1} + (1 - \hat{\alpha}) \hat{y}_{t-1} \right). \end{aligned} \tag{3.18}$$

By repeating this procedure for each $\hat{y}_{t-1}$, $\hat{y}_{t-2}$, etc., we will obtain a different form of the method:

$$\hat{y}_{t+1} = \hat{\alpha} y_t + \hat{\alpha}(1 - \hat{\alpha}) y_{t-1} + \hat{\alpha}(1 - \hat{\alpha})^2 y_{t-2} + \dots + \hat{\alpha}(1 - \hat{\alpha})^{t-1} y_1 + (1 - \hat{\alpha})^t \hat{y}_1 \tag{3.19}$$

or equivalently:

$$\hat{y}_{t+1} = \hat{\alpha} \sum_{j=0}^{t-1} (1 - \hat{\alpha})^j y_{t-j} + (1 - \hat{\alpha})^t \hat{y}_1. \tag{3.20}$$

In the form (3.20), each actual observation has a weight in front of it. For the most recent observation, it is equal to $\hat{\alpha}$, for the previous one, it is $\hat{\alpha}(1 - \hat{\alpha})$, then $\hat{\alpha}(1 - \hat{\alpha})^2$, etc. These form the geometric series or an exponential curve. Figure 3.19 shows an example with $\hat{\alpha} = 0.25$ for a sample of 30 observations.

This explains the name "exponential". The term "smoothing" comes from the idea that the parameter $\hat{\alpha}$ should be selected so that the method smooths the original time series and does not react to noise.

3.4.3 Error correction form of SES

Finally, there is an alternative form of SES, known as error correction form, which can be obtained after some simple permutations. Taking that $e_t = y_t - \hat{y}_t$ is the one step ahead forecast error, formula (3.17) can be written as:

$$\hat{y}_{t+1} = \hat{y}_t + \hat{\alpha} e_t. \tag{3.21}$$

In this form, the smoothing parameter $\hat{\alpha}$ has a different meaning: it regulates how much the model reacts to the previous forecast error. In this interpretation, it no longer needs to be restricted with (0, 1) region, but we would still typically want it to be closer to zero to filter out the noise, not to adapt to it.

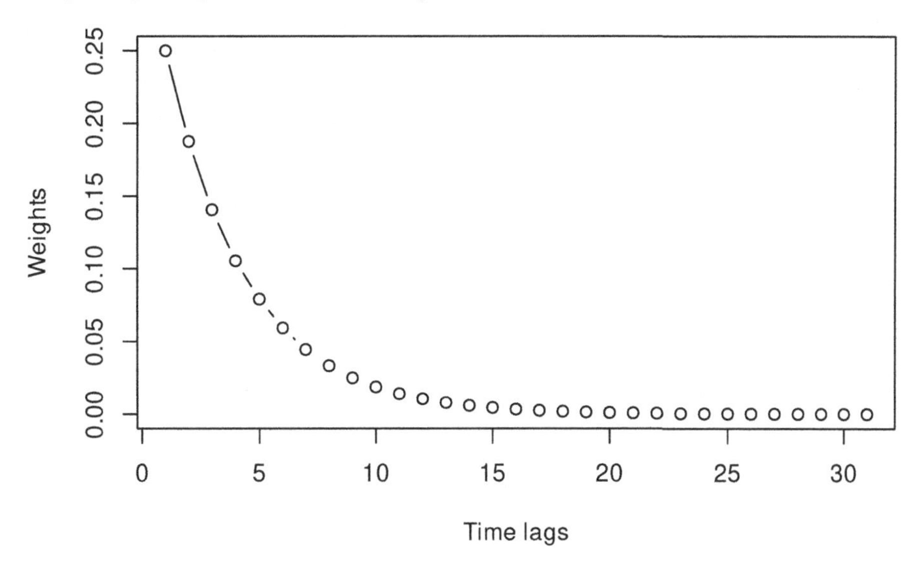

FIGURE 3.19 Example of weights distribution for $\hat{\alpha} = 0.25$.

As you see, SES is a straightforward method. It is easy to explain to practitioners, and it is very easy to implement in practice. However, this is just a forecasting method (see Section 1.4), so it provides a way of generating point forecasts but does not explain where the error comes from and how to create prediction intervals. Over the years, this was a serious limitation of the method until the introduction of state space models and ETS.

4

Introduction to ETS

Now that we know how time series can be decomposed into components, we can discuss the ETS model and its connection with Exponential Smoothing methods, focusing on the most popular of those. We do not discuss in detail how the methods were originally derived and how to work with them. Instead, we focus on the main ideas behind the conventional ETS, as formulated by Hyndman et al. (2008), and the connection between Exponential Smoothing and ETS.

The reader interested in the history of Exponential Smoothing, how it was developed, and what papers contributed to the field can refer to the reviews of Gardner (1985) and Gardner (2006). They summarise all the progress in Exponential Smoothing up until 1985 and until 2006, respectively.

4.1 ETS taxonomy

Building on the idea of time series components (from Section 3.1), we can move to the ETS taxonomy. ETS stands for "Error-Trend-Seasonality" and defines how specifically the components interact with each other. Based on the type of error, trend, and seasonality, Pegels (1969) proposed a taxonomy, which was then developed further by Hyndman et al. (2002) and refined by Hyndman et al. (2008). According to this taxonomy, error, trend, and seasonality can be:

1. Error: "Additive" (A), or "Multiplicative" (M);
2. Trend: "None" (N), or "Additive" (A), or "Additive damped" (Ad), or "Multiplicative" (M), or "Multiplicative damped" (Md);
3. Seasonality: "None" (N), or "Additive" (A), or "Multiplicative" (M).

In this taxonomy, the model (3.1) is denoted as ETS(A,A,A) while the model (3.2) is denoted as ETS(M,M,M), and (3.3) is ETS(M,A,M).

The components in the ETS taxonomy have clear interpretations: level shows average value per time period, trend reflects the change in the value, while seasonality corresponds to periodic fluctuations (e.g. increase in sales each January). Based on the the types of the components above, it is theoretically possible to devise 30 ETS models with different types of error, trend, and seasonality. Figure 4.1 shows examples of different time series with deterministic (they do not change over time) level, trend, seasonality, and with the additive error term.

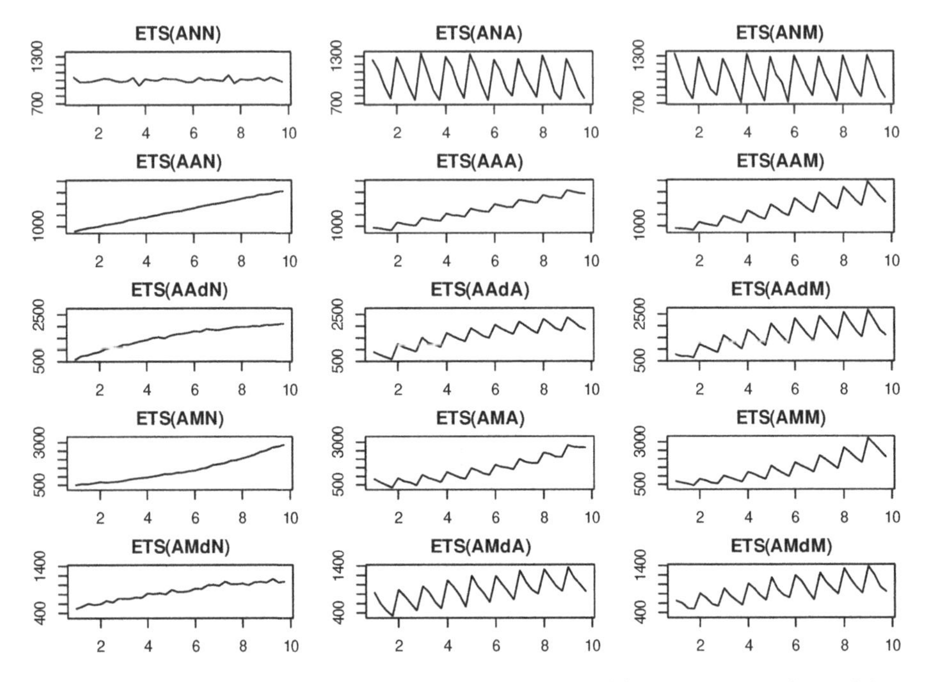

FIGURE 4.1 Time series corresponding to the additive error ETS models.

Things to note from the plots in Figure 4.1:

1. When seasonality is multiplicative, its amplitude increases with the increase of the level of the data, while with additive seasonality, the amplitude is constant. Compare, for example, ETS(A,A,A) with ETS(A,A,M): for the former, the distance between the highest and the lowest points in the first year is roughly the same as in the last year. In the case of ETS(A,A,M) the distance increases with the increase in the level of series;
2. When the trend is multiplicative, data exhibits exponential growth/decay;
3. The damped trend slows down both additive and multiplicative trends;
4. It is practically impossible to distinguish additive and multiplicative seasonality if the level of series does not change because the amplitude of seasonality will be constant in both cases (compare ETS(A,N,A) and ETS(A,N,M)).

Figure 4.2 demonstrates a similar plot for the multiplicative error models.

The plots in Figure 4.2 show roughly the same idea as the additive case,

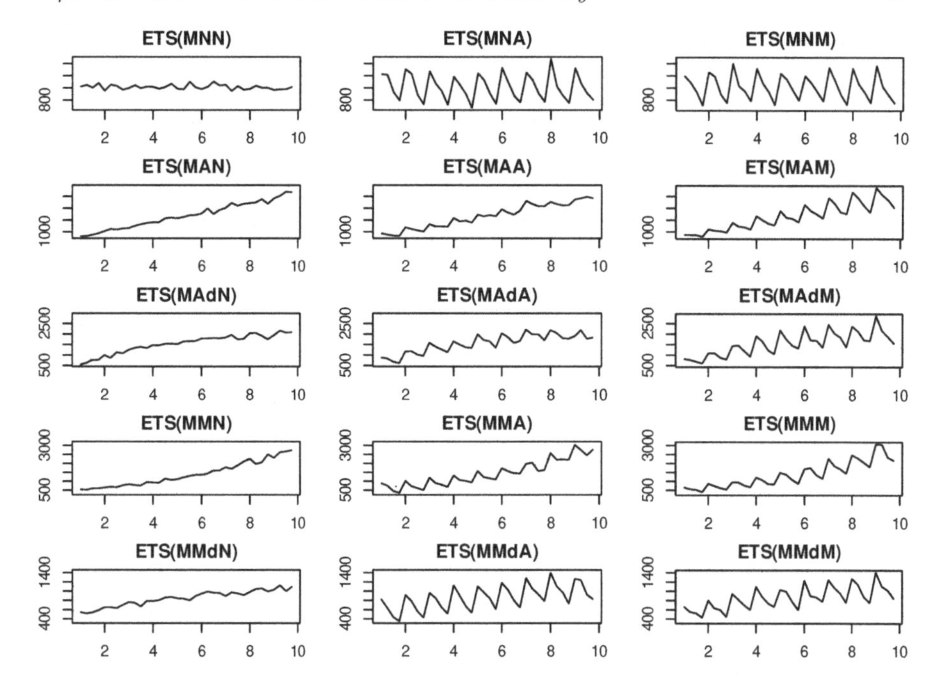

FIGURE 4.2 Time series corresponding to the multiplicative error ETS models.

the main difference being that the variance of the error increases with the increase of the level of the data – this becomes clearer on ETS(M,A,N) and ETS(M,M,N) data. This property is called heteroscedasticity in statistics, and Hyndman et al. (2008) argue that the main benefit of the multiplicative error models is in capturing this feature.

We will discuss the most important ETS family members in the following chapters. Note that not all the models in this taxonomy are sensible, and some are typically ignored entirely (this applies mainly to models with a mixture of additive and multiplicative components). ADAM implements the entire taxonomy, but we need to be aware of the potential issues with some of models, and we will discuss what to expect from them in different situations in the next chapters.

4.2 Mathematical models in the ETS taxonomy

I hope that it becomes more apparent to the reader how the ETS framework is built upon the idea of time series decomposition (from Section 3.1). By

TABLE 4.1 Additive error ETS models.

	Nonseasonal	Additive	Multiplicative
No trend	$y_t = l_{t-1} + \epsilon_t$ $l_t = l_{t-1} + \alpha\epsilon_t$	$y_t = l_{t-1} + s_{t-m} + \epsilon_t$ $l_t = l_{t-1} + \alpha\epsilon_t$ $s_t = s_{t-m} + \gamma\epsilon_t$	$y_t = l_{t-1}s_{t-m} + \epsilon_t$ $l_t = l_{t-1} + \alpha\frac{\epsilon_t}{s_{t-m}}$ $s_t = s_{t-m} + \gamma\frac{\epsilon_t}{l_{t-1}}$
Additive	$y_t = l_{t-1} + b_{t-1} + \epsilon_t$ $l_t = l_{t-1} + b_{t-1} + \alpha\epsilon_t$ $b_t = b_{t-1} + \beta\epsilon_t$	$y_t = l_{t-1} + b_{t-1} + s_{t-m} + \epsilon_t$ $l_t = l_{t-1} + b_{t-1} + \alpha\epsilon_t$ $b_t = b_{t-1} + \beta\epsilon_t$ $s_t = s_{t-m} + \gamma\epsilon_t$	$y_t = (l_{t-1} + b_{t-1})s_{t-m} + \epsilon_t$ $l_t = l_{t-1} + b_{t-1} + \alpha\frac{\epsilon_t}{s_{t-m}}$ $b_t = b_{t-1} + \beta\frac{\epsilon_t}{s_{t-m}}$ $s_t = s_{t-m} + \gamma\frac{\epsilon_t}{l_{t-1} + b_{t-1}}$
Additive damped	$y_t = l_{t-1} + \phi b_{t-1} + \epsilon_t$ $l_t = l_{t-1} + \phi b_{t-1} + \alpha\epsilon_t$ $b_t = \phi b_{t-1} + \beta\epsilon_t$	$y_t = l_{t-1} + \phi b_{t-1} + s_{t-m} + \epsilon_t$ $l_t = l_{t-1} + \phi b_{t-1} + \alpha\epsilon_t$ $b_t = \phi b_{t-1} + \beta\epsilon_t$ $s_t = s_{t-m} + \gamma\epsilon_t$	$y_t = (l_{t-1} + \phi b_{t-1})s_{t-m} + \epsilon_t$ $l_t = l_{t-1} + \phi b_{t-1} + \alpha\frac{\epsilon_t}{s_{t-m}}$ $b_t = \phi b_{t-1} + \beta\frac{\epsilon_t}{s_{t-m}}$ $s_t = s_{t-m} + \gamma\frac{\epsilon_t}{l_{t-1} + \phi b_{t-1}}$
Multiplicative	$y_t = l_{t-1}b_{t-1} + \epsilon_t$ $l_t = l_{t-1}b_{t-1} + \alpha\epsilon_t$ $b_t = b_{t-1} + \beta\frac{\epsilon_t}{l_{t-1}}$	$y_t = l_{t-1}b_{t-1} + s_{t-m} + \epsilon_t$ $l_t = l_{t-1}b_{t-1} + \alpha\epsilon_t$ $b_t = b_{t-1} + \beta\frac{\epsilon_t}{l_{t-1}}$ $s_t = s_{t-m} + \gamma\epsilon_t$	$y_t = l_{t-1}b_{t-1}s_{t-m} + \epsilon_t$ $l_t = l_{t-1}b_{t-1} + \alpha\frac{\epsilon_t}{s_{t-m}}$ $b_t = b_{t-1} + \beta\frac{\epsilon_t}{l_{t-1}s_{t-m}}$ $s_t = s_{t-m} + \gamma\frac{\epsilon_t}{l_{t-1}b_{t-1}}$
Multiplicative damped	$y_t = l_{t-1}b_{t-1}^{\phi} + \epsilon_t$ $l_t = l_{t-1}b_{t-1}^{\phi} + \alpha\epsilon_t$ $b_t = b_{t-1}^{\phi} + \beta\frac{\epsilon_t}{l_{t-1}}$	$y_t = l_{t-1}b_{t-1}^{\phi} + s_{t-m} + \epsilon_t$ $l_t = l_{t-1}b_{t-1}^{\phi} + \alpha\epsilon_t$ $b_t = b_{t-1}^{\phi} + \beta\frac{\epsilon_t}{l_{t-1}}$ $s_t = s_{t-m} + \gamma\epsilon_t$	$y_t = l_{t-1}b_{t-1}^{\phi}s_{t-m} + \epsilon_t$ $l_t = l_{t-1}b_{t-1}^{\phi} + \alpha\frac{\epsilon_t}{s_{t-m}}$ $b_t = b_{t-1}^{\phi} + \beta\frac{\epsilon_t}{l_{t-1}s_{t-m}}$ $s_t = s_{t-m} + \gamma\frac{\epsilon_t}{l_{t-1}b_{t-1}}$

introducing different components, defining their types, and adding the equations for their update, we can construct models that would work better in capturing the key features of the time series. The equations discussed in Section 3.1 represent so-called "measurement" or "observation" equations of the ETS models. But we should also consider the potential change in components over time. The "transition" or "state" equations are supposed to reflect this change: they explain how the level, trend or seasonal components evolve.

As discussed in Section 4.1, given different types of components and their interactions, we end up with 30 models in the taxonomy. Tables 4.1 and 4.2 summarise mathematically all 30 ETS models shown graphically on Figures 4.1 and 4.2, presenting formulae for measurement and transition equations.

From a statistical point of view, formulae in Tables 4.1 and 4.2 correspond to the "true models" (see Section 1.4), they explain the models underlying potential data, but when it comes to their construction and estimation, the ϵ_t is substituted by the estimated e_t (which is calculated differently depending on the error type), and time series components and smoothing parameters are also replaced by their estimates (e.g. $\hat{\alpha}$ instead of α). However, if the values of

TABLE 4.2 Multiplicative error ETS models.

	Nonseasonal	Additive	Multiplicative
No trend	$y_t = l_{t-1}(1+\epsilon_t)$ $l_t = l_{t-1}(1+\alpha\epsilon_t)$	$y_t = (l_{t-1} + s_{t-m})(1+\epsilon_t)$ $l_t = l_{t-1} + \alpha\mu_{y,t}\epsilon_t$ $s_t = s_{t-m} + \gamma\mu_{y,t}\epsilon_t$	$y_t = l_{t-1}s_{t-m}(1+\epsilon_t)$ $l_t = l_{t-1}(1+\alpha\epsilon_t)$ $s_t = s_{t-m}(1+\gamma\epsilon_t)$
Additive	$y_t = (l_{t-1}+b_{t-1})(1+\epsilon_t)$ $l_t = (l_{t-1}+b_{t-1})(1+\alpha\epsilon_t)$ $b_t = b_{t-1} + \beta\mu_{y,t}\epsilon_t$	$y_t = (l_{t-1} + b_{t-1} + s_{t-m})(1+\epsilon_t)$ $l_t = l_{t-1} + b_{t-1} + \alpha\mu_{y,t}\epsilon_t$ $b_t = b_{t-1} + \beta\mu_{y,t}\epsilon_t$ $s_t = s_{t-m} + \gamma\mu_{y,t}\epsilon_t$	$y_t = (l_{t-1}+b_{t-1})s_{t-m}(1+\epsilon_t)$ $l_t = (l_{t-1}+b_{t-1})(1+\alpha\epsilon_t)$ $b_t = b_{t-1} + \beta(l_{t-1}+b_{t-1})\epsilon_t$ $s_t = s_{t-m}(1+\gamma\epsilon_t)$
Additive damped	$y_t = (l_{t-1}+\phi b_{t-1})(1+\epsilon_t)$ $l_t = (l_{t-1}+\phi b_{t-1})(1+\alpha\epsilon_t)$ $b_t = \phi b_{t-1} + \beta\mu_{y,t}\epsilon_t$	$y_t = (l_{t-1} + \phi b_{t-1} + s_{t-m})(1+\epsilon_t)$ $l_t = l_{t-1} + \phi b_{t-1} + \alpha\mu_{y,t}\epsilon_t$ $b_t = \phi b_{t-1} + \beta\mu_{y,t}\epsilon_t$ $s_t = s_{t-m} + \gamma\mu_{y,t}\epsilon_t$	$y_t = (l_{t-1}+\phi b_{t-1})s_{t-m}(1+\epsilon_t)$ $l_t = l_{t-1} + \phi b_{t-1}(1+\alpha\epsilon_t)$ $b_t = \phi b_{t-1} + \beta(l_{t-1}+\phi b_{t-1})\epsilon_t$ $s_t = s_{t-m}(1+\gamma\epsilon_t)$
Multiplicative	$y_t = l_{t-1}b_{t-1}(1+\epsilon_t)$ $l_t = l_{t-1}b_{t-1}(1+\alpha\epsilon_t)$ $b_t = b_{t-1}(1+\beta\epsilon_t)$	$y_t = (l_{t-1}b_{t-1} + s_{t-m})(1+\epsilon_t)$ $l_t = l_{t-1}b_{t-1} + \alpha\mu_{y,t}\epsilon_t$ $b_t = b_{t-1} + \beta\frac{\mu_{y,t}}{l_{t-1}}\epsilon_t$ $s_t = s_{t-m} + \gamma\mu_{y,t}\epsilon_t$	$y_t = l_{t-1}b_{t-1}s_{t-m}(1+\epsilon_t)$ $l_t = l_{t-1}b_{t-1}(1+\alpha\epsilon_t)$ $b_t = b_{t-1}(1+\beta\epsilon_t)$ $s_t = s_{t-m}(1+\gamma\epsilon_t)$
Multiplicative damped	$y_t = l_{t-1}b_{t-1}^{\phi}(1+\epsilon_t)$ $l_t = l_{t-1}b_{t-1}^{\phi}(1+\alpha\epsilon_t)$ $b_t = b_{t-1}^{\phi}(1+\beta\epsilon_t)$	$y_t = (l_{t-1}b_{t-1}^{\phi} + s_{t-m})(1+\epsilon_t)$ $l_t = l_{t-1}b_{t-1}^{\phi} + \alpha\mu_{y,t}\epsilon_t$ $b_t = b_{t-1}^{\phi} + \beta\frac{\mu_{y,t}}{l_{t-1}}\epsilon_t$ $s_t = s_{t-m} + \gamma\mu_{y,t}\epsilon_t$	$y_t = l_{t-1}b_{t-1}^{\phi}s_{t-m}(1+\epsilon_t)$ $l_t = l_{t-1}b_{t-1}^{\phi}\,(1+\alpha\epsilon_t)$ $b_t = b_{t-1}^{\phi}\,(1+\beta\epsilon_t)$ $s_t = s_{t-m}\,(1+\gamma\epsilon_t)$

these models' parameters were known, it would be possible to produce point forecasts and conditional h steps ahead expectations from these models, which are summarised in Table 4.3 with the following elements:

- Conditional one step ahead expectation $\mu_{y,t} \equiv \mu_{y,t|t-1}$;
- Multiple steps ahead point forecast $\hat{y}_{t+h}$;
- Conditional multiple steps ahead expectation $\mu_{y,t+h|t}$.

In the case of the additive error models, the point forecasts correspond to the expectations only when the expectation of the error term is zero, i.e. $\mathrm{E}(\epsilon_t) = 0$. In contrast, in the case of the multiplicative models, the condition is changed to $\mathrm{E}(1 + \epsilon_t) = 1$.

Remark. **Not all point forecasts of ETS models correspond to conditional expectations.** This issue applies to the models with multiplicative trend and/or multiplicative seasonality. This is because the ETS model assumes that different states are correlated (they have the same source of error), and as a result, multiple steps ahead values (when h>1) of states introduce products of error terms. So, the conditional expectations in these cases might not have analytical forms ("n.c.f." in Table 4.3 stands for "No Closed Form"), and when working with these models, simulations might be required. This does not apply to the one-step-ahead forecasts, for which all the classical formulae work. This issue is discussed in Section 6.3.

The multiplicative error models have the same one step ahead expectations and point forecasts as the additive error ones. However, due to the multiplication

TABLE 4.3 Point forecasts and expectations of ETS models. n.c.f. stands for "No Closed Form".

	Nonseasonal	Additive	Multiplicative
No trend	$\mu_{y,t} = l_{t-1}$ $\hat{y}_{t+h} = l_t$ $\mu_{y,t+h\vert t} = \hat{y}_{t+h}$	$\mu_{y,t} = l_{t-1} + s_{t-m}$ $\hat{y}_{t+h} = l_t + s_{t+h-m\lceil\frac{h}{m}\rceil}$ $\mu_{y,t+h\vert t} = \hat{y}_{t+h}$	$\mu_{y,t} = l_{t-1}s_{t-m}$ $\hat{y}_{t+h} = l_t s_{t+h-m\lceil\frac{h}{m}\rceil}$ $\mu_{y,t+h\vert t} = \hat{y}_{t+h}$ only for $h \leq m$
Additive	$\mu_{y,t} = l_{t-1} + b_{t-1}$ $\hat{y}_{t+h} = l_t + hb_t$ $\mu_{y,t+h\vert t} = \hat{y}_{t+h}$	$\mu_{y,t} = l_{t-1} + b_{t-1} + s_{t-m}$ $\hat{y}_{t+h} = l_t + hb_{t-1} + s_{t+h-m\lceil\frac{h}{m}\rceil}$ $\mu_{y,t+h\vert t} = \hat{y}_{t+h}$	$\mu_{y,t} = (l_{t-1} + b_{t-1})s_{t-m}$ $\hat{y}_{t+h} = (l_t + hb_{t-1})\, s_{t+h-m\lceil\frac{h}{m}\rceil}$ $\mu_{y,t+h\vert t} = \hat{y}_{t+h}$ only for $h \leq m$
Additive damped	$\mu_{y,t} = l_{t-1} + \phi b_{t-1}$ $\hat{y}_{t+h} = l_t + \sum_{j=1}^{h} \phi^j b_t$ $\mu_{y,t+h\vert t} = \hat{y}_{t+h}$	$\mu_{y,t} = l_{t-1} + \phi b_{t-1} + s_{t-m}$ $\hat{y}_{t+h} = l_t + \sum_{j=1}^{h} \phi^j b_{t-1} + s_{t+h-m\lceil\frac{h}{m}\rceil}$ $\mu_{y,t+h\vert t} = \hat{y}_{t+h}$	$\mu_{y,t} = (l_{t-1} + \phi b_{t-1})s_{t-m}$ $\hat{y}_{t+h} = \left(l_t + \sum_{j=1}^{h} \phi^j b_t\right) s_{t+h-m\lceil\frac{h}{m}\rceil}$ $\mu_{y,t+h\vert t} = \hat{y}_{t+h}$ only for $h \leq m$
Multiplicative	$\mu_{y,t} = l_{t-1}b_{t-1}$ $\hat{y}_{t+h} = l_t b_t^h$ $\mu_{y,t+h\vert t}$ – n.c.f. for $h > 1$	$\mu_{y,t} = l_{t-1}b_{t-1} + s_{t-m}$ $\hat{y}_{t+h} = l_t b_{t-1}^h + s_{t+h-m\lceil\frac{h}{m}\rceil}$ $\mu_{y,t+h\vert t}$ – n.c.f. for $h > 1$	$\mu_{y,t} = l_{t-1}b_{t-1}s_{t-m}$ $\hat{y}_{t+h} = l_t b_{t-1}^h s_{t+h-m\lceil\frac{h}{m}\rceil}$ $\mu_{y,t+h\vert t}$ – n.c.f. for $h > 1$
Multiplicative damped	$\mu_{y,t} = l_{t-1}b_{t-1}^{\phi}$ $\hat{y}_{t+h} = l_t b_t^{\sum_{j=1}^{h} \phi^j}$ $\mu_{y,t+h\vert t}$ – n.c.f. for $h > 1$	$\mu_{y,t} = l_{t-1}b_{t-1}^{\phi} + s_{t-m}$ $\hat{y}_{t+h} = l_t b_{t-1}^{\sum_{j=1}^{h} \phi^j} + s_{t+h-m\lceil\frac{h}{m}\rceil}$ $\mu_{y,t+h\vert t}$ – n.c.f. for $h > 1$	$\mu_{y,t} = l_{t-1}b_{t-1}^{\phi}s_{t-m}$ $\hat{y}_{t+h} = l_t b_{t-1}^{\sum_{j=1}^{h} \phi^j} s_{t+h-m\lceil\frac{h}{m}\rceil}$ $\mu_{y,t+h\vert t}$ – n.c.f. for $h > 1$

by the error term, the multiple steps ahead conditional expectations between the two types of models might differ, specifically for the multiplicative trend and multiplicative seasonal models. These values do not have closed forms and can only be obtained via simulations.

Although there are 30 potential ETS models, not all of them are sensible. So, Rob Hyndman has reduced the pool of models under consideration in the `ets()` function of the `forecast` package to the following 19: ANN, AAN, AAdN, ANA, AAA, AAdA, MNN, MAN, MAdN, MNA, MAA, MAdA, MNM, MAM, MAdM, MMN, MMdN, MMM, and MMdM. In addition, the multiplicative trend models are unstable in data with outliers, so they are switched off in the `ets()` function by default, which reduces the pool of models further to the first 15.

The `es()` function from the `smooth` package implements the conventional ETS, supporting all 30 models and implementing some features, discussed in the original Hyndman et al. (2008) book (e.g. explanatory variables and cumulative over the lead time forecasts).

4.3 ETS and SES

Taking a step back, in this section we discuss one of the basic ETS models, the local level model, and the Exponential Smoothing method related to it.

4.3.1 ETS(A,N,N)

There have been several tries to develop statistical models underlying SES, and we know now that the model has underlying ARIMA(0,1,1) (Muth, 1960), local level MSOE (Multiple Source of Error, Muth, 1960) and SSOE (Single Source of Error, Snyder, 1985) models. According to Hyndman et al. (2002), the ETS(A,N,N) model also underlies the SES method. To see the connection and to get to it from SES, we need to recall two things: how in general, the actual value relates to the forecast error and the fitted value, and the error correction form of SES from Subsection 3.4.3:

$$\begin{aligned} y_t &= \hat{y}_t + e_t \\ \hat{y}_{t+1} &= \hat{y}_t + \hat{\alpha} e_t \end{aligned} . \tag{4.1}$$

In order to get to the SSOE state space model for SES, we need to substitute $\hat{y}_t = \hat{l}_{t-1}$, implying that the fitted value is equal to the level of the series:

$$\begin{aligned} y_t &= \hat{l}_{t-1} + e_t \\ \hat{l}_t &= \hat{l}_{t-1} + \hat{\alpha} e_t \end{aligned} . \tag{4.2}$$

If we now substitute the sample estimates of level, smoothing parameter, and forecast error by their population values, we will get the ETS(A,N,N), which was discussed in Section 4.2:

$$\begin{aligned} y_t &= l_{t-1} + \epsilon_t \\ l_t &= l_{t-1} + \alpha \epsilon_t \end{aligned} , \tag{4.3}$$

where, as we know from Section 3.1, l_t is the level of the data, ϵ_t is the error term, and α is the smoothing parameter. Note that we use α without the "hat" symbol, which implies that there is a "true" value of the parameter (which could be obtained if we had all the data in the world or just knew it for some reason). The main benefit of having the model (4.3) instead of just the method (3.21) is in having a flexible framework, which allows adding other components, selecting the most appropriate ones (Section 15.1), consistently estimating parameters (Chapter 11), producing prediction intervals (Section 18.3), etc. In a way, this model is the basis of ADAM.

In order to see the data that corresponds to the ETS(A,N,N) we can use the `sim.es()` function from the `smooth` package. Here are several examples with different smoothing parameters values:

```
# list with generated data
y <- vector("list",6)
# Parameters for DGP
initial <- 1000
meanValue <- 0
```

```
sdValue <- 20
alphas <- c(0.1,0.3,0.5,0.75,1,1.5)
# Go through all alphas and generate respective data
for(i in 1:length(alphas)){
  y[[i]] <- sim.es("ANN", 120, 1, 12, persistence=alphas[i],
                   initial=initial, mean=meanValue, sd=sdValue)
}
```

The generated data can be plotted the following way:

```
par(mfrow=c(3,2), mar=c(2,2,2,1))
for(i in 1:6){
  plot(y[[i]], main=paste0("alpha=",y[[i]]$persistence),
       ylim=initial+c(-500,500))
}
```

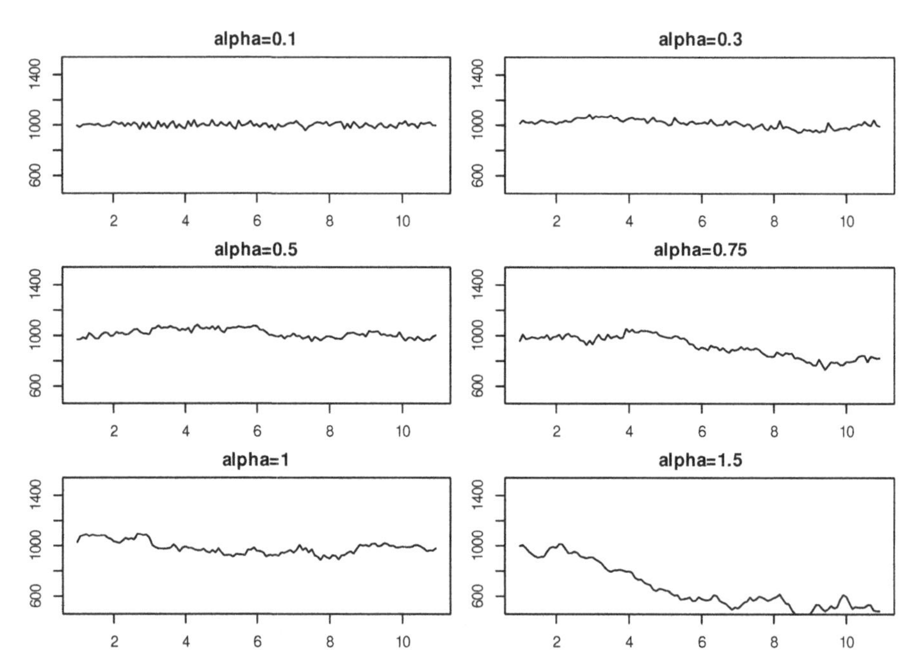

FIGURE 4.3 Local level data corresponding to the ETS(A,N,N) model with different smoothing parameters.

This simple simulation shows that the smoothing parameter in ETS(A,N,N) controls the variability in the data (Figure 4.3): the higher α is, the higher variability is and less predictable the data becomes. With the higher values of

α, the level changes faster, leading to increased uncertainty about the future values.

When it comes to the application of this model to the data, the conditional h steps ahead mean corresponds to the point forecast and is equal to the last observed level:

$$\mu_{y,t+h|t} = \hat{y}_{t+h} = l_t. \tag{4.4}$$

This holds because it is assumed (see Section 1.4.1) that $\mathrm{E}(\epsilon_t) = 0$, which implies that the conditional h steps ahead expectation of the level in the model is (from the second equation in (4.3)):

$$\mathrm{E}(l_{t+h-1}|t) = l_t + \mathrm{E}(\alpha \sum_{j=1}^{h-2} \epsilon_{t+j}|t) = l_t. \tag{4.5}$$

Here is an example of a forecast from ETS(A,N,N) with automatic parameter estimation using the `es()` function from the `smooth` package:

```
# Generate the data
y <- sim.es("ANN", 120, 1, 12, persistence=0.3, initial=1000)
# Apply ETS(A,N,N) model
esModel <- es(y$data, "ANN", h=12, holdout=TRUE)
# Produce forecasts
esModel |> forecast(h=12, interval="pred") |>
    plot(main=paste0("ETS(ANN) with alpha=",
                     round(esModel$persistence,4)))
```

As we see from Figure 4.4, the true smoothing parameter is 0.3, but the estimated one is not exactly 0.3, which is expected because we deal with an in-sample estimation. Also, notice that with such a smoothing parameter, the prediction interval widens with the increase of the forecast horizon. If the smoothing parameter were lower, the bounds would not increase, but this might not reflect the uncertainty about the level correctly. Here is an example with $\alpha = 0.01$ on the same data (Figure 4.5).

```
es(y$data, "ANN", h=12,
   holdout=TRUE, persistence=0.01) |>
    forecast(h=12, interval="pred") |>
    plot(main="ETS(ANN) with alpha=0.01")
```

Figure 4.5 shows that the prediction interval does not expand, but at the same time is wider than needed, and the forecast is biased – the model does not keep up to the fast-changing time series. So, it is essential to correctly estimate the smoothing parameters not only to approximate the data but also to produce a less biased point forecast and a more appropriate prediction interval.

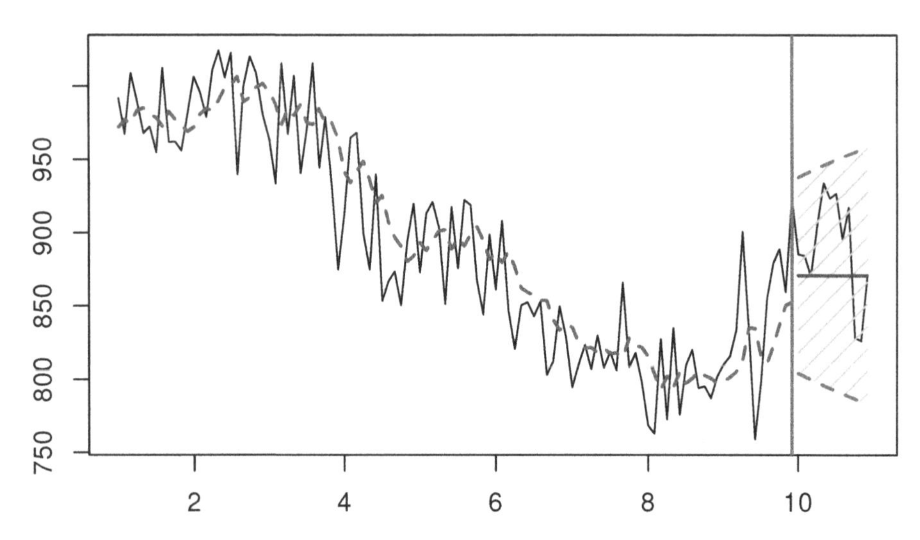

FIGURE 4.4 An example of ETS(A,N,N) applied to the data generated from the same model.

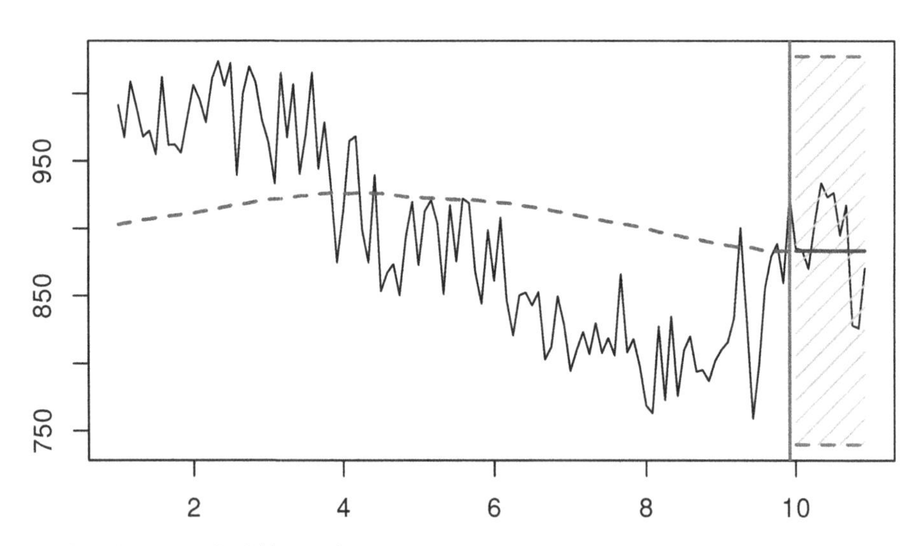

FIGURE 4.5 ETS(A,N,N) with $\hat{\alpha} = 0.01$ applied to the data generated from the same model with $\alpha = 0.3$.

4.3.2 ETS(M,N,N)

Hyndman et al. (2008) demonstrate that there is another ETS model, underlying SES. It is the model with multiplicative error, which is formulated in the following way, as mentioned in Section 4.2:

$$\begin{aligned} y_t &= l_{t-1}(1+\epsilon_t) \\ l_t &= l_{t-1}(1+\alpha\epsilon_t) \end{aligned}, \tag{4.6}$$

where $(1+\epsilon_t)$ corresponds to the ε_t discussed in Section 3.1. In order to see the connection of this model with SES, we need to revert to the estimation of the model on the data again:

$$\begin{aligned} y_t &= \hat{l}_{t-1}(1+e_t) \\ \hat{l}_t &= \hat{l}_{t-1}(1+\hat{\alpha}e_t) \end{aligned}. \tag{4.7}$$

where the one step ahead forecast is (Section 4.2) $\hat{y}_t = \hat{l}_{t-1}$ and $e_t = \frac{y_t - \hat{y}_t}{\hat{y}_t}$. Substituting these values in the second equation of (4.7) we obtain:

$$\hat{y}_{t+1} = \hat{y}_t \left(1 + \hat{\alpha}\frac{y_t - \hat{y}_t}{\hat{y}_t}\right) \tag{4.8}$$

Finally, opening the brackets, we get the SES in the form similar to (3.21):

$$\hat{y}_{t+1} = \hat{y}_t + \hat{\alpha}(y_t - \hat{y}_t). \tag{4.9}$$

This example again demonstrates the difference between a forecasting method and a model. When we use SES, we ignore the distributional assumptions, which restricts the usefulness of the method. When we work with a model, we assume a specific structure, which on the one hand, makes it more restrictive, but on the other hand, gives it additional features. The main ones in the case of ETS(M,N,N) in comparison with ETS(A,N,N) are:

1. The variance of the actual values in ETS(M,N,N) increases with the increase of the level l_t. This allows modelling a heteroscedasticity situation in the data;
2. If $(1+\epsilon_t)$ is always positive, then the ETS(M,N,N) model will always produce only positive forecasts (both point and interval). This makes this model applicable in principle to the data with low levels.

An alternative to (4.6) would be the ETS(A,N,N) model (4.3) applied to the data in logarithms (assuming that the data we work with is always positive), implying that:

$$\begin{aligned} \log y_t &= l_{t-1} + \epsilon_t \\ l_t &= l_{t-1} + \alpha\epsilon_t \end{aligned}. \tag{4.10}$$

However, to produce forecasts from (4.10), exponentiation is needed, making the application of the model more difficult than needed. The ETS(M,N,N), on the other hand, does not rely on exponentiation, making it more practical and safe in cases when the model produces very high values (e.g. `exp(1000)` returns infinity in R).

Finally, the conditional h steps ahead mean of ETS(M,N,N) corresponds to the point forecast and is equal to the last observed level, but only if $\mathrm{E}(1+\epsilon_t)=1$:

$$\mu_{y,t+h|t} = \hat{y}_{t+h} = l_t. \tag{4.11}$$

And here is an example with the ETS(M,N,N) data (Figure 4.6):

```
y <- sim.es("MNN", 120, 1, 12, persistence=0.3, initial=1000)
esModel <- es(y$data, "MNN", h=12, holdout=TRUE)
forecast(esModel, h=12, interval="pred") |>
    plot(main=paste0("ETS(MNN) with alpha=",
                     round(esModel$persistence,4)))
```

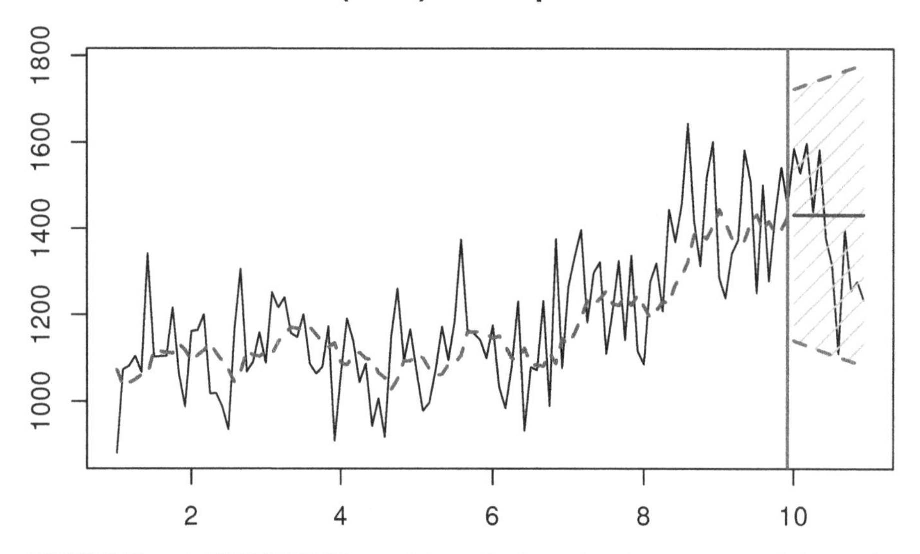

FIGURE 4.6 ETS(M,N,N) model applied to the data generated from the same model.

Conceptually, the data in Figure 4.6 looks very similar to the one from ETS(A,N,N) (Figure 4.4), but demonstrating the changing variance of the error term with the change of the level. The model itself would in general produce a wider prediction interval than its additive error counterpart, keeping the same smoothing parameter.

4.4 Several examples of ETS and related Exponential Smoothing methods

There are other Exponential Smoothing methods, which include more components, as discussed in Section 3.1. This includes but is not limited to: Holt's (Holt, 2004, originally proposed in 1957), Holt-Winter's (Winters, 1960), multiplicative trend (Pegels, 1969), damped trend (originally proposed by Roberts (1982) and then picked up by Gardner and McKenzie (1985)), damped trend Holt-Winters (Gardner & McKenzie, 1989), and damped multiplicative trend (Taylor, 2003a) methods. We will not discuss them here one by one, as we will not use them further in this monograph. Instead, we will focus on the ETS models underlying them.

We already understand that there can be different components in time series and that they can interact either in an additive or a multiplicative way, which gives us the taxonomy discussed in Section 4.1. This section considers several examples of ETS models and their relations to the conventional Exponential Smoothing methods.

4.4.1 ETS(A,A,N)

This is also sometimes known as the local trend model and is formulated similar to ETS(A,N,N), but with addition of the trend equation. It underlies **Holt's method** (Ord, Koehler, & Snyder, 1997):

$$\begin{aligned} y_t &= l_{t-1} + b_{t-1} + \epsilon_t \\ l_t &= l_{t-1} + b_{t-1} + \alpha \epsilon_t, \\ b_t &= b_{t-1} + \beta \epsilon_t \end{aligned} \tag{4.12}$$

where β is the smoothing parameter for the trend component. It has a similar idea as ETS(A,N,N): the states evolve, and the speed of their change depends on the values of α and β. The trend is not deterministic in this model: both the intercept and the slope change over time. The higher the smoothing parameters are, the more uncertain the level and the slope will be, thus, the higher the uncertainty about the future values is.

Here is an example of the data that corresponds to the ETS(A,A,N) model:

```
y <- sim.es("AAN", 120, 1, 12, persistence=c(0.3,0.1),
            initial=c(1000,20), mean=0, sd=20)
plot(y)
```

The series in Figure 4.7 demonstrates a trend that changes over time. If we need to produce forecasts for this data, we will capture the dynamics of the

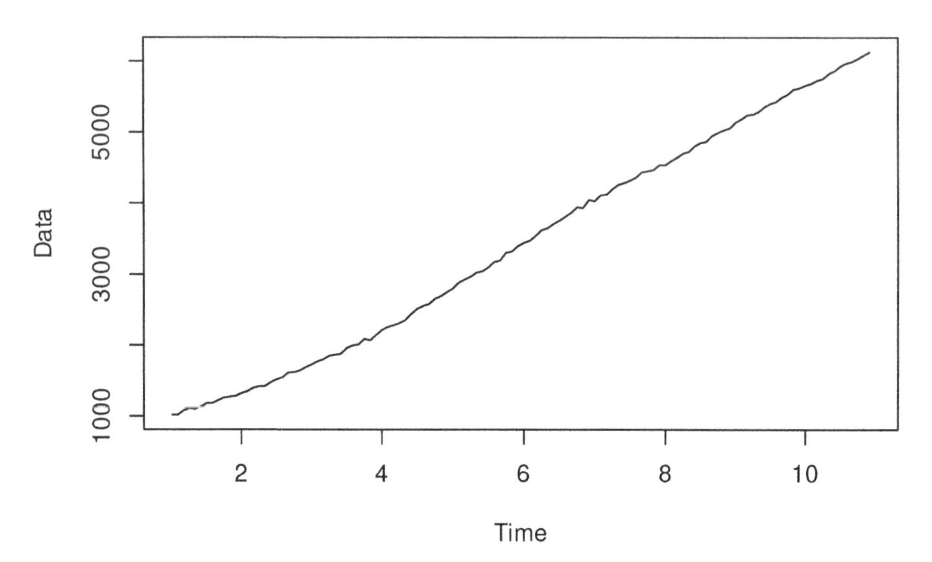

FIGURE 4.7 Data generated from an ETS(A,A,N) model.

trend component via ETS(A,A,N) and then use the last values for the several steps ahead prediction.

The point forecast h steps ahead from this model is a straight line with a slope b_t (as shown in Table 4.3 from Section 4.2):

$$\mu_{y,t+h|t} = \hat{y}_{t+h} = l_t + h b_t. \tag{4.13}$$

This becomes apparent if one takes the conditional expectations $\mathrm{E}(l_{t+h}|t)$ and $\mathrm{E}(b_{t+h}|t)$ in the second and third equations of (4.12) and then inserts them in the measurement equation. Graphically it will look as shown in Figure 4.8:

```
es(y, h=10) |>
    plot(which=7)
```

If you want to experiment with the model and see how its parameters influence the fit and forecast, you can use the following R code:

```
es(y$data, "AAN", h=10, persistence=c(0.2,0.1)) |>
    plot(which=7)
```

where `persistence` is the vector of smoothing parameters (first $\hat{\alpha}$, then $\hat{\beta}$). By changing their values, we will make the model less/more responsive to the changes in the data.

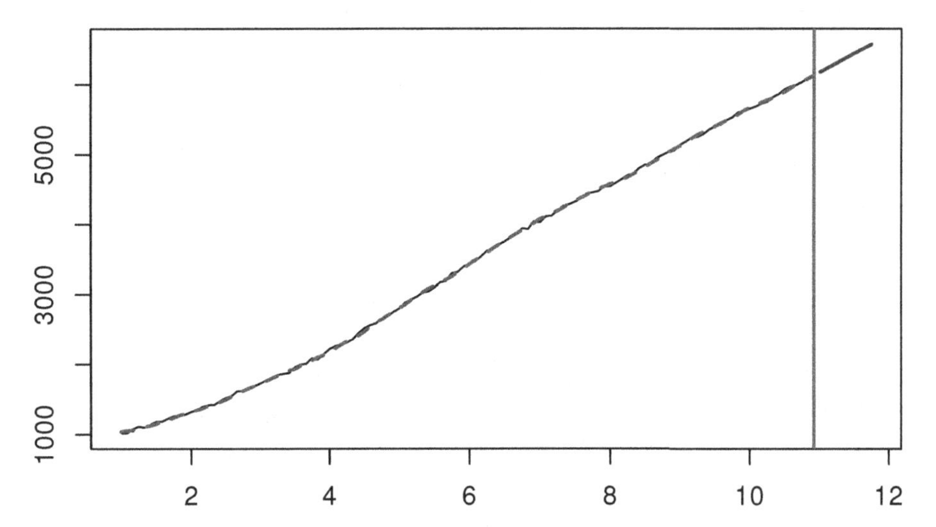

FIGURE 4.8 ETS(A,A,N) and a point forecast produced from it.

In a special case, ETS(A,A,N) corresponds to the Random Walk with drift (Subsection 3.3.4), when $\beta = 0$ and $\alpha = 1$:

$$\begin{aligned} y_t &= l_{t-1} + b_{t-1} + \epsilon_t \\ l_t &= l_{t-1} + b_{t-1} + \epsilon_t , \\ b_t &= b_{t-1} \end{aligned}$$

or

$$\begin{aligned} y_t &= l_{t-1} + b_0 + \epsilon_t \\ l_t &= l_{t-1} + b_0 + \epsilon_t \end{aligned} . \tag{4.14}$$

The connection between the two models becomes apparent, when substituting the first equation into the second one in (4.14) to obtain:

$$\begin{aligned} y_t &= l_{t-1} + b_0 + \epsilon_t \\ l_t &= y_t \end{aligned} , \tag{4.15}$$

or after inserting the second equation into the first one:

$$y_t = y_{t-1} + b_0 + \epsilon_t . \tag{4.16}$$

Finally, in another special case, when both α and β are zero, the model reverts to the Global Trend discussed in Subsection 3.3.5 and becomes:

$$\begin{aligned} y_t &= l_{t-1} + b_{t-1} + \epsilon_t \\ l_t &= l_{t-1} + b_{t-1} , \\ b_t &= b_{t-1} \end{aligned}$$

or

$$\begin{aligned} y_t &= l_{t-1} + b_0 + \epsilon_t \\ l_t &= l_{t-1} + b_0 \end{aligned} . \tag{4.17}$$

The main difference of (4.17) with the Global Trend model (3.13) is that the latter explicitly includes the trend component t and is formulated in one equation, while the former splits the equation into two parts and has an explicit level component. They both fit the data in the same way and produce the same forecasts when $a_0 = l_0$ and $a_1 = b_0$ in the Global Trend model.

Due to its flexibility, the ETS(A,A,N) model is considered as one of the good benchmarks in the case of trended time series.

4.4.2 ETS(A,Ad,N)

This is the model that underlies the **damped trend method** (Gardner & McKenzie, 1985; Roberts, 1982):

$$\begin{aligned} y_t &= l_{t-1} + \phi b_{t-1} + \epsilon_t \\ l_t &= l_{t-1} + \phi b_{t-1} + \alpha \epsilon_t, \\ b_t &= \phi b_{t-1} + \beta \epsilon_t \end{aligned} \tag{4.18}$$

where ϕ is the dampening parameter, typically lying between zero and one. If it is equal to zero, the model reduces to ETS(A,N,N), (4.3). If it is equal to one, it becomes equivalent to ETS(A,A,N), (4.12). The dampening parameter slows down the trend, making it non-linear. An example of data that corresponds to ETS(A,Ad,N) is provided in Figure 4.9.

```
y <- sim.es("AAdN", 120, 1, 12, persistence=c(0.3,0.1),
            initial=c(1000,20), phi=0.95, mean=0, sd=20)
plot(y)
```

Visually it is typically challenging to distinguish ETS(A,A,N) from ETS(A,Ad,N) data. So, some other model selection techniques are recommended (see Section 15.1).

The point forecast from this model is a bit more complicated than the one from ETS(A,A,N) (see Section 4.2):

$$\mu_{y,t+h|t} = \hat{y}_{t+h} = l_t + \sum_{j=1}^h \phi^j b_t. \tag{4.19}$$

It corresponds to the slowing down trajectory, as shown in Figure 4.10.

As can be seen in Figure 4.10, the forecast trajectory from the ETS(A,Ad,N) has a slowing down element in it. This is because of the $\phi = 0.95$ in our example.

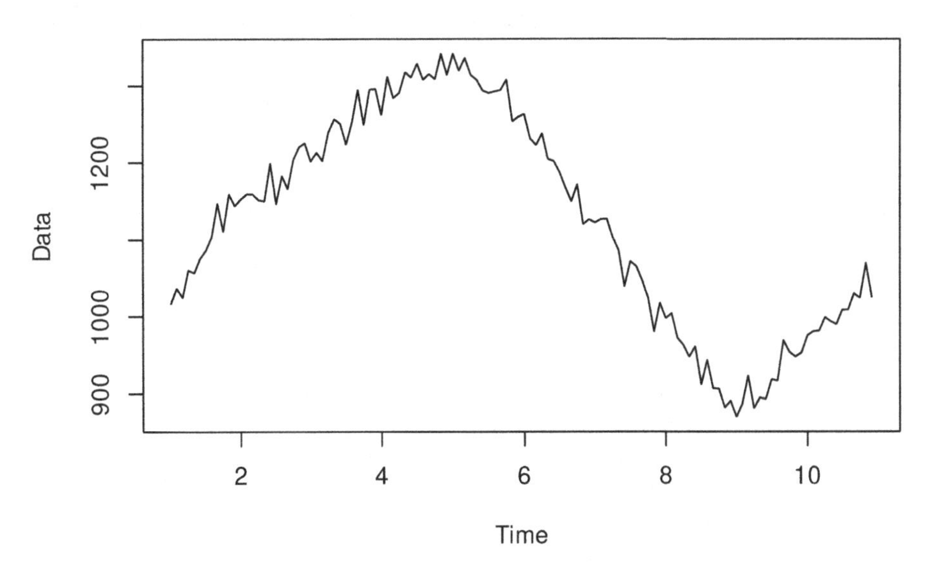

FIGURE 4.9 An example of ETS(A,Ad,N) data.

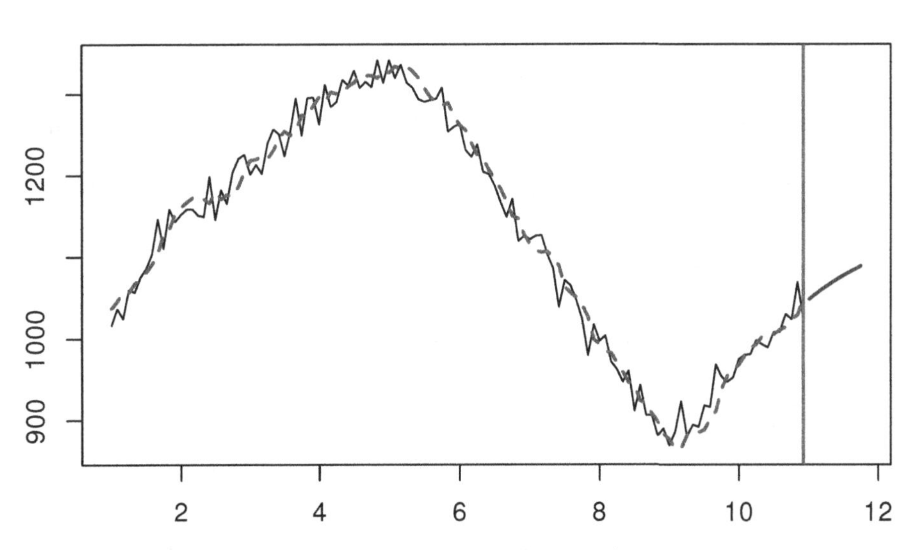

FIGURE 4.10 A point forecast from ETS(A,Ad,N).

4.4.3 ETS(A,A,M)

Finally, this is an exotic model with additive error and trend, but multiplicative seasonality. It can be considered exotic because of the misalignment of the error and seasonality. Still, we list it here, because it underlies the **Holt-Winters method** (Winters, 1960):

$$\begin{aligned} y_t &= (l_{t-1} + b_{t-1}) s_{t-m} + \epsilon_t \\ l_t &= l_{t-1} + b_{t-1} + \alpha \frac{\epsilon_t}{s_{t-m}} \\ b_t &= b_{t-1} + \beta \frac{\epsilon_t}{s_{t-m}} \\ s_t &= s_{t-m} + \gamma \frac{\epsilon_t}{l_{t-1} + b_{t-1}} \end{aligned}, \tag{4.20}$$

where s_t is the seasonal component and γ is its smoothing parameter. This is one of the potentially unstable models, which due to the mix of components, might produce unreasonable forecasts because the seasonal component might become negative, while it should always be positive. Still, it might work on the strictly positive high-level data. Figure 4.11 shows how the data for this model can look.

```
y <- sim.es("AAM", 120, 1, 4, persistence=c(0.3,0.05,0.2),
            initial=c(1000,20), initialSeason=c(0.9,1.1,0.8,1.2),
            mean=0, sd=20)
plot(y)
```

The data in Figure 4.11 exhibits an additive trend with increasing seasonal amplitude, which are the two characteristics of the model.

Finally, the point forecast from this model builds upon ETS(A,A,N), introducing seasonal component:

$$\hat{y}_{t+h} = (l_t + h b_t) s_{t+h-m\lceil\frac{h}{m}\rceil}, \tag{4.21}$$

where $\lceil\frac{h}{m}\rceil$ is the rounded up value of the fraction in the brackets, implying that the seasonal index from the previous period is used (e.g. previous January value). The point forecast from this model is shown in Figure 4.12.

Remark. The point forecasts produced from this model do not correspond to the conditional expectations. This will be discussed in Section 7.3.

Hyndman et al. (2008) argue that in ETS models, the error term should be aligned with the seasonal component because it is difficult to motivate why the amplitude of seasonality should increase with the increase of level, while the variability of the error term should stay the same. So, they recommend using

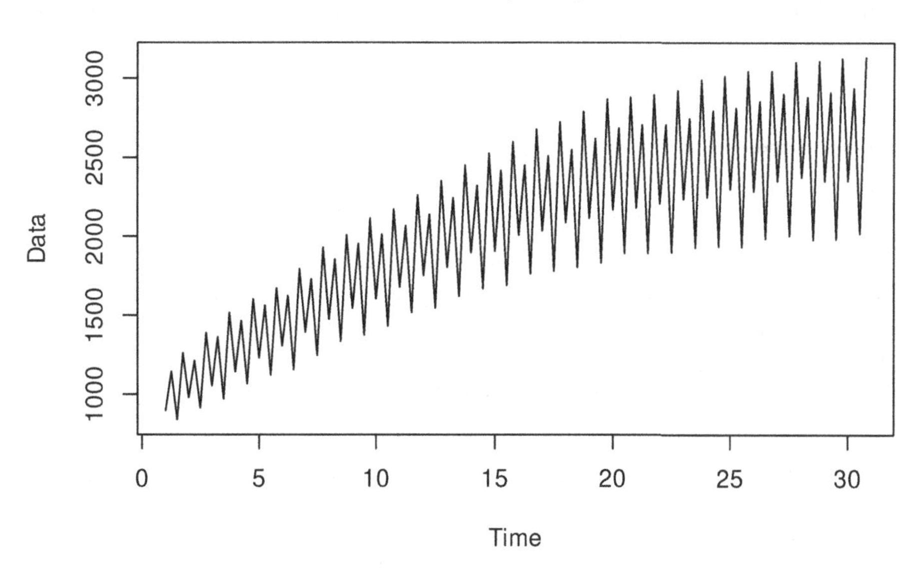

FIGURE 4.11 An example of ETS(A,A,M) data.

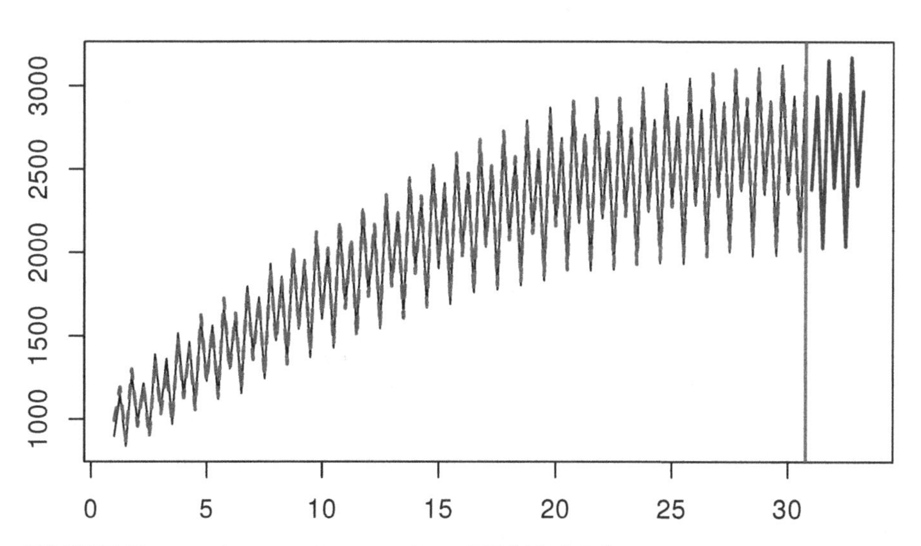

FIGURE 4.12 A point forecast from ETS(A,A,M).

ETS(M,A,M) instead of ETS(A,A,M) if you deal with positive high-volume data. This is a reasonable recommendation, but keep in mind that both models might break if you deal with the low-volume data and the trend component becomes negative.

4.5 ETS basic assumptions and principles

Several assumptions need to hold for the conventional ETS models to work properly. Some of them have already been discussed in Section 1.4.1, and we will come back to them in Chapter 14. What is important in our context is that the conventional ETS assumes that the error term ϵ_t follows the Normal distribution with zero mean and variance σ^2. There are several points related to this that need to be clarified:

1. If the mean was not equal to zero then, for example, the level models would act as models with drift (see Subsection 3.3.4). This implies that the architecture of the model should change, and the conventional ETS models cannot be efficiently applied to such data. Furthermore, correctly estimating such models would not be straightforward, because ETS exhibits a "pull to centre" effect, where the predicted value gets closer to the actual one based on the forecast error of the model. As a result, it would be challenging to capture the non-zero mean of the error term. So, **the zero mean assumption is essential** for such dynamic models as ETS. For the multiplicative error models, this translates into $\mathrm{E}(1+\epsilon_t)=1$;
2. As it is well known, the Normal distribution is defined for positive, negative, and zero values. This is not a big deal for additive models, which assume that the actual value can be anything, and it is not an issue for the multiplicative models when we deal with high-level positive data (e.g. thousands of units): in this case, the variance of the error term will be small enough, protecting it from becoming negative. However, if the level of the data is low, then the variance of the error term can be large enough for the normally distributed error to cover negative values. And if the error term $1+\epsilon_t$ becomes negative the model will break. This is a potential flaw in the conventional ETS model with the multiplicative error term. So, what the standard multiplicative error ETS model actually assumes, is that **the data we work with is strictly positive and has high-level values**.

Based on the assumption of normality of error term, the ETS model can be estimated via the maximisation of likelihood (discussed in Section 11.1), which is equivalent to the minimisation of the mean squared one step ahead forecast error e_t. Note that in order to apply the ETS models to the data, we also need to know the initial values of components, $\hat{l}_0, \hat{b}_0, \hat{s}_{-m+2}, \hat{s}_{-m+3}, \dots, \hat{s}_0$. The conventional approach is to estimate these values together with the smoothing parameters during likelihood maximisation. As a result, the optimisation might involve a large number of parameters. In addition, the variance of the error term is considered as an additional parameter in the maximum likelihood estimation, so the number of parameters for different models is (here "*" stands for any type):

1. ETS(*,N,N) – 3 parameters: $\hat{l}_0$, $\hat{\alpha}$, and $\hat{\sigma}^2$;
2. ETS(*,*,N) – 5 parameters: $\hat{l}_0$, $\hat{b}_0$, $\hat{\alpha}$, $\hat{\beta}$, and $\hat{\sigma}^2$;
3. ETS(*,*d,N) – 6 parameters: $\hat{l}_0$, $\hat{b}_0$, $\hat{\alpha}$, $\hat{\beta}$, $\hat{\phi}$, and $\hat{\sigma}^2$;
4. ETS(*,N,*) – 4+m-1 parameters: $\hat{l}_0$, $\hat{s}_{-m+2}, \hat{s}_{-m+3}, \dots, \hat{s}_0$, $\hat{\alpha}$, $\hat{\gamma}$, and $\hat{\sigma}^2$;
5. ETS(*,*,*) – 6+m-1 parameters: $\hat{l}_0$, $\hat{b}_0$, $\hat{s}_{-m+2}, \hat{s}_{-m+3}, \dots, \hat{s}_0$, $\hat{\alpha}$, $\hat{\beta}$, $\hat{\gamma}$, and $\hat{\sigma}^2$;
6. ETS(*,*d,*) – 7+m-1 parameters: $\hat{l}_0$, $\hat{b}_0$, $\hat{s}_{-m+2}, \hat{s}_{-m+3}, \dots, \hat{s}_0$, $\hat{\alpha}$, $\hat{\beta}$, $\hat{\gamma}$, $\hat{\phi}$, and $\hat{\sigma}^2$.

Remark. In the case of seasonal models, we typically make sure that the initial seasonal indices are normalised, so we only need to estimate $m-1$ of them, and the last one is calculated based on the linear combination of the others. For example, for the additive seasonality, it is equal to $-\sum_{j=1}^{m-1} s_j$ because the sum of all the indices should be equal to zero.

When it comes to selecting the most appropriate model, the conventional approach involves the application of all models to the data and then selecting the most appropriate of them based on an information criterion (Section 15.1). This was first proposed by Hyndman et al. (2002). In the case of the conventional ETS model, this relies on the likelihood value of Normal distribution used in the estimation of the model.

Finally, the assumption of normality is used to generate a prediction interval from the model. There are typically two ways of doing that:

1. Calculating the variance of multiple steps ahead forecast error and then using it for the intervals' construction (see Chapter 6 of Hyndman et al. (2008) or Sections 5.3 and 18.2);
2. Generating thousands of possible paths for the components of the series and the actual values and then taking the necessary quantiles for the prediction intervals (see Section 18.1 for details).

Typically, (1) is applied for the pure additive models, where the closed forms for the variances are known, and the assumption of normality holds for several steps ahead. In some special cases of mixed models, approximations for variances work on short horizons (see Section 6.4 of Hyndman et al., 2008). But in all the other cases, (2) should be used, despite being typically slower than (1) and producing bounds that differ slightly from run to run due to randomness.

4.6 State space form of ETS

One of the main advantages of the ETS model is its state space form, which gives it the flexibility. Hyndman et al. (2008) use the following general formulation of the model with the first equation called "measurement equation" and the second one "transition equation":

$$\begin{aligned} y_t &= w(\mathbf{v}_{t-1}) + r(\mathbf{v}_{t-1})\epsilon_t \\ \mathbf{v}_t &= f(\mathbf{v}_{t-1}) + g(\mathbf{v}_{t-1})\epsilon_t \end{aligned}, \tag{4.22}$$

where $\mathbf{v}_t$ is the state vector, containing the components of series (level, trend, and seasonal), $w(\cdot)$ is the measurement, $r(\cdot)$ is the error, $f(\cdot)$ is the transition, and $g(\cdot)$ is the persistence functions. Depending on the types of components, these functions can have different values.

Remark. Note that Hyndman et al. (2008) use $\mathbf{x}_t$ instead of $\mathbf{v}_t$. I do not use their notation because I find it confusing: x is typically used to denote explanatory variables (especially in regression context), and when we use $\mathbf{x}_t$ in ETS context, the states are sometimes perceived as related to explanatory variables. However, this is not the case. They relate more to time-varying parameters rather than exogenous variables in the regression context. This aspect is discussed on an example of a seasonal model in Section 10.5.

1. Depending on the types of trend and seasonality, $w(\mathbf{v}_{t-1})$ will be equal either to the addition or multiplication of components. The special cases were presented in Tables 4.1 and 4.2 in Section 4.2. For example, in case of ETS(M,M,M) it is: $w(\mathbf{v}_{t-1}) = l_{t-1}b_{t-1}s_{t-m}$;
2. If the error is additive, then $r(\mathbf{v}_{t-1}) = 1$, otherwise (in the case of multiplicative error) it is $r(\mathbf{v}_{t-1}) = w(\mathbf{v}_{t-1})$. For example, for ETS(M,M,M) it will be $r(\mathbf{v}_{t-1}) = l_{t-1}b_{t-1}s_{t-m}$;
3. The transition function $f(\cdot)$ will produce values depending on the types of trend and seasonality and will correspond to the first parts in Tables 4.1 and 4.2 of the transition equations (dropping the error term). This function records how components interact with each other and how they change from one observation to another (thus the term "transition"). An example is the ETS(M,M,M) model, for which the transition function will produce three values: $l_{t-1}b_{t-1}$, b_{t-1}, and s_{t-m}, respectively, for the level, trend, and seasonal components. So, if we drop the persistence function $g(\cdot)$ and the error term ϵ_t for a moment, the second equation

in (4.22) will be:

$$\begin{aligned} l_t &= l_{t-1} b_{t-1} \\ b_t &= b_{t-1} \\ s_t &= s_{t-m} \end{aligned} . \tag{4.23}$$

4. Finally, the persistence function will differ from one model to another, but in some special cases it can either be: $g(\mathbf{v}_{t-1}) = \mathbf{g}$ if all components are additive, or $g(\mathbf{v}_{t-1}) = f(\mathbf{v}_{t-1})\mathbf{g}$ if they are all multiplicative. $\mathbf{g}$ is the vector of smoothing parameters, called in the ETS context the "persistence vector". An example of persistence function is the ETS(M,M,M) model, for which it is: $l_{t-1}b_{t-1}\alpha$, $b_{t-1}\beta$, and $s_{t-m}\gamma$, respectively, for the level, trend, and seasonal components. Uniting this with the transition function (4.23) we get the equation from Table 4.2:

$$\begin{aligned} l_t &= l_{t-1} b_{t-1} + l_{t-1} b_{t-1} \alpha \epsilon_t \\ b_t &= b_{t-1} + b_{t-1} \beta \epsilon_t \\ s_t &= s_{t-m} + s_{t-m} \gamma \epsilon_t \end{aligned} , \tag{4.24}$$

which can be simplified to:

$$\begin{aligned} l_t &= l_{t-1} b_{t-1} (1 + \alpha \epsilon_t) \\ b_t &= b_{t-1} (1 + \beta \epsilon_t) \\ s_t &= s_{t-m} (1 + \gamma \epsilon_t) \end{aligned} . \tag{4.25}$$

Some of the mixed models have more complicated persistence function values. For example, for ETS(A,A,M) it is:

$$g(\mathbf{v}_{t-1}) = \begin{pmatrix} \alpha \frac{1}{s_{t-m}} \\ \beta \frac{1}{s_{t-m}} \\ \gamma \frac{1}{l_{t-1} + b_{t-1}} \end{pmatrix}, \tag{4.26}$$

which results in the state space model discussed in Subsection 4.4.3.

The compact form (4.22) is thus convenient, it underlies all the 30 ETS models discussed in the Sections 4.1 and 4.2. Unfortunately, they cannot be used directly for deriving conditional values, so they are needed just for the general understanding of ETS and can be used in programming.

4.6.1 Pure additive state space model

The more useful state space model in ETS framework is the pure additive one, which, based on the discussion above, is formulated as:

$$\begin{aligned} y_t &= \mathbf{w}' \mathbf{v}_{t-1} + \epsilon_t \\ \mathbf{v}_t &= \mathbf{F} \mathbf{v}_{t-1} + \mathbf{g} \epsilon_t \end{aligned} , \tag{4.27}$$

where $\mathbf{w}$ is the measurement vector, showing how the components form the structure, $\mathbf{F}$ is the transition matrix, showing how components interact with each other and change over time (e.g. level is equal to the previous level plus trend), and $\mathbf{g}$ is the persistence vector, containing smoothing parameters. The conditional expectation and variance can be derived based on (4.27), together with bounds on the smoothing parameters for any model that can be formulated in this way. And, as mentioned above, any pure additive ETS model can be written in the form (4.27), which means that all of them have relatively simple analytical formulae for the statistics mentioned above. For example, the h steps ahead conditional expectation and variance of the model (4.27) are (Hyndman et al., 2008, Chapter 6):

$$\begin{aligned} \mu_{y,t+h} &= \mathrm{E}(y_{t+h}|t) = \mathbf{w}'\mathbf{F}^{h-1}\mathbf{v}_t \\ \sigma^2_h &= \mathrm{V}(y_{t+h}|t) = \left(\mathbf{w}'\mathbf{F}^{j-1}\mathbf{g}\mathbf{g}'\mathbf{F}'\mathbf{w} + 1\right)\sigma^2, \end{aligned} \tag{4.28}$$

where σ^2 is the variance of the error term. The formulae in (4.28) can be used for the generation of respective moments from any pure additive ETS model. The conditional expectation can also be used for some mixed models as an approximation for the true conditional mean.

4.7 Parameters bounds

While many practitioners and academics accept that the smoothing parameters of Exponential Smoothing methods should lie between zero and one, this is not entirely true for the ETS models. There are, in fact, several possible restrictions on smoothing parameters, and it is worth discussing them separately:

1. **Classical or conventional** bounds are $\alpha, \beta, \gamma \in (0, 1)$. The idea behind them originates from the simple Exponential Smoothing method (Section 3.4), where it is logical to restrict the bounds with this region because then the smoothing parameters regulate what weight the actual value y_t will have and what weight will be assigned to the predicted one $\hat{y}_t$. Hyndman et al. (2008) showed that this condition is sometimes too loose and, in other cases, is too restrictive for some ETS models. Brenner, D'Esopo, and Fowler (1968) were some of the first to show that the bounds are wider than this region for many Exponential Smoothing methods. Still, the conventional restriction is the most often used in practice, just because it is easy to interpret.
2. **Usual or traditional** bounds are those that satisfy the set of the

following equations:

$$\begin{aligned} &\alpha \in [0,1) \\ &\beta \in [0,\alpha) \\ &\gamma \in [0,1-\alpha) \end{aligned} . \tag{4.29}$$

This set of restrictions guarantees that the weights decline over time exponentially (see Section 3.4.2), and the ETS models have the property of "averaging" the values over time. In the lower boundary condition, the model's components become deterministic, and we can say that they are calculated as the global averages of the values over time.

3. **Admissible** bounds, satisfying the stability condition. The idea here is that the most recent observation should have a higher weight than the older ones, which is regulated via the smoothing parameters. However, in this case, we do not impose the restriction of exponential decay of weights on the models, so they can oscillate or decay harmonically as long as their absolute values decrease over time. The condition is more complicated mathematically than the previous two. It will be discussed later in the monograph for the pure additive models (see Section 5.1), but here are several examples for bounds, satisfying this condition (from Chapter 10 of Hyndman et al., 2008):

- ETS(A,N,N): $\alpha \in (0,2)$;
- ETS(A,A,N): $\alpha \in (0,2); \beta \in (0,4-2\alpha)$;
- ETS(A,N,A): $\alpha \in \left(\frac{-2}{m-1}, 2-\gamma\right); \gamma \in (\max(-m\alpha,0), 2-\alpha)$;

As you see, the admissible bounds are much wider than the conventional and usual ones. In fact, in this case, smoothing parameters can become either negative or greater than one in some cases for some models, which is hard to interpret but might indicate that the data is difficult to predict. Furthermore, the admissible bounds correspond to the restrictions of the parameters for ARIMA models, underlying some of pure additive ETS models (see discussion in Section 8.4). In a way, they are more natural for the ETS models than the other two because they follow the formulation and arise naturally. However, their usage in practice has been met with mixed success, with only a handful of papers using them instead of (1) or (2) (e.g. Gardner & Diaz-Saiz, 2008, mention that they appear in some cases and Snyder, Ord, Koehler, McLaren, and Beaumont (2017) use them in their model).

5

Pure additive ADAM ETS

Now that we are familiar with the conventional ETS, we can move to the discussion of ADAM implementation of ETS, which has several important differences from the classical one. This chapter focuses on technical details of the pure additive model, discussing general formulation in algebraic form, then moving to recursive relations, which are needed to understand how to produce forecasts from the model and how to estimate it correctly (i.e. impose restrictions on the parameters). Finally, we discuss the distributional assumptions for ADAM ETS, introducing not only the Normal distribution but also showing how to use Laplace, S, Generalised Normal, Log-Normal, Gamma, and Inverse Gaussian distributions in the context.

5.1 Model formulation

The pure additive case is interesting, because this is the group of models that have closed forms for conditional moments (mean and variance) and support parametric predictive distribution for several steps ahead values. In order to understand how we can get to the general model form, we consider an example of an ETS(A,A,A) model, which, as discussed in Section 4.2, is formulated as:

$$\begin{aligned} y_t &= l_{t-1} + b_{t-1} + s_{t-m} + \epsilon_t \\ l_t &= l_{t-1} + b_{t-1} + \alpha\epsilon_t \\ b_t &= b_{t-1} + \beta\epsilon_t \\ s_t &= s_{t-m} + \gamma\epsilon_t \end{aligned} . \tag{5.1}$$

This model can be formatted in the following way:

$$\begin{aligned} y_t &= l_{t-1} & +b_{t-1} & +s_{t-m} & +\epsilon_t \\ l_t &= l_{t-1} & +b_{t-1} & +0 & +\alpha\epsilon_t \\ b_t &= 0 & +b_{t-1} & +0 & +\beta\epsilon_t \\ s_t &= 0 & +0 & +s_{t-m} & +\gamma\epsilon_t \end{aligned} . \tag{5.2}$$

To see how its elements can then be represented in the matrix form based on (5.2):

$$\begin{aligned}
y_t &= \begin{pmatrix} 1 & 1 & 1 \end{pmatrix} \begin{pmatrix} l_{t-1} \\ b_{t-1} \\ s_{t-m} \end{pmatrix} + \epsilon_t \\
\begin{pmatrix} l_t \\ b_t \\ s_t \end{pmatrix} &= \begin{pmatrix} 1 & 1 & 0 \\ 0 & 1 & 0 \\ 0 & 0 & 1 \end{pmatrix} \begin{pmatrix} l_{t-1} \\ b_{t-1} \\ s_{t-m} \end{pmatrix} + \begin{pmatrix} \alpha \\ \beta \\ \gamma \end{pmatrix} \epsilon_t
\end{aligned} . \tag{5.3}$$

I use tabulation in (5.2) to show how the matrix form is related to the general one. The positions of l_{t-1}, b_{t-1} and s_{t-m} correspond to the non-zero values in the transition matrix in (5.3). Now we can define each matrix and vector, for example:

$$\begin{aligned}
\mathbf{w} &= \begin{pmatrix} 1 \\ 1 \\ 1 \end{pmatrix}, \mathbf{F} = \begin{pmatrix} 1 & 1 & 0 \\ 0 & 1 & 0 \\ 0 & 0 & 1 \end{pmatrix}, \quad \mathbf{g} = \begin{pmatrix} \alpha \\ \beta \\ \gamma \end{pmatrix}, \\
\mathbf{v}_t &= \begin{pmatrix} l_t \\ b_t \\ s_t \end{pmatrix}, \mathbf{v}_{t-\boldsymbol{l}} = \begin{pmatrix} l_{t-1} \\ b_{t-1} \\ s_{t-m} \end{pmatrix}, \quad \boldsymbol{l} = \begin{pmatrix} 1 \\ 1 \\ m \end{pmatrix}
\end{aligned} . \tag{5.4}$$

Substituting (5.4) into (5.3), we get the general pure additive ADAM ETS model:

$$\begin{aligned}
y_t &= \mathbf{w}' \mathbf{v}_{t-\boldsymbol{l}} + \epsilon_t \\
\mathbf{v}_t &= \mathbf{F} \mathbf{v}_{t-\boldsymbol{l}} + \mathbf{g} \epsilon_t
\end{aligned} , \tag{5.5}$$

where $\mathbf{w}$ is the measurement vector, $\mathbf{F}$ is the transition matrix, $\mathbf{g}$ is the persistence vector, $\mathbf{v}_{t-\boldsymbol{l}}$ is the vector of lagged components, and $\boldsymbol{l}$ is the vector of lags. The important thing to note is that the ADAM is based on the model discussed in Section 4.6.1, but it is formulated using lags of components rather than their transition over time. This comes to the elements of the vector $\boldsymbol{l}$. Just for the comparison, the conventional ETS(A,A,A), formulated according to (4.22) would have the following transition matrix (instead of (5.4)):

$$\mathbf{F} = \begin{pmatrix} 1 & 1 & \mathbf{0}'_{m-1} & 0 \\ 0 & 1 & \mathbf{0}'_{m-1} & 0 \\ 0 & 0 & \mathbf{0}'_{m-1} & 1 \\ \mathbf{0}_{m-1} & \mathbf{0}_{m-1} & \mathbf{I}_{m-1} & \mathbf{0}_{m-1} \end{pmatrix}, \tag{5.6}$$

where $\mathbf{I}_{m-1}$ is the identity matrix of the size $(m-1) \times (m-1)$ and $\mathbf{0}_{m-1}$ is the vector of zeroes of size $m-1$. The main benefit of using the vector of lags $\boldsymbol{l}$ instead of the conventional mechanism in the transition equation is in the reduction of dimensions of matrices (the transition matrix contains 3×3 elements in the case of (5.5) instead of $(2+m) \times (2+m)$ as in the conventional ETS model). The model (5.5) is more parsimonious than the conventional one and simplifies some of the calculations, making it realistic, for example, to apply models to data with large frequency m (e.g. 24, 48, 52, 365). The main disadvantage of this approach is in the complications arising in the derivation of conditional expectation and variance, which still have closed forms, but are more cumbersome. They are discussed later in this chapter in Section 5.3.

5.2 Recursive relation

One of the useful representations of the pure additive model (5.5) is its recursive form, which can be used for further inference.

First, when we produce forecast for h steps ahead, it is important to understand what the actual value h steps ahead might be, given the information on observation t (i.e. in-sample values). In order to get to it, we first consider the model for the actual value y_{t+h}:

$$\begin{aligned} y_{t+h} &= \mathbf{w}' \mathbf{v}_{t+h-\boldsymbol{l}} + \epsilon_{t+h} \\ \mathbf{v}_{t+h} &= \mathbf{F} \mathbf{v}_{t+h-\boldsymbol{l}} + \mathbf{g} \epsilon_{t+h} \end{aligned}, \tag{5.7}$$

where $\mathbf{v}_{t+h-\boldsymbol{l}}$ is the vector of previous states, given the lagged values $\boldsymbol{l}$. Now we need to split the measurement and persistence vectors together with the transition matrix into parts for the same lags of components, leading to the following equation:

$$\begin{aligned} y_{t+h} &= (\mathbf{w}'_{m_1} + \mathbf{w}'_{m_2} + \dots + \mathbf{w}'_{m_d}) \mathbf{v}_{t+h-\boldsymbol{l}} + \epsilon_{t+h} \\ \mathbf{v}_{t+h} &= (\mathbf{F}_{m_1} + \mathbf{F}_{m_2} + \dots + \mathbf{F}_{m_d}) \mathbf{v}_{t+h-\boldsymbol{l}} + (\mathbf{g}_{m_1} + \mathbf{g}_{m_2} + \dots \mathbf{g}_{m_d}) \epsilon_{t+h} \end{aligned}, \tag{5.8}$$

where $m_1, m_2, \dots, m_d$ are the distinct lags of the model. So, for example, in the case of an ETS(A,A,A) model on quarterly data (periodicity is equal to four), $m_1 = 1$, $m_2 = 4$, leading to $\mathbf{F}_1 = \begin{pmatrix} 1 & 1 & 0 \\ 0 & 1 & 0 \\ 0 & 0 & 0 \end{pmatrix}$ and $\mathbf{F}_4 = \begin{pmatrix} 0 & 0 & 0 \\ 0 & 0 & 0 \\ 0 & 0 & 1 \end{pmatrix}$, where the split of the transition matrix is done column-wise. This split of matrices and vectors into distinct sub matrices and subvectors is needed in order to get the correct recursion and obtain the correct conditional h-steps ahead expectation and variance.

By substituting the values in the transition equation of (5.8) with their previous values until we reach t, we get:

$$\begin{aligned}
\mathbf{v}_{t+h-l} = & \mathbf{F}_{m_1}^{\lceil \frac{h}{m_1} \rceil - 1} \mathbf{v}_t + \sum_{j=1}^{\lceil \frac{h}{m_1} \rceil - 1} \mathbf{F}_{m_1}^{j-1} \mathbf{g}_{m_1} \epsilon_{t+m_1 \lceil \frac{h}{m_1} \rceil - j} + \\
& \mathbf{F}_{m_2}^{\lceil \frac{h}{m_2} \rceil - 1} \mathbf{v}_t + \sum_{j=1}^{\lceil \frac{h}{m_2} \rceil - 1} \mathbf{F}_{m_2}^{j-1} \mathbf{g}_{m_2} \epsilon_{t+m_2 \lceil \frac{h}{m_2} \rceil - j} + \\
& \dots + \\
& \mathbf{F}_{m_d}^{\lceil \frac{h}{m_d} \rceil - 1} \mathbf{v}_t + \sum_{j=1}^{\lceil \frac{h}{m_d} \rceil - 1} \mathbf{F}_{m_d}^{j-1} \mathbf{g}_{m_d} \epsilon_{t+m_d \lceil \frac{h}{m_d} \rceil - j} .
\end{aligned} \tag{5.9}$$

Inserting (5.9) in the measurement equation of (5.8), we get:

$$\begin{aligned}
y_{t+h} = & \mathbf{w}'_{m_1} \mathbf{F}_{m_1}^{\lceil \frac{h}{m_1} \rceil - 1} \mathbf{v}_t + \mathbf{w}'_{m_1} \sum_{j=1}^{\lceil \frac{h}{m_1} \rceil - 1} \mathbf{F}_{m_1}^{j-1} \mathbf{g}_{m_1} \epsilon_{t+m_1 \lceil \frac{h}{m_1} \rceil - j} + \\
& \mathbf{w}'_{m_2} \mathbf{F}_{m_2}^{\lceil \frac{h}{m_2} \rceil - 1} \mathbf{v}_t + \mathbf{w}'_{m_2} \sum_{j=1}^{\lceil \frac{h}{m_2} \rceil - 1} \mathbf{F}_{m_2}^{j-1} \mathbf{g}_{m_2} \epsilon_{t+m_2 \lceil \frac{h}{m_2} \rceil - j} + \\
& \dots + \\
& \mathbf{w}'_{m_d} \mathbf{F}_{m_d}^{\lceil \frac{h}{m_d} \rceil - 1} \mathbf{v}_t + \mathbf{w}'_{m_d} \sum_{j=1}^{\lceil \frac{h}{m_d} \rceil - 1} \mathbf{F}_{m_d}^{j-1} \mathbf{g}_{m_d} \epsilon_{t+m_d \lceil \frac{h}{m_d} \rceil - j} + \\
& \epsilon_{t+h} .
\end{aligned} \tag{5.10}$$

This recursion shows how the actual value appears based on the states on observation t, values of transition matrix and measurement and persistence vectors, and on the error term for the holdout sample. The latter is typically not known but we can usually estimate its moments (e.g. $\mathrm{E}(\epsilon_t) = 0$ and $\mathrm{V}(\epsilon_t) = \sigma^2$), which will help us in getting conditional moments for the actual value y_{t+h}.

Substituting the specific values of $m_1, m_2, \dots, m_d$ in (5.10) will simplify the equation and make it easier to understand. For example, for ETS(A,N,N), $m_1 = 1$ and all the other lags are equal to zero, so the recursion (5.10) simplifies to:

$$y_{t+h} = \mathbf{w}'_1 \mathbf{F}_1^{h-1} \mathbf{v}_t + \mathbf{w}'_1 \sum_{j=1}^{h-1} \mathbf{F}_1^{j-1} \mathbf{g}_1 \epsilon_{t+h-j} + \epsilon_{t+h}, \tag{5.11}$$

which is the recursion obtained by Hyndman et al. (2008) on page 103.

5.3 Conditional expectation and variance

Now, why is the recursion (5.10) important? This is because we can take the expectation and variance of (5.10) conditional on the values of the state vector $\mathbf{v}_t$ on the observation t and all the matrices and vectors ($\mathbf{F}$, $\mathbf{w}$, and $\mathbf{g}$), assuming that the basic model assumptions hold (error term is homoscedastic, uncorrelated, and has the expectation of zero, Subsection 1.4.1), in order to get:

$$\begin{aligned} \mu_{y,t+h} = \mathrm{E}(y_{t+h}|t) =& \sum_{i=1}^{d} \left(\mathbf{w}'_{m_i} \mathbf{F}_{m_i}^{\lceil\frac{h}{m_i}\rceil-1} \right) \mathbf{v}_{t} \\ \sigma^2_{h} = \mathrm{V}(y_{t+h}|t) =& \left(\sum_{i=1}^{d} \left(\mathbf{w}'_{m_i} \sum_{j=1}^{\lceil\frac{h}{m_i}\rceil-1} \mathbf{F}_{m_i}^{j-1} \mathbf{g}_{m_i} \mathbf{g}'_{m_i} (\mathbf{F}'_{m_i})^{j-1} \mathbf{w}_{m_i} \right) + 1 \right) \sigma^2 \end{aligned}, \tag{5.12}$$

where σ^2 is the variance of the error term. The formulae (5.12) are cumbersome, but they give the analytical solutions to the two moments for any model that can be formulated in the pure additive form (5.5). Having obtained both of them, we can construct prediction intervals, assuming, for example, that the error term follows Normal distribution (see Section 18.3 for details):

$$y_{t+h} \in \left(\mathrm{E}(y_{t+h}|t) + z_{\frac{\alpha}{2}} \sqrt{\mathrm{V}(y_{t+h}|t)}, \mathrm{E}(y_{t+h}|t) + z_{1-\frac{\alpha}{2}} \sqrt{\mathrm{V}(y_{t+h}|t)} \right), \tag{5.13}$$

where $z_{\frac{\alpha}{2}}$ is the quantile of standardised Normal distribution for the level $\frac{\alpha}{2}$. When it comes to other distributions (see Section 5.5), in order to get the conditional h steps ahead scale parameter, we can first calculate the variance use (5.12) and then using the relation between the scale and the variance for the specific distribution to get the necessary value (this is discussed in Section 5.5).

5.3.1 Example with ETS(A,N,N)

For example, for the ETS(A,N,N) model, we get:

$$\begin{aligned} \mathrm{E}(y_{t+h}|t) =& \mathbf{w}'_{1} \mathbf{F}_{1}^{h-1} \mathbf{v}_{t} \\ \mathrm{V}(y_{t+h}|t) =& \left(\mathbf{w}'_{1} \sum_{j=1}^{h-1} \mathbf{F}_{1}^{j-1} \mathbf{g}_{1} \mathbf{g}'_{1} (\mathbf{F}'_{1})^{j-1} \mathbf{w}_{1} + 1 \right) \sigma^2 \end{aligned}, \tag{5.14}$$

or by substituting $\mathbf{F} = 1$, $\mathbf{w} = 1$, $\mathbf{g} = \alpha$ and $\mathbf{v}_t = l_t$:

$$\begin{aligned} \mu_{y,t+h} =& l_{t} \\ \sigma^2_{h} =& \left((h-1) \alpha^2 + 1\right) \sigma^2 \end{aligned}, \tag{5.15}$$

which is the same conditional expectation and variance as in the Hyndman et al. (2008) monograph on page 81.

5.4 Stability and forecastability conditions

Another important aspect of the pure additive model (5.5) is the restriction on the smoothing parameters. This is related to the stability and forecastability conditions of the model, defined by Hyndman et al. (2008) in Chapter 10. The **stability** implies that the weights for observations in a dynamic model decay over time (see example with SES in Section 3.4.2). This guarantees that the newer observations will have higher weights than the older ones, thus the impact of the older information on forecasts slowly disappears with the increase of the sample size. The **forecastability** does not guarantee that the weights decay, but it guarantees that the initial value of the state vector will have a constant impact on forecasts, i.e. it will not increase in weight with the increase of the sample size. An example of the non-stable, but forecastable model is ETS(A,N,N) with $\alpha = 0$. In this case, it reverts to the global level model (Section 3.3.2), where the initial value impacts the final forecast in the same way as it does for the first observation.

In order to derive both conditions for the ADAM, we need to use a reduced form of the model by inserting the measurement equation in the transition equation via $\epsilon_t = y_t - \mathbf{w}'\mathbf{v}_{t-\boldsymbol{l}}$:

$$\begin{aligned} \mathbf{v}_t = & \mathbf{F}\mathbf{v}_{t-\boldsymbol{l}} + \mathbf{g}\left(y_t - \mathbf{w}'\mathbf{v}_{t-\boldsymbol{l}}\right) \\ = & \left(\mathbf{F} - \mathbf{g}\mathbf{w}'\right)\mathbf{v}_{t-\boldsymbol{l}} + \mathbf{g}y_t \\ = & \mathbf{D}\mathbf{v}_{t-\boldsymbol{l}} + \mathbf{g}y_t \end{aligned} . \tag{5.16}$$

The matrix $\mathbf{D} = \mathbf{F} - \mathbf{g}\mathbf{w}'$ is called the discount matrix and it shows how the weights diminish over time. It is the main part of the model that determines, whether the model will be stable/forecastable or not.

5.4.1 Example with ETS(A,N,N)

In order to better understand what we plan to discuss in this section, consider an example of an ETS(A,N,N) model, for which $\mathbf{F} = 1$, $\mathbf{w} = 1$, $\mathbf{g} = \alpha$, $\mathbf{v}_t = l_t$ and $\boldsymbol{l} = 1$. Inserting these values in (5.16), we get:

$$l_t = (1 - \alpha)\, l_{t-1} + \alpha y_t, \quad . \tag{5.17}$$

which corresponds to the formula of SES from Section 3.4. The discount matrix, in this case, is $\mathbf{D} = 1 - \alpha$. If we now substitute the values for the level on the right-hand side of the equation (5.17) by the previous values of the level, we will obtain the recursion that we have already discussed in Section 3.4.2, but now in terms of the "true" components and parameters:

$$l_t = \alpha \sum_{j=0}^{t-1} (1-\alpha)^j y_{t-j} + (1-\alpha)^t l_0 \quad . \tag{5.18}$$

The *stability* condition for ETS(A,N,N) is that the discount scalar $1-\alpha$ is less than one by absolute value. This way, the weights will decay in time because of the exponentiation in (5.18) to the power of j. This condition is satisfied when $\alpha \in (0,2)$, which is the admissible bound discussed in Section 4.7.

As for the *forecastability* condition, in this case it implies that $\lim\limits_{t\to\infty}(1-\alpha)^t l_0 = \text{const}$, which means that the effect of the initial state on future values stays the same. This is achievable, for example, when $\alpha = 0$, but is violated, when $\alpha < 0$ or $\alpha \geq 2$. So, the bounds for the smoothing parameters in the ETS(A,N,N) model, guaranteeing the forecastability of the model (i.e. making it useful), are:

$$\alpha \in [0, 2). \tag{5.19}$$

5.4.2 Coming back to the general case

In the general case, the logic is the same as with ETS(A,N,N), but it implies the usage of linear algebra. Due to our lagged formulation, the recursion becomes complicated, because the discount matrix $\mathbf{D}$ needs to be split into submatrices similar to how we did it in Section 5.2:

$$\begin{aligned} \mathbf{v}_t = & \mathbf{D}_{m_1}^{\lceil \frac{t}{m_1} \rceil} \mathbf{v}_0 + \sum_{j=0}^{\lceil \frac{t}{m_1} \rceil - 1} \mathbf{D}_{m_1}^{j} y_{t-jm_1} + \\ & \mathbf{D}_{m_2}^{\lceil \frac{t}{m_2} \rceil} \mathbf{v}_0 + \sum_{j=0}^{\lceil \frac{t}{m_2} \rceil - 1} \mathbf{D}_{m_2}^{j} y_{t-jm_2} + \\ & \dots + \\ & \mathbf{D}_{m_d}^{\lceil \frac{t}{m_d} \rceil} \mathbf{v}_0 + \sum_{j=0}^{\lceil \frac{t}{m_d} \rceil - 1} \mathbf{D}_{m_d}^{j} y_{t-jm_d} \end{aligned} , \tag{5.20}$$

where $\mathbf{D}_{m_i} = \mathbf{F}_{m_i} - \mathbf{g}_{m_i} \mathbf{w}'_{m_i}$ is the discount matrix for each lag of the model. The stability condition in this case is that the absolute values of all the non-zero eigenvalues of the discount matrices $\mathbf{D}_{m_i}$ are lower than one. This condition can be checked at the model construction stage, ensuring that the selected parameters guarantee the stability of the model. As for the forecastability, as discussed earlier, it will hold if the initial value of the state vector does not have an increasing impact on the last observed value. This is obtained by inserting (5.20) in the measurement equation of the pure additive model:

$$\begin{aligned} y_t = & \mathbf{w}'_{m_1} \mathbf{D}_{m_1}^{\lceil \frac{t-1}{m_1} \rceil} \mathbf{v}_0 + \mathbf{w}'_{m_1} \sum_{j=0}^{\lceil \frac{t-1}{m_1} \rceil - 1} \mathbf{D}_{m_1}^j y_{t-1-jm_1} + \\ & \mathbf{w}'_{m_2} \mathbf{D}_{m_2}^{\lceil \frac{t-1}{m_2} \rceil} \mathbf{v}_0 + \mathbf{w}'_{m_2} \sum_{j=0}^{\lceil \frac{t-1}{m_2} \rceil - 1} \mathbf{D}_{m_2}^j y_{t-1-jm_2} + \\ & \dots + \\ & \mathbf{w}'_{m_d} \mathbf{D}_{m_d}^{\lceil \frac{t-1}{m_d} \rceil} \mathbf{v}_0 + \mathbf{w}'_{m_d} \sum_{j=0}^{\lceil \frac{t-1}{m_d} \rceil - 1} \mathbf{D}_{m_d}^j y_{t-1-jm_d} + \epsilon_t \end{aligned} , \tag{5.21}$$

In our case the forecastability condition implies that:

$$\lim_{t \rightarrow \infty} \left(\mathbf{w}'_{m_i} \mathbf{D}_{m_i}^{\lceil \frac{t-1}{m_i} \rceil} \mathbf{v}_0 \right) = \text{const for all } i = 1, \dots, d. \tag{5.22}$$

These conditions are general but applicable to any model formulated in the pure additive form (5.5).

5.5 Distributional assumptions in pure additive ADAM

While the conventional ETS assumes that the error term follows Normal distribution, ADAM ETS proposes some flexibility, implementing the following options for the error term distribution in the additive error models:

1. Normal: $\epsilon_t \sim \mathcal{N}(0, \sigma^2)$, meaning that $y_t = \mu_{y,t} + \epsilon_t \sim \mathcal{N}(\mu_{y,t}, \sigma^2)$;
2. Laplace: $\epsilon_t \sim \mathcal{L}(0, s)$, so that $y_t = \mu_{y,t} + \epsilon_t \sim \mathcal{L}(\mu_{y,t}, s)$;
3. Generalised Normal: $\epsilon_t \sim \mathcal{GN}(0, s, \beta)$, leading to $y_t = \mu_{y,t} + \epsilon_t \sim \mathcal{GN}(\mu_{y,t}, s, \beta)$;
4. S (special case of $\mathcal{GN}$ with $\beta = 0.5$): $\epsilon_t \sim \mathcal{S}(0, s)$, implying that $y_t = \mu_{y,t} + \epsilon_t \sim \mathcal{S}(\mu_{y,t}, s)$,

where $\mu_{y,t} = \mathbf{w}'\mathbf{v}_{t-\mathbf{l}}$ is the one step ahead point forecast.

The conditional moments and stability/forecastability conditions do not change for the model with these assumptions. The main element that changes is the scale and the width of prediction intervals. Given that the scales of these distributions are linearly related to the variance, one can calculate the conditional variance as discussed in Section 5.3 and then use it in order to obtain the respective scales. Having the scales, it becomes straightforward to calculate the needed quantiles for the prediction intervals. Here are the formulae for the scales of distributions mentioned above:

1. Normal: scale is σ^2_h;
2. Laplace: $s_h = \sigma_h \sqrt{\frac{1}{2}}$;
3. Generalised Normal: $s_h = \sigma_h \sqrt{\frac{\Gamma(1/\beta)}{\Gamma(3/\beta)}}$;
4. S: $s_h = \sqrt{\sigma_h} \sqrt[4]{\frac{1}{120}}$.

The estimation of pure additive ADAM can be done via the maximisation of the likelihood of the assumed distribution (see Section 11.1), which in some cases coincides with the popular loss functions (e.g. Normal and MSE, or Laplace and MAE).

In addition, the following more exotic options for the additive error models are available in ADAM:

1. Log-Normal: $\left(1 + \frac{\epsilon_t}{\mu_{y,t}}\right) \sim \log\mathcal{N}\left(-\frac{\sigma^2}{2}, \sigma^2\right)$, implying that $y_t = \mu_{y,t}\left(1 + \frac{\epsilon_t}{\mu_{y,t}}\right) = y_t = \mu_{y,t} + \epsilon_t \sim \log\mathcal{N}\left(\log \mu_{y,t} - \frac{\sigma^2}{2}, \sigma^2\right)$. Here, σ^2 is the variance of the error term in logarithms and the $-\frac{\sigma^2}{2}$ appears due to the restriction $\mathrm{E}(\epsilon_t) = 0$;
2. Inverse Gaussian: $\left(1 + \frac{\epsilon_t}{\mu_{y,t}}\right) \sim \mathcal{IG}(1, \sigma^2)$ with $y_t = \mu_{y,t}\left(1 + \frac{\epsilon_t}{\mu_{y,t}}\right) \sim \mathcal{IG}\left(\mu_{y,t}, \frac{\sigma^2}{\mu_{y,t}}\right)$;
3. Gamma: $\left(1 + \frac{\epsilon_t}{\mu_{y,t}}\right) \sim -(\sigma^{-2}, \sigma^2)$, so that $y_t = \mu_{y,t}\left(1 + \frac{\epsilon_t}{\mu_{y,t}}\right) \sim -(\sigma^{-2}, \sigma^2 \mu_{y,t})$.

The possibility of application of these distributions arises from the reformulation of the original pure additive model (5.5) into:

$$\begin{aligned} y_t &= \mathbf{w}'\mathbf{v}_{t-\boldsymbol{l}}\left(1 + \frac{\epsilon_t}{\mathbf{w}'\mathbf{v}_{t-\boldsymbol{l}}}\right). \\ \mathbf{v}_t &= \mathbf{F}\mathbf{v}_{t-\boldsymbol{l}} + \mathbf{g}\epsilon_t \end{aligned} \tag{5.23}$$

The connection between the two formulations becomes apparent when opening the brackets in the measurement equation of (5.23). Note that in this case, the model assumes that the data is strictly positive, and while it might be possible to fit the model on the data with negative values, the calculation of the scale and the likelihood might become impossible. Using alternative losses (e.g. MSE) is a potential solution in this case.

5.6 Examples of application

5.6.1 Non-seasonal data

To see how the pure additive ADAM ETS works, we will try it out using the `adam()` function from the `smooth` package for R on Box-Jenkins sales data. We start with plotting the data:

```
plot(BJsales)
```

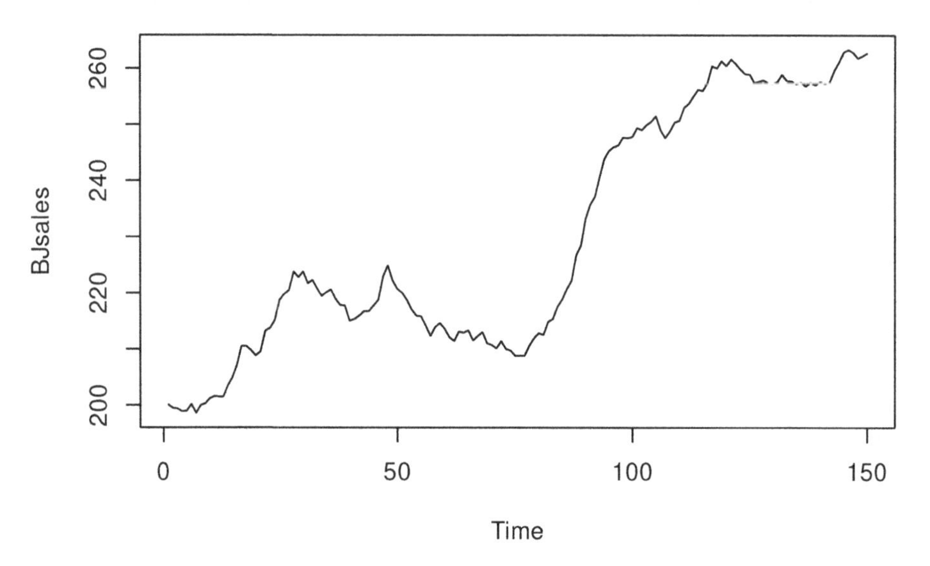

FIGURE 5.1 Box-Jenkins sales data.

The series in Figure 5.1 seem to exhibit a trend, so we will apply an ETS(A,A,N) model:

```
adamETSBJ <- adam(BJsales, "AAN")
adamETSBJ
```

```
## Time elapsed: 0.06 seconds
## Model estimated using adam() function: ETS(AAN)
## Distribution assumed in the model: Normal
## Loss function type: likelihood; Loss function value: 258.8098
## Persistence vector g:
##  alpha   beta
## 1.0000 0.2438
##
```

```
## Sample size: 150
## Number of estimated parameters: 5
## Number of degrees of freedom: 145
## Information criteria:
##       AIC     AICc      BIC     BICc
## 527.6196 528.0362 542.6728 543.7166
```

The model's output summarises which specific model was constructed, what distribution was assumed, how the model was estimated, and also provides the values of smoothing parameters. It also reports the sample size, the number of parameters, degrees of freedom, and produces information criteria (see Section 16.4 of Svetunkov, 2022). We can compare this model with the ETS(A,N,N) to see which of them performs better in terms of information criteria (e.g. in terms of AICc):

```
adam(BJsales, "ANN")
```

```
## Time elapsed: 0.03 seconds
## Model estimated using adam() function: ETS(ANN)
## Distribution assumed in the model: Normal
## Loss function type: likelihood; Loss function value: 273.2898
## Persistence vector g:
## alpha
##     1
##
## Sample size: 150
## Number of estimated parameters: 3
## Number of degrees of freedom: 147
## Information criteria:
##       AIC     AICc      BIC     BICc
## 552.5795 552.7439 561.6114 562.0233
```

In this situation the AICc for ETS(A,N,N) is higher than for ETS(A,A,N), so we should use the latter for forecasting purposes. We can produce point forecasts and a prediction interval (in this example we will construct 90% and 95% ones) and plot them (Figure 5.2):

```
forecast(adamETSBJ, h=10,
         interval="prediction", level=c(0.9,0.95)) |>
    plot(main="")
```

Notice that the bounds in Figure 5.2 are expanding fast, demonstrating that the components of the model exhibit high uncertainty, which is then reflected in the holdout sample. This is partially due to the high values of the smoothing

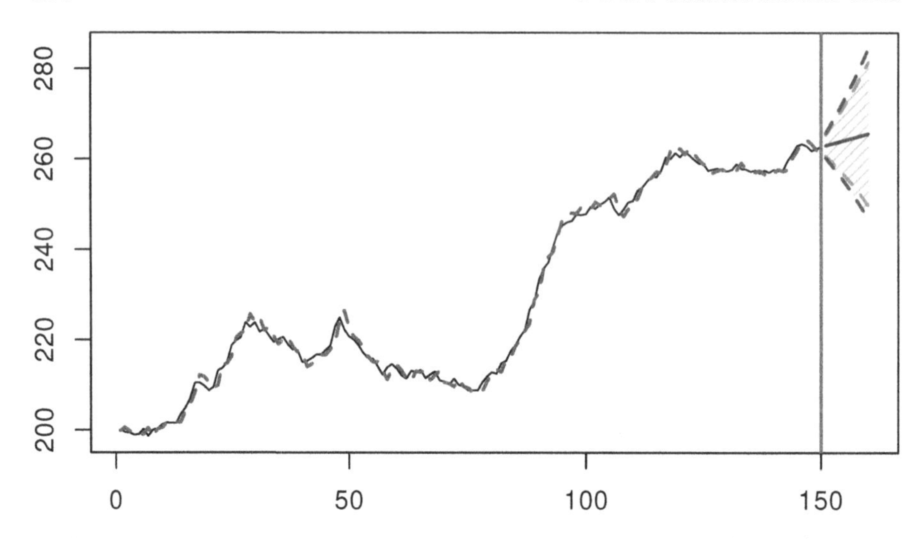

FIGURE 5.2 Forecast for Box-Jenkins sales data from ETS(A,A,N) model.

parameters of ETS(A,A,N), with $\alpha = 1$. While we typically want to have lower smoothing parameters, in this specific case this might mean that the maximum likelihood is achieved in the *admissible* bounds (i.e. data exhibits even higher variability than we expected with the usual bounds). We can try it out and see what happens:

```
adamETSBJ <- adam(BJsales, "AAN", bounds="admissible")
adamETSBJ
```

```
## Time elapsed: 0.07 seconds
## Model estimated using adam() function: ETS(AAN)
## Distribution assumed in the model: Normal
## Loss function type: likelihood; Loss function value: 258.5358
## Persistence vector g:
##  alpha   beta
## 1.0541 0.2185
##
## Sample size: 150
## Number of estimated parameters: 5
## Number of degrees of freedom: 145
## Information criteria:
##      AIC     AICc      BIC     BICc
## 527.0716 527.4883 542.1248 543.1687
```

Both smoothing parameters are now higher, which implies that the uncertainty about the future values of states is higher as well, which is then reflected in the slightly wider prediction interval (Figure 5.3):

```
forecast(adamETSBJ, h=10,
         interval="prediction", level=c(0.9,0.95)) |>
    plot(main="")
```

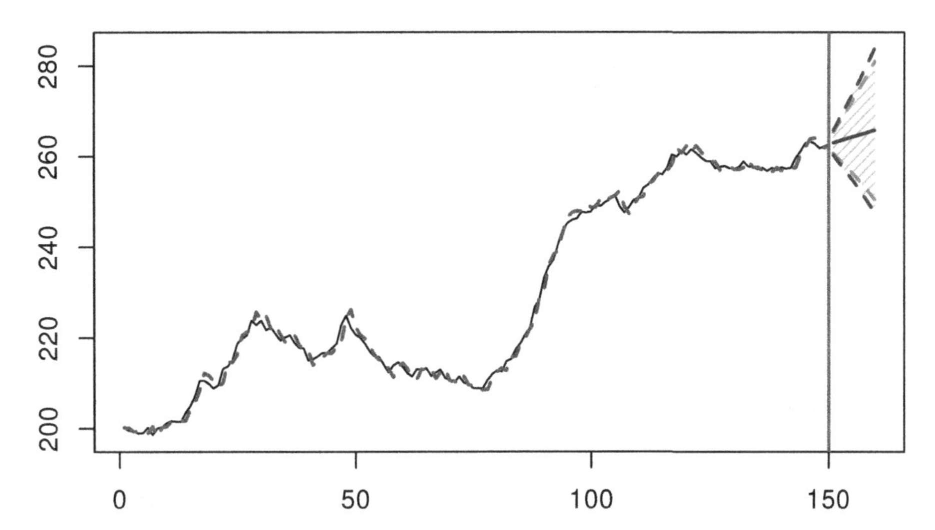

FIGURE 5.3 Forecast for Box-Jenkins sales data from an ETS(A,A,N) model with admissible bounds.

Although the values of smoothing parameters are higher than one, the model is still stable. In order to see that, we can calculate the discount matrix $\mathbf{D}$ using the objects returned by the function to reflect the formula $\mathbf{D} = \mathbf{F} - \mathbf{g}\mathbf{w}'$:

```
(adamETSBJ$transition - adamETSBJ$persistence %*%
    adamETSBJ$measurement[nobs(adamETSBJ),,drop=FALSE]) |>
    eigen(only.values=TRUE)
```

```
## $values
## [1]  0.79538429 -0.06800887
##
## $vectors
## NULL
```

Notice that the absolute values of both eigenvalues in the matrix are less than one, which means that the newer observations have higher weights than the older ones and that the absolute values of weights decrease over time, making the model stable.

If we want to test ADAM ETS with another distribution, it can be done using the respective parameter in the function (here we use Generalised Normal, estimating the shape together with the other parameters):

```
adamETSBJ <- adam(BJsales, "AAN", distribution="dgnorm")
print(adamETSBJ, digits=3)
```

```
## Time elapsed: 0.1 seconds
## Model estimated using adam() function: ETS(AAN)
## Distribution assumed in the model: Generalised Normal with shape
    =1.741
## Loss function type: likelihood; Loss function value: 258.456
## Persistence vector g:
## alpha  beta
## 1.000 0.217
##
## Sample size: 150
## Number of estimated parameters: 6
## Number of degrees of freedom: 144
## Information criteria:
##     AIC     AICc      BIC     BICc
## 528.913 529.500 546.977 548.448
```

Similar to the previous cases, we can plot the forecasts from the model:

```
forecast(adamETSBJ, h=10, interval="prediction") |>
    plot(main="")
```

The prediction interval in this case is slightly wider than in the previous one, because the Generalised Normal distribution with $\beta = 1.74$ has fatter tails than the Normal one (Figure 5.4).

5.6.2 Seasonal data

Now we will check what happens in the case of seasonal data. We use `AirPassengers` data, plotted in Figure 5.5, which apparently has multiplicative seasonality. But for demonstration purposes, we will see what happens when we use the wrong model with additive seasonality. We will withhold the last 12 observations to look closer at the performance of the ETS(A,A,A) model in this case:

```
adamETSAir <- adam(AirPassengers, "AAA", lags=12,
                   h=12, holdout=TRUE)
```

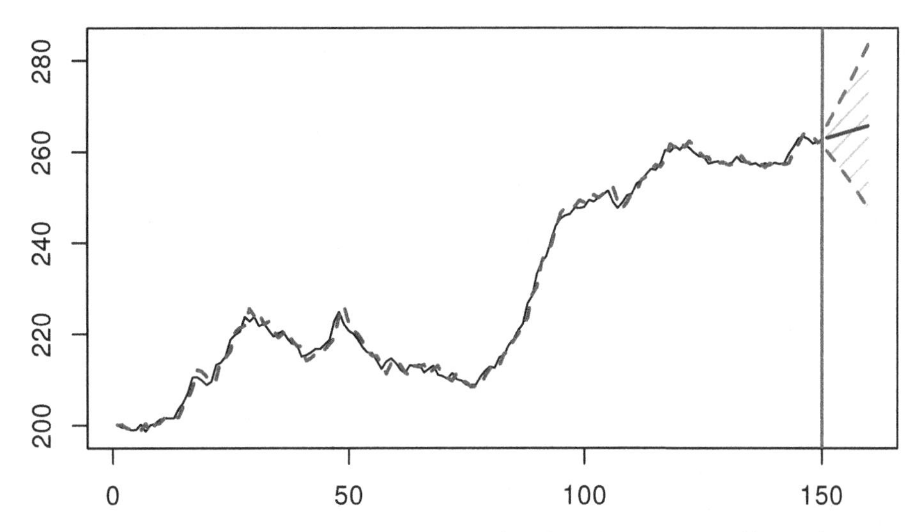

FIGURE 5.4 Forecast for Box-Jenkins sales data from an ETS(A,A,N) model with Generalised Normal distribution.

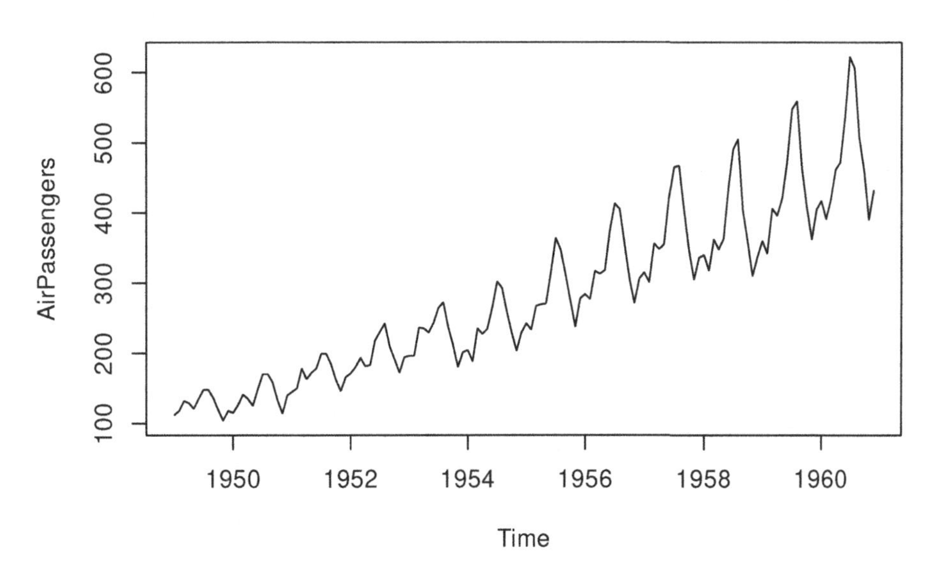

FIGURE 5.5 Air passengers data from Box-Jenkins textbook.

Remark. In this specific case, the `lags` parameter is not necessary because the function will automatically get the frequency from the `ts` object `AirPassengers`. If we were to provide a vector of values instead of the `ts` object, we would need to specify the correct lag. Note that `1` (lag for level and trend) is unnecessary; the function will always use it anyway.

Remark. In some cases, the optimiser might converge to the local minimum, so if you find the results unsatisfactory, it might make sense to reestimate the model, tuning the parameters of the optimiser (see Section 11.4 for details). Here is an example (we increase the number of iterations in the optimisation and set new starting values for the smoothing parameters):

```
adamETSAir$B[1:3] <- c(0.2,0.1,0.3)
adamETSAir <- adam(AirPassengers, "AAA", lags=12,
                   h=12, holdout=TRUE,
                   B=adamETSAir$B, maxeval=1000)
adamETSAir
```

```
## Time elapsed: 0.39 seconds
## Model estimated using adam() function: ETS(AAA)
## Distribution assumed in the model: Normal
## Loss function type: likelihood; Loss function value: 513.0026
## Persistence vector g:
##  alpha   beta  gamma
## 0.1928 0.0000 0.8072
##
## Sample size: 132
## Number of estimated parameters: 17
## Number of degrees of freedom: 115
## Information criteria:
##      AIC     AICc      BIC     BICc
## 1060.005 1065.374 1109.013 1122.119
##
## Forecast errors:
## ME: 6.216; MAE: 14.162; RMSE: 17.75
## sCE: 28.418%; Asymmetry: 51.9%; sMAE: 5.395%; sMSE: 0.457%
## MASE: 0.588; RMSSE: 0.567; rMAE: 0.186; rRMSE: 0.172
```

Notice that because we fit the seasonal additive model to the data with multiplicative seasonality, the smoothing parameter γ has become large – the seasonal component needs to be updated frequently to keep up with the changing seasonal profile. In addition, because we use the `holdout` parameter, the function also reports the error measures for the point forecasts on that part of the data. This can be useful when comparing the performance of several models on a time series. And here is how the forecast from ETS(A,A,A) looks on this data:

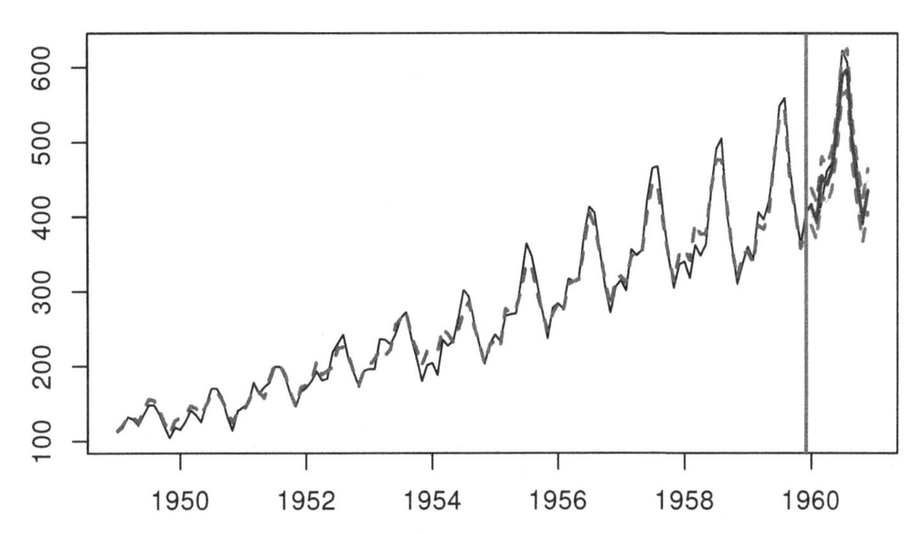

FIGURE 5.6 Forecast for air passengers data using an ETS(A,A,A) model.

Figure 5.6 demonstrates that while the fit to the data is far from perfect, due to a pure coincidence, the point forecast from this model is decent.

In order to see how the ADAM ETS decomposes the data into components, we can plot it via the `plot()` method with `which=12` parameter:

```
plot(adamETSAir, which=12)
```

We can see on the graph in Figure 5.7 that the residuals still contain some seasonality, so there is room for improvement. This probably happened because the data exhibits multiplicative seasonality rather than the additive one. For now, we do not aim to fix this issue.

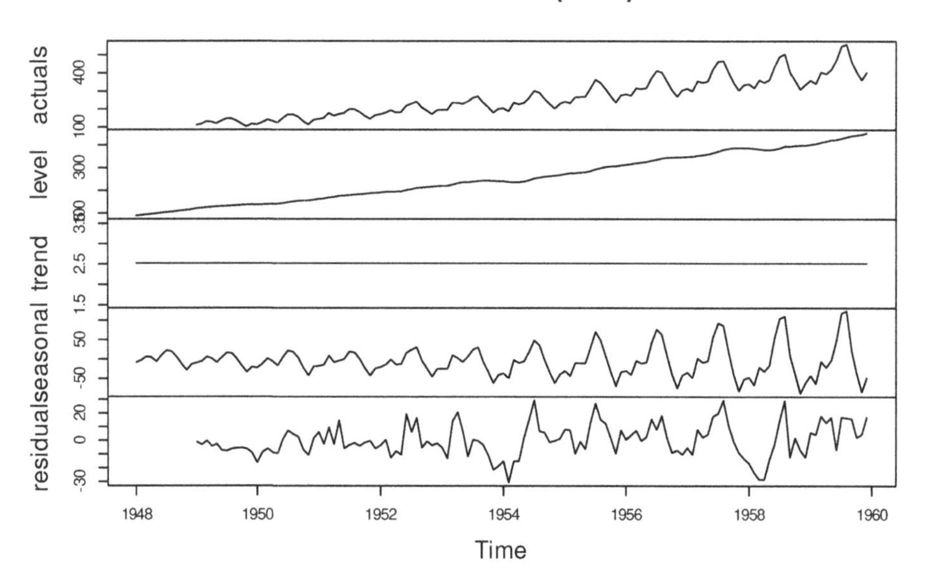

FIGURE 5.7 Decomposition of air passengers data using an ETS(A,A,A) model.

6

Pure multiplicative ADAM ETS

There is a reason why we discuss pure multiplicative ADAM ETS models separately: they are suitable for the positive data, especially when the level is low, yet they do not rely on prior data transformations (such as taking logarithms or applying a power transform), which makes them useful in a variety of contexts. However, the models discussed in this chapter are not easy to work with – they typically do not have closed forms for the conditional h steps ahead mean and variance and do not have well-defined parameter space. Furthermore, they make more sense in conjunction with positive-valued distributions, although they also work with the Normal one. All these aspects are discussed in this chapter.

6.1 Model formulation

The pure multiplicative ETS implemented in ADAM framework can be formulated using logarithms in the following way:

$$\begin{aligned} \log y_t = & \mathbf{w}' \log(\mathbf{v}_{t-\boldsymbol{l}}) + \log(1 + \epsilon_t) \\ \log \mathbf{v}_t = & \mathbf{F} \log \mathbf{v}_{t-\boldsymbol{l}} + \log(\mathbf{1}_k + \mathbf{g}\epsilon_t) \end{aligned}, \tag{6.1}$$

where $\mathbf{1}_k$ is the vector of ones, containing k elements (number of components in the model), log is the natural logarithm, applied element-wise to the vectors, and all the other objects correspond to the ones discussed in Section 5.1. An example of a pure multiplicative model is ETS(M,M,M), for which we have the following:

$$\begin{aligned} \mathbf{w} = \begin{pmatrix} 1 \\ 1 \\ 1 \end{pmatrix}, \mathbf{F} = \begin{pmatrix} 1 & 1 & 0 \\ 0 & 1 & 0 \\ 0 & 0 & 1 \end{pmatrix}, \mathbf{g} = \begin{pmatrix} \alpha \\ \beta \\ \gamma \end{pmatrix}, \\ \mathbf{v}_t = \begin{pmatrix} l_t \\ b_t \\ s_t \end{pmatrix}, \boldsymbol{l} = \begin{pmatrix} 1 \\ 1 \\ m \end{pmatrix}, \mathbf{1}_k = \begin{pmatrix} 1 \\ 1 \\ 1 \end{pmatrix} \end{aligned}. \tag{6.2}$$

By inserting these values in equation (6.1), we obtain the model in logarithms:

$$\begin{aligned} \log y_t &= \log l_{t-1} + \log b_{t-1} + \log s_{t-m} + \log \left(1 + \epsilon_t\right) \\ \log l_t &= \log l_{t-1} + \log b_{t-1} + \log(1 + \alpha\epsilon_t) \\ \log b_t &= \log b_{t-1} + \log(1 + \beta\epsilon_t) \\ \log s_t &= \log s_{t-m} + \log(1 + \gamma\epsilon_t) \end{aligned} , \tag{6.3}$$

which after exponentiation becomes equal to the one discussed in Section 4.2:

$$\begin{aligned} y_t &= l_{t-1} b_{t-1} s_{t-m} (1 + \epsilon_t) \\ l_t &= l_{t-1} b_{t-1} (1 + \alpha\epsilon_t) \\ b_t &= b_{t-1} (1 + \beta\epsilon_t) \\ s_t &= s_{t-m} (1 + \gamma\epsilon_t) \end{aligned} . \tag{6.4}$$

This example demonstrates that the model (6.1) underlies other pure multiplicative ETS models. While it can be used for some inference, it has limitations due to the $\log(\mathbf{1}_k + \mathbf{g}\epsilon_t)$ term, which introduces a non-linear transformation of the smoothing parameters and the error term (this will be discussed in more detail in next sections).

An interesting observation is that the model (6.3) will produce values similar to the model ETS(A,A,A) applied to the data in logarithms, when the values of smoothing parameters are close to zero. This becomes apparent, when recalling the limit:

$$\lim_{x \to 0} \log(1 + x) = x. \tag{6.5}$$

Based on that, the model will become close to the following one in cases of small values of smoothing parameters:

$$\begin{aligned} \log y_t &= \log l_{t-1} + \log b_{t-1} + \log s_{t-m} + \epsilon_t \\ \log l_t &= \log l_{t-1} + \log b_{t-1} + \alpha\epsilon_t \\ \log b_t &= \log b_{t-1} + \beta\epsilon_t \\ \log s_t &= \log s_{t-m} + \gamma\epsilon_t \end{aligned} , \tag{6.6}$$

which is the ETS(A,A,A) applied to the data in the logarithms. In many cases, the smoothing parameters will be small enough for the limit (6.5) to hold, so the two models will produce similar forecasts. The main benefit of (6.6) is that it has closed forms for the conditional mean and variance. However, the form (6.6) does not permit mixed components – it only supports the multiplicative ones, making it detached from the other ETS models.

6.2 Recursive relation

Similarly to how it was done for the pure additive model in Section 5.2, we can show what the recursive relation will look like for the pure multiplicative one (the logic here is the same, the main difference is in working with logarithms instead of the original values):

$$\begin{aligned}\log y_{t+h} = & \mathbf{w}'_{m_1} \mathbf{F}_{m_1}^{\lceil \frac{h}{m_1} \rceil - 1} \log \mathbf{v}_t + \mathbf{w}'_{m_1} \sum_{j=1}^{\lceil \frac{h}{m_1} \rceil - 1} \mathbf{F}_{m_1}^{j-1} \log \left(\mathbf{1}_k + \mathbf{g}_{m_1} \epsilon_{t+m_1 \lceil \frac{h}{m_1} \rceil - j} \right) + \\ & \mathbf{w}'_{m_2} \mathbf{F}_{m_2}^{\lceil \frac{h}{m_2} \rceil - 1} \log \mathbf{v}_t + \mathbf{w}'_{m_2} \sum_{j=1}^{\lceil \frac{h}{m_2} \rceil - 1} \mathbf{F}_{m_2}^{j-1} \log \left(\mathbf{1}_k + \mathbf{g}_{m_2} \epsilon_{t+m_2 \lceil \frac{h}{m_2} \rceil - j} \right) + \\ & \dots \\ & \mathbf{w}'_{m_d} \mathbf{F}_{m_d}^{\lceil \frac{h}{m_d} \rceil - 1} \log \mathbf{v}_t + \mathbf{w}'_{m_d} \sum_{j=1}^{\lceil \frac{h}{m_d} \rceil - 1} \mathbf{F}_{m_d}^{j-1} \log \left(\mathbf{1}_k + \mathbf{g}_{m_d} \epsilon_{t+m_d \lceil \frac{h}{m_d} \rceil - j} \right) + \\ & \log \left(1 + \epsilon_{t+h}\right)\end{aligned} \tag{6.7}$$

In order to see how this recursion works, we can take the example of ETS(M,N,N), for which $m_1 = 1$ and all the other lags are equal to zero:

$$y_{t+h} = \exp \left(\mathbf{w}'_1 \mathbf{F}_1^{h-1} \log \mathbf{v}_t + \mathbf{w}'_1 \sum_{j=1}^{h-1} \mathbf{F}_1^{j-1} \log \left(\mathbf{1}_k + \mathbf{g}_1 \epsilon_{t+h-j}\right) + \log \left(1 + \epsilon_{t+h}\right) \right), \tag{6.8}$$

or after inserting $\mathbf{w}_1 = 1$, $\mathbf{F}_1 = 1$, $\mathbf{v}_t = l_t$, $\mathbf{g}_1 = \alpha$, and $\mathbf{1}_k = 1$:

$$y_{t+h} = l_t \prod_{j=1}^{h-1} \left(1 + \alpha \epsilon_{t+h-j}\right) \left(1 + \epsilon_{t+h}\right). \tag{6.9}$$

This recursion is useful to understand how the states evolve, and in the case of ETS(M,N,N), it allows obtaining the conditional expectation and variance. Unfortunately, in general, for models with trend and/or seasonality, the recursion (6.7) cannot be used to calculate conditional moments, like the one for the pure additive ADAM ETS. This is discussed in the next Section 6.3.

6.3 Moments and quantiles of pure multiplicative ETS

The recursion (6.7) obtained in the previous section shows how the information on observation t influences the logarithms of states. While it is possible to calculate the expectation of the logarithm of the variable y_{t+h} based on that information under some conditions, in general, this does not allow deriving the expectation of the variable in the original scale. This is because the expectations of terms $\log(\mathbf{1}_k + \mathbf{g}_{m_i}\epsilon_{t+j})$ for different j and i are not known and are difficult to derive analytically (if possible at all). The situation does not become simpler for the conditional variance.

The only way to derive the conditional expectation and variance for the pure multiplicative models is to use the formulae from Tables 4.1 and 4.2 in Section 4.2 and manually derive the values in the original scale. This works well only for the ETS(M,N,N) model, for which it is possible to take conditional expectation and variance of the recursion (6.9) to obtain:

$$\begin{aligned} \mu_{y,t+h} = \mathrm{E}(y_{t+h}|t) = & l_t \\ \mathrm{V}(y_{t+h}|t) = & l_t^2 \left(\left(1 + \alpha^2 \sigma^2 \right)^{h-1} (1 + \sigma^2) - 1 \right), \end{aligned} \tag{6.10}$$

where σ^2 is the variance of the error term. For the other models, the conditional moments do not have general closed forms because of the product of $\log(1+\alpha\epsilon_t)$, $\log(1+\beta\epsilon_t)$, and $\log(1+\gamma\epsilon_t)$. It is still possible to derive the moments for special cases of h, but this is a tedious process. In order to see that, we demonstrate here how the recursion looks for the ETS(M,Md,M) model:

$$\begin{aligned} & y_{t+h} = l_{t+h-1} b_{t+h-1}^\phi s_{t+h-m} \left(1 + \epsilon_{t+h}\right) = l_t b_t^{\sum_{j=1}^h \phi^j} s_{t+h-m\lceil\frac{h}{m}\rceil} \\ & \times \prod_{j=1}^{h-1} \left((1 + \alpha \epsilon_{t+j}) \prod_{i=1}^{j} (1 + \beta \epsilon_{t+i})^{\phi^{j-i}} \right) \prod_{j=1}^{\lceil\frac{h}{m}\rceil} \left(1 + \gamma \epsilon_{t+j}\right) \left(1 + \epsilon_{t+h}\right) . \end{aligned} \tag{6.11}$$

The conditional expectation of the recursion (6.11) does not have a simple form, because of the difficulties in calculating the expectation of $(1+\alpha\epsilon_{t+j})(1+\beta\epsilon_{t+i})^{\phi^{j-i}}(1+\gamma\epsilon_{t+j})$. In a simple example of $h=2$ and $m>h$ the conditional expectation based on (6.11) can be simplified to:

$$\mu_{y,t+2} = l_t b_t^{\phi+\phi^2} \left(1 + \alpha \beta \sigma^2\right), \tag{6.12}$$

introducing the second moment, the variance of the error term σ^2. The case of $h=3$ implies the appearance of the third moment, the $h=4$ – the fourth, etc. This is why there are no closed forms for the conditional moments for the pure multiplicative ETS models with trend and/or seasonality. In some special cases, when smoothing parameters and the variance of the error term are all low, it is possible to use approximate formulae for some of the multiplicative models. These are discussed in Chapter 6 of Hyndman et al. (2008). In a special case when all smoothing parameters are equal to zero or when $h = 1$, the conditional expectation will coincide with the point forecast from Tables 4.1 and 4.2 in Section 4.2. But in general, the best thing that can be done in this case is the simulation of possible paths (using the formulae from the tables mentioned above) and then the calculation of mean and variance based on them.

Furthermore, it can be shown for pure multiplicative models that:

$$\hat{y}_{t+h} \leq \check{\mu}_{t+h} \leq \mu_{y,t+h}, \tag{6.13}$$

where $\mu_{y,t+h}$ is the conditional h steps ahead expectation, $\check{\mu}_{t+h}$ is the conditional h steps ahead geometric expectation (expectation in logarithms), and $\hat{y}_{t+h}$ is the point forecast (Svetunkov & Boylan, 2022). This gives an understanding that the point forecasts from pure multiplicative ETS models are always lower than geometric and arithmetic moments. If the variance of the error term is close to zero, the three elements in (6.13) will be close to each other. A similar effect will be achieved when all smoothing parameters are close to zero. Moreover, the three elements will coincide for $h = 1$ (Svetunkov & Boylan, 2022).

Finally, when it comes to conditional quantiles, the same term $\log(\mathbf{1}_k + \mathbf{g}_{m_i} \epsilon_{t+j})$ causes a different set of problems, introducing convolutions of products of random variables. To better understand this issue, we consider the persistence part of the equation for the ETS(M,N,N) model, which is:

$$\log(1 + \alpha \epsilon_t) = \log(1 - \alpha + \alpha(1 + \epsilon_t)). \tag{6.14}$$

Whatever we assume about the distribution of the variable $(1 + \epsilon_t)$, the distribution of (6.14) will be more complicated. For example, if we assume that $(1 + \epsilon_t) \sim \log\mathcal{N}(0, \sigma^2)$, then the distribution of (6.14) is something like exp-three-parameter Log-Normal distribution (Sangal & Biswas, 1970). The convolution of (6.14) for different t does not follow a known distribution, so it is not possible to calculate the conditional quantiles based on (6.7). Similar issues arise if we assume any other distribution. The problem is worsened in the case of multiplicative trend and/or multiplicative seasonality models, because then the recursion (6.7) contains several errors on the same observation (e.g. $\log(1 + \alpha \epsilon_t)$ and $\log(1 + \beta \epsilon_t)$), introducing products of random variables.

All of this means that in general in order to get adequate estimates of moments or quantiles for a pure multiplicative ETS model, we need to revert to simulations (discussed in Section 18.1).

6.4 Smoothing parameters' bounds

Similar to the pure additive ADAM ETS, it is possible to have different restrictions on smoothing parameters for pure multiplicative models. However, in this case, the classical and the usual restrictions become more reasonable from the model's point of view. In contrast, the derivation of admissible bounds becomes a challenging task. Consider the ETS(M,N,N) model, for which the level is updated using the following relation:

$$l_t = l_{t-1}(1 + \alpha\epsilon_t) = l_{t-1}(1 - \alpha + \alpha(1 + \epsilon_t)). \tag{6.15}$$

As discussed previously, the main benefit of pure multiplicative models is in dealing with positive data. So, it is reasonable to assume that $(1 + \epsilon_t) > 0$, which implies that the actual values will always be positive and that each model component should also be positive. This means that $\alpha(1 + \epsilon_t) > 0$, which implies that $(1 - \alpha + \alpha(1 + \epsilon_t)) > 1 - \alpha$ or equivalently based on (6.15) $(1 + \alpha\epsilon_t) > 1 - \alpha$ should always hold. In order for the model to make sense, the condition $(1 + \alpha\epsilon_t) > 0$ should hold as well, ensuring that the level is always positive. Connecting the two inequalities, this can be achieved when $1 - \alpha \geq 0$, meaning that $\alpha \leq 1$. Furthermore, for the level to be positive irrespective of the specific error on observation t, the smoothing parameter should be non-negative. So, in general, the bounds $[0, 1]$ guarantee that the model ETS(M,N,N) will produce positive values only. The two special cases $\alpha = 0$ and $\alpha = 1$ make sense because the level in (6.15) will be positive in both of them, implying that for the former, the model becomes equivalent to the global level, while for the latter the model is equivalent to Random Walk. Using similar logic, it can be shown that the **classical restriction** $\alpha, \beta, \gamma \in [0, 1]$ guarantees that the model will always produce positive values.

The more restrictive condition of the **usual bounds**, discussed in Section 4.7, makes sense as well, although it might be more restrictive than needed. But it has a different idea: guaranteeing that the model exhibits averaging properties.

Finally, the **admissible bounds** might still make sense for the pure multiplicative models in some cases, but the condition for parameters' bounds becomes more complicated and implies that the distribution of the error term becomes trimmed from below to satisfy the classical restrictions discussed above (this is also discussed in Akram, Hyndman, & Ord, 2009). Very crudely, the conventional restriction from pure additive models can be used to approximate the proper admissible bounds, based on the limit (6.5), but this should be used with care, given the discussion above.

From the practical point of view, the pure multiplicative models typically have low smoothing parameters, close to zero, because they rely on multiplication of components rather than on addition, so even the classical restriction might seem broad in many situations.

6.5 Distributional assumptions in pure multiplicative ETS

The conventional assumption for the error term in ETS is that $\epsilon_t \sim \mathcal{N}(0, \sigma^2)$. The condition that $\mathrm{E}(\epsilon_t) = 0$ guarantees that the conditional expectation of the model will be equal to the point forecasts when the trend and seasonal components are not multiplicative. In general, ETS works well in many cases with this assumption, mainly when the data is strictly positive, and the level of series is high (e.g. thousands of units). However, this assumption might become unhelpful when dealing with lower-level data because the models may start generating non-positive values, which contradicts the idea of pure multiplicative ETS models. Akram et al. (2009) studied the ETS models with the multiplicative error and suggested that applying ETS on data in logarithms is a better approach than just using ETS(M,Y,Y) models (here "Y" stands for a non-additive component). However, this approach sidesteps the ETS taxonomy, creating a new group of models. An alternative (also discussed in Akram et al., 2009) is to assume that the error term $1 + \epsilon_t$ follows some distribution for positive data. The authors mentioned Log-Normal, truncated Normal, and Gamma distributions but never explored them further.

Svetunkov and Boylan (2022) discussed several options for the distribution of $1 + \epsilon_t$ in ETS, including Log-Normal, Gamma, and Inverse Gaussian. Other distributions for positive data can be applied as well, but their usage might become complicated, because they need to meet condition $\mathrm{E}(1 + \epsilon_t) = 1$ in order for the expectation to coincide with the point forecasts for models with non-multiplicative trend and seasonality. For example, if the error term follows Log-Normal distribution, then this restriction implies that the location of the distribution should be non-zero: $1 + \epsilon_t \sim \log\mathcal{N}\left(-\frac{\sigma^2}{2}, \sigma^2\right)$. Using this principle the following distributions can be used for ADAM ETS:

1. Inverse Gaussian: $(1 + \epsilon_t) \sim \mathcal{IG}(1, \sigma^2)$, so that $y_t = \mu_{y,t}(1 + \epsilon_t) \sim \mathcal{IG}(\mu_{y,t}, \sigma^2)$;
2. Gamma: $(1 + \epsilon_t) \sim \Gamma(\sigma^{-2}, \sigma^2)$, so that $y_t = \mu_{y,t}(1 + \epsilon_t) \sim \Gamma(\sigma^{-2}, \sigma^2 \mu_{y,t})$;
3. Log-Normal: $(1 + \epsilon_t) \sim \log\mathcal{N}\left(-\frac{\sigma^2}{2}, \sigma^2\right)$ so that $y_t = \mu_{y,t}(1 + \epsilon_t) \sim \log\mathcal{N}(\log \mu_{y,t} - \frac{\sigma^2}{2}, \sigma^2)$.

The Maximum Likelihood Estimate (MLE) of s in Inverse Gaussian is straightforward (see Section 11.1) and is:

$$\hat{\sigma}^2 = \frac{1}{T} \sum_{t=1}^{T} \frac{e_t^2}{1 + e_t}, \tag{6.16}$$

where e_t is the estimate of the error term ϵ_t. However, when it comes to the MLE of scale parameter for the Log-Normal distribution with the aforementioned restrictions, it is more complicated and is (Svetunkov & Boylan, 2022):

$$\hat{\sigma}^2 = 2 \left(1 - \sqrt{1 - \frac{1}{T} \sum_{t=1}^{T} \log^2(1 + e_t)}\right) . \tag{6.17}$$

Finally, MLE of s in Gamma does not have a closed form. Luckily, method of moments can be used to obtain its value (Svetunkov & Boylan, 2022):

$$\hat{\sigma}^2 = \frac{1}{T} \sum_{t=1}^{T} e_t^2 . \tag{6.18}$$

This value will coincide with the variance of the error term, given the imposed restrictions on the Gamma distribution.

Even if we deal with strictly positive high level data, it is not necessary to limit the distribution exclusively with the positive ones. The following distributions can be applied as well:

1. Normal: $\epsilon_t \sim \mathcal{N}(0, \sigma^2)$, implying that $y_t = \mu_{y,t}(1+\epsilon_t) \sim \mathcal{N}(\mu_{y,t}, \mu_{y,t}^2\sigma^2)$;
2. Laplace: $\epsilon_t \sim \mathcal{L}(0, s)$, meaning that $y_t = \mu_{y,t}(1 + \epsilon_t) \sim \mathcal{L}(\mu_{y,t}, \mu_{y,t}s)$;
3. Generalised Normal: $\epsilon_t \sim \mathcal{GN}(0, s, \beta)$ and $y_t = \mu_{y,t}(1 + \epsilon_t) \sim \mathcal{GN}(\mu_{y,t}, \mu_{y,t}^{\beta}s)$;
4. S: $\epsilon_t \sim \mathcal{S}(0, s)$, so that $y_t = \mu_{y,t}(1 + \epsilon_t) \sim \mathcal{S}(\mu_{y,t}, \sqrt{\mu_{y,t}}s)$.

The MLE of scale parameters for these distributions will be calculated differently than in the case of pure additive models (these are provided in Section 11.1). For example, for the Normal distribution it is:

$$\hat{\sigma}^2 = \frac{1}{T} \sum_{t=1}^{T} \frac{y_t - \hat{\mu}_{y,t}}{\hat{\mu}_{y,t}} , \tag{6.19}$$

where the main difference from the additive error case arises from the measurement equation of the multiplicative error models:

$$y_t = \mu_{y,t}(1 + \epsilon_t), \tag{6.20}$$

implying that

$$e_t = \frac{y_t - \hat{\mu}_{y,t}}{\hat{\mu}_{y,t}} . \tag{6.21}$$

The distributional assumptions impact both the estimation of models and the prediction intervals. In the case of asymmetric distributions (such as Log-Normal, Gamma, and Inverse Gaussian), the intervals will typically be asymmetric, with the upper bound being further away from the point forecast than the lower one. Furthermore, even with the comparable estimates of scales of distributions, Inverse Gaussian distribution will typically produce wider bounds than Log-Normal and Gamma, making it a viable option for data with higher uncertainty. The width of intervals for these distributions relates to their kurtoses (Svetunkov & Boylan, 2022).

6.6 Examples of application

6.6.1 Non-seasonal data

We continue our examples with the same Box-Jenkins sales data by fitting the ETS(M,M,N) model, but this time with a holdout of ten observations:

```
adamETSBJ <- adam(BJsales, "MMN", h=10, holdout=TRUE)
adamETSBJ
```

```
## Time elapsed: 0.09 seconds
## Model estimated using adam() function: ETS(MMN)
## Distribution assumed in the model: Gamma
## Loss function type: likelihood; Loss function value: 245.3759
## Persistence vector g:
##  alpha   beta
## 1.0000 0.2412
##
## Sample size: 140
## Number of estimated parameters: 5
## Number of degrees of freedom: 135
## Information criteria:
##      AIC     AICc      BIC     BICc
## 500.7518 501.1996 515.4600 516.5664
##
## Forecast errors:
## ME: 3.217; MAE: 3.33; RMSE: 3.784
## sCE: 14.124%; Asymmetry: 91.6%; sMAE: 1.462%; sMSE: 0.028%
## MASE: 2.817; RMSSE: 2.482; rMAE: 0.925; rRMSE: 0.921
```

The output above is similar to the one we discussed in Section 5.6, so we can compare the two models using various criteria and select the most appropriate. Even though the default distribution for the multiplicative error models in ADAM is Gamma, we can compare this model with the ETS(A,A,N) via information criteria. For example, here are the AICc for the two models:

```
# ETS(M,M,N)
AICc(adamETSBJ)

## [1] 501.1996

# ETS(A,A,N)
AICc(adam(BJsales, "AAN", h=10, holdout=TRUE))

## [1] 497.2624
```

The comparison is fair because both models were estimated via likelihood, and both likelihoods are formulated correctly, without omitting any terms (e.g. the `ets()` function from the `forecast` package omits the $-\frac{T}{2}\log\left(2\pi e\frac{1}{T}\right)$ for convenience, which makes it incomparable with other models). In this example, the pure additive model is more suitable for the data than the pure multiplicative one.

Figure 6.1 shows how the model fits the data and what forecast it produces. Note that the function produces the **point forecast** in this case, which is not equivalent to the conditional expectation! The point forecast undershoots the actual values in the holdout.

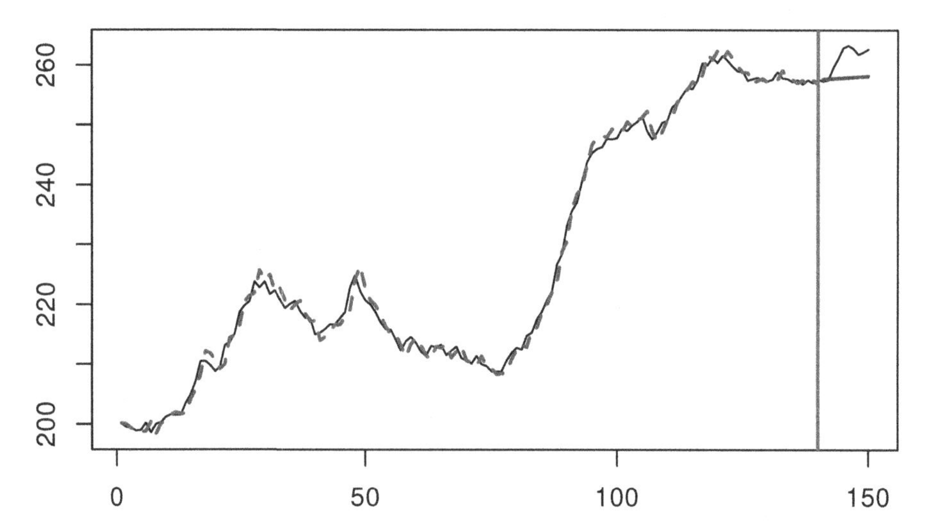

FIGURE 6.1 Model fit for Box-Jenkins sales data from ETS(M,M,N).

If we want to produce the forecasts (conditional expectation and prediction interval) from the model, we can do it, using the same command as in Section 5.6:

```
forecast(adamETSBJ, h=10,
         interval="prediction", level=0.95) |>
    plot()
```

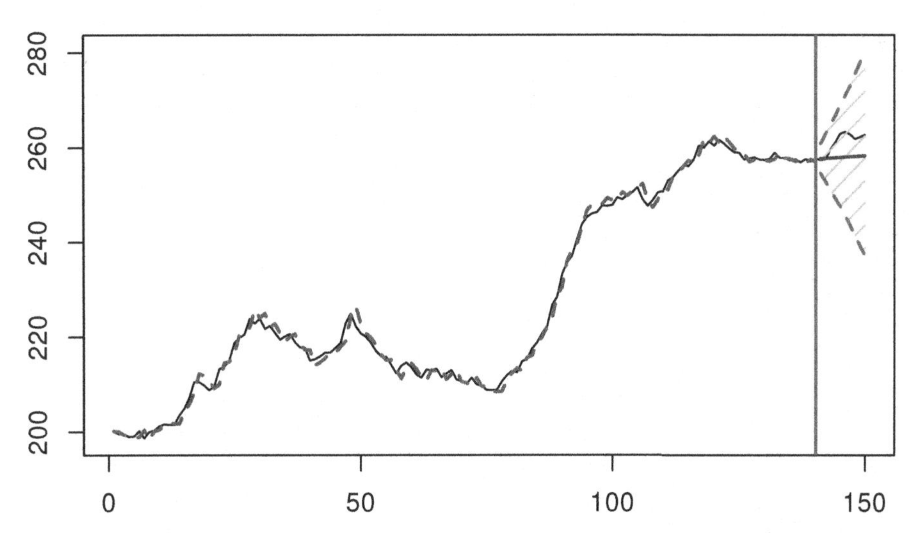

FIGURE 6.2 Forecast for Box-Jenkins sales data from ETS(M,M,N).

Note that, when we ask for "prediction" interval, the `forecast()` function will automatically decide what to use based on the estimated model: in the case of a pure additive one, it will use analytical solutions, while in the other cases, it will use simulations (see Section 18.3). The point forecast obtained from the forecast function corresponds to the conditional expectation and is calculated based on the simulations. This also means that it will differ slightly from one run of the function to another (reflecting the uncertainty in the error term). Still, the difference, in general, should be negligible for a large number of simulation paths.

The forecast with prediction interval is shown in Figure 6.2. The conditional expectation is not very different from the point forecast in this example. This is because the variance of the error term is close to zero, thus bringing the two close to each other:

```
sigma(adamETSBJ)^2
```

```
## [1] 3.928668e-05
```

We can also compare the performance of ETS(M,M,N) with Gamma distribution with the conventional ETS(M,M,N) assuming normality:

```
adamETSBJNormal <- adam(BJsales, "MMN", h=10, holdout=TRUE,
                        distribution="dnorm")
adamETSBJNormal
```

```
## Time elapsed: 0.07 seconds
## Model estimated using adam() function: ETS(MMN)
## Distribution assumed in the model: Normal
## Loss function type: likelihood; Loss function value: 245.3872
## Persistence vector g:
## alpha  beta
## 1.000 0.241
##
## Sample size: 140
## Number of estimated parameters: 5
## Number of degrees of freedom: 135
## Information criteria:
##      AIC     AICc      BIC     BICc
## 500.7745 501.2222 515.4827 516.5890
##
## Forecast errors:
## ME: 3.217; MAE: 3.33; RMSE: 3.785
## sCE: 14.126%; Asymmetry: 91.6%; sMAE: 1.462%; sMSE: 0.028%
## MASE: 2.817; RMSSE: 2.483; rMAE: 0.925; rRMSE: 0.921
```

In this specific example, the two distributions produce very similar results with almost indistinguishable estimates of parameters.

6.6.2 Seasonal data

The `AirPassengers` data used in Section 5.6 has (as we discussed) multiplicative seasonality. So, the ETS(M,M,M) model might be more suitable than the pure additive one that we used previously:

```
adamETSAir <- adam(AirPassengers, "MMM", h=12, holdout=TRUE)
```

After running the command above we might get a warning, saying that the model has a potentially explosive multiplicative trend. This happens, when the final in-sample value of the trend component is greater than one, in which case the forecast trajectory might exhibit exponential growth. Here is what we have in the output of this model:

```
adamETSAir
```

```
## Time elapsed: 0.35 seconds
## Model estimated using adam() function: ETS(MMM)
## Distribution assumed in the model: Gamma
## Loss function type: likelihood; Loss function value: 468.5176
## Persistence vector g:
##  alpha   beta  gamma
## 0.7684 0.0206 0.0000
##
## Sample size: 132
## Number of estimated parameters: 17
## Number of degrees of freedom: 115
## Information criteria:
##       AIC      AICc       BIC      BICc
##  971.0351  976.4036 1020.0428 1033.1492
##
## Forecast errors:
## ME: -5.617; MAE: 15.496; RMSE: 21.938
## sCE: -25.677%; Asymmetry: -23.1%; sMAE: 5.903%; sMSE: 0.698%
## MASE: 0.643; RMSSE: 0.7; rMAE: 0.204; rRMSE: 0.213
```

Notice that the smoothing parameter γ is equal to zero, which implies that we deal with the data with deterministic multiplicative seasonality. Comparing the information criteria (e.g. AICc) with the ETS(A,A,A) (discussed in Subsection 5.6.2), the pure multiplicative model does a better job at fitting the data than the additive one:

```
adamETSAirAdditive <- adam(AirPassengers, "AAA", lags=12,
                           h=12, holdout=TRUE)
AICc(adamETSAirAdditive)
```

```
## [1] 1130.756
```

The conditional expectation and prediction interval from this model are more adequate as well (Figure 6.3):

```
adamForecast <- forecast(adamETSAir, h=12, interval="prediction")
```

```
## Warning: Your model has a potentially explosive multiplicative
    trend. I cannot
## do anything about it, so please just be careful.
```

```
plot(adamForecast, main="")
```

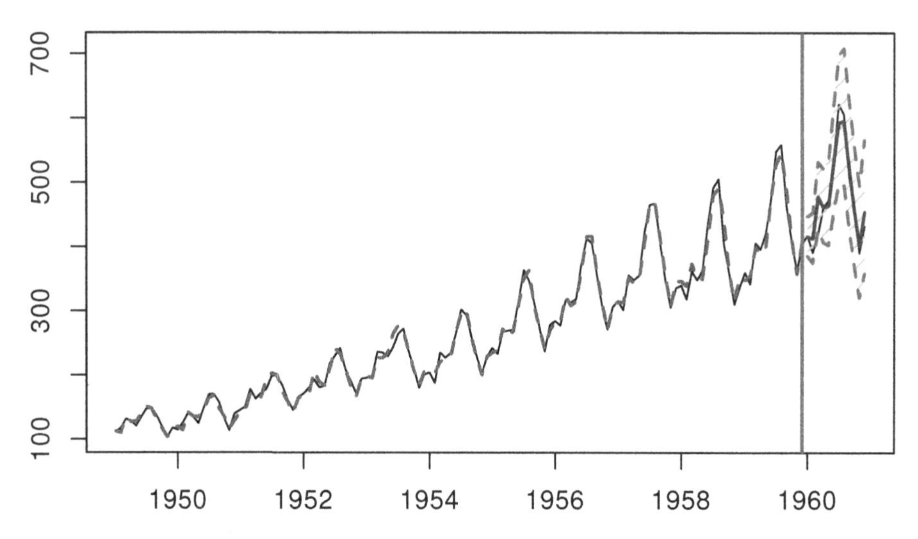

FIGURE 6.3 Forecast for air passengers data using an ETS(M,M,M) model.

If we want to calculate the error measures based on the conditional expectation, we can use the `measures()` function from the `greybox` package in the following way:

```
measures(adamETSAir$holdout,
         adamForecast$mean,
         actuals(adamETSAir))
```

```
##            ME           MAE           MSE           MPE
    MAPE
##  -5.854678539  15.372294543 476.905198437  -0.016766764
    0.033586821
##           sCE          sMAE          sMSE          MASE
    RMSSE
##  -0.267650172   0.058562812   0.006921473   0.638279108
    0.696989324
##          rMAE         rRMSE          rAME     asymmetry
    sPIS
##   0.202267033   0.212069277   0.082267146  -0.237969758
    2.183280288
```

These can be compared with the measures from the ETS(A,A,A) model:

```
measures(adamETSAir$holdout,
         adamETSAirAdditive$forecast,
         actuals(adamETSAir))
```

```
##             ME           MAE          MSE          MPE
    MAPE
##   28.36910729   37.11462699 2442.64739763   0.04881281
    0.07053896
##            sCE          sMAE         sMSE         MASE
    RMSSE
##    1.29691091    0.14139314    0.03545090   1.54105107
    1.57739510
##           rMAE         rRMSE         rAME    asymmetry
    sPIS
##    0.48835036    0.47994571    0.39862914   0.69383499
    -7.31044007
```

Comparing, for example, MSE from the two models, we can conclude that the pure multiplicative model is more accurate than the pure additive one.

We can also produce the plot of the time series decomposition according to ETS(M,M,M) (see Figure 6.4):

```
plot(adamETSAir, which=12)
```

The plot in Figure 6.4 shows that the residuals are more random for the pure multiplicative model than for the ETS(A,A,A), but there still might be some structure left. The autocorrelation and partial autocorrelation functions (discussed in Section 8.3) might help in understanding this better:

```
par(mfcol=c(2,1), mar=c(2,4,2,1))
plot(adamETSAir, which=10:11)
```

The plot in Figure 6.5 shows that there is still some correlation left in the residuals, which could be either due to pure randomness or imperfect estimation of the model. Tuning the parameters of the optimiser or selecting a different model might solve the problem.

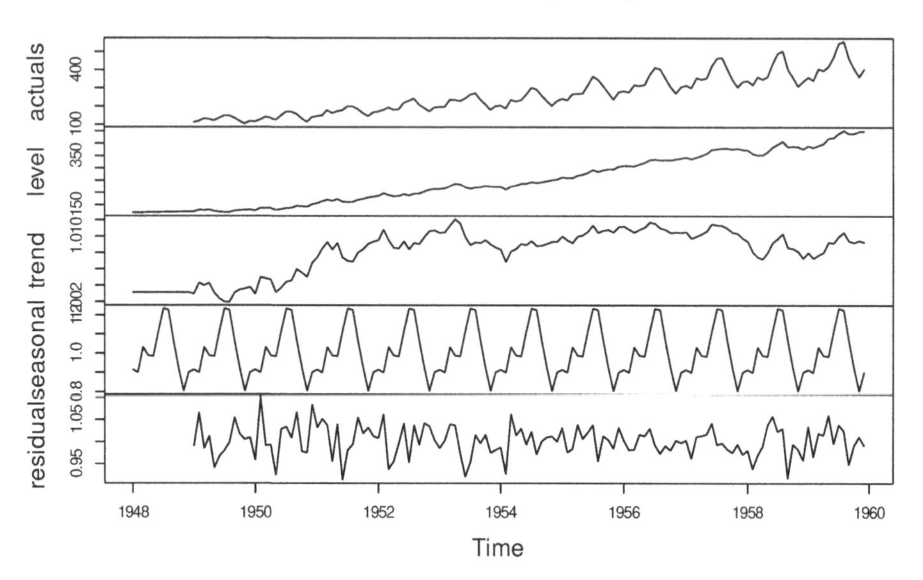

FIGURE 6.4 Decomposition of air passengers data using an ETS(M,M,M) model.

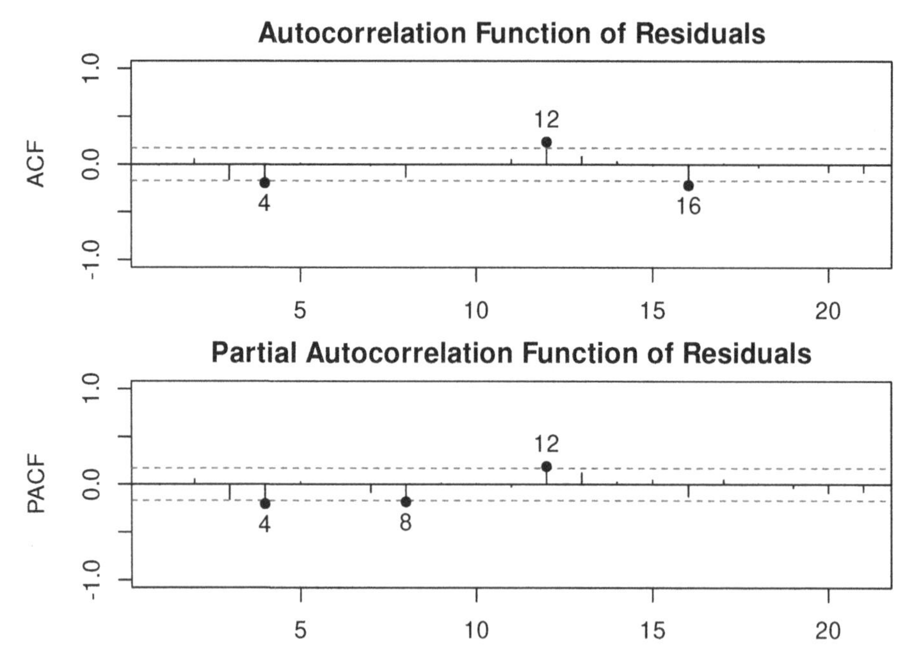

FIGURE 6.5 ACF and PACF of residuals of an ETS(M,M,M) model.

7

General ADAM ETS model

Now that we have discussed two special cases of ADAM ETS models (namely pure additive from Chapter 5 and pure multiplicative from Chapter 6 ADAM ETS), we can move to the discussion of the general model and special cases of it, including some of the mixed models. We will consider two important groups of mixed ADAM ETS models: with non-multiplicative and with multiplicative trends. They have very different properties that are worth mentioning. Finally, we will discuss the question of the normalisation of seasonal indices. This topic was studied in the literature back in the 80s when we still did not have a proper ETS model, but it has not been thoroughly studied since then. In Section 7.5, we take a more critical view towards the topic.

7.1 Model formulation

Based on the discussion in previous chapters, we can summarise the general ADAM ETS model. It is built upon the conventional model discussed in Section 4.6 but has several significant differences, the most important of which is that it is formulated using lags of components rather than the transition of them over time (this was discussed in Section 5.1 for the pure additive model). The general ADAM ETS model is formulated in the following way:

$$\begin{aligned} y_t &= w(\mathbf{v}_{t-\boldsymbol{l}}) + r(\mathbf{v}_{t-\boldsymbol{l}})\epsilon_t \\ \mathbf{v}_t &= f(\mathbf{v}_{t-\boldsymbol{l}}) + g(\mathbf{v}_{t-\boldsymbol{l}})\epsilon_t \end{aligned}, \qquad (7.1)$$

where $\mathbf{v}_{t-\boldsymbol{l}}$ is the vector of lagged components and $\boldsymbol{l}$ is the vector of lags, while all the other functions correspond to the ones used in Section 4.6, equation (4.22). This model form is mainly useful for the formulation, rather than for inference. Not only does it encompass any pure model, it also allows formulating any of the mixed ones. For example, the ETS(M,A,M) will have the following values:

$$\begin{aligned} w(\mathbf{v}_{t-\boldsymbol{l}}) &= (l_{t-1} + b_{t-1})s_{t-m}, \; r(\mathbf{v}_{t-\boldsymbol{l}}) = w(\mathbf{v}_{t-\boldsymbol{l}}), \\ f(\mathbf{v}_{t-\boldsymbol{l}}) &= \begin{pmatrix} l_{t-1} + b_{t-1} \\ b_{t-1} \\ s_{t-m} \end{pmatrix}, \; g(\mathbf{v}_{t-\boldsymbol{l}}) = \begin{pmatrix} \alpha(l_{t-1} + b_{t-1}) \\ \beta(l_{t-1} + b_{t-1}) \\ \gamma s_{t-m} \end{pmatrix}, \\ \mathbf{v}_t &= \begin{pmatrix} l_t \\ b_t \\ s_t \end{pmatrix}, \; \boldsymbol{l} = \begin{pmatrix} 1 \\ 1 \\ m \end{pmatrix}, \\ \mathbf{v}_{t-\boldsymbol{l}} &= \begin{pmatrix} l_{t-1} \\ b_{t-1} \\ s_{t-m} \end{pmatrix} \end{aligned} .$$

By inserting these values in (7.1) we will get the classical ETS(M,A,M) model, mentioned in Section 4.2:

$$\begin{aligned} y_t &= (l_{t-1} + b_{t-1})s_{t-m}(1 + \epsilon_t) \\ l_t &= (l_{t-1} + b_{t-1})(1 + \alpha\epsilon_t) \\ b_t &= b_{t-1} + (l_{t-1} + b_{t-1})\beta\epsilon_t \\ s_t &= s_{t-m}(1 + \gamma\epsilon_t) \end{aligned} . \tag{7.2}$$

The model (7.1) with different values for the functions is the basis of the `adam()` function from the `smooth` package. It is used in the C++ code to generate fitted values and/or simulate data from any ETS model.

7.2 Mixed ADAM ETS models

Hyndman et al. (2008) proposed five classes of ETS models, based on the types of their components:

1. ANN; AAN; AAdN; ANA; AAA; AAdA;
2. MNN; MAN; MAdN; MNA; MAA; MAdA;
3. MNM; MAM; MAdM;
4. MMN; MMdN; MMM; MMdM;
5. ANM; AAM; AAdM; MMA; MMdA; AMN; AMdN; AMA; AMdA; AMM; AMdM.

The idea behind this split is to distinguish the models by their complexity and the availability of analytical expressions for conditional moments. Class 1 models were discussed in Section 5.1. They have analytical expressions for conditional mean and variance; they can be applied to any data; they have simple formulae for prediction intervals.

Hyndman et al. (2008) demonstrate that models from Class 2 have closed forms for conditional expectation and variance, with the former corresponding to the point forecasts. However, the conditional distribution from these models is not Gaussian, so there are no formulae for the prediction intervals from these models. Yes, in some cases, Normal distribution might be used as a satisfactory approximation for the real one, but simulations should generally be preferred.

Class 3 models suffer from similar issues, but the situation worsens: there are no analytical solutions for the conditional mean and variance, and there are only approximations to these statistics.

Class 4 models were discussed in Section 6.1. They do not have analytical expressions for the moments, their conditional h steps ahead distributions represent a complex convolution of products of the basic ones, but they are appropriate for the positive data and become more valuable when the level of series is low, as already discussed in Section 6.1.

Finally, Class 5 models might have infinite variances, specifically on long horizons and when the data has low values. Indeed, when the level in one of these models becomes close to zero, there is an increased chance of breaking the model due to the appearance of negative values. Consider an example of the ETS(A,A,M) model, which might have a negative trend, leading to negative values, which are then multiplied by the positive seasonal indices. This would lead to unreasonable values of states in the model. That is why in practice, these models should only be used when the level of the series is high. Furthermore, some Class 5 models are very difficult to estimate and are very sensitive to the smoothing parameter values. This mainly applies to the multiplicative trend models.

The `ets()` function from the `forecast` package by default supports only Classes 1 – 4 for the reasons explained above, although it is possible to switch on the Class 5 models by setting the parameter `restrict` to `FALSE`.

To be fair, any mixed model can potentially break when the level of the series is close to zero. For example, ETS(M,A,N) can have a negative trend, which might lead to the negative level and, as a result, to the multiplication of a pure positive error term by the negative components. Estimating such a model on real data becomes a non-trivial task.

In addition, as discussed above, simulations are typically needed to get prediction intervals for models of Classes 2 – 5 and conditional mean and variance for models of Classes 4 – 5. All of this, in my opinion, means that the more useful classification of ETS models is the following (it was first proposed by Akram et al., 2009):

A) Pure additive models (Section 5.1): ANN; AAN; AAdN; ANA; AAA; AAdA;
B) Pure multiplicative models (Section 6.1): MNN; MMN; MMdN; MNM; MMM; MMdM;

C) Mixed models with non-multiplicative trend (Section 7.3): MAN; MAdN; MNA; MAA; MAdA; MAM; MAdM; ANM; AAM; AAdM;
D) Mixed models with multiplicative trend (Section 7.4: MMA; MMdA; AMN; AMdN; AMA; AMdA; AMM; AMdM.

The main idea behind the split to (C) and (D) is that the multiplicative trend makes it almost impossible to derive the formulae for the conditional moments of the distribution. So this class of models can be considered as the most challenging one.

The `adam()` function supports all 30 ETS models, but you should keep in mind the limitations of some of them discussed in this section. The `es()` function from `smooth` is a wrapper of `adam()` and as a result supports the same set of models as well.

7.3 Mixed models with non-multiplicative trend

There are two subclasses in this class of models:

1. With a non-multiplicative seasonal component (MAN, MAdN, MNA, MAA, MAdA);
2. With the multiplicative one (MAM; MAdM; ANM; AAM; AAdM).

The conditional mean for the former models coincides with the point forecasts, while the conditional variance can be calculated using the following recursive formula (Hyndman et al., 2008, page 84):

$$\begin{aligned} & \mathrm{V}(y_{t+h}|t) = (1+\sigma^2)\xi_h - \mu_{t+h|t}^2 \\ & \text{where } \xi_1 = \mu_{t+1|t}^2 \text{ and } \xi_h = \mu_{t+h|t}^2 + \sigma^2 \sum_{j=1}^{h-1} c_j^2 \xi_{h-j} \end{aligned} , \tag{7.3}$$

where σ^2 is the variance of the error term. Still, the predictive distribution from these models does not have a closed form, and as a result, in general the simulations need to be used to get the correct quantiles.

As for the second subgroup, the conditional mean corresponds to the point forecasts, when the forecast horizon is less than or equal to the seasonal frequency of the component (i.e. $h \leq m$), and there is a cumbersome formula for calculating the conditional mean to some of models in this subgroup for the $h > m$. When it comes to the conditional variance, there exists a formula for some of models in the second subgroup, but they are cumbersome as well. For all of these, the interested reader is directed to Hyndman et al. (2008), page 85.

When it comes to the parameters' bounds for the models in this group, the first subgroup of models has bounds similar to the ones for the respective additive error models (Section 5.4) because they both underlie the same Exponential Smoothing methods, but with additional restrictions, coming from the multiplicative error (Section 6.4).

1. The *traditional bounds* (aka "usual") should work fine for these models for the same reasons they work for the pure additive ones, although they might be too restrictive in some cases;
2. The *admissible bounds* for smoothing parameters for the models in this group might be too wide and violate the condition of $(1 + \alpha\epsilon_t) > 0$, which is important in order not to break the models.

The second subgroup is more challenging in terms of parameters' bounds because of the multiplication of states by the seasonal components. In general, to be on the safe side, the bounds should not be wider than [0, 1] for the smoothing parameters α and γ in these models.

Finally, some models in this group are difficult to motivate from the application point of view. For example, ETS(M,A,A) assumes that the trend and seasonal components change in units (e.g. sales increase for several units in January), while the error term reflects the percentage changes. Similarly, ETS(A,A,M) has a difficult explanation because it assumes unit changes due to the error term and percentage changes due to the seasonality.

7.4 Mixed models with multiplicative trend

This is the most challenging class of models (MMA; MMdA; AMN; AMdN; AMA; AMdA; AMM; AMdM). They do not have analytical formulae for conditional moments, and they are very sensitive to smoothing parameter values and may lead to explosive forecasting trajectories. So, to obtain the conditional expectation, variance, and prediction interval from the models of these classes, simulations should be used.

One of the issues encountered when using these models is the explosive trajectories because of the multiplicative trend. As a result, when these models are estimated, it makes sense to set the initial value of the trend to 1 or a lower value so that the optimiser does not encounter difficulties in the calculations because of the explosive behaviour in-sample.

Furthermore, the combinations of components for the models in this class make even less sense than the combinations for Class C (discussed in Section 7.3). For example, the multiplicative trend assumes either explosive growth or decay, as shown in Figure 7.1.

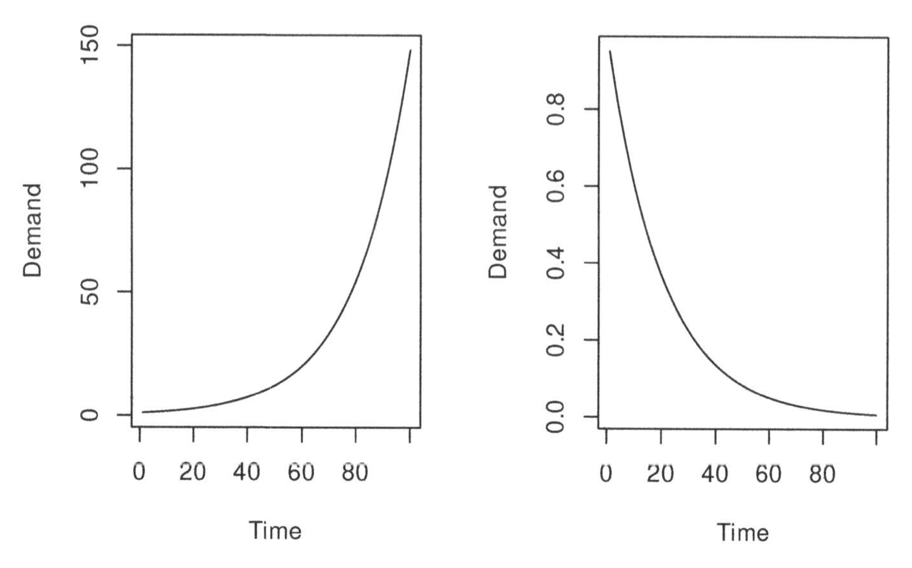

FIGURE 7.1 Plots of exponential increase and exponential decline.

However, assuming that either the seasonal component, the error term, or both will have precisely the same impact on the final value irrespective of the point of time (thus being additive) is unrealistic for this situation. The more reasonable one would be for the amplitude of seasonality to decrease together with the exponential decay of the trend and for the variance of the error term to do the same. But this means that we are talking about the pure multiplicative models (Section 6.1), not the mixed ones. There is only one situation where such mixed models could make sense: when the speed of change of the exponential trajectory is close to zero (i.e. the trend behaves similar to the linear one) and when the volume of the data is high. In this case, the mixed models might perform well and even produce more accurate forecasts than the models from the other classes.

When it comes to the bounds of the parameters, this is a mystery for the mixed models of this class. This is because the recursive relations are complicated, and calculating the discount matrix or anything like that becomes challenging. The usual bounds should still be okay, but keep in mind that these mixed models are typically not very stable and might exhibit explosive behaviour even, for example, with $\beta = 0.1$. So from my experience, the smoothing parameters in these models should be as low as possible, assuming that the initial values guarantee a working model (not breaking on some of the observations).

7.5 Normalisation of seasonal indices in ETS models

One of the ideas arising from time series decomposition (Section 3.2), inherited by the conventional ETS, is the renormalisation of seasonal indices. It comes to one of the two principles, depending on the type of seasonality:

1. If the model has additive seasonality, then the seasonal indices should add up to zero in a specific period of time, e.g. monthly indices should add up to zero over the yearly period;
2. If the model has multiplicative seasonality, then the geometric mean of seasonal indices over a specific period should be equal to one.

Condition (2) in the conventional ETS is substituted by "the arithmetic mean of multiplicative indices should be equal to one", which does not have reasonable grounds behind it. If we deal with the multiplicative effect, the geometric mean should be used, not the arithmetic one. Otherwise, we introduce bias in the model by multiplying components by indices.

While the normalisation is a natural element of the time series decomposition and works fine for the initial seasonal indices, renormalising the seasonal indices over time might not be natural for the ETS.

Hyndman et al. (2008) discuss different mechanisms for the renormalisation of seasonal indices, which, as the authors claim, are needed to make the principles (1) and (2) hold from period to period in the data. However, I argue that this is an unnatural idea for the ETS models. The indices should only be normalised during the initialisation of the model (at the moment $t = 0$), and that they should vary independently for the rest of the sample. The rationale for this comes from the model itself. To illustrate it, I will use ETS(A,N,A), but the idea can be easily applied to any other seasonal ETS model with any types of components and any number of seasonal frequencies. Just a reminder, ETS(A,N,A) is formulated as:

$$\begin{aligned} y_t &= l_{t-1} + s_{t-m} + \epsilon_t \\ l_t &= l_{t-1} + \alpha \epsilon_t \\ s_t &= s_{t-m} + \gamma \epsilon_t \end{aligned} . \tag{7.4}$$

Let's assume that this is the "true model" (as discussed in Section 1.4) for whatever data we have for whatever reason. In this case, the set of equations (7.4) tells us that the seasonal indices change over time, depending on the value of the smoothing parameter γ and each specific value of ϵ_t, which is assumed to be i.i.d. All seasonal indices s_{t+i} in a particular period (e.g. monthly indices in a year) can be written down explicitly based on (7.4):

$$
\begin{aligned}
s_{t+1} &= s_{t+1-m} + \gamma \epsilon_{t+1} \\
s_{t+2} &= s_{t+2-m} + \gamma \epsilon_{t+2} \\
&\vdots \\
s_{t+m} &= s_{t} + \gamma \epsilon_{t+m}
\end{aligned} . \tag{7.5}
$$

If this is how the data is "generated" and the seasonality evolves over time, then there is only one possibility, for the indices $s_{t+1}, s_{t+2}, \dots, s_{t+m}$ to add up to zero:

$$
s_{t+1} + s_{t+2} + \dots + s_{t+m} = 0 \tag{7.6}
$$

or after inserting (7.5) in (7.6):

$$
s_{t+1-m} + s_{t+2-m} + \dots + s_{t} + \gamma \left(\epsilon_{t+1} + \epsilon_{t+2} + \dots + \epsilon_{t+m}\right) = 0, \tag{7.7}
$$

meaning that:

1. The previous indices $s_{t+1-m}, s_{t+2-m}, \dots, s_{t}$ add up to zero **and**
2. Either:
   a. $\gamma = 0$,
   b. the sum of error terms $\epsilon_{t+1}, \epsilon_{t+2}, \dots, \epsilon_{t+m}$ is zero.

Note that we do not consider the situation $s_{t+1-m} + \dots + s_{t} = -\gamma \left(\epsilon_{t+1} + \dots + \epsilon_{t+m}\right)$ as it does not make sense. The condition (1) can be considered reasonable if the previous indices are normalised. (2.a) means that the seasonal indices do not evolve over time. However, (2.b) implies that the error term is not independent, because $\epsilon_{t+m} = -\epsilon_{t+1} - \epsilon_{t+2} - \dots - \epsilon_{t+m-1}$, which violates one of the basic assumptions of the model from Subsection 1.4.1, meaning that (7.5) cannot be considered as the "true" model anymore, as it omits some important elements. Thus renormalisation is unnatural for the ETS from the "true" model point of view.

Alternatively each seasonal index could be updated on each observation t (to make sure that the indices are renormalised). In this situation we have:

$$
\begin{aligned}
& s_t = s_{t-m} + \gamma \epsilon_t \\
& s_{t-m+1} + s_{t-m+2} + \dots + s_{t-1} + s_{t} = 0
\end{aligned},
$$

which can be rewritten as $s_{t-m} + \gamma \epsilon_t = -s_{t+1-m} - s_{t+2-m} - \dots - s_{t-1}$, meaning that:

$$
s_{t-m} + s_{t+1-m} + s_{t+2-m} + \dots + s_{t-1} = -\gamma \epsilon_t .
$$

But due to the renormalisation, the sum on the left-hand side should be equal to zero, implying that either $\gamma = 0$ or $\epsilon_t = 0$. While the former might hold in some cases (deterministic seasonality), the latter cannot hold for all $t = 1, \dots, T$ and again violates the model's assumptions. The renormalisation is thus impossible without changing the structure of the model. Hyndman et al. (2008) acknowledge that and propose in Chapter 8 some modifications for the seasonal ETS model (i.e. introducing new models), which we do not aim to discuss in this chapter.

The discussion in this section demonstrates that the renormalisation of seasonal indices is unnatural for the conventional ETS model and should not be used. This does not mean that the initial seasonal indices (corresponding to the observation $t = 0$) cannot be normalised. On the contrary, this is the desired approach as it reduces the number of estimated parameters in the model. But this means that there is no need to implement an additional mechanism of renormalisation of indices in-sample for $t > 0$.

8

Introduction to ARIMA

Another important dynamic element in ADAM is the ARIMA model (developed originally by Box & Jenkins, 1976). ARIMA stands for "AutoRegressive Integrated Moving Average", although the name does not tell much on its own and needs additional explanation, which will be provided in this chapter.

The main idea of the model is that the data might have dynamic relations over time, where the new values depend on the values on the previous observations. This becomes more obvious in the case of engineering systems and modelling physical processes. For example, Box and Jenkins (1976) use an example of a series of CO_2 output of a furnace when the input gas rate changes. In this case, the elements of the ARIMA process are natural, as the CO_2 cannot just drop to zero when the gas is switched off – it will leave the furnace in reducing quantity over time (i.e. leaving $\phi_1 \times 100\%$ of CO_2 in the next minute, where ϕ_1 is a parameter in the model).

Another example where AR processes are natural is modelling the temperature in the room, measured with five minute intervals. In this case, the temperature at 5:30 p.m. will depend on the one at 5:25 p.m.: if the temperature outside the room is lower, then the one in the room will go down slightly due to the loss of heat. Every five minutes it will go down on average by some quantity ϕ_1.

Both these examples describe the AR(1) process, and in both of them the ARIMA model can be considered a "true model" (see discussion in Section 1.4). Unfortunately, when it comes to time series in the social or business domains, it becomes very difficult to motivate ARIMA usage from the modelling point of view. For example, the demand for a product does not reproduce itself and in real life does not depend on the demand on previous observations. So, if we construct an ARIMA for such a process, we turn a blind eye to the fact that the observed time series relations in the data are most probably spurious. At best, in this case, ARIMA can be considered a very crude approximation of a complex process (demand is typically influenced by price changes, consumer behaviour, and promotional activities, etc.). Thus, whenever we work with ARIMA models in social or business domains, we should keep in mind that they are wrong even from the philosophical point of view. Nevertheless, they still can be useful (as was pointed out by one of the original authors, George Box), which is why we discuss them in this chapter. We focus our discussion on forecasting with ARIMA. A reader interested in time series analysis is directed to Box and Jenkins (1976) or to more modern editions of that textbook. ARIMA is also well explained in Chapter 6 of Ord et al. (2017).

This chapter will discuss the main theoretical properties of ARIMA processes (i.e. what would happen if the data indeed followed the specified model), moving to more practical aspects in the next chapter. We start the discussion with the non-seasonal ARIMA models, explaining what the forecasts from those models would look like, then move to the seasonal and multi-seasonal ARIMA, then discuss the classical Box-Jenkins approach for ARIMA order selection and its limitations. Finally, we explain the connection between ARIMA and ETS models.

8.1 Introduction to ARIMA

ARIMA contains several elements:

1. AR(p) – the AutoRegressive part, showing how the variable is impacted by its values on the previous observations. It contains p lags. For example, the quantity of the liquid in a vessel with an opened tap on some observation will depend on the quantity on the previous steps;
2. I(d) – the number of differences d taken in the model (I stands for "Integrated"). Working with differences rather than with the original data means that we deal with changes and rates of changes, rather than the original values. Technically, differences are needed to make data stationary (i.e. with fixed expectation and variance, although there are different definitions of the term *stationarity*, see below);
3. MA(q) – the Moving Average component, explaining how the previous white noise impacts the variable. It contains q lags. Once again, in technical systems, the idea that random error can affect the value has a relatively simple explanation. For example, when the liquid drips out of a vessel, we might not be able to predict the air fluctuations, which would impact the flow and could be perceived as elements of random noise. This randomness might, in turn, influence the quantity of liquid in a vessel on the following observation, thus introducing the MA elements in the model.

I intentionally do not provide ARIMA examples from the demand forecasting area, as these are much more difficult to motivate and explain than the examples from the more technical areas.

Before continuing our discussion, we should define the term **stationarity**. There are two definitions in the literature: one refers to "strict stationarity", while the other refers to the "weak stationarity":

- Time series is said to be **weak stationary** when its unconditional expectation and variance are constant, and the variance is finite on all observations;
- Time series is said to be **strong stationary** when its unconditional joint probability distribution does not change over time. This automatically implies that all its moments are constant (i.e. the process is also weak stationary).

The stationarity is essential in the ARIMA context and plays an important role in regression analysis. If the series is not stationary, it might be challenging to estimate its moments correctly using conventional methods. In some cases, it might be impossible to get the correct parameters (e.g. there is an infinite combination of parameters that would produce the minimum of the selected loss function). To avoid this issue, the series is transformed to ensure that the moments are finite and constant. Taking differences or detrending time series (e.g. via seasonal decomposition discussed in Section 3.2) allows making the first moment (mean) constant, while taking logarithms or doing the Box-Cox transform of the original series typically stabilises the variance, making it constant as well. After that, the model becomes easier to identify.

In contrast with ARIMA, the ETS models are almost always non-stationary and do not require the series to be stationary. We will see the connection between the two approaches in Section 8.4.

Finally, conventional ARIMA is always a pure additive model, and as a result its point forecasts coincide with the conditional expectation, and it has closed forms for the conditional moments and quantiles.

8.1.1 AR(p)

We start with a simple AR(1) model, which is written as:

$$y_t = \phi_1 y_{t-1} + \epsilon_t, \tag{8.1}$$

where ϕ_1 is the parameter of the model. This formula tells us that the value on the previous observation is carried over to the new one in the proportion of ϕ_1. Typically, the parameter ϕ_1 is restricted with the region (-1, 1) to make the model stationary, but very often in real life, ϕ_1 lies in the (0, 1) region. If the parameter is equal to one, the model becomes equivalent to Random Walk (Section 3.3.1).

The forecast trajectory (conditional expectation several steps ahead) of this model would typically correspond to the exponentially declining curve. This is because for the y_{t+h} value, AR(1) is:

$$y_{t+h} = \phi_1 y_{t+h-1} + \epsilon_{t+h} = \phi_1^2 y_{t+h-2} + \phi_1 \epsilon_{t+h-1} + \epsilon_{t+h} = \dots \phi_1^h y_t + \sum_{j=0}^{h-1} \phi_1^j \epsilon_{t+h-j}. \tag{8.2}$$

If we then take expectation conditional on the information available up until observation t, all error terms will disappear (because we assume that $\mathrm{E}(\epsilon_{t+j}) = 0$ for all j) leading to:

$$\mathrm{E}(y_{t+h}|t) = \phi_1^h y_t. \tag{8.3}$$

If $\phi_1 \in (0, 1)$, the forecast trajectory will decline exponentially.

Here is a simple example in R of a very basic forecast from AR(1) with $\phi_1 = 0.9$ (see Figure 8.1):

```
y <- vector("numeric", 20)
y[1] <- 100
phi <- 0.9
for(i in 2:length(y)){
    y[i] <- phi * y[i-1]
}
plot(y, type="l", xlab="Horizon", ylab="Forecast")
```

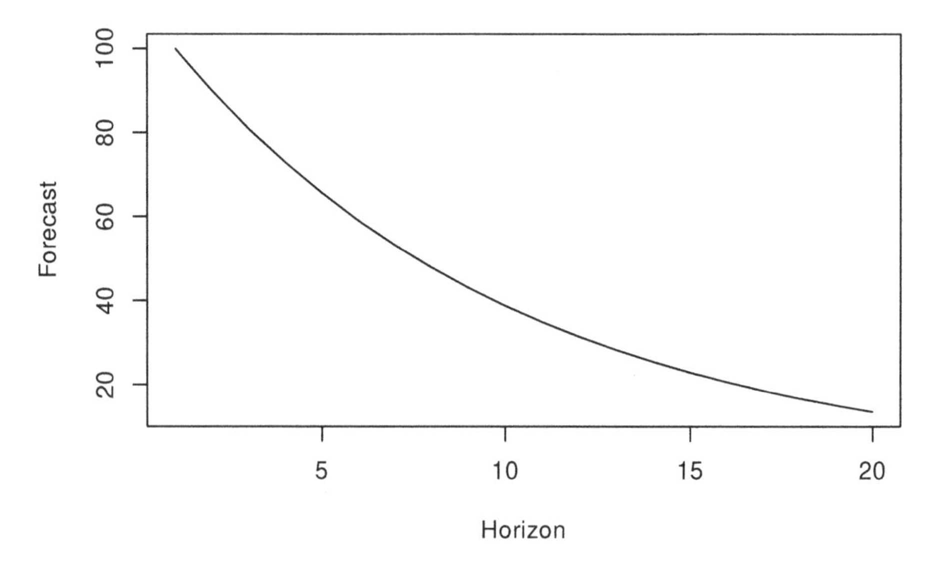

FIGURE 8.1 Forecast trajectory for AR(1) with $\phi_1 = 0.9$.

If, for some reason, we get $\phi_1 > 1$, then the trajectory will exhibit an exponential increase, becoming explosive, implying non-stationary behaviour. The model, in this case, becomes very difficult to work with, even if the parameter is close to one. So it is typically advised to restrict the parameter with the stationarity region (we will discuss this in more detail later in this chapter).

In general, it is possible to imagine the situation, when the value at the moment of time t would depend on several previous values, so the model AR(p) would be needed:

$$y_t = \phi_1 y_{t-1} + \phi_2 y_{t-2} + \dots + \phi_p y_{t-p} + \epsilon_t, \tag{8.4}$$

where ϕ_i is the parameter for the i-th lag of the model. So, the model assumes that the data on the recent observation is influenced by the p previous observations. The more lags we introduce in the model, the more complicated the forecasting trajectory becomes, potentially demonstrating harmonic behaviour. Here is an example of a point forecast from an AR(3) model $y_t = 0.9y_{t-1} - 0.7y_{t-2} + 0.6y_{t-3} + \epsilon_t$ (Figure 8.2):

```
y <- vector("numeric", 30)
y[1:3] <- c(100, 75, 30)
phi <- c(0.9,-0.7,0.6)
for(i in 4:30){
    y[i] <- phi[1] * y[i-1] + phi[2] * y[i-2] + phi[3] * y[i-3]
}
plot(y, type="l", xlab="Horizon", ylab="Forecast")
```

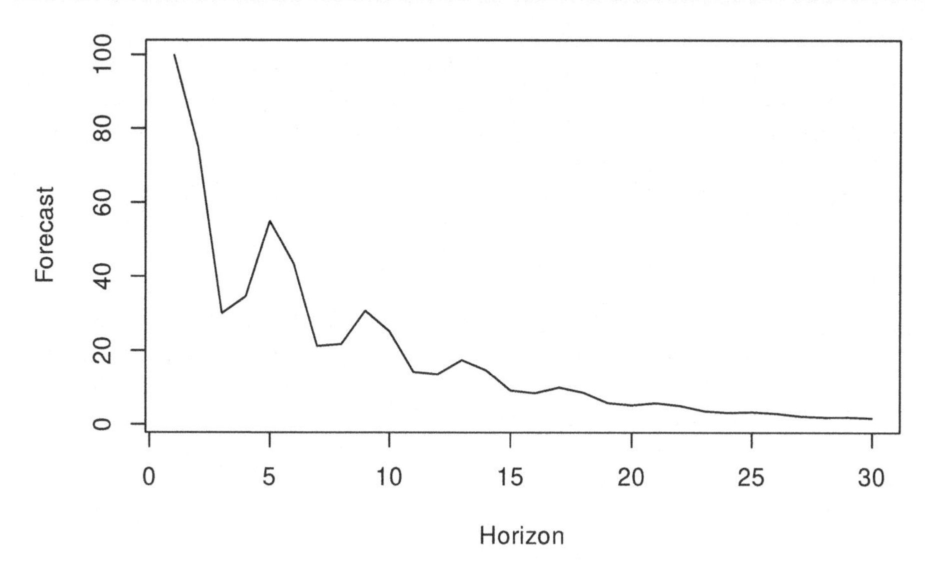

FIGURE 8.2 Forecast trajectory for an AR(3) model.

No matter what the forecast trajectory of the AR model is, it will asymptotically converge to zero as long as the model is stationary.

An example of an AR(3) time series generated using the `sim.ssarima()` function from the `smooth` package and a forecast for it via `msarima()` is shown in Figure 8.3.

As can be seen from Figure 8.3, the series does not exhibit any distinct characteristics so that it would be possible to identify the order of the model

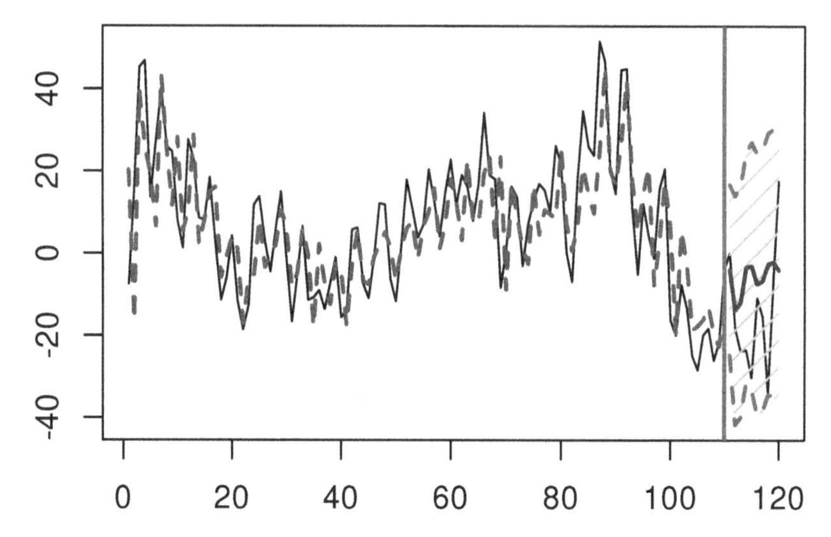

FIGURE 8.3 An example of an AR(3) series and a forecast for it.

just by looking at it. The only thing that we can probably say is that there is some structure in it: it has some periodic fluctuations and some parts of series are consistently above zero, while the others are consistently below. The latter is an indicator of autocorrelation in data. We do not discuss how to identify the order of AR in this section, but we will come back to it in Section 8.3.

8.1.2 MA(q)

Before discussing the "Moving Averages" model, we should acknowledge that the name is quite misleading and that the model has *nothing to do* with Centred Moving Averages used in time series decomposition (Section 3.2) or Simple Moving Averages (discussed in Section 3.3.3) used in forecasting. The idea of the simplest MA(1) model can be summarised in the following mathematical way:

$$y_t = \theta_1 \epsilon_{t-1} + \epsilon_t, \tag{8.5}$$

where θ_1 is the parameter of the model, typically lying between (-1, 1), showing what part of the error is carried out to the next observation. Because of the conventional assumption that the error term has a zero mean ($\mathrm{E}(\epsilon_t) = 0$), the forecast trajectory of this model is just a straight line coinciding with zero starting from the $h = 2$. But for the one step ahead point forecast we have:

$$\mathrm{E}(y_{t+1}|t) = \theta_1 \mathrm{E}(\epsilon_t|t) + \mathrm{E}(\epsilon_{t+1}|t) = \theta_1 \epsilon_t. \tag{8.6}$$

Starting from $h = 2$ there are no observable error terms ϵ_t, so all the values past that are equal to zero:

$$\mathrm{E}(y_{t+2}|t) = \theta_1 \mathrm{E}(\epsilon_{t+1}|t) + \mathrm{E}(\epsilon_{t+2}|t) = 0. \tag{8.7}$$

So, the forecast trajectory for the MA(1) model drops to zero, when $h > 1$.

More generally, the MA(q) model is written as:

$$y_t = \theta_1 \epsilon_{t-1} + \theta_2 \epsilon_{t-2} + \dots + \theta_q \epsilon_{t-q} + \epsilon_t, \tag{8.8}$$

where θ_i is the parameters for the i-th lag of the error term, typically restricted with the so-called invertibility region (discussed in the next section). In this case, the model implies that the recent observation is influenced by several errors on previous observations (your mistakes in the past will haunt you in the future). The more lags we introduce, the more complicated the model becomes. As for the forecast trajectory, it will reach zero when $h > q$.

An example of an MA(3) time series and a forecast for it is shown in Figure 8.4.

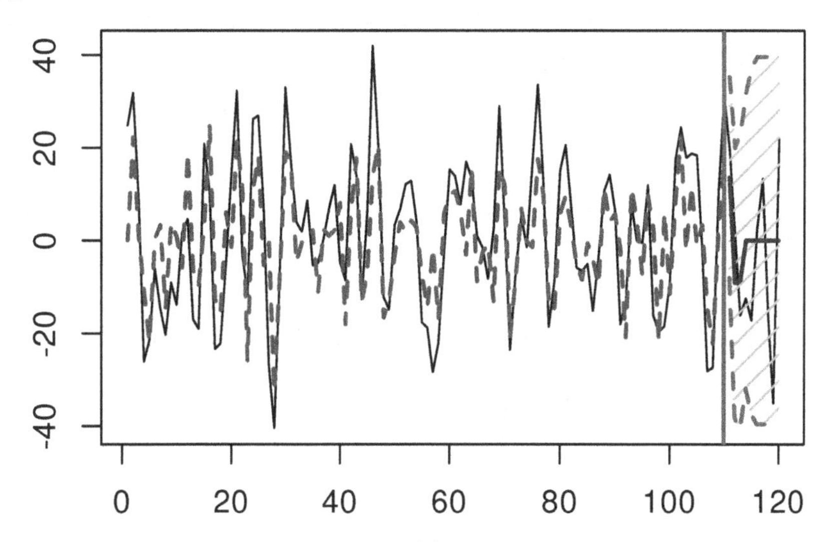

FIGURE 8.4 An example of an MA(3) series and forecast for it.

Similarly to how it was with AR(3), Figure 8.4 does not show anything specific that could tell us that this is an MA(3) process. The proper identification of MA orders will be discussed in Section 8.3.

8.1.3 ARMA(p,q)

Connecting the models (8.4) and (8.8), we get the more complicated model, ARMA(p,q):

$$y_t = \phi_1 y_{t-1} + \phi_2 y_{t-2} + \dots + \phi_p y_{t-p} + \theta_1 \epsilon_{t-1} + \theta_2 \epsilon_{t-2} + \dots + \theta_q \epsilon_{t-q} + \epsilon_t, \tag{8.9}$$

which has the properties of the two models discussed above. The forecast trajectory from this model will have a combination of trajectories for AR and MA for $h \leq q$ and then will correspond to AR(p) for $h > q$.

To simplify the work with ARMA models, the equation (8.9) is typically rewritten by moving all terms with y_t to the left-hand side:

$$y_t - \phi_1 y_{t-1} - \phi_2 y_{t-2} - \dots - \phi_p y_{t-p} = \epsilon_t + \theta_1 \epsilon_{t-1} + \theta_2 \epsilon_{t-2} + \dots + \theta_q \epsilon_{t-q}. \quad (8.10)$$

Furthermore, in order to make this even more compact, the backshift operator B can be introduced. It just shows by how much the subscript of the variable is shifted back in time:

$$y_t B^i = y_{t-i}. \quad (8.11)$$

Using (8.11), the ARMA model can be written as:

$$y_t(1 - \phi_1 B - \phi_2 B^2 - \dots - \phi_p B^p) = \epsilon_t(1 + \theta_1 B + \theta_2 B^2 + \dots + \theta_q B^q). \quad (8.12)$$

Finally, we can also introduce the AR and MA polynomial functions (corresponding to the elements in the brackets of (8.12)) to make the model even more compact:

$$\begin{aligned} \varphi^p(B) &= 1 - \phi_1 B - \phi_2 B^2 - \dots - \phi_p B^p \\ \vartheta^q(B) &= 1 + \theta_1 B + \theta_2 B^2 + \dots + \theta_q B^q. \end{aligned} \quad (8.13)$$

Inserting the functions (8.13) in (8.12) leads to the compact presentation of the ARMA model:

$$y_t \varphi^p(B) = \epsilon_t \vartheta^q(B). \quad (8.14)$$

The model (8.14) can be considered a compact form of (8.9). It is more difficult to understand and interpret but easier to work with from a mathematical point of view. In addition, this form permits introducing additional elements, which will be discussed later in this chapter.

Figure 8.5 shows an example of an ARMA(3,3) series with a forecast from it.

Similarly to the AR(3) and MA(3) examples above, the process is not visually distinguishable from other ARMA processes.

Coming back to the ARMA model (8.9), as discussed earlier, it assumes convergence of forecast trajectory to zero, the speed of which is regulated by its parameters. This implies that the data has the mean of zero, and ARMA should be applied to somehow pre-processed data so that it is stationary and varies around zero. This means that if you work with non-stationary and/or with non-zero mean data, the pure AR/MA or ARMA will be inappropriate – some prior transformations will be required.

8.1.4 ARMA with constant

One of the simpler ways to deal with the issue with zero forecasts is to introduce the constant (or intercept) in ARMA:

$$y_t = a_0 + \phi_1 y_{t-1} + \phi_2 y_{t-2} + \dots + \phi_p y_{t-p} + \theta_1 \epsilon_{t-1} + \theta_2 \epsilon_{t-2} + \dots + \theta_p \epsilon_{t-p} + \epsilon_t \quad (8.15)$$

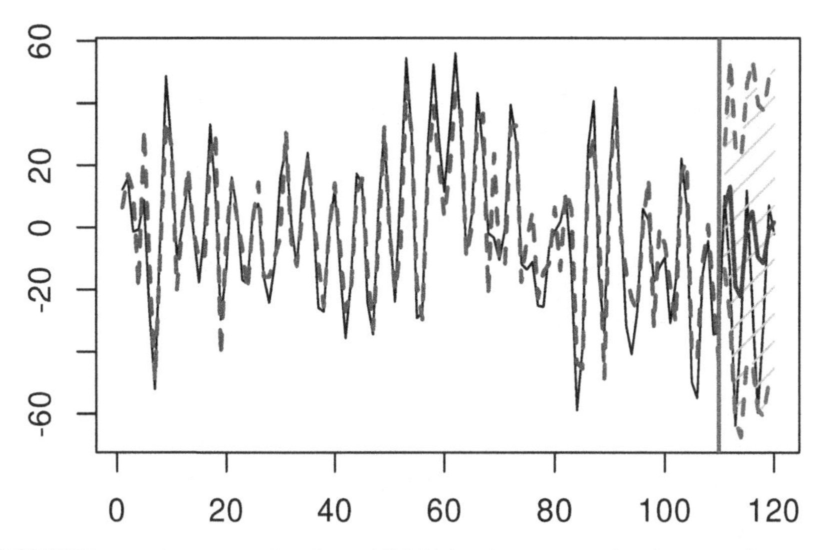

FIGURE 8.5 An example of an ARMA(3,3) series and a forecast for it.

or

$$y_t \varphi^p(B) = a_0 + \epsilon_t \vartheta^q(B), \tag{8.16}$$

where a_0 is the constant parameter, which in this case also works as the unconditional mean of the series. The forecast trajectory in this case would converge to a non-zero number, but with some minor differences from the trajectory of ARMA without constant. For example, in case of ARMA(1,1) with constant we will have:

$$y_t = a_0 + \phi_1 y_{t-1} + \theta_1 \epsilon_{t-1} + \epsilon_t. \tag{8.17}$$

The conditional expectation of y_{t+h} for $h = 1$ and $h = 2$ can be written as (based on the discussions in previous sections):

$$\begin{aligned} & \mathrm{E}(y_{t+1}|t) = a_0 + \phi_1 y_t + \theta_1 \epsilon_t \\ & \mathrm{E}(y_{t+2}|t) = a_0 + \phi_1 \mathrm{E}(y_{t+1}|t) = a_0 + \phi_1 a_0 + \phi_1^2 y_t + \phi_1 \theta_1 \epsilon_t \end{aligned}, \tag{8.18}$$

or in general for the horizon h:

$$\mathrm{E}(y_{t+h}|t) = \sum_{j=1}^h a_0 \phi_1^{j-1} + \phi_1^h y_t + \phi_1^{h-1} \theta_1 \epsilon_t. \tag{8.19}$$

So, the forecast trajectory from this model dampens out, similar to the ETS(A,Ad,N) model (Subsection 4.4.2), and the rate of dampening is regulated by the value of ϕ_1. The following simple example demonstrates this point (see Figure 8.6; I drop the MA(1) part because it does not change the shape of the curve, but only shifts it):

```
y <- vector("numeric", 20)
y[1] <- 100
phi <- 0.9
for(i in 2:length(y)){
    y[i] <- 100 + phi * y[i-1]
}
plot(y, type="l", xlab="horizon", ylab="Forecast")
```

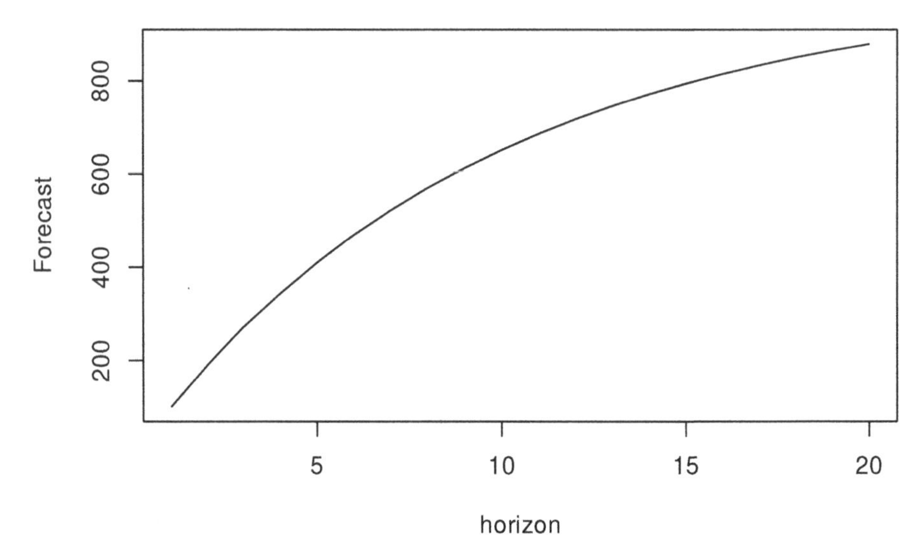

FIGURE 8.6 Forecast trajectory for an ARMA(1,1) model with constant.

The more complicated ARMA(p,q) models with p>1 will have more complex trajectories with potential harmonics, but the idea of dampening in the AR(p) part of the model stays.

An example of time series generated from ARMA(3,3) with constant is provided in Figure 8.7.

Finally, as alternative to adding a_0, each actual value of y_t can be centred via $y'_t = y_t - \bar{y}$, where $\bar{y}$ is the in-sample mean, making sure that the mean of y'_t is zero and ARMA can be applied to the y'_t data instead of y_t. However, this approach introduces additional steps, but the result on stationary data is typically the same as adding the constant.

8.1.5 I(d)

Based on the previous discussion, we can conclude that ARMA cannot be efficiently applied to non-stationary data. So, if we deal with one, we need to make it stationary. The conventional way of doing that is by taking differences

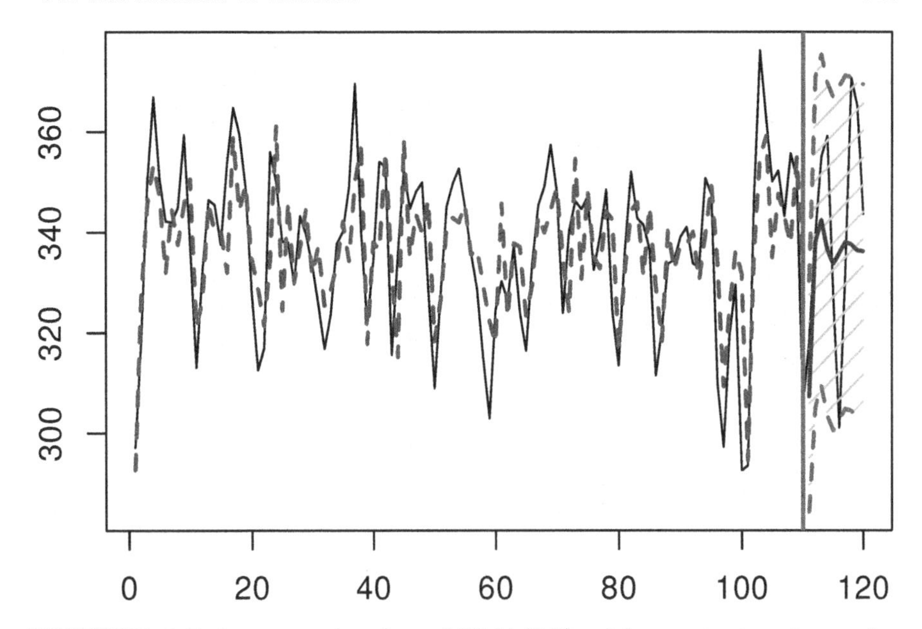

FIGURE 8.7 An example of an ARMA(3,3) with constant series and a forecast for it.

in the data. The logic behind this is straightforward: if the data is not stationary, the mean somehow changes over time. This can be, for example, due to a trend in the data. In this case, we should be talking about the change of variable y_t rather than the variable itself. So we should work on the following data instead:

$$\Delta y_t = y_t - y_{t-1} = y_t(1 - B), \tag{8.20}$$

if the first differences have a constant mean. The simplest model with differences is I(1), which is also known as the **Random Walk** (see Section 3.3.1):

$$\Delta y_t = \epsilon_t. \tag{8.21}$$

It can also be reformulated in a simpler, more interpretable form by inserting (8.20) in (8.21) and regrouping elements:

$$y_t = y_{t-1} + \epsilon_t. \tag{8.22}$$

The model (8.22) can also be perceived as AR(1) with $\phi_1 = 1$. This is a non-stationary model (meaning that the unconditional mean of y_t is not constant) and the point forecast from it corresponds to the one from the Naïve method (see Section 3.3.1) with a straight line equal to the last observed actual value (again, assuming that $\mathrm{E}(\epsilon_t) = 0$ and that the other basic assumptions from Section 1.4.1 hold):

$$\mathrm{E}(y_{t+h}|t) = \mathrm{E}(y_{t+h-1}|t) + \mathrm{E}(\epsilon_{t+h}|t) = y_t. \tag{8.23}$$

Visually, the Random Walk data and the forecast from it are shown in Figure 8.8.

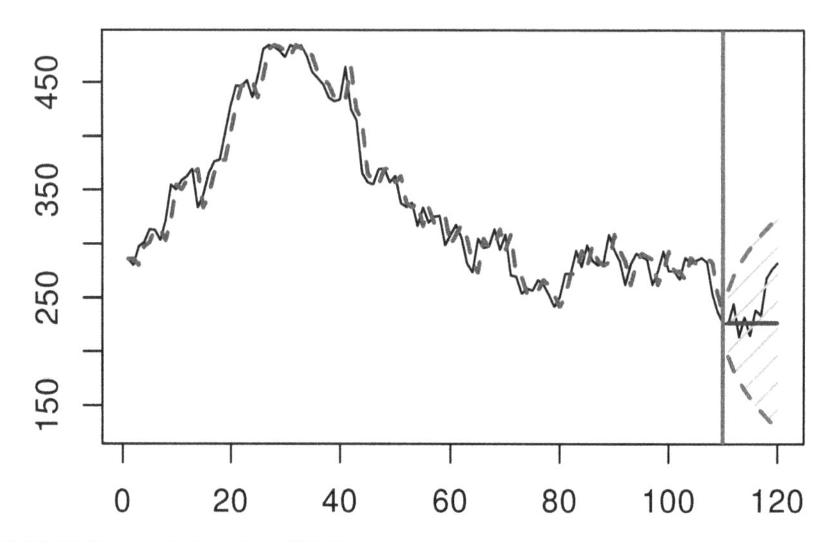

FIGURE 8.8 A Random Walk data example.

Another simple model that relies on differences of the data is called **Random Walk with drift** (which was discussed in Section 3.3.4) and is formulated by adding constant a_0 to the right-hand side of equation (8.21):

$$\Delta y_t = a_0 + \epsilon_t. \tag{8.24}$$

This model has some similarities with the global level model, which is formulated via the actual value rather than differences (see Section 3.3.2):

$$y_t = a_0 + \epsilon_t.$$

Using a similar regrouping as with the Random Walk, we can obtain a simpler form of (8.24):

$$y_t = a_0 + y_{t-1} + \epsilon_t, \tag{8.25}$$

which is, again, equivalent to the AR(1) model with $\phi_1 = 1$, but this time with a constant. The term "drift" appears because a_0 acts as an additional element, showing the tendency in the data: if it is positive, the model will exhibit a positive trend; if it is negative, the trend will be negative. This can be seen for the conditional mean, for example, for the case of $h = 2$:

$$\mathrm{E}(y_{t+2}|t) = \mathrm{E}(a_0) + \mathrm{E}(y_{t+1}|t) + \mathrm{E}(\epsilon_{t+2}|t) = a_0 + \mathrm{E}(a_0 + y_t + \epsilon_t|t) = 2a_0 + y_t, \tag{8.26}$$

or in general for the horizon h:

$$\mathrm{E}(y_{t+h}|t) = ha_0 + y_t. \tag{8.27}$$

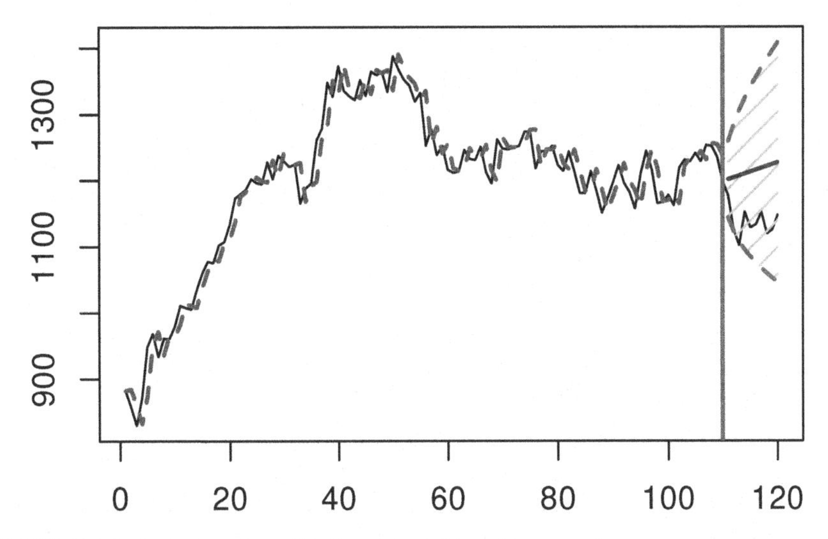

FIGURE 8.9 A Random Walk with drift with $a_0 = 2$.

Visually, the data and the forecast from it are shown in Figure 8.9.

In a manner similar to (8.20), we can also introduce second differences of the data (differences of differences) if we suspect that the change of variable over time is not stationary, which would be written as:

$$\Delta^2 y_t = \Delta y_t - \Delta y_{t-1} = y_t - y_{t-1} - y_{t-1} + y_{t-2}, \tag{8.28}$$

or using the backshift operator:

$$\Delta^2 y_t = y_t(1 - 2B + B^2) = y_t(1 - B)^2. \tag{8.29}$$

In fact, we can introduce higher level differences if we want to (but typically we should not) based on the idea of (8.29):

$$\Delta^d = (1 - B)^d. \tag{8.30}$$

Using (8.30), the I(d) model is formulated as:

$$\Delta^d y_t = \epsilon_t. \tag{8.31}$$

8.1.6 ARIMA(p,d,q)

Finally, having made the data stationary via the differences, we can introduce ARMA elements (8.14) to it which would be done on the differenced data, instead of the original y_t:

$$y_t \Delta^d(B) \varphi^p(B) = \epsilon_t \vartheta^q(B), \tag{8.32}$$

or in a more general form (8.12) with (8.30):

$$y_t(1-B)^d(1-\phi_1 B - \dots - \phi_p B^p) = \epsilon_t(1+\theta_1 B + \dots + \theta_q B^q), \tag{8.33}$$

which is an ARIMA(p,d,q) model. This model allows producing trends with some values of differences and also inherits the trajectories from both AR(p) and MA(q). This implies that the point forecasts from the model can exhibit complicated trajectories, depending on the values of p, d, q, and the model's parameters.

The model (8.33) is difficult to interpret in a general form, but opening the brackets and moving all elements but y_t to the right-hand side helps understanding each specific model.

Figure 8.10 demonstrates how data would look if it was generated from ARIMA(1,1,2) and how the forecasts would look if the same model ARIMA(1,1,2) was applied to the data. Because of the AR(1) term in the model, its forecast trajectory is dampening, similar to how it was done in the ETS(A,Ad,N) model.

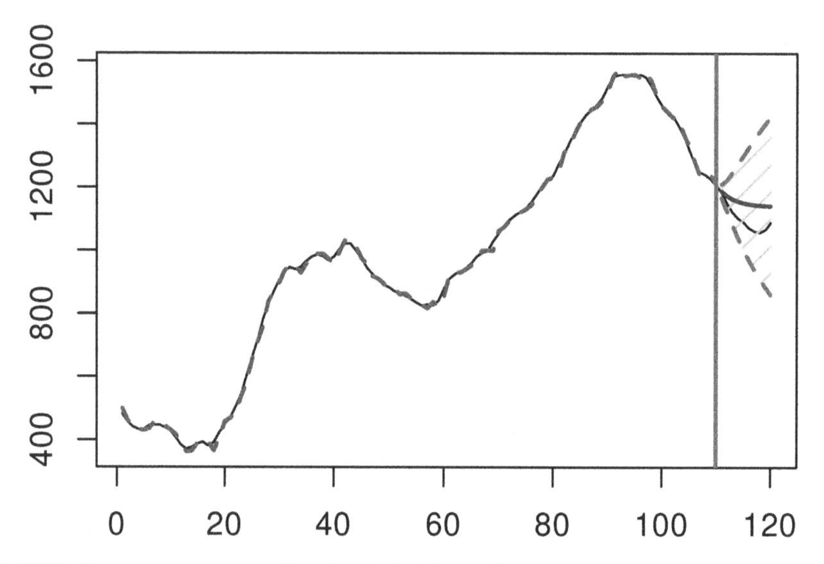

FIGURE 8.10 An example of ARIMA(1,1,2) data and a forecast for it.

8.1.7 parameters' bounds

ARMA models have two conditions that need to be satisfied for them to be useful and to work appropriately:

1. Stationarity;
2. Invertibility.

Condition (1) has already been discussed in Section 8.1. It is imposed on the model's AR parameters, ensuring that the forecast trajectories do not exhibit explosive behaviour (in terms of both mean and variance). (2) is equivalent to the stability condition in ETS (Section 5.4) and refers to the MA parameters: it guarantees that the old observations do not have an increasing impact on the recent ones. The term "invertibility" comes from the idea that any MA process can be represented as an infinite AR process via the inversion of the parameters as long as the parameters lie in some specific bounds. For example, an MA(1) model, which is written as:

$$y_t = \epsilon_t + \theta_1 \epsilon_{t-1} = \epsilon_t (1 + \theta_1 B), \tag{8.34}$$

can be rewritten as:

$$y_t (1 + \theta_1 B)^{-1} = \epsilon_t, \tag{8.35}$$

or in a slightly easier to digest form (based on (8.34) and the idea that $\epsilon_t = y_t - \theta_1 \epsilon_{t-1}$, implying that $\epsilon_{t-1} = y_{t-1} - \theta_1 \epsilon_{t-2}$):

$$y_t = \theta_1 y_{t-1} - \theta_1^2 \epsilon_{t-2} + \epsilon_t = \theta_1 y_{t-1} - \theta_1^2 y_{t-2} + \theta_1^3 \epsilon_{t-2} + \epsilon_t = \sum_{j=1}^{\infty} -1^{j-1} \theta_1^j y_{t-j} + \epsilon_t. \tag{8.36}$$

The recursion in (8.36) shows that the recent actual value y_t depends on the previous infinite number of values of y_{t-j} for $j = \{1, \dots, \infty\}$. The parameter θ_1, in this case, is exponentiated to the power j and leads to the exponential distribution of weights in this infinite series (looks similar to SES from Section 3.4, doesn't it?). The *invertibility* condition makes sure that those weights decline over time with the increase of j so that the older observations do not have an increasing impact on the most recent y_t.

There are different ways to check both conditions, the conventional one is by calculating the roots of the polynomial equations:

$$\begin{aligned} \varphi^p(B) = 0 \text{ for AR} \\ \vartheta^q(B) = 0 \text{ for MA'} \end{aligned} \tag{8.37}$$

or expanding the functions in (8.37) and substituting B with a variable x (for convenience):

$$\begin{aligned} 1 - \phi_1 x - \phi_2 x^2 - \dots - \phi_p x^p = 0 \text{ for AR} \\ 1 + \theta_1 x + \theta_2 x^2 + \dots + \theta_q x^q = 0 \text{ for MA} \end{aligned}. \tag{8.38}$$

Solving the first equation for x in (8.38), we get p roots (some of them might be complex numbers). For the model to be stationary, all the roots must be greater than one by absolute value. Similarly, if all the roots of the second equation in (8.38) are greater than one by absolute value, then the model is invertible (aka stable).

Calculating roots of polynomials is a difficult task, so there are simpler special cases for both conditions that guarantee that the more complicated ones are satisfied:

$$\begin{aligned} 0 < \sum_{j=1}^{p} \phi_j < 1 \\ 0 < \sum_{j=1}^{q} \theta_j < 1 \end{aligned} . \tag{8.39}$$

But note that the condition (8.39) is rather restrictive and is not generally applicable for all ARIMA models. It can be used to skip the check of the more complicated condition (8.38) if it is satisfied by a set of estimated parameters.

Finally, in a special case with an AR(p) model with $0 < \phi_j < 1$ for all j and $\sum_{j=1}^{p} \phi_j = 1$, we end up with the moving weighted average, which is a non-stationary model. This becomes apparent from the connection between Simple Moving Average and AR processes (Svetunkov & Petropoulos, 2018).

8.2 Seasonal ARIMA

8.2.1 Single seasonal ARIMA

When it comes to the actual data, we typically have relations between consecutive observations and between observations happening with some fixed seasonal lags. In the ETS framework, the latter relations are taken care of by seasonal indices, repeating every m observations. In the ARIMA framework, this is done by introducing lags in the model elements. For example, seasonal AR(P)$_m$ with lag m can be written similar to AR(p), but with some minor modifications:

$$y_t = \phi_{m,1} y_{t-m} + \dots + \phi_{m,P} y_{t-p.m.} + \varepsilon_t, \tag{8.40}$$

where $\phi_{m,i}$ is the parameter for the lagged actual value in the model, and ε_t is the error term of the seasonal AR model. We use the underscore "m" to show that the parameters here refer to the seasonal part of the model. The idea of the model (8.40) on the example of monthly data is that the current observation is influenced by a similar value, the same month a year ago, then the same month two years ago, etc. This is hard to justify from the theoretical point of view (demand two years ago impacts demand this year?), but this model allows capturing complex relations in the data.

Similarly to seasonal AR(P), we can have seasonal MA(Q)$_m$:

$$y_t = \theta_{m,1} \varepsilon_{t-m} + \dots + \theta_{m,Q} \varepsilon_{t-Qm} + \varepsilon_t, \tag{8.41}$$

where $\theta_{m,i}$ is the parameter for the lagged error term in the model. This model is even more difficult to justify than the MA(q) because it is difficult to explain how the white noise the same month last year can impact the actual value this year. Still, this is a useful instrument for forecasting purposes.

Finally, we have the seasonal differences, $I(D)_m$, which are easier to present using the backshift operator:

$$y_t(1 - B^m)^D = \varepsilon_t. \tag{8.42}$$

The seasonal differences allow dealing with the seasonality that changes its amplitude from year to year, i.e. model the multiplicative seasonality via ARIMA by making the seasonality itself stationary.

A special case of an I(D) model is $I(1)_m$, which is a seasonal **Random Walk**, underlying the Seasonal Naïve method from Section 3.3.6:

$$y_t(1 - B^m) = \varepsilon_t, \tag{8.43}$$

or equivalently:

$$y_t = y_{t-m} + \varepsilon_t. \tag{8.44}$$

Combining (8.40), (8.41), and (8.42) we get a pure seasonal ARIMA$(P,D,Q)_m$ model, similar to the ARIMA(p,d,q):

$$y_t(1-B^m)^D(1-\phi_{m,1}B^m-\dots-\phi_{m,P}B^{p.m.}) = \varepsilon_t(1+\theta_{m,1}B^m+\dots+\theta_{m,Q}B^{Qm}), \tag{8.45}$$

or if we introduce the polynomial functions for seasonal AR and MA and use notations similar to (8.29):

$$y_t\Delta^D(B^m)\varphi^P(B^m) = \varepsilon_t\vartheta^Q(B^m), \tag{8.46}$$

where

$$\begin{aligned} \Delta^D(B^m) &= (1 - B^m)^D \\ \varphi^P(B^m) &= 1 - \phi_{m,1}B^m - \dots - \phi_{m,P}B^{p.m.} \\ \vartheta^Q(B^m) &= 1 + \theta_{m,1}B^m + \dots + \theta_{m,Q}B^{Qm}. \end{aligned} \tag{8.47}$$

Now that we have taken care of the seasonal part of the model, we should not forget that there is a non-seasonal one. If it exists in the data, then ε_t would not be just a white noise, but could be modelled using a non-seasonal ARIMA(p,d,q):

$$\varepsilon_t\Delta^d(B)\varphi^p(B) = \epsilon_t\vartheta^q(B), \tag{8.48}$$

implying that:

$$\varepsilon_t = \epsilon_t \frac{\vartheta^q(B)}{\Delta^d(B)\varphi^p(B)}. \tag{8.49}$$

Inserting (8.49) into (8.46), we get the final SARIMA$(p,d,q)(P,D,Q)_m$ model in the compact form after regrouping the polynomials:

$$y_t\Delta^D(B^m)\varphi^P(B^m)\Delta^d(B)\varphi^p(B) = \epsilon_t\vartheta^Q(B^m)\vartheta^q(B). \tag{8.50}$$

The equation (8.50) does not tell us much about what specifically happens in the model. It just shows how different elements interact with each other in it. To understand, what SARIMA means, we need to consider a specific order of model and see what impacts the current actual value. For example, here is what we will have in the case of SARIMA(1,0,1)(1,0,1)$_4$ (i.e. applied to stationary quarterly data):

$$y_t \Delta^0(B^4)\varphi^1(B^4)\Delta^0(B)\varphi^1(B) = \epsilon_t \vartheta^1(B^4)\vartheta^1(B). \tag{8.51}$$

Inserting the values of polynomials (8.47), (8.30), and (8.13) in (8.51), we get:

$$y_t(1 - \phi_{4,1}B^4)(1 - \phi_1 B) = \epsilon_t(1 + \theta_{4,1}B^4)(1 + \theta_1 B), \tag{8.52}$$

which is slightly easier to understand but still does not explain how past values impact the present one. So, we open the brackets and move all the elements except for y_t to the right-hand side of the equation to get:

$$y_t = \phi_1 y_{t-1} + \phi_{4,1} y_{t-4} - \phi_1 \phi_{4,1} y_{t-5} + \theta_1 \epsilon_{t-1} + \theta_{4,1}\epsilon_{t-4} + \theta_1\theta_{4,1}\epsilon_{t-5} + \epsilon_t. \tag{8.53}$$

So, now we see that SARIMA(1,0,1)(1,0,1)$_4$ implies that the present value is impacted by the value in the previous quarter ($\phi_1 y_{t-1} + \theta_1 \epsilon_{t-1}$), the value last year ($\phi_{4,1} y_{t-4} + \theta_{4,1}\epsilon_{t-4}$) on the same quarter, and the value from last year on the previous quarter ($-\phi_1\phi_{4,1}y_{t-5} + \theta_1\theta_{4,1}\epsilon_{t-5}$), which introduces a much more complicated interaction than any ETS model can. However, this complexity is obtained with a minimum number of parameters: we have three lagged actual values and three lagged error terms, but we only have four parameters to estimate, not six. Thus the SARIMA model mentioned above can be considered as parsimonious for modelling this specific situation.

The more complicated SARIMA models would have even more complicated interactions, making it more challenging to interpret, but all of that comes with a benefit of having a parsimonious model with just $p + q + P + Q$ parameters to estimate.

When it comes to forecasting from such models as SARIMA(1,0,1)(1,0,1)$_4$, the trajectories would have elements of the classical ARMA model, discussed in Section 8.1.3, converging to zero as long as there is no constant and the model is stationary. The main difference would be in having the seasonal element. Here is an R example of a prediction for such a model for $h > m + 1$ (see Figure 8.11; the MA part is dropped because the expectation of the error term is assumed to be equal to zero):

```
y <- vector("numeric", 100)
y[1:5] <- c(97,87,85,94,95)
phi <- c(0.6,0.8)
for(i in 6:length(y)){
    y[i] <- phi[1] * y[i-1] + phi[2] * y[i-4] -
```

```
        phi[1] * phi[2] * y[i-5]
}
plot(y, type="l", xlab="horizon", ylab="Forecast")
```

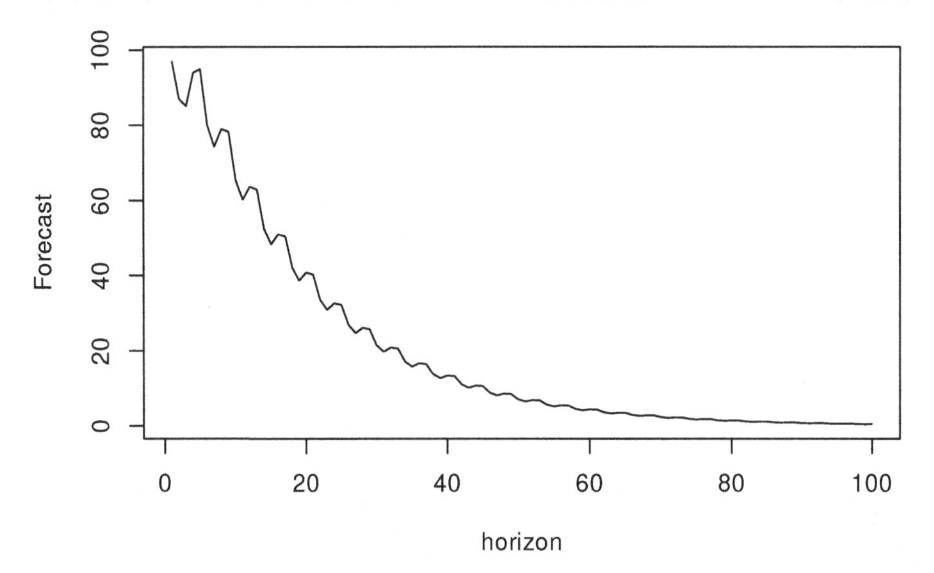

FIGURE 8.11 Forecast trajectory for a SARIMA(1,0,1)(1,0,1)$_4$.

As we see from Figure 8.11, the values converge to zero due to $0 < \phi_1 < 1$ and the seasonality disappears because $0 < \phi_{4,1} < 1$ as well. So, this is the forecast implied by the SARIMA without differences. If the differences are introduced, then the model would produce non-stationary and seasonally non-stationary trajectories.

8.2.2 SARIMA with constant

In addition, it is possible to add the constant term to the SARIMA model, and it will have a more complex effect on the forecast trajectory, depending on the order of the model. In the case of zero differences, the effect will be similar to ARMA with constant (Section 8.1.4), introducing the dampening trajectory. Here is an example (see Figure 8.12):

```
y <- vector("numeric", 100)
y[1:5] <- c(97,87,85,94,95)
phi <- c(0.6,0.8)
for(i in 6:length(y)){
    y[i] <- phi[1] * y[i-1] + phi[2] * y[i-4] -
      phi[1] * phi[2] * y[i-5] + 8
```

```
}
plot(y, type="l", xlab="horizon", ylab="Forecast")
```

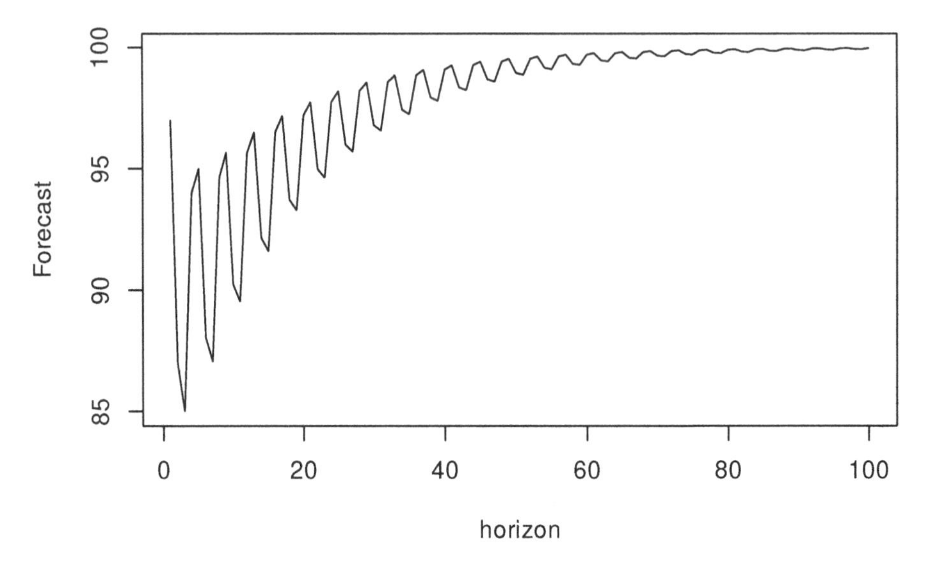

FIGURE 8.12 Forecast trajectory for a SARIMA model with constant.

In case of the model with the differences, the constant would have a two-fold effect: working as a drift for the non-seasonal part and increasing the amplitude of seasonality for the seasonal one. Here is an example from SARIMA(1,0,0)(1,1,0)$_4$ with constant:

$$y_t(1 - \phi_{4,1}B^4)(1 - \phi_1 B)(1 - B^4) = \epsilon_t + a_0, \tag{8.54}$$

which can be reformulated as (after opening brackets and moving elements to the right-hand side):

$$y_t = \phi_1 y_{t-1} + (1+\phi_{4,1})y_{t-4} - (1+\phi_{4,1})\phi_1 y_{t-5} - \phi_{4,1}y_{t-8} + \phi_1\phi_{4,1}y_{t-9} + a_0 + \epsilon_t. \tag{8.55}$$

This formula can then be used to see, how the trajectory from such a model will look:

```
y <- vector("numeric", 100)
y[1:9] <- c(96,87,85,94,97,88,86,95,98)
phi <- c(0.6,0.8)
for(i in 10:length(y)){
    y[i] <- phi[1] * y[i-1] + (1+phi[2]) * y[i-4] -
      (1+ phi[2]) *phi[1] * y[i-5] - phi[2] * y[i-8] +
      phi[1] * phi[2] * y[i-9] + 0.1
```

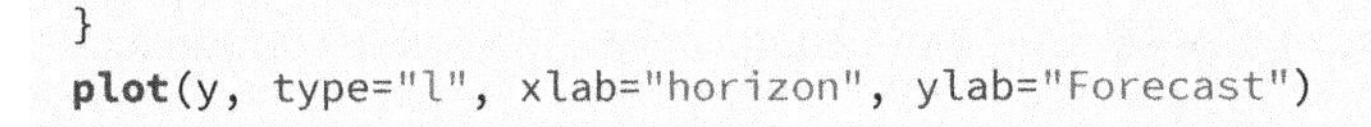

```
}
plot(y, type="l", xlab="horizon", ylab="Forecast")
```

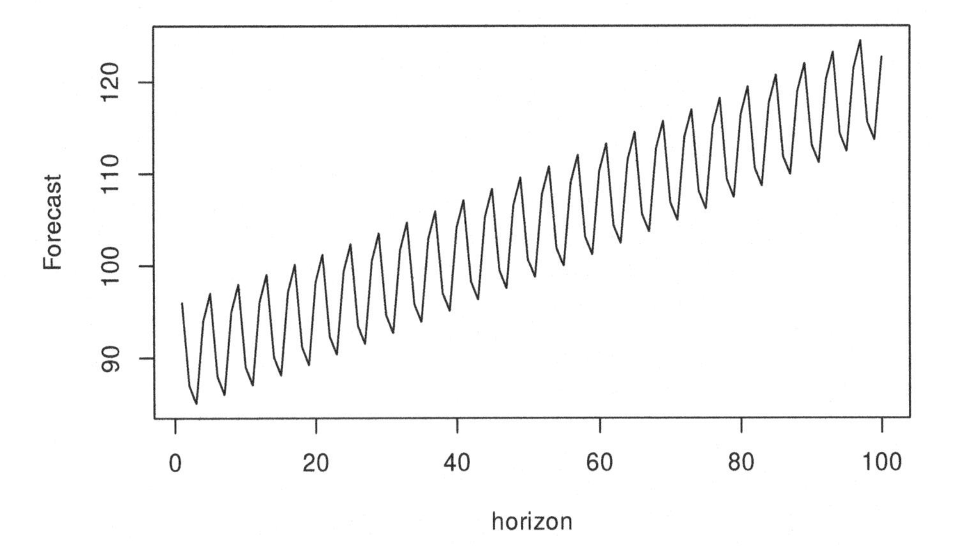

FIGURE 8.13 Forecast trajectory for a SARIMA model with differences and a constant.

As we see from Figure 8.13, the trajectory exhibits a drift, coming from the non-seasonal part of the model and a stable seasonality (the amplitude of which does not converge to zero anymore). More complex behaviours for the future trajectories can be obtained with higher orders of seasonal and non-seasonal parts of the SARIMA model.

8.2.3 Multiple Seasonal ARIMA

Using the same approach as the conventional SARIMA, we can introduce more terms (similar to how Taylor, 2003b, did it) with several seasonal frequencies. For example, we can have an hour of the day, a day of the week, and a week of the year frequencies in the data. Given that we work with the hourly data in this situation, we should introduce three seasonal ARIMA elements with seasonalities $m_1 = 24$, $m_2 = 24 \times 7 = 168$, and $m_3 = 24 \times 7 \times 365 = 61,320$. In this example, we would have AR, I, and MA polynomials for each seasonal part of the model, introducing a triple seasonal ARIMA, which is not even easy to formulate in the compact form. This type of model with multiple seasonal components can be called "Multiple Seasonal ARIMA", MSARIMA, which in general can be written as:

$$y_t \Delta^{D_n}(B^{m_n}) \varphi^{P_n}(B^{m_n}) \dots \Delta^{D_0}(B^{m_0}) \varphi^{P_0}(B^{m_0}) = \epsilon_t \vartheta^{Q_n}(B^{m_n}) \dots \vartheta^{Q_0}(B^{m_0}), \tag{8.56}$$

where n is the number of seasonal cycles, and $D_0 = d$, $P_0 = p$, $Q_0 = q$, and $m_0 = 1$ for convenience. The slightly more compact and even less comprehensible form of (8.56) is:

$$y_t \prod_{j=0}^{n} \Delta^{D_j}(B^{m_j})\varphi^{P_j}(B^{m_j}) = \epsilon_t \prod_{j=0}^{n} \vartheta^{Q_j}(B^{m_j}). \tag{8.57}$$

Conceptually, the model (8.57) is neat, as it captures all the complex relations in the data, but it is not easy to understand and work with, not to mention the potential estimation and order selection problems. To understand what the forecast from such a model can be, we would need to take a special case, multiply the polynomials, and move all the past elements on the right-hand side, leaving only y_t on the left-hand side, as we did with the SARIMA example above. It is worth noting that the `msarima()` function from the `smooth` package implements the model (8.57), although not in this form, but in the state space form, discussed in Chapter 9.

8.2.4 Parameters' bounds for MSARIMA

When it comes to parameters' bounds of SARIMA, the logic stays similar to the process discussed for the case of the non-seasonal model in Section 8.1.7, with the only difference being that instead of analysing the polynomials of a specific part of a model, we need to consider the product of all polynomials. So, the *stationarity* condition for the MSARIMA is for all the roots of the following polynomial to be greater than one by absolute value (lie outside the unit circle):

$$\prod_{j=0}^{n} \varphi^{P_j}(B^{m_j}) = 0, \tag{8.58}$$

while the invertibility condition is for all the roots of the following polynomial to lie outside the unit circle:

$$\prod_{j=0}^{n} \vartheta^{Q_j}(B^{m_j}) = 0. \tag{8.59}$$

Both of these conditions are difficult to check, especially for high frequencies m_j: the polynomial equation of order n has n complex roots, so if you fit a Multiple Seasonal ARIMA on hourly data, where the maximum frequency is $24 \times 7 \times 365 = 61,320$, then the equation will have at least 61,320 roots (this number will increase if there are lower frequencies or non-seasonal orders of the model). Finding all of them is not a trivial task even for modern computers (for example, the `polyroot()` function from the `base` package cannot handle this). So, when considering ARIMA on high-frequency data with high seasonal frequency values, it might make sense to find other ways of checking the stationarity and stability conditions. The `msarima()` and the `adam()` functions in the `smooth` package use the state space form of ARIMA (discussed in Section 9.1) and rely on slightly different principles of checking the same conditions. They do that more efficiently than in the case of the conventional approach of finding the roots of polynomials (8.58) and (8.59).

8.3 Box-Jenkins approach

Now that we are more or less familiar with the idea of ARIMA models, we can move to practicalities. As it might become apparent from the previous sections, one of the issues with the model is the identification of orders p, d, q, P_j, D_j, Q_j etc. Back in the 20th century, when computers were slow, this was a challenging task, so George Box and Gwilym Jenkins (Box & Jenkins, 1976) developed a methodology for identifying and estimating ARIMA models. While there are more efficient ways of order selection for ARIMA nowadays, some of their principles are still used in time series analysis and in forecasting. We briefly outline the idea in this section, not purporting to give a detailed explanation of the approach.

8.3.1 Identifying stationarity

Before doing any time series analysis, we need to make the data stationary, which is done via the differences in the context of ARIMA (Section 8.1.5). But before doing anything, we need to understand whether the data is stationary or not in the first place: over-differencing typically is harmful to the model and would lead to misspecification issues. At the same time, in the case of under-differencing, it might not be possible to identify the model correctly.

There are different ways of understanding whether the data is stationary or not. The simplest of them is just looking at the data: in some cases, it becomes apparent that the mean of the data changes or that there is a trend in the data, so the conclusion would be relatively straightforward. If it is not stationary, then taking differences and analysing the differenced data again would be the next step to ensure that the second differences are not needed.

The more formal approach would be to conduct statistical tests, such as ADF (the `adf.test()` from the `tseries` package) or KPSS (the `kpss.test()` from the `tseries` package). Note that they test different hypotheses:

1. In the case of ADF, it is:
    - H_0: the data is **not** stationary;
    - H_1: the data is stationary;
2. In the case of KPSS:
    - H_0: the data is stationary;

- H_1: the data is **not** stationary.

I do not intent to discuss how the tests are conducted and what they imply in detail, and the interested reader is referred to the original papers of Dickey and Fuller (1979) and Kwiatkowski, Phillips, Schmidt, and Shin (1992). It should suffice to say that ADF is based on estimating parameters of the AR model and then testing the hypothesis for those parameters, while KPSS includes the component of Random Walk in a model (with potential trend) and checks whether the variance of that component is zero or not. Both tests have their advantages and disadvantages and sometimes might contradict each other. No matter what test you choose, do not forget what testing a statistical hypothesis means (see, for example, Section 8.1 of Svetunkov, 2022): if you fail to reject H_0, it does not mean that it is true.

Remark. Even if you select the test-based approach, the procedure should still be iterative: test the hypothesis, take differences if needed, test the hypothesis again, etc. This way, we can determine the order of differences I(d).

When you work with seasonal data, the situation becomes more complicated. Yes, you can probably spot seasonality by visualising the data, but it is not easy to conclude whether the seasonal differences are needed. In this case, the Canova-Hansen test (the `ch.test()` in the `uroot` package) can be used to formally test the hypothesis similar to the one in the KPSS test, but applied to the seasonal differences.

Only after making sure that the data is stationary can we move to the identification of AR and MA orders.

8.3.2 Autocorrelation function (ACF)

At the core of the Box-Jenkins approach, lies the idea of autocorrelation and partial autocorrelation functions. **Autocorrelation** is the correlation (see Section 9.3 of Svetunkov, 2022) of a variable with itself from a different period of time. Here is an example of an autocorrelation coefficient for lag 1:

$$\rho(1) = \frac{\sigma_{y_t,y_{t-1}}}{\sigma_{y_t}\sigma_{y_{t-1}}} = \frac{\sigma_{y_t,y_{t-1}}}{\sigma^2_{y_t}}, \tag{8.60}$$

where $\rho(1)$ is the "true" autocorrelation coefficient, $\sigma_{y_t,y_{t-1}}$ is the covariance between y_t and y_{t-1}, while σ_{y_t} and $\sigma_{y_{t-1}}$ are the "true" standard deviations of y_t and y_{t-1}. Note that $\sigma_{y_t} = \sigma_{y_{t-1}}$, because we are talking about one and the same variable. This is why we can simplify the formula to get the one on the right-hand side of (8.60). The formula (8.60) corresponds to the classical correlation coefficient, so this interpretation is the same as for the classical one: the value of $\rho(1)$ shows the closeness of the lagged relation to linear. If it is close to one, then this means that variable has a strong linear relation with itself on the previous observation. It obviously does not tell you anything about the causality, just shows a technical relation between variables, even if it is spurious in real life.

Using the formula (8.60), we can calculate the autocorrelation coefficients for other lags as well, just substituting y_{t-1} with y_{t-2}, y_{t-3}, ..., $y_{t-\tau}$ etc. In a way, $\rho(\tau)$ can be considered a function of a lag τ, which is called the "Autocorrelation function" (ACF). If we know the order of the ARIMA process we deal with, we can plot the values of ACF on the y-axis by changing the τ on the x-axis. Box and Jenkins (1976) discuss different theoretical functions for several special cases of ARIMA, which we do not plan to repeat here fully. But, for example, they show that if you deal with an AR(1) process, then the $\rho(1) = \phi_1$, $\rho(2) = \phi_1^2$, etc. This can be seen on the example of $\rho(1)$ by calculating the covariance for AR(1):

$$\begin{aligned} \sigma_{y_t,y_{t-1}} =& \mathrm{cov}(y_t, y_{t-1}) = \mathrm{cov}(\phi_1 y_{t-1} + \epsilon_t, y_{t-1}) = \\ & \mathrm{cov}(\phi_1 y_{t-1}, y_{t-1}) = \phi_1 \sigma^2_{y_t}, \end{aligned} \tag{8.61}$$

which when inserted in (8.60) leads to $\rho(1) = \phi_1$. The ACF for AR(1) with a positive ϕ_1 will have the shape shown in Figure 8.14 (on the example of $\phi_1 = 0.9$).

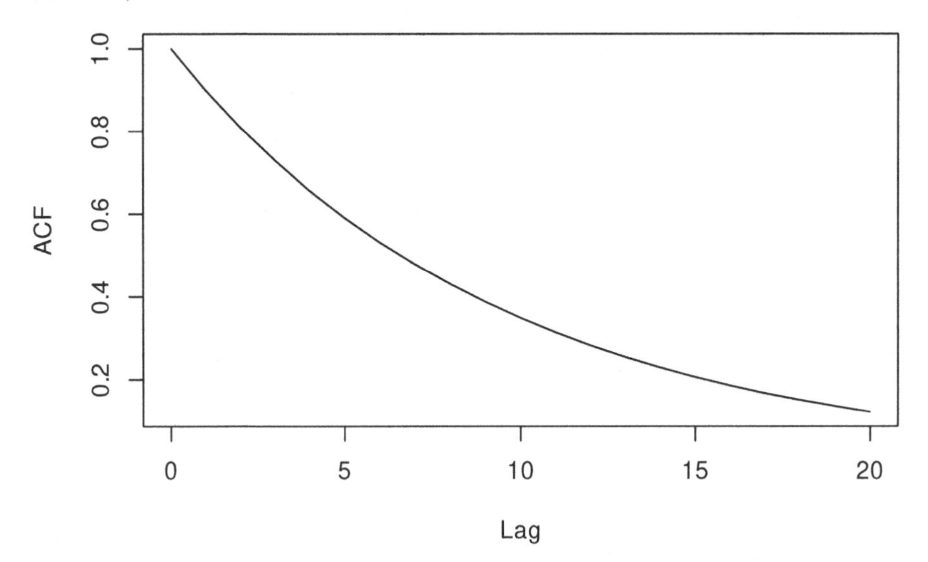

FIGURE 8.14 ACF for an AR(1) model.

Note that $\rho(0) = 1$ just because the value is correlated with itself, so lag 0 is typically dropped as not being useful. The declining shape of the ACF tells us that if y_t is correlated with y_{t-1}, then the correlation between y_{t-1} and y_{t-2} will be exactly the same, also implying that y_t is somehow correlated with y_{t-2}, even if there is no true relation between them. It is difficult to say anything for the AR process based on ACF exactly because of this temporal relation of the variable with itself.

On the other hand, ACF can be used to judge the order of an MA(q) process. For example, if we consider MA(1) (Section 8.1.2), then the $\rho(1)$ will depend on the following covariance:

$$\begin{aligned} \sigma_{y_t,y_{t-1}} =& \mathrm{cov}(y_t, y_{t-1}) = \mathrm{cov}(\theta_1 \epsilon_{t-1} + \epsilon_t, \theta_1 \epsilon_{t-2} + \epsilon_{t-1}) = \\ & \mathrm{cov}(\theta_1 \epsilon_{t-1}, \epsilon_{t-1}) = \theta_1 \sigma^2, \end{aligned} \tag{8.62}$$

where σ^2 is the variance of the error term, which in case of MA(1) is equal to $\sigma^2_{y_t}$, because $\mathrm{E}(y_t) = 0$. However, the covariance between the higher lags will be equal to zero for the pure MA(1) (given that the usual assumptions from Section 1.4.1 hold). Box and Jenkins (1976) showed that for the moving averages, ACF tells more about the order of the model than for the autoregressive one: **ACF will drop rapidly right after the specific lag q for the MA(q) process**.

When it comes to seasonal models, in the case of seasonal AR(P), ACF will decrease exponentially from season to season (e.g. you would see a decrease on lags 4, 8, 12, etc. for SAR(1) and $m = 4$), while in case of seasonal MA(Q), ACF would drop abruptly, starting from the lag $(Q + 1)m$ (so, the subsequent seasonal lag from the one that the process has, e.g. on lag 8, if we deal with SMA(1) with $m = 4$).

8.3.3 Partial autocorrelation function (PACF)

The other instrument useful for time series analysis with respect to ARIMA is called "partial ACF". The idea of this follows from ACF directly. As we have spotted, if the process we deal with follows AR(1), then $\rho(2) = \phi_1^2$ just because of the temporal relation. In order to get rid of this temporal effect, when calculating the correlation between y_t and y_{t-2} we could remove the correlation $\rho(1)$ in order to get the clean effect of y_{t-2} on y_t. This type of correlation is called "partial", denoting it as $\varrho(\tau)$. There are different ways to do that. One of the simplest is to estimate the following regression model:

$$y_t = a_1 y_{t-1} + a_2 y_{t-2} + \dots + a_\tau y_{t-\tau} + \epsilon_t, \tag{8.63}$$

where a_i is the parameter for the i-th lag of the model. In this regression, all the relations between y_t and $y_{t-\tau}$ are captured separately, so the last parameter a_τ is clean of all the temporal effects discussed above. We then can use the value $\varrho(\tau) = a_\tau$ as the coefficient, showing this relation. In order to obtain the PACF, we would need to construct and estimate regressions (8.63) for each lag $\tau = \{1, 2, \dots, p\}$ and get the respective parameters a_1, a_2, ..., a_p, which would correspond to $\varrho(1)$, $\varrho(2)$, ..., $\varrho(p)$.

Just to show what this implies, we consider calculating PACF for an AR(1) process. In this case, the true model is:

$$y_t = \phi_1 y_{t-1} + \epsilon_t.$$

For the first lag we estimate exactly the same model, so that $\varrho(1) = \phi_1$. For the second lag we estimate the model:

$$y_t = a_1 y_{t-1} + a_2 y_{t-2} + \epsilon_t.$$

But we know that for AR(1), $a_2 = 0$, so when estimated in population, this would result in $\varrho(2) = 0$ (in the case of a sample, this would be a value very close to zero). If we continue with other lags, we will come to the same conclusion: for all lags $\tau > 1$ for the AR(1), we will have $\varrho(\tau) = 0$. This is one of the properties of PACF: **if we deal with an AR(p) process, then PACF drops rapidly to zero right after the lag** p.

When it comes to the MA(q) process, PACF behaves differently. In order to understand how it would behave, we take an example of an MA(1) model, which is formulated as:

$$y_t = \theta_1 \epsilon_{t-1} + \epsilon_t.$$

As it was discussed in Section 8.1.7, the MA process can be also represented as an infinite AR (see (8.36) for derivation):

$$y_t = \sum_{j=1}^{\infty} -1^{j-1} \theta_1^j y_{t-j} + \epsilon_t.$$

If we construct and estimate the regression (8.63) for any lag τ for such a process we will get $\varrho(\tau) = -1^{\tau-1}\theta_1^\tau$. This would correspond to the exponentially decreasing curve (if the parameter θ_1 is positive, this will be an alternating series of values), similar to the one we have seen for the AR(1) and ACF. More generally, PACF will decline exponentially for an MA(q) process, starting from the $\varrho(q) = \theta_q$.

When it comes to seasonal ARIMA models, the behaviour of PACF would resemble one of the non-seasonal ones, but with lags, multiple to the seasonality m. e.g., for the SAR(1) process with $m = 4$, the $\varrho(4) = \phi_{4,1}$, while $\varrho(8) = 0$.

8.3.4 Summary

Summarising the discussions in this section, we can conclude that:

1. For an AR(p) process, ACF will decrease either exponentially or alternatingly (depending on the parameters' values), starting from the lag p;
2. For an AR(p) process, PACF will drop abruptly right after the lag p;
3. For an MA(q) process, ACF will drop abruptly right after the lag q;
4. For an MA(q) process, PACF will decline either exponentially or alternatingly (based on the specific values of parameters), starting from the lag q.

These rules are tempting to use when determining the appropriate order of the ARMA model. This is what Box and Jenkins (1976) propose to do, but in an iterative way, analysing the ACF/PACF of residuals of models obtained on previous steps. The authors also recommend identifying the seasonal orders before moving to the non-seasonal ones.

However, these rules are not necessarily bi-directional and might not work in practice, e.g. *if we deal with MA(q), ACF drops abruptly right after the lag q, but if ACF drops abruptly after the lag q, then this does not necessarily mean that we deal with MA(q).* The former follows directly from the assumed "true" model, while the latter refers to the identification of the model on the data, and there can be different reasons for the ACF to behave in the way it does. The logic here is similar to the following:

Example 8.1. All birds have wings. Sarah has wings. Thus, Sarah is a bird.

Here is Sarah:

FIGURE 8.15 Sarah by Yegor Kamelev.

This slight discrepancy led to issues in ARIMA identification over the years. We now understand that we should not rely entirely on the Box-Jenkins approach when identifying ARIMA orders. There are more appropriate methods for order selection, which can be used in the context of ARIMA (see, for example, Hyndman & Khandakar, 2008), and we will discuss some of them in Chapter 15. Still, ACF and PACF could be helpful in diagnostics. They might show you whether anything important is missing in the model. But they should not be used on their own. They are helpful together with other additional instruments (see discussion in Section 14.5).

8.3.5 Examples of ACF/PACF for several ARMA processes

Building upon the behaviours of ACF/PACF for AR and MA processes, we consider several examples of data generated from AR/MA. This might provide some guidelines of how the orders of ARMA can be identified in practice and show how challenging this is in some situations.

Figure 8.16 shows the plot of the data and ACF/PACF for an AR(1) process with $\phi_1 = 0.9$. As discussed earlier, the ACF in this case declines exponentially, starting from lag 1, while the PACF drops right after the lag 1. The plot also shows how the fitted values and the forecast would look like if the order of the model was correctly selected and the parameters were known.

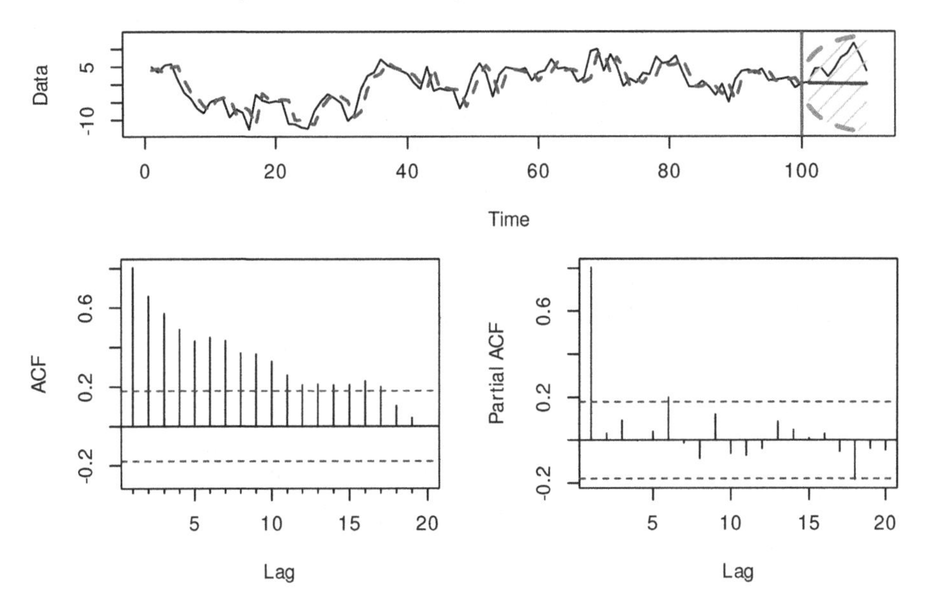

FIGURE 8.16 AR(1) process with $\phi_1 = 0.9$.

Remark. You might notice that there are some values of the PACF lying outside the standard 95% confidence bounds in the plots above. This does not mean that we need to include the respective AR elements (AR(6) and AR(18)), because, by the definition of confidence interval, we can expect 5% of coefficients of correlation to lie outside the bounds, so we can ignore them.

A similar idea is shown in Figure 8.17 for an AR(2) process with $\phi_1 = 0.9$ and $\phi_2 = -0.4$. In the Figure, the ACF oscillates slightly around the origin, while the PACF drops to zero right after the lag 2. These are the characteristics that we have discussed earlier in this section.

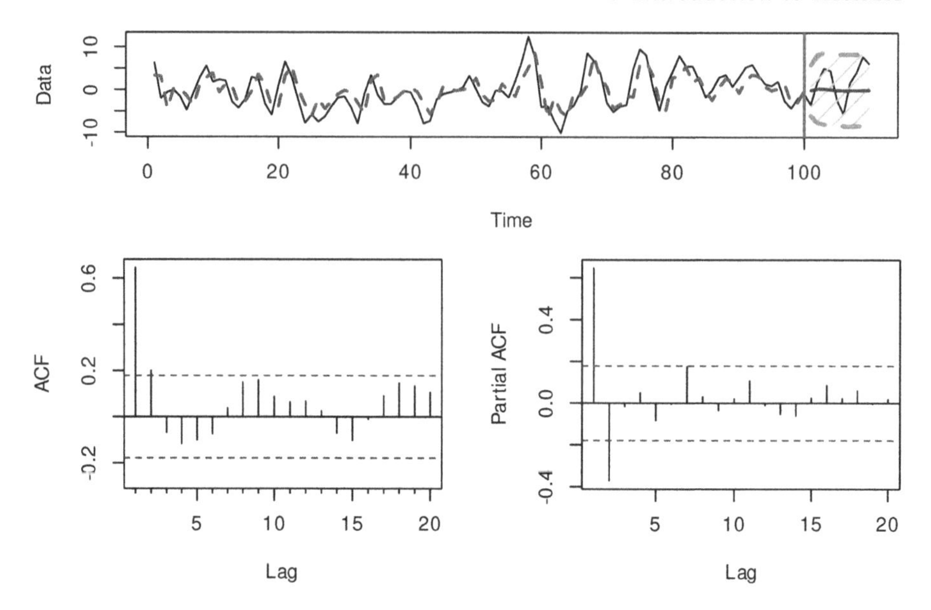

FIGURE 8.17 AR(2) process with $\phi_1 = 0.9$ and $\phi_2 = -0.4$.

Similar plots but with different behaviours of ACF and PACF can be shown for MA(1) and MA(2) processes. Figure 8.18 shows how the MA(2) process looks like with $\theta_1 = -0.9$ and $\theta_2 = 0.8$. As we can see, the correct diagnostics of the MA order in this case is already challenging: while the ACF drops to zero after lag 2, the PACF seems to contain some values outside of the interval on lags 3 and 4. So, using the Box and Jenkins (1976) approach, we would be choosing between an MA(2) and AR(1)/AR(4) models.

To make things even more complicated, we present similar plots for an ARMA(2,2) model in Figure 8.19.

Based on the ACF/PACF in Figure 8.19, we can conclude that the process is not fundamentally distinguishable from AR(2) and/or MA(2). In order to correctly identify the order in this situation, Box and Jenkins (1976) recommend doing the identification sequentially, by fitting an AR model, then analysing the residuals and selecting an appropriate MA order. Still, even this iterative process is challenging and does not guarantee that the correct order will be selected. This is one of the reasons why modern ARIMA order selection methods do not fully rely on the Box-Jenkins approach and involve selection based on information criteria (Hyndman & Khandakar, 2008, Svetunkov and Boylan (2020)).

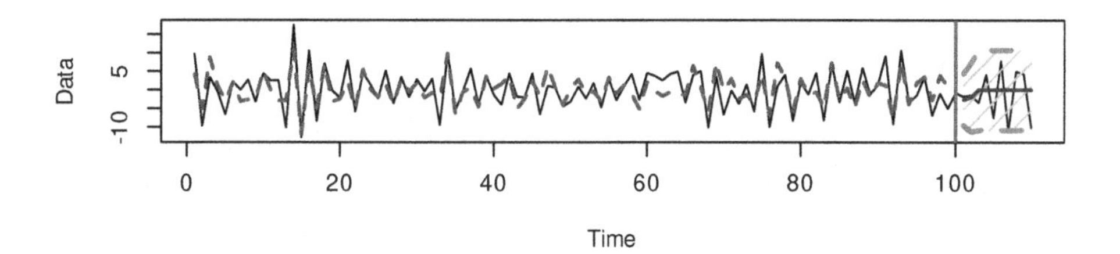

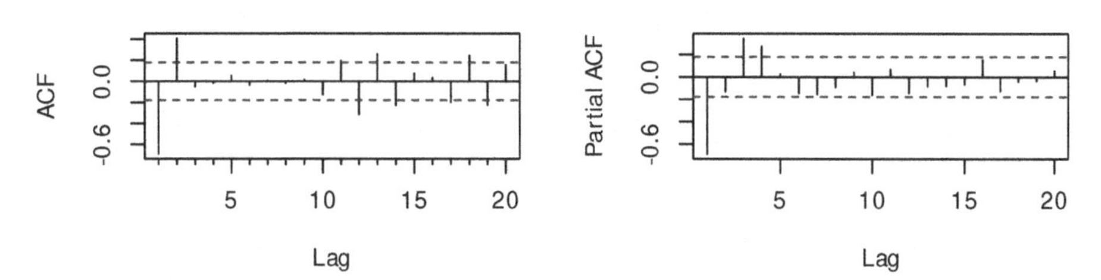

FIGURE 8.18 MA(2) process with $\theta_1 = -0.9$ and $\theta_2 = 0.8$.

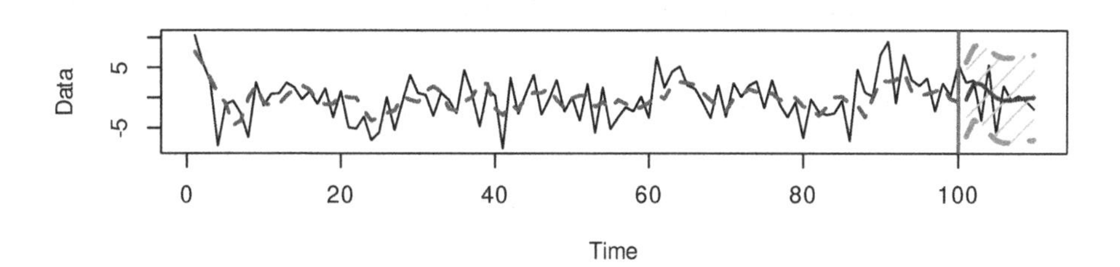

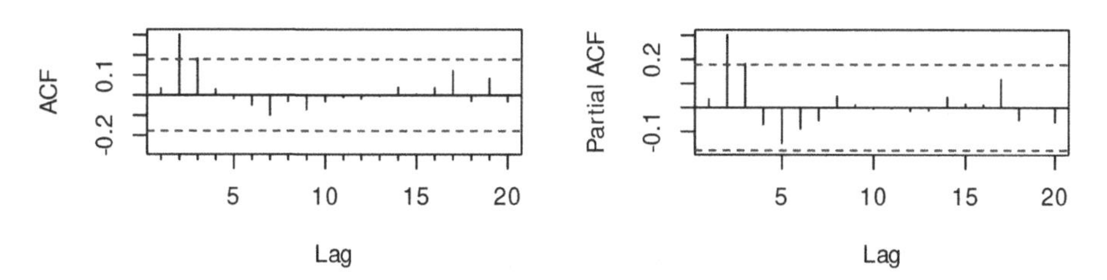

FIGURE 8.19 ARMA(2,2) process with $\phi_1 = 0.9, \phi_2 = -0.4$, $\theta_1 = -0.9$ and $\theta_2 = 0.8$.

8.4 ARIMA and ETS

Box and Jenkins (1976) showed in their textbook that several Exponential Smoothing methods could be considered special cases of the ARIMA model. Because of that, statisticians have thought for many years that ARIMA is a superior model and paid no attention to Exponential Smoothing. It took many years, many papers, and a lot of effort (Fildes, Hibon, Makridakis, & Meade, 1998; Makridakis et al., 1982; Makridakis & Hibon, 2000) to show that this is not correct and that if you are interested in forecasting, then Exponential Smoothing, being a simpler model, typically does a better job than ARIMA. It was only after Ord et al. (1997) that statisticians have started considering ETS as a separate model with its own properties. Furthermore, it seems that some of the conclusions from the previous competitions mainly apply to the Box-Jenkins approach (for example, see Makridakis & Hibon, 1997), pointing out that selecting the correct order of ARIMA models is a much more challenging task than statisticians had thought before.

Still, there is a connection between ARIMA and ETS models, which can benefit both models, so it is worth discussing this in a separate section of the monograph.

8.4.1 ARIMA(0,1,1) and ETS(A,N,N)

Muth (1960) was one of the first authors who showed that Simple Exponential Smoothing (Section 3.4) has an underlying ARIMA(0,1,1) model. This becomes apparent, when we study the error correction form of SES:

$$\hat{y}_t = \hat{y}_{t-1} + \hat{\alpha} e_{t-1}.$$

Recalling that $e_t = y_t - \hat{y}_t$, this equation can be rewritten as:

$$y_t = y_{t-1} - e_{t-1} + \hat{\alpha} e_{t-1} + e_t,$$

or after regrouping elements:

$$y_t - y_{t-1} = e_t + (\hat{\alpha} - 1) e_{t-1}.$$

Finally, using the backshift operator for ARIMA, substituting the estimated values by their "true" ones, we get the ARIMA(0,1,1) model:

$$y_t(1 - B) = \epsilon_t(1 + \theta_1 B),$$

where $\theta_1 = \alpha - 1$. This relation was one of the first hints that α in SES should lie in a wider interval: based on the fact that $\theta_1 \in (-1, 1)$, the smoothing parameter $\alpha \in (0, 2)$. This is the same region we get when we deal with the admissible bounds of the ETS(A,N,N) model (Section 4.3). This connection between the parameters of ARIMA(0,1,1) and ETS(A,N,N) is useful on its own because we can transfer the properties of ETS to ARIMA. For example, we know that the level in ETS(A,N,N) will change slowly when α is close to zero. Similar behaviour would be observed in ARIMA(0,1,1) with θ_1 close to -1. In addition, we know that ETS(A,N,N) reverts to Random Walk, when $\alpha = 1$, which corresponds to $\theta_1 = 0$. So, the closer θ_1 is to zero, the more abrupt behaviour the ARIMA model exhibits. In cases of $\theta_1 > 0$, the model's behaviour becomes even more uncertain. In a way, this relation gives us the idea of what to expect from more complicated ARIMA(p,d,q) models when the parameters for the moving average are negative – the model should typically behave smoother. However, this might differ from one model to another, depending on the MA order.

The main conceptual difference between ARIMA(0,1,1) and ETS(A,N,N) is that the latter still makes sense, when $\alpha = 0$, while in the case of ARIMA(0,1,1), the condition $\theta_1 = -1$ is unacceptable. The global level model with $\theta_1 = -1$ corresponds to just a different model, ARIMA(0,0,0) with constant.

Finally, the connection between the two models tells us that if we have the ARIMA(0,1,q) model, this model would be suitable for the data called "level" in the ETS framework. The length of q would define the distribution of the weights in the model. The specific impact of each MA parameter on the actual values would differ, depending on the order q and values of parameters. The forecast from the ARIMA(0,1,q) would be a straight line parallel to the x-axis for $h \geq q$.

In order to demonstrate the connection between the two models we consider the following example in R using functions `sim.es()`, `es()`, and `msarima()` from `smooth` package:

```
# Generate data from ETS(A,N,N) with alpha=0.2
y <- sim.es("ANN", obs=120, persistence=0.2)
# Estimate ETS(A,N,N)
esModel <- es(y$data, "ANN")
# Estimate ARIMA(0,1,1)
msarimaModel <- msarima(y$data, c(0,1,1), initial="optimal")
```

Given the the two models in smooth have the same initialisation mechanism, they should be equivalent. The result might differ slightly only because of the optimisation routine in the two functions. The values of their losses and information criteria should be similar:

```
# Loss values
```

```
setNames(c(esModel$lossValue, msarimaModel$lossValue),
         c("ETS(A,N,N)","ARIMA(0,1,1)"))
```

```
##   ETS(A,N,N) ARIMA(0,1,1)
##     496.6615     496.6624
```

```
# AIC
setNames(c(AIC(esModel), AIC(msarimaModel)),
         c("ETS(A,N,N)","ARIMA(0,1,1)"))
```

```
##   ETS(A,N,N) ARIMA(0,1,1)
##     999.3231     999.3249
```

In addition, their parameters should be related based on the formula discussed above. The following two lines should produce similar values:

```
# Smoothing parameter and theta_1
setNames(c(esModel$persistence, msarimaModel$arma$ma+1),
         c("ETS(A,N,N)","ARIMA(0,1,1)"))
```

```
##   ETS(A,N,N) ARIMA(0,1,1)
##   0.08215179   0.08519610
```

Finally, the fit and the forecasts from the two models should be exactly the same if the parameters are linearly related (Figure 8.20):

```
par(mfcol=c(2,1), mar=c(2,2,2,1))
plot(esModel,7)
plot(msarimaModel,7)
```

We expect the ETS(A,N,N) and ARIMA(0,1,1) models to be similar in this example because they are estimated using the respective functions `es()` and `msarima()`, which are implemented in the same way, using the same framework. If the framework, initialisation, construction, or estimation would be different, then the relation between the applied models might be not exact but approximate.

8.4.2 ARIMA(0,2,2) and ETS(A,A,N)

Nerlove and Wage (1964) showed that there is an underlying ARIMA(0,2,2) for the Holt's method (Subsection 4.4.1), although they do not say that explicitly in their paper. Skipping the derivations, the relation between Holt's method

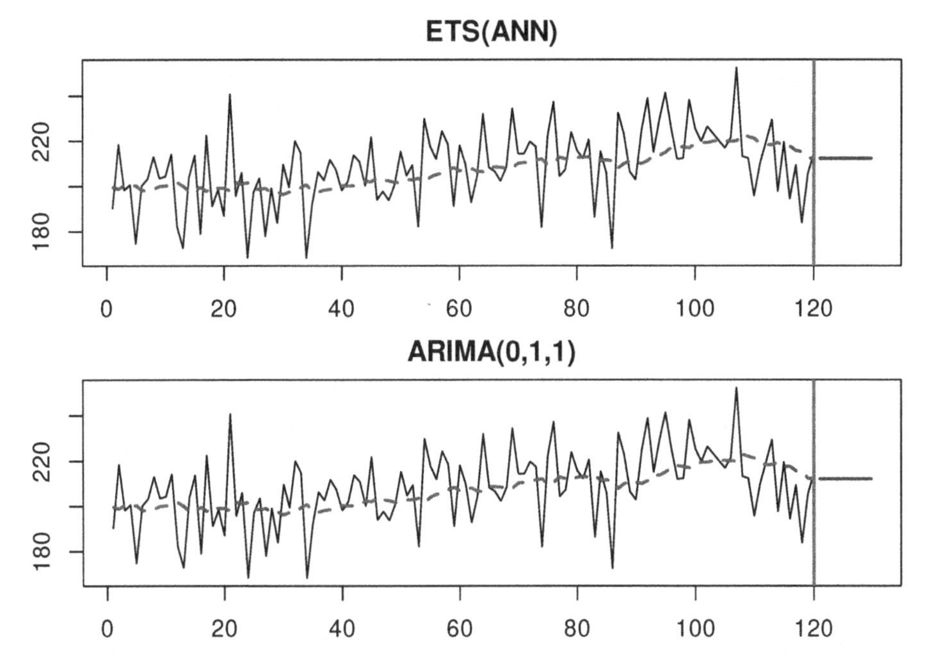

FIGURE 8.20 ETS(A,N,N) and ARIMA(0,1,1) models producing the same fit and forecast trajectories.

and the ARIMA model is expressed in the following two equations about their parameters (in the form of ARIMA discussed in this monograph):

$$\begin{aligned} \theta_1 &= \alpha + \beta - 2 \\ \theta_2 &= 1 - \alpha \end{aligned} .$$

We also know from Section 4.4 that Holt's method has an underlying ETS(A,A,N) model. Thus, there is a connection between this model and ARIMA(0,2,2). This means that ARIMA(0,2,2) will produce linear trajectories for the data and that the MA parameters of the model regulate the speed of the update of values. Because of the second difference, ARIMA(0,2,q) will produce a straight line as a forecasting trajectory for any $h \geq q$.

Similarly to the ARIMA(0,1,1) vs ETS(A,N,N), one of the important differences between the models is that the boundary values for parameters are not possible for ARIMA(0,2,2): $\alpha = 0$ and $\beta = 0$ are possible in ETS, but the respective $\theta_1 = 2$ and $\theta_2 = -1$ in ARIMA are not.

Furthermore, the model that corresponds to the situation, when $\alpha = 1$ and $\beta = 0$ can only be formulated as ARIMA(0,1,0) with drift (discussed in Section 8.1.5). The Global Trend ARIMA can only appear in the boundary case with

$\theta_1 = -2$ and $\theta_2 = 1$, implying the following model:

$$y_t(1-B)^2 = \epsilon_t - 2\epsilon_{t-1} + \epsilon_{t-2} = \epsilon_t(1-B)^2,$$

which tells us that in ARIMA framework, the Global Trend model is only available as a Global Mean on second differences of the data.

Finally, the ETA(A,A,N) and ARIMA(0,2,2) will fit the data similarly and produce the exact forecasts as long as they are constructed, initialised, and estimated in the same way.

8.4.3 ARIMA(1,1,2) and ETS(A,Ad,N)

Roberts (1982) proposed damped trend Exponential Smoothing method (Section 4.4.2), showing that it is related to the ARIMA(1,1,2) model, with the following connection between the parameters of the two:

$$\begin{aligned} \theta_1 &= \alpha - 1 + \phi(\beta - 1) \\ \theta_2 &= \phi(1-\alpha) \\ \phi_1 &= \phi \end{aligned} .$$

At the same time, the damped trend method has underlying ETS(A,Ad,N), thus the two models are connected. Recalling that ETS(A,Ad,N) reverts to ETS(A,A,N), when $\phi = 1$, we can see a similar property in ARIMA: when $\phi_1 = 1$, the model should be reformulated as ARIMA(0,2,2) instead of ARIMA(1,1,2). Given the direct connection between the dampening parameters and the AR(1) parameter of the two models, we can conclude that AR(1) defines the forecasting trajectory's dampening effect. We have already noticed this in Section 8.1.4. However, we should acknowledge that the dampening only happens when $\phi_1 \in (0,1)$. The case of $\phi_1 > 1$ is unacceptable in the ARIMA framework and is not very useful in the case of ETS, producing explosive exponential trajectories. The case of $\phi_1 \in (-1,0)$ is possible but is less useful in practice, as the trajectory will oscillate.

The lesson to learn from the connection between the two models is that the AR(p) part of ARIMA can act as a dampening element for the forecasting trajectories, although the specific shape would depend on the value of p and the values of parameters.

8.4.4 ARIMA and other ETS models

The pure additive seasonal ETS models (Section 5.1) also have a connection with ARIMA, but the resulting models are not parsimonious. For example, ETS(A,A,A) is related to SARIMA(0,1,m+1)(0,1,0)$_m$ (Chatfield, 1977; McKenzie, 1976) with some restrictions on parameters. If we were to work with SARIMA and wanted to model the seasonal time series, we would probably apply SARIMA(0,1,1)(0,1,1)$_m$ instead of this larger model.

When it comes to pure multiplicative (Chapter 6) and mixed (Section 7.2) ETS models, there are no appropriate ARIMA analogues for them. For example, Chatfield (1977) showed that there are no ARIMA models for the Exponential Smoothing with the multiplicative seasonal component. This makes ETS distinct from ARIMA. The closest one can get to a pure multiplicative ETS model is the ARIMA applied to logarithmically transformed data when the smoothing parameters of ETS are close to zero, coming from the limit (6.5).

8.4.5 ETS + ARIMA

Finally, based on the discussion above, it is possible to have a combination of ETS and ARIMA, but not all combinations would be meaningful and helpful. For example, fitting a combination of ETS(A,N,N)+ARIMA(0,1,1) is not a good idea due to the connection of the two models (Subsection 8.4.1). However, doing ETS(A,N,N) and adding an ARIMA(1,0,0) component would make sense – the resulting model would exhibit the dampening trend as discussed in Section 8.4.3 but would have fewer parameters to estimate than ETS(A,Ad,N). Gardner (1985) pointed out that using AR(1) with Exponential Smoothing methods improves forecasting accuracy, so this combination of the two models is potentially beneficial for ETS. In the next chapter, we will discuss how specifically the two models can be united in one framework.

8.5 Examples of application

8.5.1 Non-seasonal data

Using the time series from the Box-Jenkins textbook (Box & Jenkins, 1976), we fit the ARIMA model to the data but based on our judgment rather than their approach. Just a reminder, here is how the data looks (series `BJsales`, Figure 8.21):

It seems to exhibit the trend in the data, so we can consider ARIMA(1,1,2), ARIMA(0,2,2), and ARIMA(1,1,1) models. We do not consider models with drift in this example, because they would imply the same slope over time for the whole series, which does not seem to be the case here. We use the `msarima()` function from the `smooth` package, which is a wrapper for the `adam()` function:

```
adamARIMABJ <- vector("list",3)
# ARIMA(1,1,2)
adamARIMABJ[[1]] <- msarima(BJsales, orders=c(1,1,2),
                            h=10, holdout=TRUE)
```

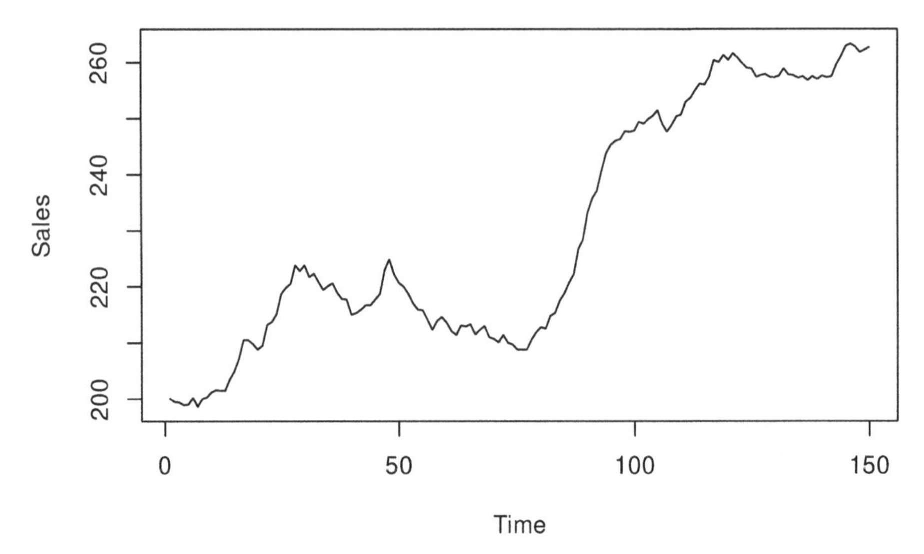

FIGURE 8.21 Box-Jenkins sales data.

```
# ARIMA(0,2,2)
adamARIMABJ[[2]] <- msarima(BJsales, orders=c(0,2,2),
                            h=10, holdout=TRUE)
# ARIMA(1,1,1)
adamARIMABJ[[3]] <- msarima(BJsales, orders=c(1,1,1),
                            h=10, holdout=TRUE)
names(adamARIMABJ) <- c("ARIMA(1,1,2)", "ARIMA(0,2,2)",
                        "ARIMA(1,1,1)")
```

Comparing information criteria (we will use AICc) of the three models, we can select the most appropriate one:

```
sapply(adamARIMABJ, AICc)
```

```
## ARIMA(1,1,2) ARIMA(0,2,2) ARIMA(1,1,1)
##     503.8630     497.0121     505.7502
```

Remark. Note that this comparison is possible in `adam()`, `ssarima()`, and `msarima()` because the implemented ARIMA is formulated in state space form, sidestepping the issue of the conventional ARIMA (where taking differences reduces the sample size).

Based on this comparison, it looks like the ARIMA(0,2,2) is the most appropriate model (among the three) for the data. Here is how the fit and the forecast from the model looks (Figure 8.22):

```
plot(adamARIMABJ[[2]], which=7)
```

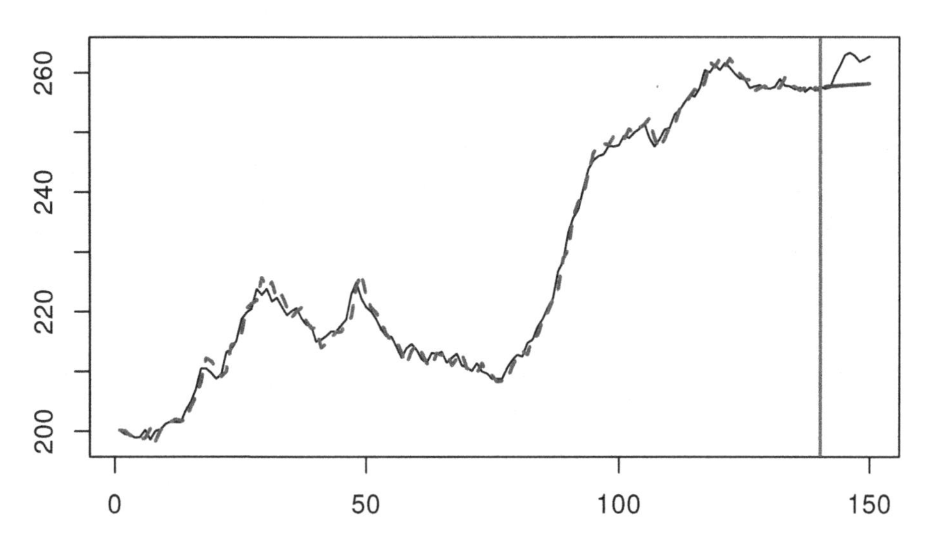

FIGURE 8.22 BJSales series and ARIMA(0,2,2).

Comparing this model with the ETS(A,A,N), we will see a slight difference because the two models are initialised and estimated differently:

```
adam(BJsales, "AAN", h=10, holdout=TRUE)
```

```
## Time elapsed: 0.07 seconds
## Model estimated using adam() function: ETS(AAN)
## Distribution assumed in the model: Normal
## Loss function type: likelihood; Loss function value: 243.4073
## Persistence vector g:
##  alpha   beta
## 1.0000 0.2392
##
## Sample size: 140
## Number of estimated parameters: 5
## Number of degrees of freedom: 135
## Information criteria:
##      AIC     AICc      BIC     BICc
```

```
## 496.8146 497.2624 511.5228 512.6292
##
## Forecast errors:
## ME: 3.229; MAE: 3.341; RMSE: 3.797
## sCE: 14.178%; Asymmetry: 91.7%; sMAE: 1.467%; sMSE: 0.028%
## MASE: 2.826; RMSSE: 2.491; rMAE: 0.928; rRMSE: 0.924
```

If we are interested in a more classical Box-Jenkins approach, we can always analyse the residuals of the constructed model and try improving it further. Here is an example of ACF and PACF of the residuals of the ARIMA(0,2,2):

```
par(mfcol=c(1,2))
plot(adamARIMABJ[[2]], which=c(10,11), main="")
```

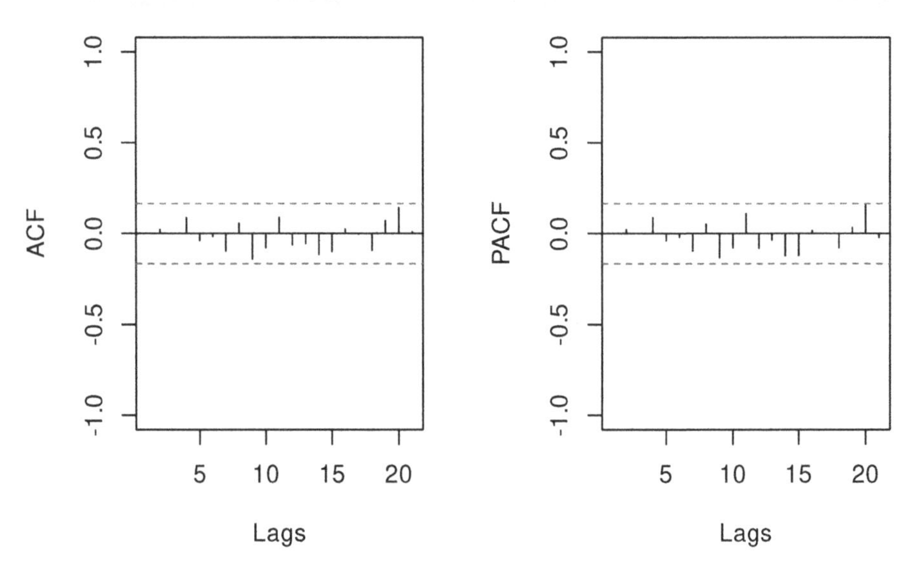

FIGURE 8.23 ACF and PACF of the ARIMA(0,2,2) model on BJSales data

As we see from the plot in Figure 8.23, all autocorrelation coefficients lie inside the confidence interval, implying that there are no significant AR/MA lags to include in the model.

8.5.2 Seasonal data

Similarly to the previous cases, we use Box-Jenkins AirPassengers data, which exhibits a multiplicative seasonality and a trend. We will model this using SARIMA$(0,2,2)(0,1,1)_{12}$, SARIMA$(0,2,2)(1,1,1)_{12}$, and SARIMA$(0,2,2)(1,1,0)_{12}$ models, which are selected to see what type of seasonal ARIMA is more appropriate to the data:

```
adamSARIMAAir <- vector("list",3)
# SARIMA(0,2,2)(0,1,1)[12]
adamSARIMAAir[[1]] <- msarima(AirPassengers, lags=c(1,12),
                              orders=list(ar=c(0,0), i=c(2,1),
                                          ma=c(2,1)),
                              h=12, holdout=TRUE)
# SARIMA(0,2,2)(1,1,1)[12]
adamSARIMAAir[[2]] <- msarima(AirPassengers, lags=c(1,12),
                              orders=list(ar=c(0,1), i=c(2,1),
                                          ma=c(2,1)),
                              h=12, holdout=TRUE)
# SARIMA(0,2,2)(1,1,0)[12]
adamSARIMAAir[[3]] <- msarima(AirPassengers, lags=c(1,12),
                              orders=list(ar=c(0,1), i=c(2,1),
                                          ma=c(2,0)),
                              h=12, holdout=TRUE)
names(adamSARIMAAir) <- c("SARIMA(0,2,2)(0,1,1)[12]",
                          "SARIMA(0,2,2)(1,1,1)[12]",
                          "SARIMA(0,2,2)(1,1,0)[12]")
```

Note that now that we have a seasonal component, we need to provide the SARIMA lags: 1 and $m = 12$ and specify `orders` differently – as a list with values for AR, I, and MA orders separately. This is done because the SARIMA implemented in `msarima()` and `adam()` supports multiple seasonalities (e.g. you can have `lags=c(1,24,24*7)` if you want). The resulting information criteria of models are:

```
sapply(adamSARIMAAir, AICc)
```

```
## SARIMA(0,2,2)(0,1,1)[12] SARIMA(0,2,2)(1,1,1)[12] SARIMA(0,2,2)
    (1,1,0)[12]
##                 1065.427                 1083.727
    1082.625
```

It looks like the second model is slightly better than the other two, so we will use it in order to produce forecasts (see Figure 8.24):

```
forecast(adamSARIMAAir[[2]], h=12,
         interval="prediction") |>
    plot(main="")
```

This model is directly comparable with ETS models, so here is, for example, the AICc value of ETS(M,A,M) on the same data:

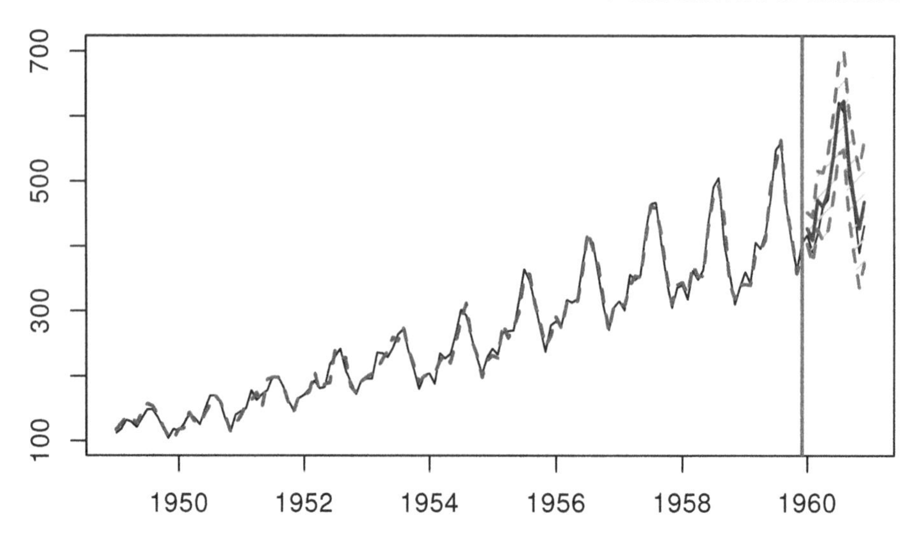

FIGURE 8.24 Forecast of AirPassengers data from a SARIMA$(0,2,2)(1,1,1)_{12}$ model.

```
AICc(adamETSAir <- adam(AirPassengers, "MAM",
                        h=12, holdout=TRUE))
```

```
## [1] 973.9646
```

It is lower than for the SARIMA model, which means that ETS(M,A,M) is more appropriate for the data in terms of information criteria than SARIMA$(0,2,2)(1,1,1)_{12}$. We can also investigate if there is a way to improve ETS(M,A,M) by adding some ARMA components (Figure 8.25):

```
par(mfcol=c(1,2))
plot(adamETSAir, which=c(10,11), main="")
```

Judging by the plots in Figure 8.25, significant correlation coefficients exist for some lags. Still, it is not clear whether they appear due to randomness or not. Just to check, we will see if adding SARIMA$(0,0,0)(0,0,1)_{12}$ helps (reduces AICc) in this case:

```
adam(AirPassengers, "MAM", h=12, holdout=TRUE,
     order=list(ma=c(0,1)), lags=c(1,12)) |>
    AICc()
```

```
## [1] 1012.416
```

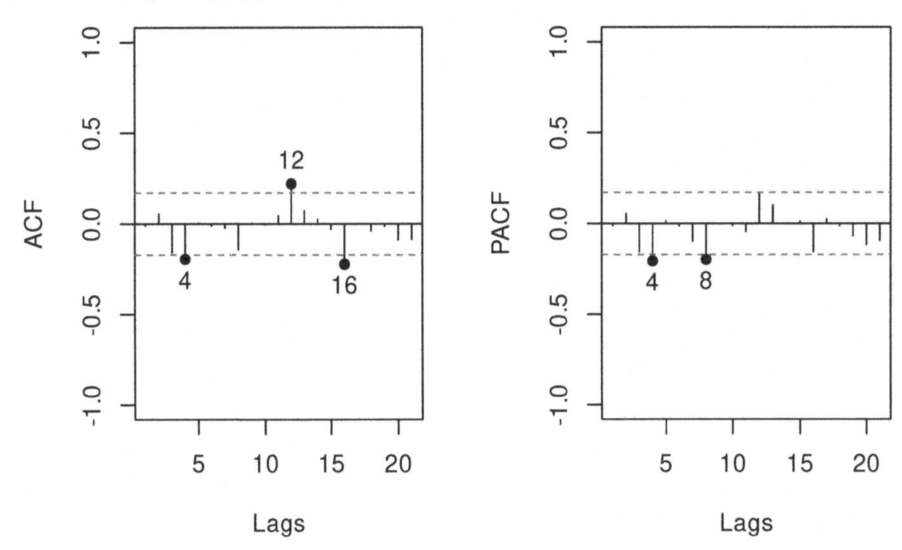

FIGURE 8.25 ACF and PACF of ETS(M,A,M) on AirPassengers data.

As we see, the increased complexity does not decrease the AICc (probably because now we need to estimate 13 parameters more than in just ETS(M,A,M)), so we should not add the SARIMA component. We could try adding other SARIMA elements to see if they improve the model, but we do not aim to find the best model here. The interested reader is encouraged to do that as an additional exercise.

9

ADAM ARIMA

There are different ways to formulate and implement ARIMA. The one discussed in Chapter 8 is the conventional way. The model, in that case, can be estimated directly, assuming that its initialisation happens at some point before the Big Bang: the conventional ARIMA assumes that there is no starting point of the model. The idea is that we observe a specific piece of data from a population without any beginning or end. Obviously, this assumption is idealistic and does not necessarily agree with reality (imagine the series of infinitely lasting sales of Siemens S45 mobile phones. Do you even remember such a thing?).

But besides the conventional formulation, there are also state space forms of ARIMA, the most relevant to our topic being the one implemented in SSOE form (Chapter 11 of Hyndman et al., 2008). Svetunkov and Boylan (2020) adapted this state space model for supply chain forecasting, developing an order selection mechanism, sidestepping the hypotheses testing and focusing on information criteria. However, the main limitation of that approach is that the resulting ARIMA model works very slow on the high frequency data with several seasonal patterns (because the model was formulated based on the conventional SSOE). Luckily, the SSOE used in ADAM (introduced in Section 5.1) addresses this issue. This model is already implemented in the `msarima()` function of the `smooth` package and was also used as the basis for the ADAM ARIMA.

In this chapter, we discuss the state space ADAM ARIMA for both pure additive and pure multiplicative cases. We then explore the conditional moments from the model and parameter space and move to the distributional assumptions of the model (including the conditional distributions). We conclude the chapter with the discussion of the implications of the ETS+ARIMA model. The latter has not been discussed in the literature and might make the model unidentifiable, so an analyst using the combination should be cautious.

9.1 State space ARIMA

9.1.1 An example of state space ARIMA

In order to understand how the state space ADAM ARIMA can be formulated, we consider an arbitrary example of SARIMA(1,1,2)(0,1,0)$_4$:

$$y_t(1-\phi_1 B)(1-B)(1-B^4) = \epsilon_t(1+\theta_1 B+\theta_2 B^2),$$

which can be rewritten in the expanded form after opening the brackets:

$$y_t(1-\phi_1 B - B + \phi_1 B^2 - B^4 + \phi_1 B^5 + B^5 - \phi_1 B^6) = \epsilon_t(1+\theta_1 B+\theta_2 B^2),$$

and after moving all the lagged values to the right-hand side as:

$$y_t = (1+\phi_1)y_{t-1} - \phi_1 y_{t-2} + y_{t-4} - (1+\phi_1)y_{t-5} + \phi_1 y_{t-6} + \theta_1 \epsilon_{t-1} + \theta_2 \epsilon_{t-2} + \epsilon_t.$$

Now we can define the states of the model for each of the indices $t-j$:

$$\begin{aligned} v_{1,t-1} &= (1+\phi_1)y_{t-1} + \theta_1 \epsilon_{t-1} \\ v_{2,t-2} &= -\phi_1 y_{t-2} + \theta_2 \epsilon_{t-2} \\ v_{3,t-3} &= 0 \\ v_{4,t-4} &= y_{t-4} \\ v_{5,t-5} &= -(1+\phi_1)y_{t-5} \\ v_{6,t-6} &= \phi_1 y_{t-6} \end{aligned} . \tag{9.1}$$

In our example all the MA parameters are zero for $j > 2$, which is why they disappear from the states above. Furthermore, there are no elements for lag 3, so that state can be dropped. The measurement equation of the ARIMA model in this situation can be written as:

$$y_t = \sum_{j=1,2,4,5,6} v_{j,t-j} + \epsilon_t,$$

based on which the actual value on some lag i can also be written as:

$$y_{t-i} = \sum_{j=1,2,4,5,6} v_{j,t-j-i} + \epsilon_{t-i}. \tag{9.2}$$

Inserting (9.2) in (9.1) and shifting the lags from $t-i$ to t in every equation, we get the transitions equation of state space ARIMA:

$$\begin{aligned}
v_{1,t} &= (1+\phi_1) \sum_{j=1,2,4,5,6} v_{j,t-j} + (1+\phi_1+\theta_1)\epsilon_t \\
v_{2,t} &= -\phi_1 \sum_{j=1,2,4,5,6} v_{j,t-j} + (-\phi_1+\theta_2)\epsilon_t \\
v_{4,t} &= \sum_{j=1,2,4,5,6} v_{j,t-j} + \epsilon_t \\
v_{5,t} &= -(1+\phi_1) \sum_{j=1,2,4,5,6} v_{j,t-j} - (1+\phi_1)\epsilon_t \\
v_{6,t} &= \phi_1 \sum_{j=1,2,4,5,6} v_{j,t-j} + \phi_1\epsilon_t
\end{aligned} .$$

This model can then be applied to the data, and forecasts can be produced similarly to how it was done for the pure additive ETS model (see Section 5.1). Furthermore, it can be shown that any ARIMA model can be written in the compact form (5.5), meaning that the same principles as for ETS can be applied to ARIMA and that the two models can be united in one framework.

9.1.2 Additive ARIMA

In a more general case, in order to develop the state space ARIMA, we will use the Multiple Seasonal ARIMA, discussed in Subsection 8.2.3:

$$y_t \prod_{j=0}^{n} \Delta^{D_j}(B^{m_j}) \varphi^{P_j}(B^{m_j}) = \epsilon_t \prod_{j=0}^{n} \vartheta^{Q_j}(B^{m_j}),$$

This model can be represented in an easier to digest form by expanding the polynomials and moving all the previous values to the right-hand side. In a general case we will have:

$$y_t = \sum_{j=1}^{K} \eta_j y_{t-j} + \sum_{j=1}^{K} \psi_j \epsilon_{t-j} + \epsilon_t, \tag{9.3}$$

where each element η_j and ψ_j can be called the parameter of polynomial. In our example with SARIMA(1,1,2)(0,1,0)$_4$ in the previous subsection they were:

$$\begin{aligned}
\eta_1 &= 1 + \phi_1 \\
\eta_2 &= -\phi_1 \\
\eta_3 &= 0 \\
\eta_4 &= 1 \\
\eta_5 &= -(1 + \phi_1) \\
\eta_6 &= \phi_1 \\
\psi_1 &= \theta_1 \\
\psi_2 &= \theta_2
\end{aligned}$$

In the equation (9.3), K is the order of the highest polynomial, calculated as $K = \max\left(\sum_{j=0}^n (P_j + D_j)m_j, \sum_{j=0}^n Q_j m_j\right)$. If, for example, the MA order is higher than the sum of ARI orders, then polynomials $\eta_i = 0$ for $i > \sum_{j=0}^n (P_j + D_j)m_j$. The same holds for the opposite situation of the sum of ARI orders being higher than the MA orders, where $\psi_i = 0$ for all $i > \sum_{j=0}^n Q_j m_j$. Using this idea we could define states for each of the previous elements:

$$v_{i,t-i} = \eta_i y_{t-i} + \theta_i \epsilon_{t-i}, \tag{9.4}$$

leading to the following model based on (9.4) and (9.3):

$$y_t = \sum_{j=1}^K v_{j,t-j} + \epsilon_t. \tag{9.5}$$

This can be considered a measurement equation of the state space ARIMA. Now if we consider the previous values of y_t based on (9.5), for y_{t-i}, it will be equal to:

$$y_{t-i} = \sum_{j=1}^K v_{j,t-j-i} + \epsilon_{t-i}. \tag{9.6}$$

The value (9.6) can then be inserted into (9.4), to get the set of transition equations for all $i = 1, 2, \dots, K$:

$$v_{i,t-i} = \eta_i \sum_{j=1}^K v_{j,t-j-i} + (\eta_i + \psi_i)\epsilon_{t-i}. \tag{9.7}$$

This leads to the SSOE state space model based on (9.6) and (9.7):

$$\begin{aligned}
y_t &= \sum_{j=1}^K v_{j,t-j} + \epsilon_t \\
v_{i,t} &= \eta_i \sum_{j=1}^K v_{j,t-j} + (\eta_i + \psi_i)\epsilon_t \text{ for each } i = \{1, 2, \dots, K\}
\end{aligned}, \tag{9.8}$$

which can be formulated in the conventional form as a pure additive ADAM (Section 5.1):

$$\begin{aligned} y_t &= \mathbf{w}' \mathbf{v}_{t-\boldsymbol{l}} + \epsilon_t \\ \mathbf{v}_t &= \mathbf{F} \mathbf{v}_{t-\boldsymbol{l}} + \mathbf{g} \epsilon_t \end{aligned},$$

with the following values for matrices:

$$\begin{aligned} \mathbf{F} = \begin{pmatrix} \eta_1 & \eta_1 & \dots & \eta_1 \\ \eta_2 & \eta_2 & \dots & \eta_2 \\ \vdots & \vdots & \ddots & \vdots \\ \eta_K & \eta_K & \dots & \eta_K \end{pmatrix}, \mathbf{w} = \begin{pmatrix} 1 \\ 1 \\ \vdots \\ 1 \end{pmatrix}, \\ \mathbf{g} = \begin{pmatrix} \eta_1 + \psi_1 \\ \eta_2 + \psi_2 \\ \vdots \\ \eta_K + \psi_K \end{pmatrix}, \mathbf{v}_t = \begin{pmatrix} v_{1,t} \\ v_{2,t} \\ \vdots \\ v_{K,t} \end{pmatrix}, \quad \boldsymbol{l} = \begin{pmatrix} 1 \\ 2 \\ \vdots \\ K \end{pmatrix} \end{aligned}. \tag{9.9}$$

I should point out that the states in this model do not have any specific meaning, they just represent a combination of lagged actual values and error terms. Furthermore, there are zero states in this model, corresponding to zero polynomials of ARI and MA. These can be dropped to make the model even more compact.

In general, state space ARIMA looks more complicated than the original one in the conventional form, but it brings the model to the same ground as ETS in ADAM (Chapter 5), making them directly comparable via information criteria and allowing us to easily combine the two models, not to mention comparing ARIMA of any order with another ARIMA (e.g. with different orders of integration) or introduce multiple seasonality and explanatory variables. Several examples of ARIMA models in ADAM framework are provided in Subsection 9.1.5.

9.1.3 State space ARIMA with constant

If we want to add the constant to the model (similar to how it was done in Section 8.1.4), we need to modify the equation (9.3):

$$y_t = \sum_{j=1}^K \eta_j y_{t-j} + \sum_{j=1}^K \theta_j \epsilon_{t-j} + a_0 + \epsilon_t. \tag{9.10}$$

This then leads to the appearance of the new state:

$$v_{K+1,t} = a_0, \tag{9.11}$$

and modified measurement equation:

$$y_t = \sum_{j=1}^{K+1} v_{j,t-j} + \epsilon_t, \tag{9.12}$$

with the following transition equations:

$$\begin{aligned} v_{i,t} &= \eta_i \sum_{j=1}^{K+1} v_{j,t-j} + (\eta_i + \theta_i)\epsilon_t, \text{ for } i = \{1, 2, \dots, K\} \\ v_{K+1,t} &= v_{K+1,t-1}. \end{aligned} \tag{9.13}$$

The state space equations (9.12) and (9.13) lead to the following matrices:

$$\begin{aligned} \mathbf{F} = \begin{pmatrix} \eta_1 & \dots & \eta_1 & \eta_1 \\ \eta_2 & \dots & \eta_2 & \eta_2 \\ \vdots & \vdots & \ddots & \vdots \\ \eta_K & \dots & \eta_K & \eta_K \\ 0 & \dots & 0 & 1 \end{pmatrix}, \mathbf{w} = \begin{pmatrix} 1 \\ 1 \\ \vdots \\ 1 \\ 1 \end{pmatrix}, \\ \mathbf{g} = \begin{pmatrix} \eta_1 + \theta_1 \\ \eta_2 + \theta_2 \\ \vdots \\ \eta_K + \theta_K \\ 0 \end{pmatrix}, \mathbf{v}_t = \begin{pmatrix} v_{1,t} \\ v_{2,t} \\ \vdots \\ v_{K,t} \\ v_{K+1,t} \end{pmatrix}, \quad \boldsymbol{l} = \begin{pmatrix} 1 \\ 2 \\ \vdots \\ K \\ 1 \end{pmatrix} \end{aligned}. \tag{9.14}$$

Remark. Note that the constant term introduced in this model has a changing meaning, depending on the order of differences of the model. For example, if $D_j = 0$ for all j, it acts as an intercept, while for the $D_0 = d = 1$, it will act as a drift.

9.1.4 Multiplicative ARIMA

In order to connect ARIMA with ETS, we also need to define cases for multiplicative models. This implies that the error term $(1 + \epsilon_t)$ is multiplied by components of the model. The state space ARIMA in this case can be formulated using logarithms in the following way:

$$\begin{aligned} y_t &= \exp\left(\sum_{j=1}^{K} \log v_{j,t-j} + \log(1 + \epsilon_t)\right) \\ \log v_{i,t} &= \eta_i \sum_{j=1}^{K} \log v_{j,t-j} + (\eta_i + \theta_i) \log(1 + \epsilon_t) \text{ for each } i = \{1, 2, \dots, K\} \end{aligned}. \tag{9.15}$$

The model (9.15) can be written in the following more general form:

$$\begin{aligned} y_t &= \exp\left(\mathbf{w}' \log \mathbf{v}_{t-\boldsymbol{l}} + \log(1 + \epsilon_t)\right) \\ \log \mathbf{v}_t &= \mathbf{F} \log \mathbf{v}_{t-\boldsymbol{l}} + \mathbf{g} \log(1 + \epsilon_t) \end{aligned}, \tag{9.16}$$

where $\mathbf{w}$, $\mathbf{F}$, $\mathbf{v}_t$, $\mathbf{g}$, and $\boldsymbol{l}$ are defined as before for the pure additive ARIMA (Section 9.1). This model is equivalent to applying ARIMA to log-transformed data but at the same time shares some similarities with the pure multiplicative ETS from Section 6.1. The main advantage of this formulation is that this model has analytical solutions for the conditional moments and has well-defined h steps ahead conditional distribution if the distribution of $\log(1+\epsilon_t)$ supports convolutions. This simplifies substantially the work with the model in contrast with the pure multiplicative ETS.

To distinguish the additive ARIMA from the multiplicative one, we will use the notation "Log-ARIMA" in this book, pointing out what such model is equivalent to (applying ARIMA to the log-transformed data).

Finally, it is worth mentioning that due to the logarithmic transform, the Log-ARIMA model would be suitable for the cases of time-varying heteroscedasticity, similar to the multiplicative error ETS models.

9.1.5 Several examples of state space ARIMA in ADAM

There are several important special cases of ARIMA model that are often used in practice. We provide their state space formulations in this subsection.

9.1.5.1 ARIMA(0,1,1)

$$(1-B)y_t = (1+\theta_1 B)\epsilon_t,$$

or

$$y_t = y_{t-1} + \theta_1 \epsilon_{t-1} + \epsilon_t,$$

which is equivalent to:

$$\begin{aligned} y_t &= v_{1,t-1} + \epsilon_t \\ v_{1,t} &= v_{1,t-1} + (1+\theta_1)\epsilon_t \end{aligned} . \tag{9.17}$$

9.1.5.2 ARIMA(0,1,1) with drift

$$(1-B)y_t = a_0 + (1+\theta_1 B)\epsilon_t,$$

or

$$y_t = y_{t-1} + a_0 + \theta_1 \epsilon_{t-1} + \epsilon_t,$$

which is in state space:

$$\begin{aligned} y_t &= v_{1,t-1} + v_{2,t-1} + \epsilon_t \\ v_{1,t} &= v_{1,t-1} + v_{2,t-1} + (1+\theta_1)\epsilon_t, \\ v_{2,t} &= v_{2,t-1} \end{aligned} \tag{9.18}$$

where $v_{2,0} = a_0$.

9.1.5.3 ARIMA(0,2,2)

$$\begin{aligned} & (1-B)^2 y_t = (1+\theta_1 B + \theta_2 B^2)\epsilon_t, \\ & \text{or} \\ & y_t = 2y_{t-1} - y_{t-2} + \theta_1 \epsilon_{t-1} + \theta_2 \epsilon_{t-2} + \epsilon_t . \end{aligned}$$

In ADAM, this is formulated as:

$$\begin{aligned} & y_t = v_{1,t-1} + v_{2,t-2} + \epsilon_t \\ & v_{1,t} = 2(v_{1,t-1} + v_{2,t-2}) + (2+\theta_1)\epsilon_t \\ & v_{2,t} = -(v_{1,t-1} + v_{2,t-2}) + (-1+\theta_2)\epsilon_t \end{aligned} . \tag{9.19}$$

9.1.5.4 ARIMA(1,1,2)

$$\begin{aligned} & (1-B)(1-\phi_1 B) y_t = (1+\theta_1 B + \theta_2 B^2)\epsilon_t, \\ & \text{or} \\ & y_t = (1+\phi_1) y_{t-1} - \phi_1 y_{t-2} + \theta_1 \epsilon_{t-1} + \theta_2 \epsilon_{t-2} + \epsilon_t, \end{aligned}$$

which is equivalent to:

$$\begin{aligned} & y_t = v_{1,t-1} + v_{2,t-2} + \epsilon_t \\ & v_{1,t} = (1+\phi_1)(v_{1,t-1} + v_{2,t-2}) + (1+\phi_1+\theta_1)\epsilon_t \\ & v_{2,t} = -\phi_1(v_{1,t-1} + v_{2,t-2}) + (-\phi_1+\theta_2)\epsilon_t \end{aligned} . \tag{9.20}$$

9.1.5.5 Log-ARIMA(0,1,1)

This model is equivalent to ARIMA applied to the $\log y_t$. It can be written as:

$$\begin{aligned} & (1-B)\log y_t = (1+\theta_1 B)\log(1+\epsilon_t), \\ & \text{or} \\ & \log y_t = \log y_{t-1} + \theta_1 \log(1+\epsilon_{t-1}) + \log(1+\epsilon_t). \end{aligned}$$

In ADAM, it becomes:

$$\begin{aligned} & y_t = \exp(\log v_{1,t-1} + \log(1+\epsilon_t)) \\ & \log v_{1,t} = \log v_{1,t-1} + (1+\theta_1)\log(1+\epsilon_t) \end{aligned} . \tag{9.21}$$

9.2 Recursive relation

Both additive and multiplicative ARIMA models can be written in the recursive form, similar to pure additive ETS (see Section 5.2). For the pure additive ARIMA it is:

$$y_{t+h} = \sum_{i=1}^{K} \mathbf{w}_i' \mathbf{F}_i^{\lceil \frac{h}{i} \rceil - 1} \mathbf{v}_t + \mathbf{w}_i' \sum_{j=1}^{\lceil \frac{h}{i} \rceil - 1} \mathbf{F}_i^{j-1} \mathbf{g}_i \epsilon_{t+i\lceil \frac{h}{i} \rceil - j} + \epsilon_{t+h}, \tag{9.22}$$

while for the pure multiplicative one:

$$\log y_{t+h} = \sum_{i=1}^{K} \mathbf{w}_i' \mathbf{F}_i^{\lceil \frac{h}{i} \rceil -1} \log \mathbf{v}_t + \mathbf{w}_i' \sum_{j=1}^{\lceil \frac{h}{i} \rceil -1} \mathbf{F}_i^{j-1} \mathbf{g}_i \log(1+\epsilon_{t+i\lceil \frac{h}{i} \rceil -j}) + \log(1+\epsilon_{t+h}), \tag{9.23}$$

where i corresponds to each lag of the model from 1 to K, $\mathbf{w}_i$ is the measurement vector, $\mathbf{g}_i$ is the persistence vector, both including only i-th elements, and $\mathbf{F}_i$ is the transition matrix, including only the i-th column. Based on these recursions, point forecasts can be produced from the additive and multiplicative ARIMA models, which will be, respectively:

$$\hat{y}_{t+h} = \sum_{i=1}^{K} \mathbf{w}_i' \mathbf{F}_i^{\lceil \frac{h}{i} \rceil -1} \mathbf{v}_t, \tag{9.24}$$

and:

$$\hat{y}_{t+h} = \exp \left(\sum_{i=1}^{K} \mathbf{w}_i' \mathbf{F}_i^{\lceil \frac{h}{i} \rceil -1} \log \mathbf{v}_t \right). \tag{9.25}$$

Remark. Similarly to the multiplicative ETS, the point forecasts of Log-ARIMA will not necessarily coincide with the conditional expectations. In the most general case they will correspond to the conditional geometric means. For some distributions, they can be used to get the arithmetic ones.

Based on the recursions (9.22) and (9.23), we can calculate conditional moments of ADAM ARIMA.

9.2.1 Conditional moments of ADAM ARIMA

In the case of the pure additive ARIMA, the moments correspond to the ones for ETS, discussed in Section 5.1 and follow directly from (9.22):

$$\begin{aligned} \mu_{y,t+h} = \mathrm{E}(y_{t+h}|t) = & \sum_{i=1}^{K} \left(\mathbf{w}_i' \mathbf{F}_i^{\lceil \frac{h}{i} \rceil -1} \right) \mathbf{v}_t \\ \sigma^2_h = \mathrm{V}(y_{t+h}|t) = & \left(\sum_{i=1}^{K} \left(\mathbf{w}_i' \sum_{j=1}^{\lceil \frac{h}{i} \rceil -1} \mathbf{F}_i^{j-1} \mathbf{g}_i \mathbf{g}_i' (\mathbf{F}_i')^{j-1} \mathbf{w}_i \right) + 1 \right) \sigma^2 \end{aligned}.$$

When it comes to the multiplicative ARIMA, the conditional moments would depend on the assumed distribution and might become quite complicated. Here is an example of the conditional logarithmic mean for Log-Normal distribution, assuming that $(1+\epsilon_t) \sim \log\mathcal{N}\left(\frac{\sigma^2}{2}, \sigma^2\right)$ based on (9.23):

$$\mu_{\log y,t+h} = \mathrm{E}(\log y_{t+h}|t) = \sum_{i=1}^{d} \left(\mathbf{w}_{m_i}' \mathbf{F}_{m_i}^{\lceil \frac{h}{m_i} \rceil -1} \right) \log \mathbf{v}_t - \left(\mathbf{w}_i' \sum_{j=1}^{\lceil \frac{h}{i} \rceil -1} \mathbf{F}_i^{j-1} \mathbf{g}_i + 1 \right) \frac{\sigma^2}{2}. \tag{9.26}$$

Note that the conditional logarithmic variance of the model will be the same for all Log-ARIMA models, independent of the distributional assumptions:

$$\sigma^2_{\log y,h} = \mathrm{V}(\log y_{t+h}|t) = \left(\sum_{i=1}^d \left(\mathbf{w}'_{m_i} \sum_{j=1}^{\lceil\frac{h}{m_i}\rceil-1} \mathbf{F}_{m_i}^{j-1} \mathbf{g}_{m_i} \mathbf{g}'_{m_i} (\mathbf{F}'_{m_i})^{j-1} \mathbf{w}_{m_i}\right) + 1\right) \sigma^2. \quad (9.27)$$

The obtained logarithmic moments can then be used to get the ones in the original scale, after using the connection between the moments in Log-Normal distribution. The conditional expectation and variance in this case can be calculated as:

$$\begin{aligned} \mu_{y,t+h} = \mathrm{E}(y_{t+h}|t) = \exp\left(\mu_{\log y,t+h} + \frac{\sigma^2_{\log y,h}}{2}\right) \\ \sigma^2_h = \mathrm{V}(y_{t+h}|t) = \left(\exp\left(\sigma^2_{\log y,h}\right) - 1\right) \exp\left(2 \times \mu_{\log y,t+h} + \sigma^2_{\log y,h}\right). \end{aligned} \quad (9.28)$$

Inserting the values (9.26) and (9.27) in (9.28), we will get the analytical solutions for the two moments.

If some other distributions are assumed in the model, then the conditional logarithmic mean would change because the variable $\log(1+\epsilon_t)$ would follow a different distribution with a different mean. For example:

1. Gamma: if $(1+\epsilon_t) \sim -(\sigma^{-2}, \sigma^2)$, then $\log(1+\epsilon_t) \sim \exp-(\sigma^{-2}, \sigma^2)$, which is exponential Gamma distribution, which has the following logarithmic mean: $\mathrm{E}(\log(1+\epsilon_t)) = \psi\left(\sigma^{-2}\right) + 2\log(\sigma)$;
2. Inverse Gaussian: if $(1+\epsilon_t) \sim \mathcal{IG}(1, \sigma^2)$, then $\log(1+\epsilon_t) \sim \exp\mathcal{IG}(1, \sigma^2)$, which is exponential Inverse Gaussian distribution, which does not have a simple formula for the logarithmic mean (but it can be calculated based on its connection with Generalised $\mathcal{IG}$ and formulae provided in Sichel, Dohm, & Kleingeld, 1997).

After that, similarly to how it was done for Log-Normal distribution above, the connection between the logarithmic and normal moments should be used to get the conditional expectation and variance. If these relations are not available for the distribution or are too complicated, then simulations can be used to obtain the numeric approximations (see discussion in Subsection 18.1.1).

Finally, we should remark that the formulae for the conditional moments in Log-ARIMA are complicated mainly because of the distributional assumptions inherited from ETS. However, this allows the construction of more complicated models, some of which are discussed in Section 9.4.

9.2.2 Parameters' bounds

Finally, modifying the recursions (9.22) and (9.23), we can get the stability condition for the parameters, similar to the one for the pure additive ETS from Section 5.4. The advantage of the pure multiplicative ARIMA formulated in the form (9.16) is that the adequate stability condition can be obtained in contrast with the pure multiplicative ETS models. It will be the same as the pure additive ARIMA and/or ETS. The ARIMA model will be **stable**, when the absolute values of all non-zero eigenvalues of the discount matrices $\mathbf{D}_i$ are lower than one, given that:

$$\mathbf{D}_i = \mathbf{F}_i - \mathbf{g}_i \mathbf{w}'_i. \tag{9.29}$$

Hyndman et al. (2008) show that the stability condition for SSOE models corresponds to the invertibility condition of ARIMA (Section 8.2.4), so the model can either be checked via the discount matrix (9.29) or via the MA polynomials (8.59).

When it comes to **stationarity**, state space ARIMA is always non-stationary if the differences $D_j \neq 0$ for any j. So, there needs to be a different mechanism for the stationarity check. The simplest thing to do would be to expand the AR(P_j) polynomials, ignoring all I(D_j), fill in the transition matrix $\mathbf{F}$ and then calculate its eigenvalues. If they are lower than one by absolute value, the model is stationary. The same condition can be checked via the roots of the polynomial of AR(P_j) (8.58). However, the eigenvalues approach is more computationally efficient, and I recommend using it instead of the conventional polynomials calculation, especially in case of Multiple Seasonal ARIMA.

If both stability and stationarity conditions for ARIMA are satisfied, we will call the bounds that the AR/MA parameters form "admissible", similar to how they are called in ETS. Note that ARIMA has no "usual" or "traditional" bounds.

9.3 Distributional assumptions of ADAM ARIMA

Following the same idea as in pure additive (Section 5.5) and pure multiplicative (Section 6.5) ETS models, we can have state space ARIMA with different distributional assumptions, but with the distributions aligning more appropriately with the types of models. For additive ARIMA:

1. Normal: $\epsilon_t \sim \mathcal{N}(0, \sigma^2)$;
2. Laplace: $\epsilon_t \sim \mathcal{L}(0, s)$;
3. Generalised Normal: $\epsilon_t \sim \mathcal{GN}(0, s, \beta)$;
4. S: $\epsilon_t \sim \mathcal{S}(0, s)$.

For multiplicative ARIMA:

1. Inverse Gaussian: $(1+\epsilon_t) \sim \mathcal{IG}(1, \sigma^2)$;
2. Log-Normal: $(1+\epsilon_t) \sim \mathrm{log}\mathcal{N}\left(-\frac{\sigma^2}{2}, \sigma^2\right)$;
3. Gamma: $(1+\epsilon_t) \sim -(\sigma^{-2}, \sigma^2)$.

The restrictions imposed on the model parameters are inherited from the ADAM ETS.

9.3.1 Conditional distributions

When it comes to conditional distribution of variables, additive ADAM ARIMA with the assumptions above has closed forms for all of them. For example, if we work with additive ARIMA, then according to recursive relation (9.22) from Section 9.2, the h steps ahead value follows the same distribution but with different conditional mean and variance. For example, if $\epsilon_t \sim \mathcal{GN}(0, s, \beta)$, then $y_{t+h} \sim \mathcal{GN}(\mu_{y,t+h}, s_h, \beta)$, where s_h is the conditional h steps ahead scale, found from the connection between variance and scale in Generalised Normal distribution via:

$$s_h = \sqrt{\frac{\sigma^2_h \Gamma(1/\beta)}{\Gamma(3/\beta)}}.$$

Using similar principles, we can calculate scale parameters for the other distributions.

When it comes to the multiplicative models, the conditional distribution has the closed form in case of Log-Normal (it is Log-Normal as well), but does not have it in case of Inverse Gaussian and Gamma. In the former case, the logarithmic moments can be directly used to define the parameters of distribution, i.e. if $(1+\epsilon_t) \sim \log\mathcal{N}\left(-\frac{\sigma^2}{2}, \sigma^2\right)$, then $y_{t+h} \sim \log\mathcal{N}\left(\mu_{\log y,t+h}, \sigma^2_{\log y,h}\right)$. In the other cases, simulations need to be used in order to get the quantile, cumulative and density functions.

9.4 ETS + ARIMA

Coming back to the topic of ETS and ARIMA, we can now look at it from the point of view of the SSOE state space model.

9.4.1 Pure additive models

A pure additive ETS + ARIMA model can be formulated in the general form, which we have already discussed several times in this monograph (Section 5.1):

$$\begin{aligned} y_t &= \mathbf{w}' \mathbf{v}_{t-\boldsymbol{l}} + \epsilon_t \\ \mathbf{v}_t &= \mathbf{F} \mathbf{v}_{t-\boldsymbol{l}} + \mathbf{g} \epsilon_t \end{aligned},$$

but now the matrices and vectors of the model contain ETS and ARIMA components, stacked one after another.

For example, if we want to construct ETS(A,N,A)+ARIMA(2,0,0), we can formulate this model as:

$$\begin{aligned} y_t &= l_{t-1} + s_{t-m} + v_{1,t-1} + v_{2,t-2} + \epsilon_t \\ l_t &= l_{t-1} + \alpha \epsilon_t \\ s_t &= s_{t-m} + \gamma \epsilon_t \\ v_{1,t} &= \phi_1 v_{1,t-1} + \phi_1 v_{2,t-2} + \phi_1 \epsilon_t \\ v_{2,t} &= \phi_1 v_{1,t-1} + \phi_2 v_{2,t-2} + \phi_2 \epsilon_t \end{aligned}, \tag{9.30}$$

where ϕ_i is the i-th parameter of the AR(2) part of the model. This model represented in the conventional additive SSOE state space form leads to the following matrices and vectors:

$$\begin{aligned} \mathbf{w} &= \begin{pmatrix} 1 \\ 1 \\ 1 \\ 1 \end{pmatrix}, \mathbf{F} = \begin{pmatrix} 1 & 0 & 0 & 0 \\ 0 & 1 & 0 & 0 \\ 0 & 0 & \phi_1 & \phi_1 \\ 0 & 0 & \phi_2 & \phi_2 \end{pmatrix}, \quad \mathbf{g} = \begin{pmatrix} \alpha \\ \gamma \\ \phi_1 \\ \phi_2 \end{pmatrix}, \\ \mathbf{v}_t &= \begin{pmatrix} l_t \\ s_t \\ v_{1,t} \\ v_{2,t} \end{pmatrix}, \mathbf{v}_{t-\boldsymbol{l}} = \begin{pmatrix} l_{t-1} \\ s_{t-m} \\ v_{1,t-1} \\ v_{2,t-2} \end{pmatrix} \quad \boldsymbol{l} = \begin{pmatrix} 1 \\ m \\ 1 \\ 2 \end{pmatrix} \end{aligned}. \tag{9.31}$$

So, in this formulation, the states of ETS and ARIMA are independent and form a combination of models only in the measurement equation. In a way, this model becomes similar to fitting sequentially ETS to the data and then ARIMA to the residuals, but estimating both elements at the same time. This simultaneous estimation is supposed to remove the bias in the estimates of parameters, which might appear in the sequential procedure.

ADAM introduces the flexibility necessary for fitting any ETS+ARIMA combination, but not all combinations make sense. For example, here how ETS(A,N,N)+ARIMA(0,1,1) would look like:

$$\begin{aligned} y_t &= l_{t-1} + v_{1,t-1} + \epsilon_t \\ l_t &= l_{t-1} + \alpha \epsilon_t \\ v_{1,t} &= v_{1,t-1} + (1 + \theta_1) \epsilon_t \end{aligned}. \tag{9.32}$$

In the transition part of the model (9.32), the two equations duplicate each other because they have exactly the same mechanism of update of states. In fact, as we know from Section 8.4.1, ETS(A,N,N) and ARIMA(0,1,1) are equivalent, when $\alpha = 1 + \theta_1$. If we estimate this model, then we are duplicating the state, splitting it into two parts with some arbitrary weights. This becomes apparent if we insert the transition equations in the measurement one, obtaining:

$$\begin{aligned} y_t = & l_{t-2} + \alpha \epsilon_{t-1} + v_{1,t-2} + (1 + \theta_1) \epsilon_{t-1} + \epsilon_t = \\ & l_{t-2} + v_{1,t-2} + (1 + \theta_1 + \alpha) \epsilon_{t-1} + \epsilon_t \end{aligned}, \tag{9.33}$$

which leads to an infinite combination of values of parameters θ and α that would produce exactly the same model fit. So, the model (9.32) does not have unique parameters and thus is **unidentifiable**. This means that we cannot reach the "true model" based on ETS(A,N,N)+ARIMA(0,1,1) and thus the model selection via information criteria becomes inappropriate. Furthermore, the estimates of parameters of such model might become biased, inefficient and inconsistent due to the "infinite combination" issue mentioned above.

In some other cases, some parts of the model might be duplicated (not the whole), making the overall model unidentifiable, so it makes sense to switch to either ETS or ARIMA, depending on circumstances. For example, if we have ETS(A,A,N)+ARIMA(0,2,3), then some parts of the models will be duplicated (because ETS(A,A,N) is equivalent to ARIMA(0,2,2)), so it would be more reasonable to switch to pure ARIMA(0,2,3) instead. On the other hand, if we deal with ETS(A,Ad,N)+ARIMA(0,1,2), then dropping the ARIMA part would be more appropriate (due to the relation between ETS(A,Ad,N) and ARIMA(1,1,2)).

These examples show that, when using ETS+ARIMA, model building needs to be done with care, not to get an unreasonable model that cannot be identified. As a general recommendation, keep the ETS and ARIMA connection in mind (see Section 8.4), when deciding, what to construct. And here is a short list of guidelines of what to do in some special cases:

1. For ETS(A,N,N)+ARIMA(0,1,q):
    - use ARIMA(0,1,q) in case of $q > 1$;
    - use ETS(A,N,N) in case of $q \leq 1$;
2. For ETS(A,A,N)+ARIMA(0,2,q):
    - use ARIMA(0,2,q) in case of $q > 2$;
    - use ETS(A,A,N) in case of $q \leq 2$;
3. For ETS(A,Ad,N)+ARIMA(p,1,q):
    - use ARIMA(p,1,q), when either $p > 1$ or $q > 2$;
    - use ETS(A,Ad,N), when $p \leq 1$ and $q \leq 2$.

Regarding seasonal models, the relation between ETS and ARIMA is more complex. It is highly improbable to get to equivalent ARIMA models, so we can neglect the rules for the seasonal part and focus on making sure that the three rules above hold for the non-seasonal part of the model.

9.4.2 Pure multiplicative models

In the most general case the pure multiplicative ETS+ARIMA model can be written as (based on (6.1) and (9.16)):

$$\begin{aligned} y_t &= \exp\left(\mathbf{w}'_E \log \mathbf{v}_{E,t-\boldsymbol{l}_E} + \mathbf{w}'_A \log \mathbf{v}_{A,t-\boldsymbol{l}_A} + \log(1+\epsilon_t)\right) \\ \log \mathbf{v}_{E,t} &= \mathbf{F}_E \log \mathbf{v}_{E,t-\boldsymbol{l}_E} + \log(\mathbf{1}_k + \mathbf{g}_E \epsilon_t) \\ \log \mathbf{v}_{A,t} &= \mathbf{F}_A \log \mathbf{v}_{A,t-\boldsymbol{l}_A} + \mathbf{g}_A \log(1+\epsilon_t) \end{aligned}, \qquad (9.34)$$

where the subscript reflects, which part corresponds to which model: "E" – ETS, "A" – ARIMA. The formulation (9.34) demonstrates that the ETS+ARIMA might not have the same issues as the pure additive one. This is because the multiplicative ETS (Section 6.1) and multiplicative ARIMA (Subsection 9.1.4) are formulated differently. Consider an example of ETS(M,N,N)+Log-ARIMA(0,1,1), which is formulated as:

$$\begin{aligned} y_t &= l_{t-1} v_{1,t-1} (1 + \epsilon_t) \\ l_t &= l_{t-1} (1 + \alpha \epsilon_t) \\ \log v_{1,t} &= \log v_{1,t-1} + (1 + \theta_1) \log(1 + \epsilon_t) \end{aligned} . \tag{9.35}$$

The last equation in (9.35) can be rewritten as $v_{1,t} = v_{1,t-1}(1 + \epsilon_t)^{(1+\theta_1)}$, demonstrating the difference between the transition equation of ETS(M,N,N) and multiplicative ARIMA(0,1,1). Still, the two models will be similar in cases when α is close to zero and (respectively) θ is close to -1. So this combination of models should be treated with care, along with other potentially similar combinations. The following combinations of the two models can be considered as potentially unidentifiable under some conditions:

1. ETS(M,N,N)+Log-ARIMA(0,1,1);
2. ETS(M,M,N)+Log-ARIMA(0,2,2);
3. ETS(M,Md,N)+Log-ARIMA(1,1,2).

In addition, the recommendations discussed for the pure additive ETS+ARIMA can be applied here for the pure multiplicative ETS+ARIMA to guarantee that the resulting model is identifiable in all cases.

Finally, mixing additive ETS with multiplicative ARIMA or multiplicative ETS with additive ARIMA does not make sense from the modelling point of view. It only complicates the model building process, so we do not consider these exotic cases in this book, although they are theoretically possible.

9.5 Examples of application

Building upon the example with `AirPassengers` data from Section 8.5.2, we will construct several multiplicative ARIMA models and see which one is the most appropriate for the data. As a reminder, the best additive ARIMA model was SARIMA$(0,2,2)(1,1,1)_{12}$, which had AICc of 1065.427. We will do something similar here, using Log-Normal distribution, thus working with Log-ARIMA. To understand what model can be used in this case, we can take the logarithm of data and see what happens with the components of time series:

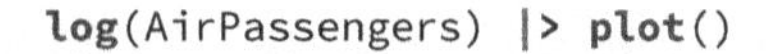

```
log(AirPassengers) |> plot()
```

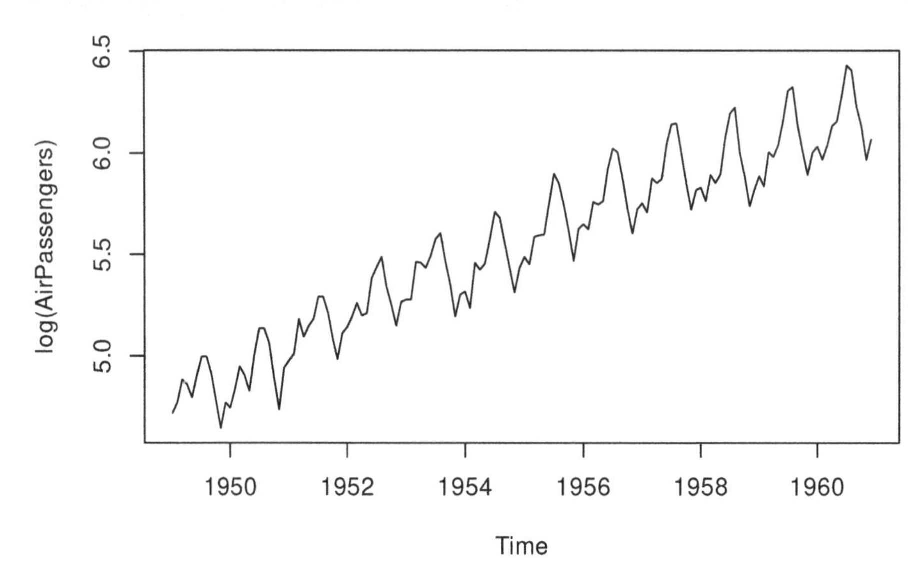

We still have the trend in the data, and the seasonality now corresponds to the additive rather than the multiplicative (as expected). While we might still need the second differences for the non-seasonal part of the model, taking the first differences for the seasonal should suffice because the logarithmic transform will take care of the expanding seasonal pattern in the data. So we can test several models with different options for ARIMA orders:

```
adamLogSARIMAAir <- vector("list",3)
# logSARIMA(0,1,1)(0,1,1)[12]
adamLogSARIMAAir[[1]] <-
  adam(AirPassengers, "NNN", lags=c(1,12),
       orders=list(ar=c(0,0), i=c(1,1), ma=c(1,1)),
       h=12, holdout=TRUE, distribution="dlnorm")
# logSARIMA(0,2,2)(0,1,1)[12]
adamLogSARIMAAir[[2]] <-
  adam(AirPassengers, "NNN", lags=c(1,12),
       orders=list(ar=c(0,0), i=c(2,1), ma=c(2,2)),
       h=12, holdout=TRUE, distribution="dlnorm")
# logSARIMA(1,1,2)(0,1,1)[12]
adamLogSARIMAAir[[3]] <-
  adam(AirPassengers, "NNN", lags=c(1,12),
       orders=list(ar=c(1,0), i=c(1,1), ma=c(2,1)),
       h=12, holdout=TRUE, distribution="dlnorm")
names(adamLogSARIMAAir) <- c("logSARIMA(0,1,1)(0,1,1)[12]",
```

```
                          "logSARIMA(0,2,2)(0,1,1)[12]",
                          "logSARIMA(1,1,2)(0,1,1)[12]")
```

The thing that is different between the models is the non-seasonal part. Using the connection with ETS (discussed in Section 8.4), the first model should work on local level data, the second should be optimal for the local trend series, and the third one is placed somewhere in between the two. We can compare the models using AICc:

```
sapply(adamLogSARIMAAir, AICc)
```

```
## logSARIMA(0,1,1)(0,1,1)[12] logSARIMA(0,2,2)(0,1,1)[12]
##                    991.4084                   1159.7709
## logSARIMA(1,1,2)(0,1,1)[12]
##                   1012.8962
```

It looks like the logSARIMA(0,1,1)(0,1,1)$_{12}$ is more appropriate for the data. In order to make sure that we did not miss anything, we analyse the residuals of this model (Figure 9.1):

```
par(mfcol=c(2,1), mar=c(2,2,2,1))
plot(adamLogSARIMAAir[[1]], which=10:11)
```

We can see that there are no significant coefficients on either the ACF or PACF, so there is nothing else to improve in this model (we discuss this in more detail in Section 14.5). We can then produce a forecast from the model and see how it performed on the holdout sample (Figure 9.2):

```
forecast(adamLogSARIMAAir[[1]], h=12, interval="prediction") |>
    plot(main=paste0(adamLogSARIMAAir[[1]]$model,
                     " with Log-Normal distribution"))
```

The ETS model closest to the logSARIMA(0,1,1)(0,1,1)$_{12}$ would probably be ETS(M,M,M), because the former has both seasonal and non-seasonal differences (see discussion in Subsection 8.4.4):

```
adamETSAir <- adam(AirPassengers, "MMM", h=12, holdout=TRUE)
adamETSAir
```

```
## Time elapsed: 0.29 seconds
## Model estimated using adam() function: ETS(MMM)
## Distribution assumed in the model: Gamma
## Loss function type: likelihood; Loss function value: 468.5176
```

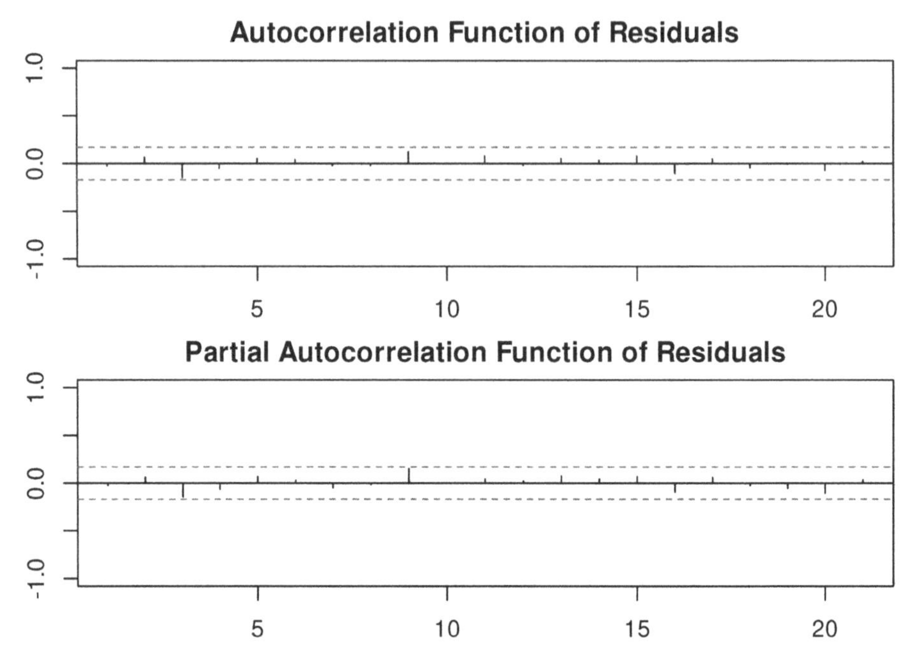

FIGURE 9.1 ACF and PACF of logSARIMA(0,1,1)(0,1,1)$_{12}$.

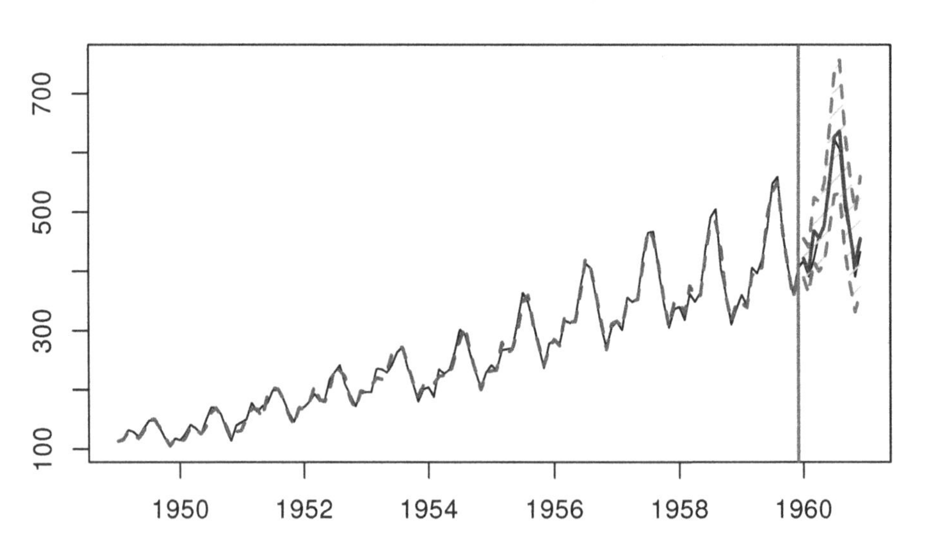

FIGURE 9.2 Forecast from logSARIMA(0,1,1)(0,1,1)$_{12}$.

```
## Persistence vector g:
##  alpha   beta  gamma
## 0.7684 0.0206 0.0000
##
## Sample size: 132
## Number of estimated parameters: 17
## Number of degrees of freedom: 115
## Information criteria:
##       AIC      AICc       BIC      BICc
##  971.0351  976.4036 1020.0428 1033.1492
##
## Forecast errors:
## ME: -5.617; MAE: 15.496; RMSE: 21.938
## sCE: -25.677%; Asymmetry: -23.1%; sMAE: 5.903%; sMSE: 0.698%
## MASE: 0.643; RMSSE: 0.7; rMAE: 0.204; rRMSE: 0.213
```

Comparing information criteria, ETS(M,M,M) should be preferred to Log-ARIMA, but in terms of accuracy on the holdout, Log-ARIMA is more accurate than ETS on this data:

```
adamLogSARIMAAir[[1]]
```

```
## Time elapsed: 0.5 seconds
## Model estimated using adam() function: SARIMA(0,1,1)[1](0,1,1)[12]
## Distribution assumed in the model: Log-Normal
## Loss function type: likelihood; Loss function value: 477.339
## ARMA parameters of the model:
## MA:
##  theta1[1] theta1[12]
##    -0.2287    -0.4981
##
## Sample size: 132
## Number of estimated parameters: 16
## Number of degrees of freedom: 116
## Information criteria:
##       AIC      AICc       BIC      BICc
##  986.6780  991.4084 1032.8028 1044.3517
##
## Forecast errors:
## ME: -14.783; MAE: 15.664; RMSE: 20.506
## sCE: -67.583%; Asymmetry: -92.9%; sMAE: 5.967%; sMSE: 0.61%
## MASE: 0.65; RMSSE: 0.654; rMAE: 0.206; rRMSE: 0.199
```

If we decide to stick with the information theory approach, we should use ETS(M,M,M). If we are more inclined towards empirical selection, we would need to apply the models in the rolling origin fashion (Section 2.4), collect a distribution of errors, and then decide, which one to choose.

10

Explanatory variables in ADAM

In real life, the need for explanatory variables arises when there are some external factors that have relation with the response variable and impact the final forecasts and their accuracy. Examples of such variables in the demand forecasting context include price changes, promotional activities, temperature, etc. In some cases, the changes in these factors would not substantially influence the demand, but this does not apply universally to the problem. If we omit this information from the model, this will be damaging for both point forecasts and prediction intervals (see discussion in Chapter 15 of Svetunkov, 2022).

While the inclusion of explanatory variables in the context of ARIMA models is a relatively well-studied topic (for example, this was discussed by Box & Jenkins, 1976), in the case of ETS, there is only Chapter 9 in Hyndman et al. (2008) and a handful of papers. Koehler, Snyder, Ord, and Beaumont (2012) discuss the mechanism of detection and approximation of outliers via an ETSX model (ETS with explanatory variables). The authors show that if an outlier appears at the end of the series, it will seriously impact the final forecast and needs to be modelled correctly. However, if it appears either in the middle or at the beginning of the series, the impact on the final forecast is typically negligible. Kourentzes and Petropoulos (2016) used ETSX successfully for promotional modelling, demonstrating that it outperforms the conventional ETS in terms of point forecasts accuracy in cases when promotions happen. So, the inclusion of explanatory variables in dynamic models is not just a nice feature, but in some situations is a necessity, which helps improve the forecasting accuracy.

In ADAM, the state-space model (7.1) can be easily extended by including additional components and explanatory variables. This chapter discusses the main aspects of ADAM with explanatory variables, how it is formulated, and how the more advanced models can be built upon it. Furthermore, the parameters for these additional components can either be fixed (static) or change over time (dynamic). We discuss both in the following sections. We also show that the stability and forecastability conditions, discussed in Section 5.4 for the pure additive ETS model, will be different in the case of the ETSX model and that the classical definitions should be updated to cater for the introduction of the explanatory variables. We also briefly discuss the inclusion of categorical variables in the ETSX model and show that the seasonal ETS models can be considered as special cases of ADAM ETSX with dummy variables.

Furthermore, we will use the term "**deterministic**" explanatory variable to denote the situations when the values of variables are known in advance or can be controlled by us. An example is the price of a product or a promotion that we decide to have. On the contrary, we will use the term "**stochastic**" explanatory variable for the cases, when its future value is not known and is beyond our control. An example of this variable is the temperature, which we cannot control and do not know for sure in advance. Usage of deterministic variables in dynamic models might differ from the usage of the stochastic ones.

As a final note, we will carry out the discussion of the topic in this Chapter on the example of ADAM ETSX, keeping in mind that the same principles will hold for ADAM ARIMAX because the two are formulated in the same framework. We will call the more general dynamic model (encompassing ETS and/or ARIMA) with explanatory variables "ADAMX" in this and further chapters.

10.1 ADAMX: Model formulation

As discussed previously, there are two types of error terms in ADAM:

1. Additive, discussed in Chapter 5 in the case of ETS and Chapter 9 for ARIMA;
2. Multiplicative, covered in Chapter 6 for ETS and in Subsection 9.1.4 for ARIMA.

The inclusion of explanatory variables in ADAMX is determined by the type of the error, so that in case of (1) the measurement equation of the model is:

$$y_t = a_{0,t-1} + a_{1,t-1} x_{1,t} + a_{2,t-1} x_{2,t} + \dots + a_{n,t-1} x_{n,t} + \epsilon_t, \tag{10.1}$$

where $a_{0,t-1}$ is the point value based on all ETS components (for example, $a_{0,t-1} = l_{t-1}$ in case of ETS(A,N,N)), $x_{i,t}$ is the i-th explanatory variable, $a_{i,t-1}$ is its parameter, and n is the number of explanatory variables. We will denote the estimated parameters of such models as $\hat{a}_{i,t-1}$. In the simple case, the transition equation for such a model would imply that the parameters $a_{i,t}$ do not change over time and are equal to some fixed value:

$$a_{i,t} = a_{i,t-1} = \dots = a_{i,0} \text{ for all } i = 1, \dots, n. \tag{10.2}$$

Various complex mechanisms for the states update can be proposed instead of (10.2), but we do not discuss them at this point. Typically, the initial values of parameters would be estimated at the optimisation stage, either based on likelihood or some other loss function, so the index t can be dropped, substituting $a_{i,t} = a_i$ for all $i = 1, \dots, n$.

When it comes to the multiplicative error model, it should be formulated differently. The most straight forward would be to formulate the model in logarithms in order to linearise it:

$$\log y_t = \log a_{0,t-1} + a_{1,t-1}x_{1,t} + a_{2,t-1}x_{2,t} + \dots + a_{n,t-1}x_{n,t} + \log(1+\epsilon_t). \quad (10.3)$$

Remark. If a log-log model is required, all that needs to be done, is to substitute $x_{i,t}$ with $\log x_{i,t}$.

The model (10.3) aligns with both pure multiplicative ETS and ARIMA, discussed, respectively, in Chapter 6 and in Subsection 9.1.4.

The compact form of the ADAMX model implies that the explanatory variables $x_{i,t}$ are included in the measurement vector $\mathbf{w}_t$, making it change over time. The parameters are then moved to the state vector, and a diagonal matrix is added to the existing transition matrix to reflect the updating mechanism (10.2). Finally, the persistence vector for the parameters of explanatory variables should contain zeroes, because for now we assume that the parameters do not change over time. The pure additive state space model, in that case, can be represented as:

$$\begin{aligned} y_t &= \mathbf{w}'_t \mathbf{v}_{t-\boldsymbol{l}} + \epsilon_t \\ \mathbf{v}_t &= \mathbf{F} \mathbf{v}_{t-\boldsymbol{l}} + \mathbf{g} \epsilon_t \end{aligned}, \quad (10.4)$$

while the pure multiplicative models is:

$$\begin{aligned} y_t &= \exp\left(\mathbf{w}'_t \log \mathbf{v}_{t-\boldsymbol{l}} + \log(1+\epsilon_t)\right) \\ \log \mathbf{v}_t &= \mathbf{F} \log \mathbf{v}_{t-\boldsymbol{l}} + \log(\mathbf{1}_k + \mathbf{g}\epsilon_t) \end{aligned}. \quad (10.5)$$

So, the only thing that changes in these models in comparison with the conventional ones in Chapters 5 and 6 is the time varying measurement vector $\mathbf{w}'_t$ instead of the fixed one. For example, in the case of ETSX(A,Ad,A) we will have:

$$\mathbf{F} = \begin{pmatrix} 1 & \phi & 0 & 0 & \dots & 0 \\ 0 & \phi & 0 & 0 & \dots & 0 \\ 0 & 0 & 1 & 0 & \dots & 0 \\ 0 & 0 & 0 & 1 & \dots & 0 \\ \vdots & \vdots & \vdots & \vdots & \ddots & \vdots \\ 0 & 0 & 0 & 0 & \dots & 1 \end{pmatrix}, \mathbf{w}_t = \begin{pmatrix} 1 \\ \phi \\ 1 \\ x_{1,t} \\ \vdots \\ x_{n,t} \end{pmatrix}, \quad \mathbf{g} = \begin{pmatrix} \alpha \\ \beta \\ \gamma \\ 0 \\ \vdots \\ 0 \end{pmatrix}, \\ \mathbf{v}_t = \begin{pmatrix} l_t \\ b_t \\ s_t \\ a_{1,t} \\ \vdots \\ a_{n,t} \end{pmatrix}, \quad \boldsymbol{l} = \begin{pmatrix} 1 \\ 1 \\ m \\ 1 \\ \vdots \\ 1 \end{pmatrix} \tag{10.6}$$

which is equivalent to the set of equations:

$$\begin{aligned} & y_t = l_{t-1} + \phi b_{t-1} + s_{t-m} + a_{1,t-1} x_{1,t} + \dots + a_{n,t-1} x_{n,t} + \epsilon_t \\ & l_t = l_{t-1} + \phi b_{t-1} + \alpha \epsilon_t \\ & b_t = \phi b_{t-1} + \beta \epsilon_t \\ & s_t = s_{t-m} + \gamma \epsilon_t \\ & a_{1,t} = a_{1,t-1} \\ & \vdots \\ & a_{n,t} = a_{n,t-1} \end{aligned} . \tag{10.7}$$

Alternatively, the state, measurement, and persistence vectors and transition matrix can be split, each into two parts, separating the ETS and X parts in the state space equations:

$$\begin{aligned} & y_t = \mathbf{w}' \mathbf{v}_{t-\boldsymbol{l}} + \mathbf{x}'_t \mathbf{a}_{t-1} + \epsilon_t \\ & \mathbf{v}_{1,t} = \mathbf{F} \mathbf{v}_{t-\boldsymbol{l}} + \mathbf{g} \epsilon_t \\ & \mathbf{a}_t = \mathbf{a}_{t-1} \end{aligned} , \tag{10.8}$$

where $\mathbf{w}$, $\mathbf{F}$, $\mathbf{g}$ and $\mathbf{v}_t$ contain the elements of the conventional components of ADAM, and $\mathbf{a}_t$ is the vector of parameters for the explanatory variables.

When all the smoothing parameters of the ETS part of the model are equal to zero, the ETSX reverts to a deterministic model, becoming just a multiple linear regression. For example, in case of ETSX(A,N,N) with $\alpha = 0$ we get:

$$\begin{aligned} & y_t = l_{t-1} + a_{1,t-1} x_{1,t} + \dots + a_{n,t-1} x_{n,t} + \epsilon_t \\ & l_t = l_{t-1} \\ & a_{1,t} = a_{1,t-1} \\ & \vdots \\ & a_{n,t} = a_{n,t-1} \end{aligned} , \tag{10.9}$$

where $l_t = a_0$ is the intercept of the model. (10.9) can be rewritten then in the conventional way, dropping the transition part of the state space model:

$$y_t = a_0 + a_1 x_{1,t} + \dots + a_n x_{n,t} + \epsilon_t. \tag{10.10}$$

In the case of models with trend and/or seasonal components, the model becomes equivalent to the regression with deterministic trend and/or seasonality. This means that, in general, ADAMX implies that we are dealing with a regression with a time-varying intercept, where the principles of this variability are defined by the ADAM components (e.g. intercept can vary seasonally). Similar properties are obtained with the multiplicative error model. The main difference is that the specific impact of explanatory variables on the response variable will vary with the intercept changes. The model, in this case, combines the strengths of the multiplicative regression and the dynamic model, where the variability of the response variable changes with the change of the baseline model (ADAM ETS and/or ADAM ARIMA in this case).

10.2 Conditional expectation and variance of ADAMX

10.2.1 ADAMX with deterministic explanatory variables

The conventional ETS and ARIMA models have a severe limitation, which will be discussed in Chapter 16: it assumes that the model's parameters are known, i.e. there is no variability in them and that the in-sample estimates are fixed no matter how the sample size changes. While in the case of point forecasts, this is not important, this affects the conditional variance and prediction intervals, which appear too narrow in many studies (e.g. Athanasopoulos, Hyndman, Song, & Wu, 2011). In case of regression, this limitation is lifted: the uncertainty of parameters in it translates to the final uncertainty in the confidence/prediction interval. ADAMX, having both dynamic (ETS/ARIMA) and static (regression) parts, has the limitation of the former, which can be resolved only with more complicated approaches (Section 16.5). As a result, the conditional mean and variance of the conventional ADAMX assume that the parameters $a_0, \dots a_n$ are known, leading to the following formulae in the case of the pure additive model, based on what was discussed in Section 5.3:

$$\begin{aligned} \mu_{y,t+h} = \mathrm{E}(y_{t+h}|t) = & \sum_{i=1}^{d} \left(\mathbf{w}'_{m_i,t} \mathbf{F}_{m_i}^{\lceil \frac{h}{m_i} \rceil -1}\right) \mathbf{v}_{t} \\ \mathrm{V}(y_{t+h}|t) = & \left(\sum_{i=1}^{d} \left(\mathbf{w}'_{m_i,t} \sum_{j=1}^{\lceil \frac{h}{m_i} \rceil -1} \mathbf{F}_{m_i}^{j-1} \mathbf{g}_{m_i} \mathbf{g}'_{m_i} (\mathbf{F}'_{m_i})^{j-1} \mathbf{w}_{m_i,t} \right) + 1 \right) \sigma^2 \end{aligned}, \tag{10.11}$$

the main difference from the moments of the conventional model (from Section 5.3) being the index t in the measurement vector $\mathbf{w}_t$. As an example, here is how the two statistics will look in the case of ETSX(A,N,N):

$$\begin{aligned} \mu_{y,t+h} = \mathrm{E}(y_{t+h}|t) = & l_t + \sum_{i=1}^{n} a_i x_{i,t+h} \\ \mathrm{V}(y_{t+h}|t) = & \left((h-1)\alpha^2 + 1\right) \sigma^2 \end{aligned}, \tag{10.12}$$

where the variance ignores the potential variability rising from the explanatory variables because of the limitations discussed above. The formulae assume that the future values of explanatory variables $x_{i,t}$ are known (the variable is deterministic). As a result, the prediction and confidence intervals for the ADAMX would typically be narrower than expected and would only be adequate in cases of large samples, where the Law of Large Numbers would start working (Section 6.1 of Svetunkov, 2022), reducing the variance of parameters (this is assuming that the typical assumptions of the model from Subsection 1.4.1 hold).

10.2.2 ADAMX with stochastic explanatory variables

Note that the ADAMX works well in cases when the future values of $x_{i,t+h}$ are known. It is a realistic assumption when we control the explanatory variables (e.g. prices and promotions for our product). But when the variables are out of our control, they need to be forecasted somehow. In this case we are assuming that each $x_{i,t}$ is a stochastic variable with some dynamic conditional one step ahead expectation $\mu_{x_{i,t}}$ and a one step ahead variance $\sigma^2_{x_{i,1}}$.

Remark. In this case, we treat the available explanatory variables as models on their own, not just as values given to us from above. This assumption of randomness will change the conditional moments of the model.

Here is what we will have for the moments of the model in the case of ETSX(A,N,N) (given that the typical assumptions from Subsection 1.4.1

hold):

$$
\begin{aligned}
\mu_{y,t+h} = \mathrm{E}(y_{t+h}|t) = l_t + \sum_{i=1}^n a_i \mu_{x_{i,t+h}} \\
\mathrm{V}(y_{t+h}|t) = \left((h-1)\alpha^2 + 1\right) \sigma^2 + \sum_{i=1}^n a_i^2 \sigma^2_{x_{i,h}} + 2 \sum_{i=1}^{n-1} \sum_{j=i+1}^{n} a_i a_j \sigma_{x_{i,h},x_{j,h}}
\end{aligned} , \tag{10.13}
$$

where $\sigma^2_{x_{i,h}}$ is the conditional variance of x_i h steps ahead, $\sigma_{x_{i,h},x_{j,h}}$ is the h steps ahead covariance between the explanatory variables $x_{i,t+h}$ and $x_{j,t+h}$, both conditional on the information available at the observation t. Similarly, if we are interested in one step ahead point forecast from the model, it should take the randomness of explanatory variables into account and become:

$$
\begin{aligned}
\mu_{y,t|t-1} = & \mathrm{E}\left(l_{t-1} + \sum_{i=1}^n a_i x_{i,t} + \epsilon_t \middle| t-1 \right) = \\
= & l_{t-1} + \sum_{i=1}^n a_i \mu_{x_{i,t}}
\end{aligned} . \tag{10.14}
$$

So, in the case of ADAMX with random explanatory variables, the model should be constructed based on the expectations of those variables, not the random values themselves. This explains, for example, why Athanasopoulos et al. (2011) found that some models with predicted explanatory variables work better than the models with the original variables. This means that when estimating the model, such as ETS(A,N,N), the following should be constructed:

$$
\begin{aligned}
& \hat{y}_t = \hat{l}_{t-1} + \sum_{i=1}^n \hat{a}_{i,t-1} \hat{x}_{i,t} \\
& e_t = y_t - \hat{y}_t \\
& \hat{l}_t = \hat{l}_{t-1} + \hat{\alpha} e_t \\
& \hat{a}_{i,t} = \hat{a}_{i,t-1} \text{ for each } i \in \{1, \dots, n\}
\end{aligned} , \tag{10.15}
$$

where $\hat{x}_{i,t}$ is the in-sample conditional one step ahead mean for the explanatory variable x_i.

Finally, as discussed previously, the conditional moments for the pure multiplicative and mixed models do not generally have closed forms, implying that the simulations need to be carried out. The situation becomes more challenging in the case of random explanatory variables because that randomness needs to be introduced in the model itself and propagated throughout the time series. This is not a trivial task, which has not been resolved yet.

10.3 Dynamic X in ADAMX

Remark. The model discussed in this section assumes particular dynamics of parameters, aligning with what the conventional ETS assumes: regression parameters are correlated with the states of the model. It does not treat parameters as independent as, for example, MSOE state space models do, which makes this model restrictive in its application. But this type of model works well with categorical variables, as I show later in this section.

As discussed in Section 10.1, the parameters of the explanatory variables in ADAMX can be assumed to be constant over time or can be assumed to vary according to some mechanism. The most reasonable one in the SSOE framework relies on the same error for different components of the model because this mechanism aligns with the model itself. Osman and King (2015) proposed one such mechanism, relying on the differences of the data. The primary motivation of their approach was to make the dynamic ETSX model stable, which is a challenging task. However, this mechanism relies on the assumption of non-stationarity of the explanatory variables, which does not always make sense (for example, it is unreasonable in the case of promotional data). An alternative approach discussed in this section is the one initially proposed by Svetunkov (1985), based on the stochastic approximation mechanism and further developed in Svetunkov and Svetunkov (2014).

We start with the following linear regression model:

$$y_t = a_{0,t-1} + a_{1,t-1} x_{1,t} + \dots + a_{n,t-1} x_{n,t} + \epsilon_t, \tag{10.16}$$

where all parameters vary over time and $a_{0,t}$ represents the value from the conventional additive error ETS model (e.g. level of series, i.e. $a_{0,t} = l_t$). The updating mechanism for the parameters in this case is straight forward and relies on the ratio of the error term and the respective explanatory variables:

$$a_{i,t} = a_{i,t-1} + \left \lbrace \begin{aligned} & \delta_i \frac{\epsilon_t}{x_{i,t}} \text{ for each } i \in \{1, \dots, n\}, \text{ if } x_{i,t} \neq 0 \\ & 0 \text{ otherwise} \end{aligned} \right. , \tag{10.17}$$

where δ_i is the smoothing parameter of the i-th explanatory variable. The same model can be represented in the state space form, based on the equations, similar to (10.4):

$$\begin{aligned} & y_{t} = \mathbf{w}'_t \mathbf{v}_{t-\boldsymbol{l}} + \epsilon_t \\ & \mathbf{v}_{t} = \mathbf{F} \mathbf{v}_{t-\boldsymbol{l}} + \mathbf{z}_t \mathbf{g} \epsilon_t \end{aligned}, \tag{10.18}$$

where $\mathbf{z}_t = \mathrm{diag}\left(\mathbf{w}_t\right)^{-1} = \left(\mathbf{I}_{k+n} \odot (\mathbf{w}_t \mathbf{1}'_{k+n})\right)^{-1}$ is the diagonal matrix consisting of inverses of explanatory variables, $\mathbf{I}_{k+n}$ is the identity matrix for

k components and n explanatory variables, and $\odot$ is Hadamard product for element-wise multiplication. This is the inverse of the diagonal matrix based on the measurement vector, for which those values that cannot be inverted (due to division by zero) are substituted by zeroes in order to reflect the condition in (10.17). In addition to what (10.4) contained, we add smoothing parameters δ_i in the persistence vector $\mathbf{g}$ for each of the explanatory variables.

If the error term is multiplicative, then the model changes to:

$$\begin{aligned} y_t &= \exp\left(a_{0,t-1} + a_{1,t-1}x_{1,t} + \dots + a_{n,t-1}x_{n,t} + \log(1+\epsilon_t)\right) \\ a_{i,t} &= a_{i,t-1} + \begin{cases} \delta_i \frac{\log(1+\epsilon_t)}{x_{i,t}} & \text{for each } i \in \{1, \dots, n\}, \text{ if } x_{i,t} \neq 0 \\ 0 & \text{otherwise} \end{cases} . \end{aligned} \quad (10.19)$$

The formulation (10.19) differs from the conventional pure multiplicative ETS model because the smoothing parameter δ_i is not included inside the error term $1 + \epsilon_t$, which simplifies some derivations and makes the model easier to work with (it has some similarities to logARIMA from Subsection 9.1.4).

Note that if it is suspected that the explanatory variables exhibit non-stationarity and are not cointegrated with the response variable, then their differences can be used instead of $x_{i,t}$ in (10.18) and (10.19). In this case, the additive model would coincide with the one proposed by Osman and King (2015). However, the decision of taking the differences for the different parts of the model should be made based on each specific situation, not across the whole set of variables. Here is an example of the ETSX(A,N,N) model with differenced explanatory variables:

$$\begin{aligned} y_t &= a_{0,t-1} + a_{1,t-1}\Delta x_{1,t} + \dots + a_{n,t-1}\Delta x_{n,t} + \epsilon_t, \\ a_{i,t} &= a_{i,t-1} + \begin{cases} \delta_i \frac{\epsilon_t}{\Delta x_{i,t}} & \text{for each } i \in \{1, \dots, n\}, \text{ if } \Delta x_{i,t} \neq 0 \\ 0 & \text{otherwise} \end{cases} , \end{aligned} \quad (10.20)$$

where $\Delta x_{i,t} = x_{i,t} - x_{i,t-1}$ is the differences of the i-th exogenous variable.

Finally, to distinguish the ADAMX with static parameters from the one with dynamic ones, we will use the letters "S" and "D" in the names of models. So, the model (10.9) can be called ETSX(A,N,N){S}, while the model (10.19), assuming that $a_{0,t-1} = l_{t-1}$, would be called ETSX(M,N,N){D}. We use curly brackets to split the ETS states from the type of X. Furthermore, given that the model with static regressors is assumed in many contexts to be the default one, the ETSX(*,*,*){S} model can also be denoted as just ETSX(*,*,*).

10.3.1 Recursion for dynamic ADAMX

Similarly to how it was discussed in Subsection 10.2.2, we can have two cases in the dynamic model: (1) deterministic explanatory variables, (2) stochastic explanatory variables. For illustrative purposes, we will use a non-seasonal model for which the lag vector $\boldsymbol{l}$ contains ones only, keeping in mind that other pure additive models can be easily used instead with slight changes in formulae. The cases of non-additive ETS models are complicated and are not discussed in this monograph. So, as discussed previously, the model can be written in the following general way:

$$\begin{aligned} y_t &= \mathbf{w}'_t \mathbf{v}_{t-1} + \epsilon_t \\ \mathbf{v}_t &= \mathbf{F} \mathbf{v}_{t-1} + \mathbf{z}_t \mathbf{g} \epsilon_t \end{aligned}. \tag{10.21}$$

Based on this model, we can get the recursive relation for h steps ahead, similar to how it was done in Section 5.2:

$$\begin{aligned} y_{t+h} &= \mathbf{w}'_{t+h} \mathbf{v}_{t+h-1} + \epsilon_{t+h} \\ \mathbf{v}_{t+h-1} &= \mathbf{F} \mathbf{v}_{t+h-2} + \mathbf{z}_{t+h-1} \mathbf{g} \epsilon_{t+h-1} \end{aligned}, \tag{10.22}$$

where the second equation can be expressed via matrices and vectors using the values available on observations from t to $t+h-1$:

$$\mathbf{v}_{t+h-1} = \mathbf{F}^{h-1} \mathbf{v}_t + \sum_{j=1}^{h-1} \mathbf{F}^{h-1-j} \mathbf{z}_{t+j} \mathbf{g} \epsilon_{t+j}. \tag{10.23}$$

Substituting the equation (10.23) in the measurement equation of (10.22) leads to the final recursion:

$$y_{t+h} = \mathbf{w}'_{t+h} \mathbf{F}^{h-1} \mathbf{v}_t + \mathbf{w}'_{t+h} \sum_{j=1}^{h-1} \mathbf{F}^{h-1-j} \mathbf{z}_{t+j} \mathbf{g} \epsilon_{t+j} + \epsilon_{t+h}, \tag{10.24}$$

which can be used for the derivation of moments of ADAMX{D}.

10.3.2 Conditional moments for deterministic explanatory variables in ADAMX{D}

Based on the recursion (10.24), we can calculate the conditional mean and variance for the model. First, we assume that the explanatory variables are controlled by an analyst, and are known for all $j = 1, \dots, h$, which leads to:

$$\begin{aligned} \mu_{y,t+h} =& \mathrm{E}(y_{t+h}|t) = \mathbf{w}'_{t+h} \mathbf{F}^{h-1} \mathbf{v}_t \\ \mathrm{V}(y_{t+h}|t) =& \left(\left(\mathbf{w}'_{t+h} \sum_{j=1}^{h-1} \mathbf{F}^{h-1-j} \mathbf{z}_{t+j} \mathbf{g} \right)^2 + 1 \right) \sigma^2 \end{aligned}. \tag{10.25}$$

The formulae for conditional moments in this case look similar to the ones from the pure additive ETS model in Section 5.3 with the only difference being that the element $\mathbf{z}_{t+j}\mathbf{g}$ is in general not equal to zero for the parameters of the explanatory variables.

10.3.3 Conditional mean for stochastic explanatory variables in ADAMX{D}

In the case of stochastic explanatory variables, the conditional expectation is straightforward and is similar to the one in the static ADAMX model:

$$\mu_{y,t+h} = \mathrm{E}(y_{t+h}|t) = \boldsymbol{\mu}'_{w,t+h}\mathbf{F}^{h-1}\mathbf{v}_t, \tag{10.26}$$

where $\boldsymbol{\mu}'_{w,t+h}$ is the vector of conditional h steps ahead expectations for each element in the $\mathbf{w}_{t+h}$. In the case of ETS components, the vector would contain ones. However, when it comes to conditional variance, it becomes more complicated because it introduces complex interactions between variances of different variables and the error term of the model. As a result, it would be easier to get the correct variance based on simulations, assuming that the explanatory variables and the error term change according to some assumed models instead of deriving analytical expressions.

10.4 Stability and forecastability conditions of ADAMX

It can be shown that any **static** ADAMX is not *stable* (as it was defined for pure additive ETS models in Section 5.4), meaning that the weights of such models do not decline to zero over time. To see this, we can draw an analogy with a deterministic model, discussed in the context of pure additive ETS (from Section 5.4). For example, we have already discussed that when $\alpha = 0$, the ETS(A,N,N) model becomes equivalent to the global level, loses the stability condition, but still can be forecastable. It becomes a simple but still useful and efficient model:

$$y_t = a_0 + \epsilon_t. \tag{10.27}$$

Similarly, the X part of ADAMX will always be unstable, but can be useful. For example, with $\alpha = 0$, ETSX(A,N,N) reverts back to the linear regression:

$$y_t = a_0 + a_1 x_{1,t} + \dots + a_n x_{n,t} + \epsilon_t, \tag{10.28}$$

which is not stable, because its weights do not decline over time, and the very first observation (a set of initial states with model parameters) impacts the final forecast. This does not make the model inappropriate in any way, but according to the conventional approach to ETS, if the dynamic part of the model is stable, but the overall model does not pass the stability check just because of the X part, then the whole model will be considered unstable and potentially dangerous to use. This is absurd. Following the same logic, we would need to avoid regression models in forecasting because they are not stable. Furthermore, there are no issues constructing ARIMAX models, but Osman and King (2015) argue that there are some with ETSX, which does not make sense if we recall the connection between ETS and ARIMA (discussed in Section 8.4). This only means that the stability/forecastability conditions should be checked for the dynamic part of the model (ETS or ARIMA) separately, ignoring the X part. Technically, this implies creating a separate transition matrix, persistence and measurement vectors, and calculating the discount matrix for the ETS/ARIMA part to check already discussed stability and forecastability conditions (Section 5.4).

When it comes to the **dynamic** ADAMX, the situation changes because now the smoothing parameters for the model coefficients determine how weights decline over time. It can be shown based on (5.10) that the values of the state vector on the observation t can be calculated via the recursion (here we provide a formula for the non-seasonal case, keeping in mind that in the case of the seasonal one, the derivation and the main message will be similar):

$$\mathbf{v}_t = \prod_{j=1}^{t-1} \mathbf{D}_{t-j} \mathbf{v}_0 + \sum_{j=0}^{t-1} \prod_{i=0}^{j} \mathbf{D}_{t-i} y_{t-j}, \tag{10.29}$$

where $\mathbf{D}_t = \mathbf{F} - \mathbf{z}_t \mathbf{g} \mathbf{w}'_t$ is the time varying discount matrix. The main issue in the case of dynamic ADAMX is that the stability condition varies over time together with the values of explanatory variables in $\mathbf{z}_t$. So, it is not possible to derive the stability condition for the general case. In order to make sure that the model is stable, we need for all eigenvalues of each $\mathbf{D}_j$ for all $j = \{1, \dots, t\}$ to lie in the unit circle.

Alternatively, we can introduce a new condition. We say that the model is **stable on average** if the eigenvalues of the geometric mean $\bar{\mathbf{D}} = \sqrt[t]{\prod_{j=1}^{t} \mathbf{D}_j}$ all lie in the unit circle. This way, some of the observations might have a higher impact on the final value, but they will be cancelled out by those with much lower weights in the product in (10.29). This condition can be checked during the model estimation, similar to how the conventional stability condition is checked.

As for the **forecastability** condition, for the ADAMX{D} it should be (based on the same logic as in Section 5.4):

$$\lim_{t \rightarrow \infty} \left(\mathbf{w}'_t \prod_{j=1}^{t-1} \mathbf{D}_{t-j} \mathbf{v}_0 \right) = \text{const}. \tag{10.30}$$

However, this condition will always be violated for the ADAMX models, just because the explanatory variables in $\mathbf{w}_t$ have their own variability and typically do not converge to a stable value with the increase of the sample size. So, if a forecastability condition needs to be checked for either ADAMX{D} or ADAMX{S}, it should be checked separately for the dynamic part of the model, dropping the X part.

10.5 Dealing with categorical variables in ADAMX

When dealing with categorical variables in a regression context, they are typically expanded to a set of dummy variables. So, for example, a variable "promotions" that can be "light", "medium", and "heavy" for different observations t would be expanded to three dummy variables, `promoLight`, `promoMedium`, and `promoHeavy`, each one of which is equal to 1, when the respective promotion type happens and equal to zero otherwise. When including these variables in the model, we would typically drop one of them (it is sometimes called pivot variable) and have a model with two dummy variables of a type:

$$y_t = a_0 + a_1 x_{1,t} + \dots + a_n x_{n,t} + d_1 promoLight_t + d_2 promoMedium_t + \epsilon_t, \tag{10.31}$$

where d_i is the parameter for the i-th dummy variable. The same procedure can be done in the context of ADAMX. The logic will be exactly the same for ADAMX{S}, but when it comes to the dynamic model, the parameters will have time indeces, and there can be different ways of formulating the model. Here is the first one:

$$\begin{aligned}
y_t = & \quad a_{0,t-1} + a_{1,t-1} x_{1,t} + \dots + a_{n,t-1} x_{n,t} + d_{1,t-1} promoLight_t + \\
& \qquad d_{2,t-1} promoMedium_t + \epsilon_t \\
a_{i,t} = & \quad a_{i,t-1} + \begin{cases} \delta_i \frac{\log(1+\epsilon_t)}{x_{i,t}} & \text{for each } i \in \{1, \dots, n\}, \text{ if } x_{i,t} \neq 0 \\ 0 \text{ otherwise} \end{cases} \\
d_{1,t} = & \quad d_{1,t-1} + \begin{cases} \delta_{n+1} \epsilon_t, & \text{if } promoLight_t \neq 0 \\ 0 \text{ otherwise} \end{cases} \\
d_{2,t} = & \quad d_{2,t-1} + \begin{cases} \delta_{n+2} \epsilon_t, & \text{if } promoMedium_t \neq 0 \\ 0 \text{ otherwise} \end{cases}
\end{aligned} . \tag{10.32}$$

Here we assume that each specific category of the variable promotion changes over time on its own with its own smoothing parameters δ_{n+1} and δ_{n+2}. Alternatively, we can assume that they have the same smoothing parameters, implying that the changes of the parameters are similar throughout different categories of the variable:

$$\begin{aligned}
d_{1,t} = d_{1,t-1} + \begin{cases} \delta_{n+1} \epsilon_t, & \text{if } promoLight_t \neq 0 \\ 0 \text{ otherwise} \end{cases} \\
d_{2,t} = d_{2,t-1} + \begin{cases} \delta_{n+1} \epsilon_t, & \text{if } promoMedium_t \neq 0 \\ 0 \text{ otherwise} \end{cases}
\end{aligned} . \tag{10.33}$$

The rationale for such restriction is that we might expect the adaptation mechanism to apply to the promo variable as a whole, not to its specific values. Indeed, in the example above, the variable of interest is `promo`, not the `promoLight` or `promoMedium`. Doing that will reduce the number of parameters and might simplify the estimation.

The mechanism (10.33) also becomes useful in connecting the ETSX and the conventional seasonal ETS model. Consider an example with quarterly data with no trend, and a categorical variable `quarterOfYear`, which can be `First`, `Second`, `Third`, and `Fourth`, depending on the specific observation. For convenience, I will call the parameters for the dummy variables, created from this categorical variable as $s_{1,t}$, $s_{2,t}$, $s_{3,t}$ and $s_{4,t}$. Based on (10.33), the model can then be formulated as:

$$\begin{aligned} y_t =& l_{t-1} + s_{1,t} quarterOfYear_{1,t} + s_{2,t} quarterOfYear_{2,t} + \\ & s_{3,t} quarterOfYear_{3,t} + s_{4,t} quarterOfYear_{4,t} + \epsilon_t \\ l_t =& l_{t-1} + \alpha \epsilon_t \\ s_{i,t} =& s_{i,t-1} + \begin{cases} \delta \epsilon_t \text{ for each } i \in \{1, \dots, 4\}, \text{ if } quarterOfYear_{i,t} \neq 0 \\ 0 \text{ otherwise} \end{cases} \end{aligned} . \tag{10.34}$$

We intentionally added all four dummy variables in (10.34) to separate the seasonal effect from the level component. While in the case of the classical regression, this does not make much sense, in the ETSX we avoid the trap of dummy variables due to the dynamic update of components and/or parameters (see discussion in Section 14.9). Having done that, we have just formulated the conventional ETS(A,N,A) model using a set of dummy variables and one smoothing parameter, the difference being that the latter relies on the lag of component:

$$\begin{aligned} y_t &= l_{t-1} + s_{t-4} + \epsilon_t \\ l_t &= l_{t-1} + \alpha \epsilon_t \\ s_t &= s_{t-4} + \gamma \epsilon_t \end{aligned} . \tag{10.35}$$

So, this comparison shows on one hand that the mechanism of ADAMX{D} is natural for the ADAM, and on the other hand that using the same smoothing parameter for different values of a categorical variable can be a reasonable idea, especially in cases when we can assume that all categories of the variable should evolve together.

10.6 Examples of application

For the example in this section, we will use the data of Road Casualties in Great Britain 1969–84, the `Seatbelts` dataset in the `datasets` package for R, which contains several variables (the description is provided in the documentation for the data and can be accessed via the `?Seatbelts` command).

The variable of interest, in this case is `drivers`, and the dataset contains more variables than needed, so we will restrict the data with `drivers`, `kms` (distance driven), `PetrolPrice`, and `law` – the latter three seem to influence the number of injured/killed drivers in principle:

```
SeatbeltsData <- Seatbelts[,c("drivers","kms","PetrolPrice","law")]
```

The dynamics of these variables over time is shown in Figure 10.1.

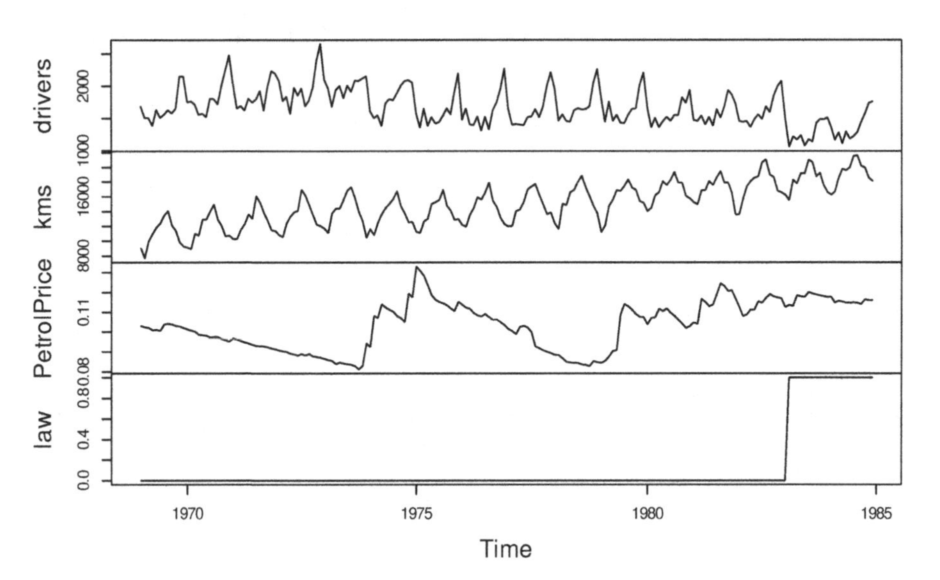

FIGURE 10.1 The time series dynamics of variables from Seatbelts dataset.

Apparently, the `drivers` variable exhibits seasonality but does not seem to have a trend. The type of seasonality is challenging to determine, but we will assume that it is multiplicative for now. A simple ETS(M,N,M) model applied to the data will produce the following (we will withhold the last 12 observations for the forecast evaluation, Figure 10.2):

```
adamETSMNMSeat <- adam(SeatbeltsData[,"drivers"], "MNM",
                       h=12, holdout=TRUE)
forecast(adamETSMNMSeat, h=12, interval="prediction") |>
    plot()
```

This simple model already does a fine job fitting the data and producing forecasts. However, the forecast is biased and is consistently lower than needed because of the sudden drop in the level of series, which can only be explained

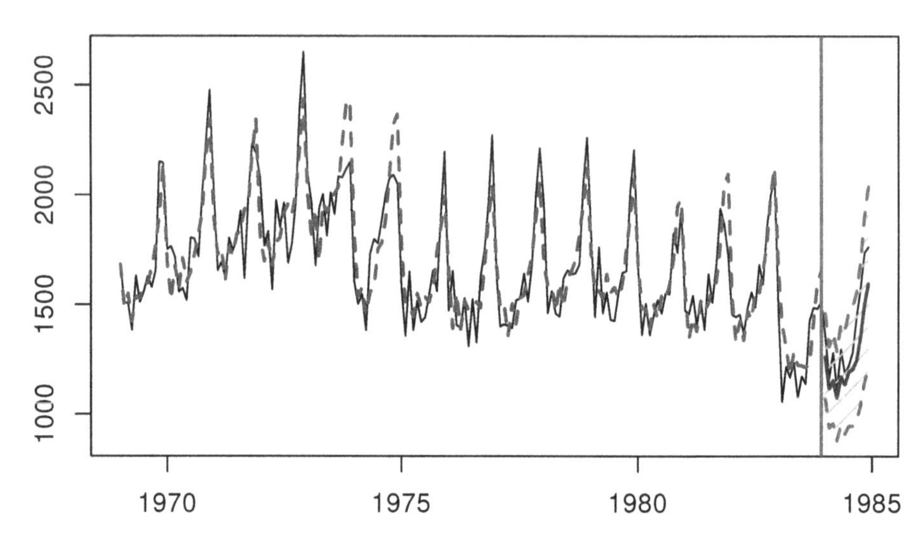

FIGURE 10.2 The actual values for drivers and a forecast from an ETS(M,N,M) model.

by the introduction of the new law in the UK in 1983, making the seatbelts compulsory for drivers. Due to the sudden drop, the smoothing parameter for the level of series is higher than needed, leading to wider intervals and less accurate forecasts. Here is the output of the model:

```
adamETSMNMSeat
```

```
## Time elapsed: 0.25 seconds
## Model estimated using adam() function: ETS(MNM)
## Distribution assumed in the model: Gamma
## Loss function type: likelihood; Loss function value: 1125.509
## Persistence vector g:
##  alpha  gamma
## 0.4133 0.0000
##
## Sample size: 180
## Number of estimated parameters: 15
## Number of degrees of freedom: 165
## Information criteria:
##      AIC     AICc      BIC     BICc
## 2281.019 2283.945 2328.913 2336.512
##
## Forecast errors:
```

```
## ME: 117.9; MAE: 117.9; RMSE: 137.596
## sCE: 83.695%; Asymmetry: 100%; sMAE: 6.975%; sMSE: 0.663%
## MASE: 0.684; RMSSE: 0.611; rMAE: 0.504; rRMSE: 0.542
```

In order to further explore the data we will produce the scatterplots and boxplots between the variables using the `spread()` function from the `greybox` package (Figure 10.3):

```
spread(SeatbeltsData)
```

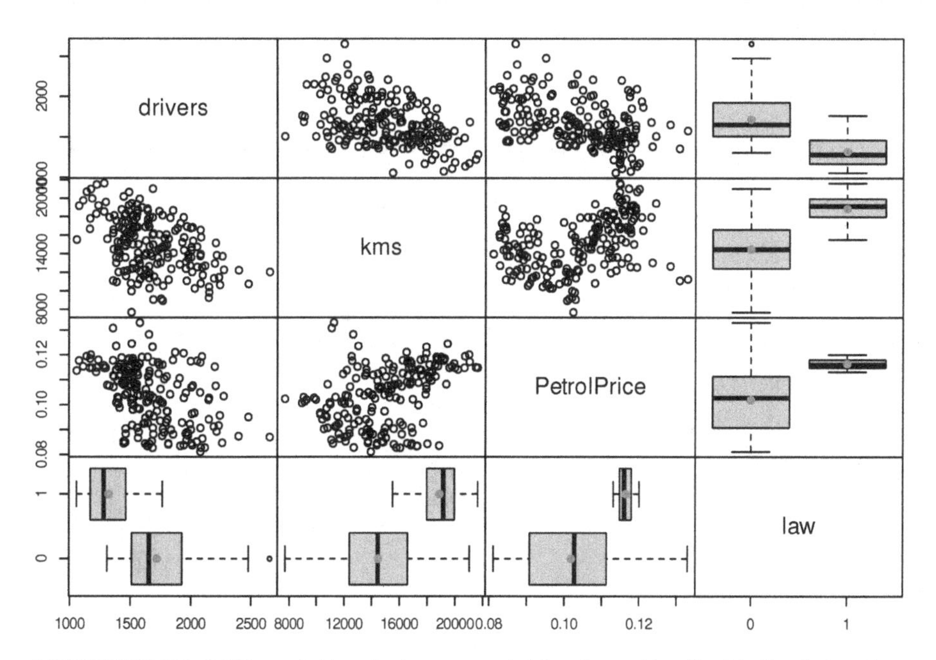

FIGURE 10.3 The relation between variables from the Seatbelts dataset.

Figure 10.3 shows a negative relation between `kms` and `drivers`: the higher the distance driven, the lower the total of car drivers killed or seriously injured. A similar relation is observed between the petrol price and drivers (when the prices are high, people tend to drive less, thus causing fewer incidents). Interestingly, the increase of both variables causes the variance of the response variable to decrease (heteroscedasticity effect). Using a multiplicative error model and including the variables in logarithms, in this case, might address this potential issue. Note that we do not need to take the logarithm of `drivers`, as we already use the model with multiplicative error. Finally, the legislation of a new law seems to have caused a decrease in the number of causalities. To have a better model in terms of explanatory and predictive power, we should include all three variables. This is how we can do that using `adam()`:

```
adamETSXMNMSeat <- adam(SeatbeltsData, "MNM", h=12, holdout=TRUE,
                        formula=drivers~log(kms)+log(PetrolPrice)+law)
```

The parameter `formula` in general is not compulsory. It can either be substituted by "`formula=drivers~.`" or dropped completely – in the latter case, the function would fit the model of the first variable in the matrix from everything else. We need the formula in our case because we introduce log-transformations of some explanatory variables.

```
forecast(adamETSXMNMSeat, h=12, interval="prediction") |>
    plot()
```

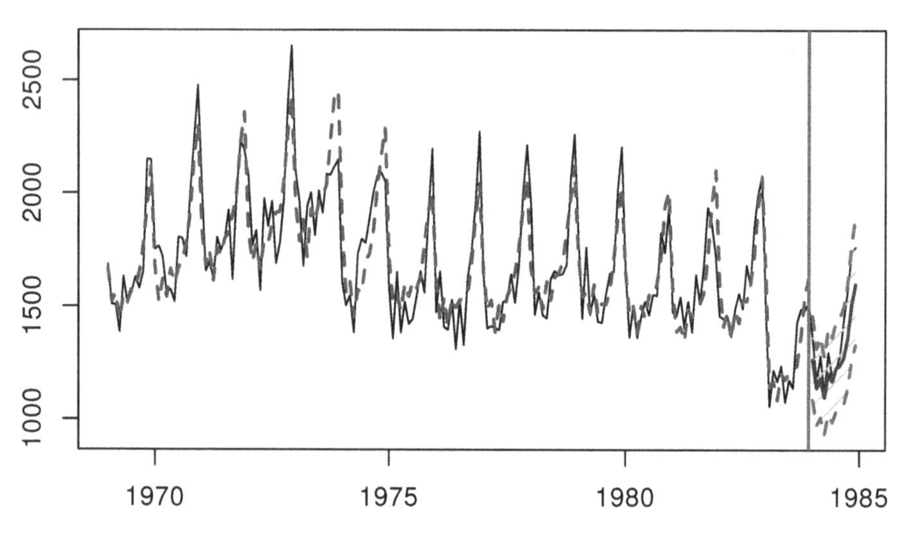

FIGURE 10.4 The actual values for drivers and a forecast from an ETSX(M,N,M) model.

Figure 10.4 shows the forecast from the second model, which is slightly more accurate. More importantly, the prediction interval is narrower than in the simple ETS(M,N,M) because now the model takes the external information into account. Here is the summary of the second model:

```
adamETSXMNMSeat
```

```
## Time elapsed: 0.88 seconds
## Model estimated using adam() function: ETSX(MNM)
## Distribution assumed in the model: Gamma
```

```
## Loss function type: likelihood; Loss function value: 1114.042
## Persistence vector g (excluding xreg):
##  alpha  gamma
## 0.2042 0.0000
##
## Sample size: 180
## Number of estimated parameters: 18
## Number of degrees of freedom: 162
## Information criteria:
##      AIC     AICc      BIC     BICc
## 2264.085 2268.333 2321.558 2332.589
##
## Forecast errors:
## ME: 96.28; MAE: 97.238; RMSE: 123.401
## sCE: 68.347%; Asymmetry: 97.1%; sMAE: 5.752%; sMSE: 0.533%
## MASE: 0.564; RMSSE: 0.548; rMAE: 0.416; rRMSE: 0.486
```

Note that the smoothing parameter α has reduced from 0.41 to 0.2. This led to the reduction in error measures. For example, based on RMSE, we can conclude that the model with explanatory variables is more precise than the simple univariate ETS(M,N,M). Still, we could try introducing the update of the parameters for the explanatory variables to see how it works (it might be unnecessary for this data):

```
adamETSXMNMDSeat <- adam(SeatbeltsData, "MNM", h=12, holdout=TRUE,
                         formula=drivers~log(kms)+log(PetrolPrice)+law,
                         regressors="adapt", maxeval=10000)
```

Remark. Given the complexity of the estimation task, the default number of iterations needed for the optimiser to find the minimum of the loss function might not be sufficient. This is why I introduced `maxeval=10000` in the code above, increasing the number of maximum iterations to 10,000. In order to see how the optimiser worked out, you can add `print_level=41` in the code above.

In our specific case, the difference between the ETSX and ETSX{D} models is infinitesimal in terms of the accuracy of final forecasts and prediction intervals. Here is the output of the model:

```
adamETSXMNMDSeat
```

```
## Time elapsed: 2.38 seconds
## Model estimated using adam() function: ETSX(MNM){D}
## Distribution assumed in the model: Gamma
```

```
## Loss function type: likelihood; Loss function value: 1113.523
## Persistence vector g (excluding xreg):
##  alpha  gamma
## 0.1153 0.0000
##
## Sample size: 180
## Number of estimated parameters: 21
## Number of degrees of freedom: 159
## Information criteria:
##      AIC     AICc      BIC     BICc
## 2269.046 2274.894 2336.098 2351.282
##
## Forecast errors:
## ME: 96.005; MAE: 96.767; RMSE: 123.024
## sCE: 68.151%; Asymmetry: 97.4%; sMAE: 5.724%; sMSE: 0.53%
## MASE: 0.561; RMSSE: 0.546; rMAE: 0.414; rRMSE: 0.484
```

We can spot that the error measures of the dynamic model are slightly lower than the ones from the static one (e.g., compare RMSE of models). The information criteria are slightly lower as well. So based on all of this, we should use the dynamic model for forecasting and analytical purposes. In order to see the set of smoothing parameters for the explanatory variables in this model, we can use the command:

```
round(adamETSXMNMDSeat$persistence,4)
```

```
##  alpha  gamma delta1 delta2 delta3
## 0.1153 0.0000 0.0056 0.0764 0.0000
```

And see how the states/parameters of the model change over time (Figure 10.5):

```
plot(adamETSXMNMDSeat, 12, main="")
```

As we see in Figure 10.5, the effects of variables change over time. This mainly applies to the `PetrolPrice` variable, the smoothing parameter for which is the highest among all δ_i parameters.

To see the initial effects of the explanatory variables on the number of incidents with drivers, we can look at the parameters for those variables:

```
adamETSXMNMDSeat$initial$xreg
```

```
##         log.kms. log.PetrolPrice.              law
##       -0.0771078       -0.2973723       -0.2381336
```

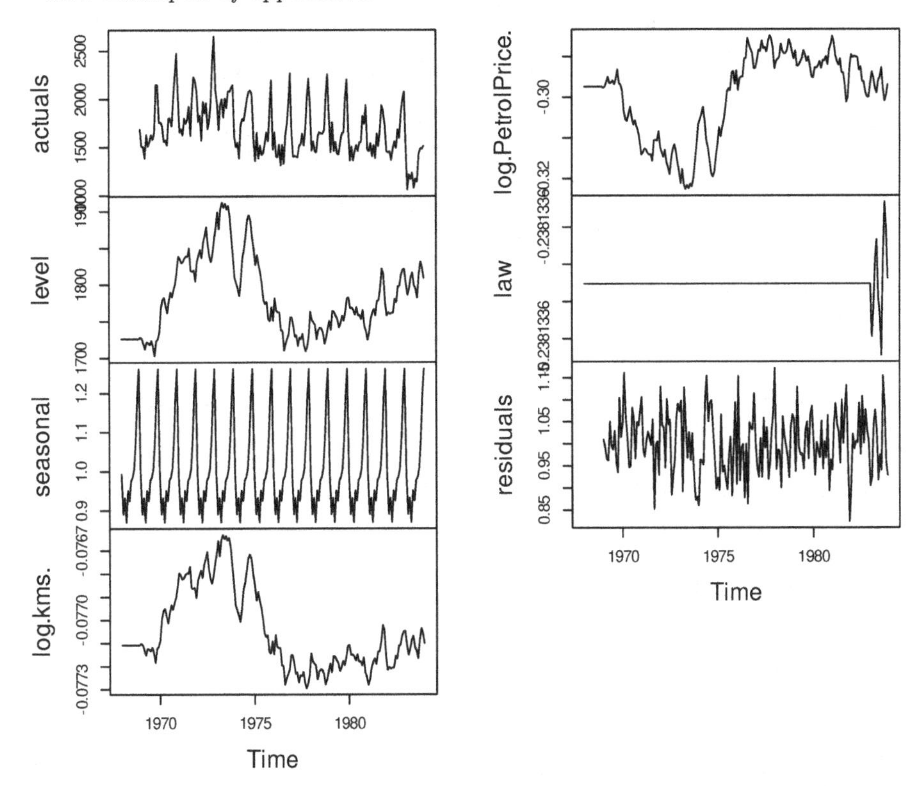

FIGURE 10.5 Dynamic of states of the ETSX(M,N,M){D} model.

Based on that, we can conclude that the introduction of the law reduced on average the number of incidents by approximately 24%, while the increase of the petrol price by 1% leads on average to a decrease in the number of incidents by 0.3%. Finally, the distance negatively impacts the incidents as well, reducing it on average by 0.1% for each 1% increase in the distance. This is the standard interpretation of parameters, which we can use based on the estimated model (see, for example, discussion in Section 11.3 of Svetunkov, 2022). We will discuss how to do further analysis using ADAM in Chapter 16, introducing the standard errors and confidence intervals for the parameters.

Finally, `adam()` has some shortcuts when a matrix of variables (not a data frame!) is provided with no formula, assuming that the necessary expansion has already been done. This leads to the decrease in computational time of the function and becomes especially useful when working on large samples of data. Here is an example with ETSX(M,N,N):

```
# Create matrix for the model
SeatbeltsDataExpanded <-
```

```
  ts(model.frame(drivers~log(kms)+log(PetrolPrice)+law,
                 SeatbeltsData),
     start=start(SeatbeltsData), frequency=frequency(SeatbeltsData))
# Fix the names of variables
colnames(SeatbeltsDataExpanded) <-
  make.names(colnames(SeatbeltsDataExpanded))
# Apply the model
adamETSXMNMExpandedSeat <- adam(SeatbeltsDataExpanded, "MNM",
                                lags=12, h=12, holdout=TRUE)
```

11

Estimation of ADAM

Now that we have discussed the properties of ETS, ARIMA, ETSX, and ARIMAX models, we need to understand how to estimate them. As mentioned earlier, when we apply a model to the data, we assume that it is suitable and we need to see how it fits the data and produces forecasts to assess this suitability. In this case, all the model parameters are substituted by their estimates (observed in the sample), and the error term becomes an estimate of the true one. In general, this means that the state space model (7.1) is substituted by:

$$\begin{aligned} y_t &= w(\hat{\mathbf{v}}_{t-\boldsymbol{l}}) + r(\hat{\mathbf{v}}_{t-\boldsymbol{l}}) e_t \\ \hat{\mathbf{v}}_t &= f(\hat{\mathbf{v}}_{t-\boldsymbol{l}}) + \hat{g}(\hat{\mathbf{v}}_{t-\boldsymbol{l}}) e_t \end{aligned}, \tag{11.1}$$

implying that the initial values of components and the smoothing parameters of the model are estimated. An example is the ETS(A,A,A) model applied to the data:

$$\begin{aligned} \hat{y}_t &= \hat{l}_{t-1} + \hat{b}_{t-1} + \hat{s}_{t-m} \\ e_t &= y_t - \hat{y}_t \\ \hat{l}_t &= \hat{l}_{t-1} + \hat{b}_{t-1} + \hat{\alpha} e_t \\ \hat{b}_t &= \hat{b}_{t-1} + \hat{\beta} e_t \\ \hat{s}_t &= \hat{s}_{t-m} + \hat{\gamma} e_t \end{aligned}, \tag{11.2}$$

where the initial values $\hat{l}_0, \hat{b}_0$, and $\hat{s}_{-m+2}, ...\hat{s}_0$ are estimated and then influence all the future values of components via the recursion (11.2) and $e_t = y_t - \hat{y}_t$ is the one step ahead in-sample forecast error, also known in statistics as the residual of the model. The set of equations (11.2) allows constructing the model by applying equations one by one (you can even do that in MS Excel by creating five columns for the respective five equations).

The estimation itself does not happen on its own, a complicated process of minimisation/maximisation of the pre-selected loss function by changing the values of parameters is involved. Typically, there is no analytical solution for estimates of ADAM parameters because of the model's recursive nature. As a result, numerical optimisation is used to obtain the estimates of parameters. The results of the estimation will differ depending on:

1. The assumed distribution;
2. The used loss function;

3. The initial values of parameters that are fed to the optimiser;
4. The parameters of the optimiser (such as sensitivity, number of iterations etc.);
5. The sample size;
6. The number of parameters to estimate and;
7. The restrictions imposed on parameters.

The aspects above are covered in this chapter.

11.1 Maximum Likelihood Estimation

Maximum Likelihood Estimation (MLE) is one of the popular approaches for estimating parameters of a statistical model. It relies on assumptions about the distribution of the response variable and uses either a probability density or a cumulative distribution function, depending on the assumed model, and aims to maximise the likelihood that the observations can be described by the model with specific parameters. An interested reader can find a detailed explanation of the likelihood approach in Chapter 16 of Svetunkov (2022).

Remark. Contrary to common belief, the MLE itself does not assume that a true model follows a specific distribution. It just tries to fit the predefined distribution to the available data.

MLE of the ADAM relies on the distributional assumptions of each specific model and might differ from one model to another (see assumptions in Sections 5.5, 6.5, and 9.3). There are several options for the `distribution` parameter supported by the `adam()` function in the `smooth` package, here we briefly discuss how the estimation is done for each one of them. We start with the additive error model, for which the assumptions, log-likelihoods, and MLE of scales are provided in Table 11.1.

The likelihoods are derived based on the probability density functions, by taking the logarithms of their products for all in-sample observations. For

TABLE 11.1 Likelihood approach for additive error models. $\mathcal{N}$ is the Normal, $\mathcal{L}$ is the Laplace, $\mathcal{S}$ is the S, $\mathcal{GN}$ is the Generalised Normal, $\mathcal{IG}$ is the Inverse Gaussian, Γ is the Gamma, and $\log\mathcal{N}$ is the Log-Normal distribution.

	Assumption	log-likelihood	MLE of scale
Normal	$\epsilon_t \sim \mathcal{N}(0, \sigma^2)$	$-\frac{T}{2}\log(2\pi\sigma^2) - \frac{1}{2}\sum_{t=1}^T \frac{\epsilon_t^2}{\sigma^2}$	$\hat{\sigma}^2 = \frac{1}{T}\sum_{t=1}^T e_t^2$
Laplace	$\epsilon_t \sim \mathcal{L}(0, s)$	$-T\log(2s) - \sum_{t=1}^T \frac{\|\epsilon_t\|}{s}$	$\hat{s} = \frac{1}{T}\sum_{t=1}^T \|e_t\|$
S	$\epsilon_t \sim \mathcal{S}(0, s)$	$-T\log(4s^2) - \sum_{t=1}^T \frac{\sqrt{\|\epsilon_t\|}}{s}$	$\hat{s} = \frac{1}{2T}\sum_{t=1}^T \sqrt{\|e_t\|}$
Generalised Normal	$\epsilon_t \sim \mathcal{GN}(0, s, \beta)$	$T\log\beta - T\log(2s\Gamma\left(\beta^{-1}\right))- \sum_{t=1}^T \frac{\|\epsilon_t\|^\beta}{s}$	$\hat{s} = \sqrt[\beta]{\frac{\beta}{T}\sum_{t=1}^T \|e_t\|^\beta}$
Inverse Gaussian	$1 + \frac{\epsilon_t}{\mu_{y,t}} \sim \mathcal{IG}(1, \sigma^2)$	$-\frac{T}{2}\log\left(2\pi\sigma^2\right) - \frac{1}{2}\sum_{t=1}^T\left(\left(\frac{y_t}{\mu_{y,t}}\right)^3\right) - \frac{3}{2}\sum_{t=1}^T \log y_t - \frac{1}{2\sigma^2}\sum_{t=1}^T \frac{\epsilon_t^2}{\mu_{y,t} y_t}$	$\hat{\sigma}^2 = \frac{1}{T}\sum_{t=1}^T \frac{e_t^2}{\hat{\mu}_{y,t} y_t}$
Gamma	$1 + \frac{\epsilon_t}{\mu_{y,t}} \sim -(\sigma^{-2}, \sigma^2)$	$-T\log\Gamma\left(\sigma^{-2}\right) - \frac{T}{\sigma^2}\log\sigma^2 + \frac{1}{\sigma^2}\sum_{t=1}^T \log\left(\frac{y_t/\mu_{y,t}}{\exp(y_t/\mu_{y,t})}\right) - \sum_{t=1}^T \log y_t$	$\hat{\sigma}^2 = \frac{1}{T}\sum_{t=1}^T \left(\frac{e_t}{\mu_{y,t}}\right)^2$ *
Log-Normal	$1 + \frac{\epsilon_t}{\mu_{y,t}} \sim \log\mathcal{N}\left(-\frac{\sigma^2}{2}, \sigma^2\right)$	$-\frac{T}{2}\log\left(2\pi\sigma^2\right) - \sum_{t=1}^T \log y_t - \frac{1}{2\sigma^2}\sum_{t=1}^T \left(\log\left(\frac{y_t}{\mu_{y,t}}\right) + \frac{\sigma^2}{2}\right)^2$	$\hat{\sigma}^2 = 2\left(1 - \sqrt{1 - \frac{1}{T}\sum_{t=1}^T \log^2\left(\frac{y_t}{\hat{\mu}_{y,t}}\right)}\right)$

example, this is what we get for the Normal distribution:

$$
\begin{aligned}
\mathcal{L}(\boldsymbol{\theta}, \sigma^2 | \mathbf{y}) &= \prod_{t=1}^T \left(\frac{1}{\sqrt{2\pi\sigma^2}} \exp \left(-\frac{\left(y_t - \mu_t\right)^2}{2\sigma^2} \right) \right) \\
\mathcal{L}(\boldsymbol{\theta}, \sigma^2 | \mathbf{y}) &= \frac{1}{\left(\sqrt{2\pi\sigma^2}\right)^T} \exp \left(\sum_{t=1}^T -\frac{\epsilon_t^2}{2\sigma^2} \right) \\
\ell(\boldsymbol{\theta}, \sigma^2 | \mathbf{y}) &= \log \mathcal{L}(\boldsymbol{\theta}, \sigma^2 | \mathbf{y}) \\
\ell(\boldsymbol{\theta}, \sigma^2 | \mathbf{y}) &= -\frac{T}{2} \log(2\pi\sigma^2) - \frac{1}{2} \sum_{t=1}^T \frac{\epsilon_t^2}{\sigma^2}
\end{aligned} , \quad (11.3)
$$

where $\mathbf{y}$ is the vector of all in-sample actual values, $\mathcal{L}(\boldsymbol{\theta}, \sigma^2|\mathbf{y})$ is the likelihood value, while $\ell(\boldsymbol{\theta}, \sigma^2|\mathbf{y})$ is the log-likelihood. As for the scale, it is obtained by solving the equation after taking the derivative of the log-likelihood (11.3) with respect to σ^2 and setting it equal to zero. We do not discuss the concentrated log-likelihoods (obtained after inserting the estimated scale in the respective log-likelihood function), because they are not useful in understanding of how the model is estimated, but knowing how to calculate scale helps, because it simplifies the model estimation process.

Remark. MLE of scale parameter for Gamma distribution (formulated in ADAM) does not have a closed-form. While there are no proofs for it, it seems that the maximum of the likelihood of Gamma distribution is achieved, when

$\hat{\sigma}^2 = \frac{1}{T}\sum_{t=1}^T \left(\frac{e_t}{\mu_{y,t}}\right)^2$, which corresponds to the estimate based on method of moments.

While other distributions can be used in ADAM ETS (for example, Logistic distribution or Student's t), we do not discuss them here. In all cases in Table 11.1, the assumptions imply that the actual value follows the same distributionn as the error term (due to additivity of the model) but with a different location and/or scale. For example, for the Normal distribution, we have:

$$\begin{aligned} & \epsilon_t \sim \mathcal{N}(0, \sigma^2) \\ & \text{or} \\ & y_t = \mu_{y,t} + \epsilon_t \sim \mathcal{N}(\mu_{y,t}, \sigma^2) \end{aligned} . \tag{11.4}$$

When it comes to the multiplicative error models, the likelihoods become slightly different. For example, when we assume that $\epsilon_t \sim \mathcal{N}(0, \sigma^2)$ in the multiplicative error model, this implies that:

$$y_t = \mu_{y,t}(1 + \epsilon_t) \sim \mathcal{N}(\mu_{y,t}, \mu_{y,t}^2 \sigma^2). \tag{11.5}$$

As a result, the log-likelihoods would have the $\mu_{y,t}$ part in the formulae:

$$\ell(\boldsymbol{\theta}, \sigma^2 | \mathbf{y}) = -\frac{1}{2} \sum_{t=1}^T \log(2 \pi \mu_{y,t}^2 \sigma^2) - \frac{1}{2} \sum_{t=1}^T \frac{\epsilon_t^2}{\sigma^2}. \tag{11.6}$$

Remark. The part $\frac{1}{2}\sum_{t=1}^T \frac{\epsilon_t^2}{\sigma^2}$ does not contain $\mu_{y,t}$ because when inserted in the formula of Normal distribution, the numerator (based on (11.5)) is: $y_t - \mu_{y,t} = \mu_{y,t}\epsilon_t$. After inserting it in the formula for the Normal distribution that part becomes: $\frac{1}{2}\sum_{t=1}^T \frac{(\mu_{y,t}\epsilon_t)^2}{\mu_{y,t}^2\sigma^2}$ which after cancellations leads to (11.6).

Similar logic is applicable to Laplace ($\mathcal{L}$), Generalised Normal ($\mathcal{GN}$), and S ($\mathcal{S}$) distributions. From the practical point of view, these assumptions imply that the scale (and variance) of the distribution of y_t changes together with the level of the series. When it comes to the Inverse Gaussian ($\mathcal{IG}$), Gamma (Γ), and Log-Normal (log$\mathcal{N}$), the assumptions are imposed on the $1 + \epsilon_t$ part of the model and the respective likelihoods do not involve the expectation $\mu_{y,t}$, but the formulation still implies that the variance of the data increases together with the increase of the level of data.

All the likelihoods for the multiplicative error models are summarised in Table 11.2.

TABLE 11.2 Likelihood approach for multiplicative error models. $\mathcal{N}$ is the Normal, $\mathcal{L}$ is the Laplace, $\mathcal{S}$ is the S, $\mathcal{GN}$ is the Generalised Normal, $\mathcal{IG}$ is the Inverse Gaussian, Γ is the Gamma, and $\log\mathcal{N}$ is the Log-Normal distribution.

	Assumption	log-likelihood	MLE of scale
Normal	$\epsilon_t \sim \mathcal{N}(0, \sigma^2)$	$-\frac{T}{2}\log(2\pi\sigma^2) - \frac{1}{2}\sum_{t=1}^{T}\frac{\epsilon_t^2}{\sigma^2} - \sum_{t=1}^{T}\log\|\mu_{y,t}\|$	$\hat{\sigma}^2 = \frac{1}{T}\sum_{t=1}^{T} e_t^2$
Laplace	$\epsilon_t \sim \mathcal{L}(0, s)$	$-T\log(2s) - \sum_{t=1}^{T}\frac{\|\epsilon_t\|}{s} - \sum_{t=1}^{T}\log \mu_{y,t}$	$\hat{s} = \frac{1}{T}\sum_{t=1}^{T}\|e_t\|$
S	$\epsilon_t \sim \mathcal{S}(0, s)$	$-T\log(4s^2) - \sum_{t=1}^{T}\frac{\sqrt{\|\epsilon_t\|}}{s} - \sum_{t=1}^{T}\log \mu_{y,t}$	$\hat{s} = \frac{1}{2T}\sum_{t=1}^{T}\sqrt{\|e_t\|}$
Generalised Normal	$\epsilon_t \sim \mathcal{GN}(0, s, \beta)$	$T\log\beta - T\log\left(2s\Gamma\left(\beta^{-1}\right)\right) - \sum_{t=1}^{T}\frac{\|\epsilon_t\|^\beta}{s} - \sum_{t=1}^{T}\log \mu_{y,t}$	$\hat{s} = \sqrt[\beta]{\frac{\beta}{T}\sum_{t=1}^{T}\|e_t\|^\beta}$
Inverse Gaussian	$1+\epsilon_t \sim \mathcal{IG}(1, \sigma^2)$	$-\frac{T}{2}\log\left(2\pi\sigma^2\right) - \frac{1}{2}\sum_{t=1}^{T}(1+\epsilon_t)^3 - \frac{3}{2}\sum_{t=1}^{T}\log y_t - \frac{1}{2\sigma^2}\sum_{t=1}^{T}\frac{\epsilon_t^2}{1+\epsilon_t}$	$\hat{\sigma}^2 = \frac{1}{T}\sum_{t=1}^{T}\frac{e_t^2}{1+e_t}$
Gamma	$1+\epsilon_t \sim -(\sigma^{-2}, \sigma)$	$-T\log\Gamma\left(\sigma^{-2}\right) - \frac{T}{\sigma^2}\log\sigma^2 + \frac{1}{\sigma^2}\sum_{t=1}^{T}\log\left(\frac{1+\epsilon_t}{\exp(1+\epsilon_t)}\right) - \sum_{t=1}^{T}\log y_t$	$\hat{\sigma}^2 = \frac{1}{T}\sum_{t=1}^{T} e_t^2$ *
Log-Normal	$1+\epsilon_t \sim \log\mathcal{N}\left(-\frac{\sigma^2}{2}, \sigma^2\right)$	$-\frac{T}{2}\log\left(2\pi\sigma^2\right) - \sum_{t=1}^{T}\log y_t - \frac{1}{2\sigma^2}\sum_{t=1}^{T}\left(\log(1+\epsilon_t) + \frac{\sigma^2}{2}\right)^2$	$\hat{\sigma}^2 = 2\left(1 - \sqrt{1 - \frac{1}{T}\sum_{t=1}^{T}\log^2(1+e_t)}\right)$

When it comes to practicalities, the optimiser in the `adam()` function calculates the scale from Tables 11.1 and 11.2 on each iteration and then uses it in the log-likelihood based on the respective distribution function. For additive error models:

1. Normal, $\mathcal{N}$ – `dnorm(x=actuals, mean=fitted, sd=scale, log=TRUE)` from the `stats` package;
2. Laplace, $\mathcal{L}$ – `dlaplace(q=actuals, mu=fitted, scale=scale, log=TRUE)` from the `greybox` package;
3. S, $\mathcal{S}$ – `ds(q=actuals, mu=fitted, scale=scale, log=TRUE)` from the `greybox` package;
4. Generalised Normal, $\mathcal{GN}$ – `dgnorm(x=actuals, mu=fitted, alpha=scale, beta=beta, log=TRUE)` implemented in the `greybox` package based on the `gnorm` package (the version on CRAN is outdated);
5. Inverse Gaussian, $\mathcal{IG}$ – `dinvgauss(x=actuals, mean=fitted, dispersion=scale/fitted, log=TRUE)` from the `statmod` package;
6. Log-Normal, $\log\mathcal{N}$ – `dlnorm(x=actuals, meanlog=fitted-scale^2/2, sdlog=scale, log=TRUE)` from the `stats` package;
7. Gamma, Γ – `dgamma(x=actuals, shape=1/scale, scale=scale*fitted, log=TRUE)` from the `stats` package.

And for multiplicative error models:

1. Normal, $\mathcal{N}$ – `dnorm(x=actuals, mean=fitted, sd=scale*fitted, log=TRUE)`;
2. Laplace, $\mathcal{L}$ – `dlaplace(q=actuals, mu=fitted, scale=scale*fitted, log=TRUE)`;
3. S, $\mathcal{S}$ – `ds(q=actuals, mu=fitted, scale=scale*sqrt(fitted), log=TRUE)`;
4. Generalised Normal, $\mathcal{GN}$ – `dgnorm(x=actuals, mu=fitted, alpha=scale*fitted^beta, beta=beta, log=TRUE)`;
5. Inverse Gaussian, $\mathcal{IG}$ – `dinvgauss(x=actuals, mean=fitted, dispersion=scale/fitted, log=TRUE)`;
6. Log-Normal, $\log\mathcal{N}$ – `dlnorm(x=actuals, meanlog=fitted-scale^2/2, sdlog=scale, log=TRUE)`;
7. Gamma, Γ – `dgamma(x=actuals, shape=1/scale, scale=scale*fitted, log=TRUE)`.

Remark. In cases of $\mathcal{GN}$, an additional parameter (namely β) is needed. If the user does not provide it, it will be estimated together with the other parameters via the maximisation of respective likelihoods. Note however that the estimate of β can be inefficient if its true value is lower than 1.

The MLE of ADAM makes different models comparable with each other irrespective of the types of components and distributional assumptions. As a result, model selection based on information criteria can be done using the `auto.adam()` function from the `smooth` package, which will select the most appropriate distribution for ADAM.

11.1.1 An example in R

The `adam()` function in the `smooth` package has the parameter `distribution`, which allows selecting between several options discussed in this chapter, based on the respective density functions. Here is a brief example in R with ADAM ETS(M,A,M) applied to the `AirPassengers` data with several distributions:

```
adamETSMAMAir <- vector("list",5)
adamETSMAMAir[[1]] <- adam(AirPassengers, "MAM", h=12, holdout=TRUE,
                           distribution="dnorm")
adamETSMAMAir[[2]] <- adam(AirPassengers, "MAM", h=12, holdout=TRUE,
                           distribution="dlaplace")
adamETSMAMAir[[3]] <- adam(AirPassengers, "MAM", h=12, holdout=TRUE,
                           distribution="dgnorm")
adamETSMAMAir[[4]] <- adam(AirPassengers, "MAM", h=12, holdout=TRUE,
                           distribution="dinvgauss")
```

```
adamETSMAMAir[[5]] <- adam(AirPassengers, "MAM", h=12, holdout=TRUE,
                           distribution="dgamma")
```

In this case, the function will select the most appropriate ETS model for each distribution. We can see what was selected in each case and compare the models using information criteria:

```
sapply(adamETSMAMAir, AICc) |>
    setNames(c("dnorm","dlaplace","dgnorm","dinvgauss","dgamma"))
```

```
##     dnorm  dlaplace    dgnorm dinvgauss    dgamma
## 971.5172  977.4446  974.9190  973.5000  973.9646
```

We could compare the performance of models in detail, but for demonstration purposes, it should suffice to say that among the four models considered above, based on the AICc value, the model with the Normal distribution should be selected. This process of fit and selection can be automated using the `auto.adam()` function, which accepts the vector of distributions to test and by default would consider `distribution=c("default", "dnorm", "dlaplace", "ds", "dgnorm", "dlnorm", "dinvgauss", "dgamma")` (where `default` is Normal for additive error and Gamma for the multiplicative error model):

```
auto.adam(AirPassengers, "MAM", h=12, holdout=TRUE)
```

This command should return the ADAM ETS(M,A,M) model with the most appropriate distribution, selected based on the AICc.

11.2 Non MLE-based loss functions

An alternative approach for estimating ADAM is using the conventional loss functions, such as MSE, MAE, etc. In this case, the model selection using information criteria would not work, but this might not matter when you have already decided what model to use and want to improve it. Alternatively, you can use cross validation (e.g. rolling origin from Section 2.4) to select a model estimated using a non MLE-based loss function. But in some special cases, the minimisation of these losses would give the same results as the maximisation of some likelihood functions:

- MSE is minimised by mean, and its minimum corresponds to the maximum likelihood of Normal distribution (see discussion in Kolassa, 2016);

- MAE is minimised by median, and its minimum corresponds to the maximum likelihood of Laplace distribution (Schwertman, Gilks, & Cameron, 1990);
- RMSLE is minimised by the geometric mean, and its minimum corresponds to the maximum likelihood of Log-Normal distribution (this is based on the fact that the maximum of the distribution coincides with the Normal one applied to the log-transformed data).

The main difference between using these losses and maximising respective likelihoods is in the number of estimated parameters: the latter implies that the scale is estimated together with the other parameters, while the former does not consider it and in a way provides it for free.

The assumed distribution does not necessarily depend on the used loss and vice versa. For example, we can assume that the actuals follow the Inverse Gaussian distribution, but estimate the model via minimisation of MAE. The estimates of parameters in this case might not be as efficient as in the case of MLE, but it is still possible to do.

When it comes to different types of models, the forecast error (which is used in respective losses) depends on the error type:

- Additive error: $e_t = y_t - \hat{\mu}_{y,t}$;
- Multiplicative error: $e_t = \frac{y_t - \hat{\mu}_{y,t}}{\hat{\mu}_{y,t}}$.

This follows directly from the respective ETS models.

11.2.1 MSE and MAE

MSE and MAE have been discussed in Section 2.1, but in the context of forecasts evaluation rather than estimation. If they are used for the latter, then the formulae (2.1) and (2.2) will be amended to:

$$\mathrm{MSE} = \sqrt{\frac{1}{T} \sum_{j=1}^T \left(e_t \right)^2} \tag{11.7}$$

and

$$\mathrm{MAE} = \frac{1}{T} \sum_{j=1}^T \left| e_t \right| , \tag{11.8}$$

where the specific formula for e_t would depend on the error type as shown above. The main difference between the two estimators is in what they are minimised by: MSE is minimised by mean, while MAE is minimised by the median. This means that models estimated using MAE will typically be more conservative (i.e. in the case of ETS, have lower smoothing parameters). Gardner (2006) recommended using MAE in cases of outliers in the data, as the ETS model is supposed to become less reactive than in the case of MSE. Note that in case of intermittent demand, MAE should be in general avoided due to the issues discussed in Section 2.1.

11.2.2 HAM

Along with the discussed MSE and MAE, there is also HAM – "Half Absolute Moment" (Svetunkov, Kourentzes, & Svetunkov, 2023):

$$\mathrm{HAM} = \frac{1}{T} \sum_{j=1}^{T} \sqrt{|e_t|}, \tag{11.9}$$

the minimum of which corresponds to the maximum likelihood of S distribution. The idea of this estimator is to minimise the errors that happen very often, close to the location of the distribution. It will typically ignore the outliers and focus on the most frequently appearing values. As a result, if used for the integer values, the minimum of HAM would correspond to the mode of that distribution if it is unimodal. I do not have a proof of this property, but it becomes apparent, given that the square root in (11.9) would reduce the influence of all values lying above one and increase the values of everything that lies between (0, 1) (e.g. $\sqrt{0.16} = 0.4$, but $\sqrt{16} = 4$).

Similar to HAM, one can calculate other fractional losses, which would be even less sensitive to outliers and more focused on the frequently appearing values, e.g. by using the $\sqrt[\alpha]{|e_t|}$ with $\alpha > 1$. This would then correspond to the maximum of Generalised Normal distribution with shape parameter $\beta = \frac{1}{\alpha}$.

11.2.3 LASSO and RIDGE

It is also possible to use LASSO (Tibshirani, 1996) and RIDGE for the estimation of ADAM (James et al., 2017, give a good overview of these losses with examples in R). This was studied in detail for ETS by Pritularga, Svetunkov, and Kourentzes (2023). The losses can be formulated in ADAM as:

$$\begin{aligned} \mathrm{LASSO} =& (1-\lambda) \sqrt{\frac{1}{T} \sum_{j=1}^{T} e_t^2} + \lambda \sum |\hat{\theta}| \\ \mathrm{RIDGE} =& (1-\lambda) \sqrt{\frac{1}{T} \sum_{j=1}^{T} e_t^2} + \lambda \sqrt{\sum \hat{\theta}^2}, \end{aligned} \tag{11.10}$$

where θ is the vector of all parameters in the model except for initial states of ETS and ARIMA components (thus, this includes all smoothing parameters, dampening parameter, AR, MA, and the initial parameters for the explanatory variables) and λ is the regularisation parameter. The idea of these losses is in the shrinkage of parameters. If $\lambda = 0$, then the losses become equivalent to the MSE, and when $\lambda = 1$, the optimiser would minimise the values of parameters, ignoring the MSE part. Pritularga et al. (2023) argues that the initial states of the model do not need to be shrunk (they will be handled by the optimiser automatically) and that the dampening parameter should shrink to one instead of zero (thus enforcing local trend model). Following the same idea, we introduce the following modifications of parameters in the loss in ADAM:

1. The smoothing parameters should shrink towards zero, implying that the respective states will change slower over time (the model becomes less stochastic);
2. Damping parameter is modified to shrink towards one, enforcing no dampening of the trend via $\hat{\phi}' = 1 - \hat{\phi}$;
3. All AR parameters are forced to shrink to one: $\hat{\phi}_i' = 1 - \hat{\phi}_i$ for all i. This means that ARIMA shrinks towards non-stationarity (this idea comes from the connection of ETS and ARIMA, discussed in Section 8.4);
4. All MA parameters shrink towards zero, removing their impact on the model;
5. If there are explanatory variables and the error term of the model is *additive*, then the respective parameters are divided by the standard deviations of the respective variables. In the case of the *multiplicative* error term, nothing is done because the parameters would typically be close to zero anyway (see a Section 10.1).
6. Finally, in order to make λ slightly more meaningful, in the case of an *additive* error model, we also divide the MSE part of the loss by $\mathrm{V}(\Delta y_t)$, where $\Delta y_t = y_t - y_{t-1}$. This sort of scaling helps in cases when there is a trend in the data. We do not do anything for the *multiplicative* error models because, typically, the error is already small.

Remark. There is no good theory behind the shrinkage of AR and MA parameters. More research is required to understand how to shrink them properly.

Remark. The `adam()` function does not select the most appropriate `lambda` and will set it equal to zero if the user does not provide it. It is up to the user to try out different `lambda` and select the one that minimises a chosen error measure. This can be done, for example, using rolling origin procedure (Section 2.4).

11.2.4 Custom losses

It is also possible to use other non-standard loss functions for ADAM estimation. The `adam()` function allows doing that via the parameter `loss`. For example, we could estimate an ETS(A,A,N) model on the `BJsales` data using an absolute cubic loss (note that the parameters `actual`, `fitted`, and `B` are compulsory for the function):

```
lossFunction <- function(actual, fitted, B){
  return(mean(abs(actual-fitted)^3));
}
adam(BJsales, "AAN", loss=lossFunction, h=10, holdout=TRUE)
```

where `actual` is the vector of actual values y_t, `fitted` is the estimate of the one-step-ahead point forecast $\hat{y}_t$, and B is the vector of all estimated parameters, $\hat{\theta}$. The syntax above allows using more advanced estimators, such as, for example, M-estimators (Barrow, Kourentzes, Sandberg, & Niklewski, 2020).

11.2.5 Examples in R

`adam()` has two parameters, one regulating the assumed `distribution`, and another one, regulating how the model will be estimated, and what `loss` will be used for these purposes. Here are examples with combinations of different losses and the Inverse Gaussian distribution for ETS(M,A,M) on `AirPassengers` data. We start with likelihood, MSE, MAE, and HAM:

```
adamETSMAMAir <- vector("list",6)
names(adamETSMAMAir) <-
  c("likelihood", "MSE", "MAE", "HAM", "LASSO", "Huber")
adamETSMAMAir[[1]] <- adam(AirPassengers, "MAM",h=12, holdout=TRUE,
                           distribution="dinvgauss",
                           loss="likelihood")
adamETSMAMAir[[2]] <- adam(AirPassengers, "MAM", h=12, holdout=TRUE,
                           distribution="dinvgauss",
                           loss="MSE")
adamETSMAMAir[[3]] <- adam(AirPassengers, "MAM", h=12, holdout=TRUE,
                           distribution="dinvgauss",
                           loss="MAE")
adamETSMAMAir[[4]] <- adam(AirPassengers, "MAM", h=12, holdout=TRUE,
                           distribution="dinvgauss",
                           loss="HAM")
```

In these cases, the models assuming the same distribution for the error term are estimated using likelihood, MSE, MAE, and HAM. Their smoothing parameters should differ, with MSE producing fitted values closer to the mean, MAE – closer to the median, and HAM – closer to the mode (but not exactly the mode) of the distribution.

In addition, we introduce ADAM ETS(M,A,M) estimated using LASSO with arbitrarily selected $\lambda = 0.9$:

```
adamETSMAMAir[[5]] <- adam(AirPassengers, "MAM", h=12, holdout=TRUE,
                           distribution="dinvgauss",
                           loss="LASSO", lambda=0.9)
```

And, finally, we estimate the same model using a custom loss, which in this case is Huber loss (Huber, 1992) with a threshold of 1.345:

```
# Huber loss with a threshold of 1.345
lossFunction <- function(actual, fitted, B){
    errors <- actual-fitted;
    return(sum(errors[errors<=1.345]^2) +
               sum(abs(errors)[errors>1.345]));
}
adamETSMAMAir[[6]] <- adam(AirPassengers, "MAM", h=12, holdout=TRUE,
                           distribution="dinvgauss",
                           loss=lossFunction)
```

Now we can compare the performance of the six models. First, we can compare the smoothing parameters:

```
sapply(adamETSMAMAir,"[[","persistence") |>
    round(3)
```

```
##       likelihood   MSE   MAE   HAM LASSO Huber
## alpha      0.776 0.775 0.888 0.909 0.002 0.664
## beta       0.000 0.000 0.019 0.017 0.002 0.001
## gamma      0.000 0.000 0.000 0.008 0.000 0.001
```

What we observe in this case is that LASSO has the lowest smoothing parameter α because it is shrunk directly in the model estimation. Also note that likelihood and MSE give similar values. They both rely on squared errors, but not in the same way because the likelihood of Inverse Gaussian distribution has some additional elements (see Section 11.1).

Unfortunately, we do not have information criteria for models 2 – 6 in this case because the likelihood function is not maximised with these losses, so it's not possible to compare them via the in-sample statistics. But we can compare their holdout sample performance:

```
round(sapply(adamETSMAMAir,"[[","accuracy"),
      3)[c("ME","MAE","MSE"),]
```

```
##     likelihood     MSE     MAE     HAM   LASSO    Huber
## ME      12.021   9.347   5.048   5.356   6.791   47.426
```

```
## MAE      22.792  20.698  17.059  17.389  18.116   51.732
## MSE     752.472 680.667 524.190 551.972 542.632 3552.266
```

And we can also produce forecasts and plot them for the visual inspection (Figure 11.1):

```
adamETSMAMAirForecasts <- lapply(adamETSMAMAir, forecast,
                                 h=12, interval="empirical")
layout(matrix(c(1:6),3,2,byrow=TRUE))
for(i in 1:6){
    plot(adamETSMAMAirForecasts[[i]],
         main=paste0("ETS(MAM) estimated using ",
                     names(adamETSMAMAir)[i]))
}
```

What we observe in Figure 11.1 is that different losses led to different forecasts and prediction intervals (we used the empirical ones, discussed in Subsection 18.3.5). What can be done to make this practical is the rolling origin evaluation (Section 2.4) for different losses and then comparing forecast errors between them to select the most accurate approach.

11.3 Multistep losses

Another family of losses that can be used to estimate ADAM is the multistep losses. The idea behind them is to produce the point forecast for h steps ahead from each observation in-sample and then calculate a measure based on that, which the optimiser will then minimise to find the most suitable values of parameters. There is a lot of literature on this topic. Svetunkov, Kourentzes, and Killick (2023) studied them in detail, showing that their usage implies shrinkage of smoothing parameters in ETS models. In this section, we will discuss the most popular multistep losses, see what they imply, and make a connection between them and predictive likelihoods from the ADAM.

11.3.1 MSE_h – MSE for h steps ahead

One of the simplest estimators is the MSE_h – mean squared h steps ahead error:

$$\mathrm{MSE}_h = \frac{1}{T-h} \sum_{t=1}^{T-h} e_{t+h|t}^2, \tag{11.11}$$

where $e_{t+h|t}$ is the conditional h steps ahead forecast error on the observation $t+h$ produced from the point at time t. In the case of the additive error model,

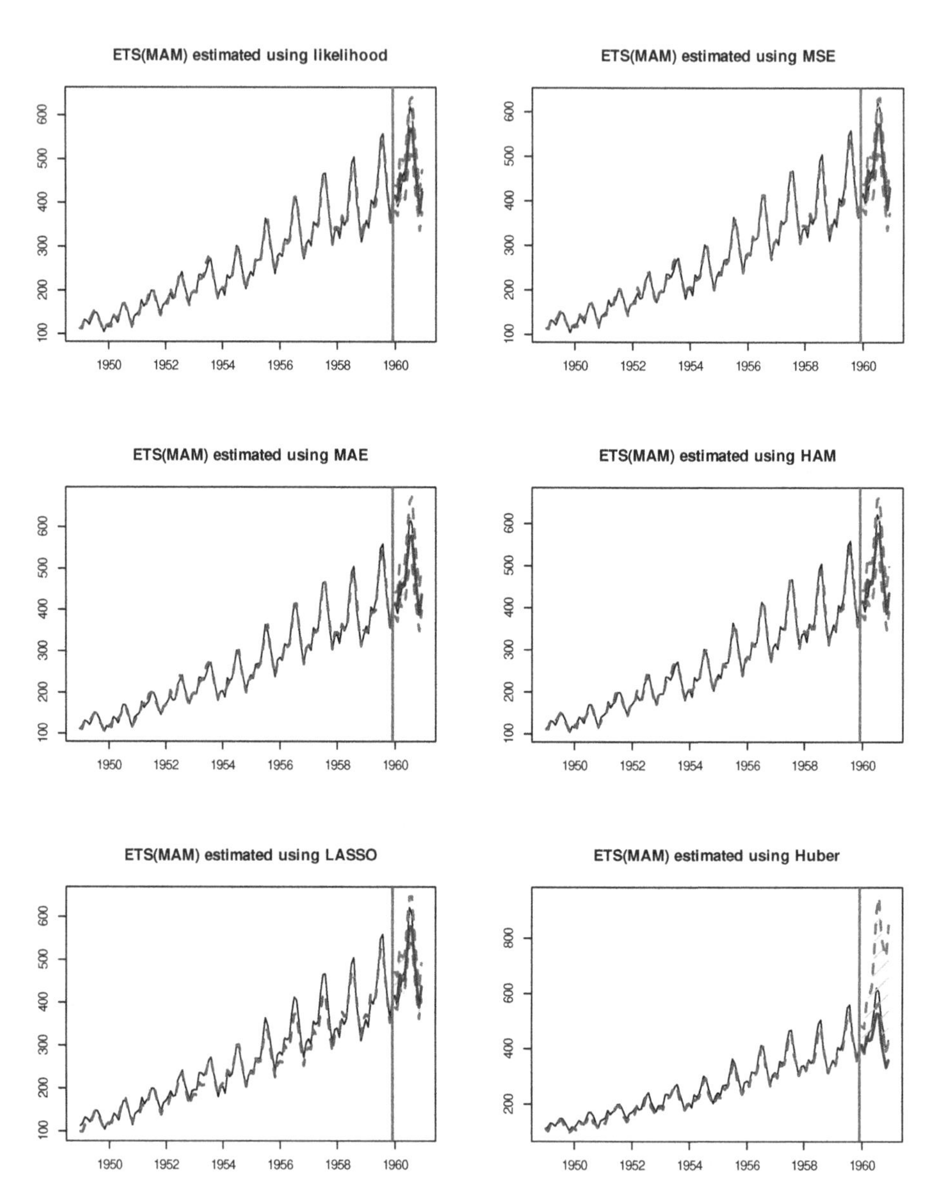

FIGURE 11.1 Forecasts from an ETS(M,A,M) model estimated using different loss functions.

it is calculated as $e_{t+h|t} = y_{t+h} - \hat{y}_{t+h}$, while in the case of the multiplicative one it is $e_{t+h|t} = \frac{y_{t+h}-\hat{y}_{t+h}}{\hat{y}_{t+h}}$. This estimator is sometimes used to fit a model several times, for each horizon from 1 to h steps ahead, resulting in h different values of parameters for each $j = 1, \dots, h$. The estimation process, in this case, becomes at least h times more complicated than estimating one model but is reported to result in increased accuracy (see for example Kourentzes, Li, & Strauss, 2019). Svetunkov, Kourentzes, and Killick (2023) show that MSE_h is proportional to the h steps ahead forecast error variance $V(y_{t+h}|t) = \sigma^2_h$. This implies that the minimisation of (11.11) leads to the minimisation of the variance σ^2_h and in turn to the minimisation of both one step ahead MSE and a combination of smoothing parameters of a model. This becomes more obvious in the case of pure additive ETS (Section 5.1), where the analytical formulae for variance(from Section 5.3) are available. In the case of ETS, the parameters are shrunk towards zero, making the model deterministic. The effect is softened on large samples when the ratio $\frac{T-h}{T-1}$ becomes close to one. In the case of ARIMA, the shrinkage mechanism is similar, making models closer to the deterministic ones. However, shrinkage direction in ARIMA is more complicated than in ETS and differs from one model to another. The shrinkage strength is proportional to the forecast horizon h and is weakened with the increase of the sample size.

Svetunkov, Kourentzes, and Killick (2023) demonstrate that the minimum of MSE_h corresponds to the maximum of the predictive likelihood based on the Normal distribution, assuming that $\epsilon_t \sim N(0, \sigma^2)$. The log-likelihood in this case is:

$$\ell_{\mathrm{MSE}_h}(\boldsymbol{\theta}, \sigma^2_h|\mathbf{y}) = -\frac{T-h}{2}\left(\log(2\pi) + \log \sigma^2_h\right) - \frac{1}{2}\sum_{t=1}^{T-h}\left(\frac{\eta^2_{t+h|t}}{\sigma^2_h}\right), \quad (11.12)$$

where $\eta_{t+h|t} \sim N(0, \sigma^2_h)$ is the h steps ahead forecast error, conditional on the information available at time t, $\boldsymbol{\theta}$ is the vector of all estimated parameters of the model, and $\mathbf{y}$ is the vector of y_{t+h} for all $t = 1, .., T-h$. The MLE of the scale parameter in (11.12) coincides with the MSE_h:

$$\hat{\sigma}^2_h = \frac{1}{T-h}\sum_{t=1}^{T-h} e^2_{t+h|t}, \quad (11.13)$$

where $e_{t+h|t}$ is the in-sample estimate of the η_{t+h}. The formula (11.12) can be used for the calculation of information criteria and in turn for the model selection in cases, when MSE_h is used for the model estimation.

Svetunkov, Kourentzes, and Killick (2023) demonstrate that (11.11) is more efficient than the conventional MSE_1 when the true smoothing parameters are close to zero and is less efficient otherwise. On smaller samples, MSE_h produces biased estimates of parameters due to shrinkage. This can still be considered an advantage if you are interested in forecasting and do not want the smoothing parameters to vary substantially from one sample to another.

11.3.2 TMSE – Trace MSE

An alternative to MSE_h is to in-sample produce 1 to h steps ahead forecasts and calculate the respective forecast errors. Then, based on that, we can calculate the overall measure, which we will call "Trace MSE":

$$\mathrm{TMSE} = \sum_{j=1}^{h} \frac{1}{T-h} \sum_{t=1}^{T-h} e_{t+j|t}^2 = \sum_{j=1}^{h} \mathrm{MSE}_j. \tag{11.14}$$

The benefit of this estimator is in minimising the error for the whole 1 to h steps ahead in one model – there is no need to construct h models, minimising MSE_j for $j = 1, ..., h$. However, this comes with a cost: typically, short-term forecast errors have lower MSE than the longer-term ones, so if we sum their squares up, we are mixing different values, and the minimisation will be done mainly for the ones on the longer horizons.

TMSE does not have a related predictive likelihood, so it is difficult to study its properties. Still, the simulations show that it tends to produce less biased and more efficient estimates of parameters than MSE_h, especially in cases of higher smoothing parameters. Kourentzes, Trapero, and Barrow (2019) showed that TMSE performs well compared to the conventional MSE_1 and MSE_h in terms of forecasting accuracy and does not take as much computational time as the estimation of h models using MSE_h.

11.3.3 GTMSE – Geometric Trace MSE

An estimator that addresses the issues of TMSE, is the GTMSE, which is derived from a so-called General Predictive Likelihood (GPL by Clements & Hendry, 1998; Svetunkov, Kourentzes, & Killick, 2023). The word "Geometric" sums up how the value is calculated:

$$\mathrm{GTMSE} = \sum_{j=1}^{h} \log \left(\frac{1}{T-h} \sum_{t=1}^{T-h} e_{t+j|t}^2 \right) = \sum_{j=1}^{h} \log \mathrm{MSE}_j. \tag{11.15}$$

Logarithms in the formula (11.15) bring the MSEs on different horizons to the similar level so that both short term and long term errors are minimised with similar power. As a result, the shrinkage effect in this estimator is milder than in MSE_h and TMSE, and the estimates of parameters are less biased and more efficient on smaller samples. It still has the benefits of other multistep estimators, shrinking the parameters towards zero. Although it is possible to derive a predictive likelihood that would be maximised when GTMSE is minimised, it relies on unrealistic assumptions of independence of multistep forecast errors (they are always correlated as long as smoothing parameters are not zero, Svetunkov, Kourentzes, & Killick, 2023).

11.3.4 MSCE – Mean Squared Cumulative Error

This estimator aligns the loss function with a specific inventory decision: ordering based on the lead time h:

$$\mathrm{MSCE} = \frac{1}{T-h} \sum_{t=1}^{T-h} \left(\sum_{j=1}^{h} e_{t+j|t} \right)^2 . \tag{11.16}$$

Kourentzes, Li, and Strauss (2019) demonstrated that it produced more accurate forecasts in cases of intermittent demand and leads to fewer revenue losses. Svetunkov, Kourentzes, and Killick (2023) showed that the shrinkage effect is much stronger in this estimator than in the others discussed in this section. In addition, it is possible to derive a predictive log-likelihood related to this estimator (assuming normality of the error term):

$$\ell_{\mathrm{MSCE}}(\theta, \varsigma_h^2 | \mathbf{y}_c) = -\frac{T-h}{2} \left(\log(2\pi) + \log \varsigma_h^2\right) - \frac{1}{2} \sum_{t=1}^{T-h} \left(\frac{\left(\sum_{j=1}^{h} \eta_{t+j|t}\right)^2}{2\varsigma_h^2} \right), \tag{11.17}$$

where $\mathbf{y}_c$ is the cumulative sum of actual values, the vector of $y_{c,t} = \sum_{j=1}^{h} y_{t+j}$ for all $t = 1, \dots, T-h$, and ς_h^2 is the variance of the cumulative error term, the MLE of which is equal to (11.16). Having the likelihood (11.17), permits the model selection and combination using information criteria (Section 16.4 Svetunkov, 2022) and also means that the parameters estimated using MSCE will be asymptotically consistent and efficient.

11.3.5 GPL – General Predictive Likelihood

Finally, Svetunkov, Kourentzes, and Killick (2023) studied the General Predictive Likelihood for a normally distributed variable from Clements and Hendry (1998), p.77, the logarithmic version of which can be written as:

$$\ell_{\mathrm{GPL}_h}(\theta, \mathbf{\Sigma} | \mathbf{Y}) = -\frac{T-h}{2} \left(h \log(2\pi) + \log |\mathbf{\Sigma}|\right) - \frac{1}{2} \sum_{t=1}^{T} \left(\mathbf{E_t}' \mathbf{\Sigma}^{-1} \mathbf{E_t} \right), \tag{11.18}$$

where $\mathbf{\Sigma}$ is the conditional covariance matrix for the vector of variables $\mathbf{y}_t = \begin{pmatrix} y_{t+1|t} & y_{t+2|t} & \dots & y_{t+h|t} \end{pmatrix}$, $\mathbf{Y}$ is the matrix consisting of $\mathbf{y}_t$ for all $t = 1, \dots, T-h$, and $\mathbf{E_t}' = \begin{pmatrix} \eta_{t+1|t} & \eta_{t+2|t} & \dots & \eta_{t+h|t} \end{pmatrix}$ is the vector of 1 to h steps ahead forecast errors. Svetunkov, Kourentzes, and Killick (2023) showed that the maximisation of the likelihood (11.18) is equivalent to minimisation of the generalised variance of the error term, $|\hat{\mathbf{\Sigma}}|$, where:

$$\hat{\mathbf{\Sigma}} = \frac{1}{T-h} \sum_{t=1}^{T-h} \mathbf{E_t} \mathbf{E_t}' = \begin{pmatrix} \hat{\sigma}_1^2 & \hat{\sigma}_{1,2} & \dots & \hat{\sigma}_{1,h} \\ \hat{\sigma}_{1,2} & \hat{\sigma}_2^2 & \dots & \hat{\sigma}_{2,h} \\ \vdots & \vdots & \ddots & \vdots \\ \hat{\sigma}_{1,h} & \hat{\sigma}_{2,h} & \dots & \hat{\sigma}_h^2 \end{pmatrix}, \tag{11.19}$$

where $\hat{\sigma}_{i,j}$ is the covariance between i-th and j-th steps ahead forecast errors. Svetunkov, Kourentzes, and Killick (2023) show that this estimator encompassess all the other estimators discussed in this section: minimising MSE_h is equivalent to minimising the $\hat{\sigma}^2_h$; minimising TMSE is equivalent to minimising the trace of the matrix $\hat{\mathbf{\Sigma}}$; minimising GTMSE is the same as minimising the determinant of $\hat{\mathbf{\Sigma}}$ but with the restriction that all off-diagonal elements are equal to zero; minimising MSCE produces the same results as minimising the sum of all elements in $\hat{\mathbf{\Sigma}}$. However, the maximum of GPL is equivalent to the maximum of the conventional one step ahead likelihood for a Normal model in case when all the basic assumptions (discussed in Subsection 1.4.1) hold. In other cases, they would be different, but it is still not clear, whether the difference would be favouring the conventional likelihood of the GPL. Nonetheless, GPL, being the likelihood, guarantees that the estimates of parameters will be efficient and consistent and permits model selection and combination via information criteria.

When it comes to models with a multiplicative error term, the formula of GPL (11.18) will need to be amended by analogy with the log-likelihood of Normal distribution in the same situation (Table 11.2):

$$\begin{aligned} \ell_{\mathrm{GPL}_h}(\theta, \mathbf{\Sigma} | \mathbf{Y}) = & -\frac{T-h}{2} \left(h \log(2\pi) + \log |\mathbf{\Sigma}| \right) - \frac{1}{2} \sum_{t=1}^{T} \left(\mathbf{E_t'} \mathbf{\Sigma}^{-1} \mathbf{E_t} \right) \\ & - \sum_{t=1}^{T-h} \sum_{j=1}^{h} \log |\mu_{y,t+j|t}|, \end{aligned} \tag{11.20}$$

where the term $\sum_{t=1}^{T-h} \sum_{j=1}^{h} \log |\mu_{y,t+j|t}|$ appears because we assume that the actual value h steps ahead follows multivariate Normal distribution with the conditional expectation $\mu_{y,t+j|t}$. Note that this is only a very crude approximation, as the conditional distribution for h steps ahead is not defined for multiplicative error models. So, when dealing with GPL, it is recommended to use pure additive models only.

11.3.6 Other multistep estimators

It is also possible to derive multistep estimators based on MAE, HAM, and other error measures. `adam()` unofficially supports the following multistep losses by analogy with MSE_h, TMSE, and MSCE discussed in this section:

1. MAE_h;
2. TMAE;
3. MACE;
4. HAM_h;
5. THAM;
6. CHAM.

When calculating likelihoods based on these losses, `adam()` will assume Laplace distribution for (1) – (3) and S distribution for (4) – (6) if the user does not specify the `distribution` parameter.

11.3.7 An example in R

In order to see how different estimators perform, we will apply an ETS(A,A,N) model to Box-Jenkins sales data, setting forecast horizon $h = 10$:

```
adamETSAANBJ <- vector("list",6)
names(adamETSAANBJ) <- c("MSE","MSEh","TMSE","GTMSE","MSCE","GPL")
for(i in 1:length(adamETSAANBJ)){
    adamETSAANBJ[[i]] <- adam(BJsales, "AAN", h=10, holdout=TRUE,
                              loss=names(adamETSAANBJ)[i])
}
```

We can compare the smoothing parameters of these models to see how the shrinkage effect worked in different estimators:

```
round(sapply(adamETSAANBJ,"[[","persistence"),5)
```

```
##             MSE MSEh TMSE   GTMSE MSCE GPL
## alpha 1.00000    1    1 1.00000    1   1
## beta  0.23915    0    0 0.14617    0   0
```

The table above shows that β is close to zero for the estimators that impose harder shrinkage on parameters: MSE_h, TMSE, and MSCE. MSE does not shrink the parameters, while GTMSE has a mild shrinkage effect. While the models estimated using these losses are in general not comparable in-sample (although MSE, MSE_h, MSCE, and GPL could be compared via information criteria if they are scaled appropriately), they are comparable on the holdout via the error measures:

```
round(sapply(adamETSAANBJ,"[[","accuracy"),5)[c("ME","MAE","MSE"),]
```

```
##          MSE    MSEh    TMSE    GTMSE    MSCE     GPL
## ME   3.22900 1.06479 1.05233  3.44962 1.04604 0.95515
## MAE  3.34075 1.41264 1.40153  3.53730 1.39593 1.31495
## MSE 14.41862 2.89067 2.85880 16.26344 2.84288 2.62394
```

In this case, ETS(A,A,N) estimated using GPL produced a more accurate forecast than the other estimators. Repeating the experiment on many samples and selecting the approach that produces more accurate forecasts would allow choosing the most appropriate one for the specific model on specific data.

11.4 Initialisation of ADAM

To construct a model, we need to initialise it, defining the values of initial states of the model. In the case of the non-seasonal model, this means estimating values of $\mathbf{v}_0$. In the case of the seasonal one, it is $\mathbf{v}_{-m+1}, \dots, \mathbf{v}_0$, where m is the seasonal lag. There are different ways of doing that, but here we only discuss the following three:

1. Optimisation of initials;
2. Backcasting;
3. Provided values.

The first option implies that the values of initial states are found in the same procedure as the other parameters of the model. This is what Hyndman et al. (2002) suggested doing. (2) means that the initials are refined iteratively when the model is fit to the data from observation $t = 1$ to $t = T$ and then going backwards to get values for $t = 0$. Finally, (3) is when a user knows initials and provides them to the model.

As a side note, we assume in ADAM that the model is initialised at the moment just before $t = 1$. We do not believe that it was initialised at some point before the Big Bang (as ARIMA typically does), and we do not initialise it at the start of the sample. This way, we make all models in ADAM comparable, making them work on the same sample, no matter how many differences are taken or how many seasonal components we define.

11.4.1 Optimisation vs backcasting

In the case of **optimisation**, all the model parameters are estimated together. This includes (depending on the type of model, in the order used in optimiser):

- Smoothing parameters of ETS;
- Smoothing parameters for the regression part of the model (from Section 10.3);
- Dampening parameter of ETS;
- Parameters of ARIMA: first AR(p), then MA(q);
- Initials of ETS;
- Initials of ARIMA;
- Initial values for parameters of explanatory variables;
- Constant/drift for ARIMA;
- Other additional parameters that are needed by assumed distributions.

The more complex the selected model is, the more parameters we will need to estimate, and all of this will happen in one and the same iterative process in the optimiser:

1. Choose parameters;
2. Fit the model;
3. Calculate loss function;
4. Compare the loss with the previous one;
5. Update the parameters based on (4);
6. Go to (2) and repeat until a specific criterion is met.

The user can specify the stopping criteria. There are several options accepted by the optimiser of `adam()`:

1. Maximum number of iterations (`maxeval`), which by default is equal to $40 \times k$ in case of ETS/ARIMA and $100 \times k$ for the model with explanatory variables, where k is the number of all estimated parameters;
2. The relative precision of the optimiser (`xtol_rel`) with the default value of 10^{-6}, which regulates the relative change of parameters;
3. The absolute precision of the optimiser (`xtol_abs`) with the default value of 10^{-8}, which regulates the absolute change of parameters;
4. The stopping criterion in case of the relative change in the loss function (`ftol_rel`) with the default value of 10^{-8}.

All these parameters are explained in more detail in the documentation of the `nloptr()` function from the `nloptr` package for R (Johnson, 2021), which handles the estimation of ADAM. `adam()` accepts several other stopping criteria, which can be found in the documentation of the function.

Remark. `adam()` can also print the results of the optimisation via the `print_level` parameter, which is defined in the same way as in the `nloptr()` function, but with an additional option of `print_level=41`, which will print the results of the optimisation, without producing step-by-step outputs.

The mechanism explained above can slow down substantially if a complex model is constructed, and it might take a lot of time and manual tuning of parameters to get to the optimum. In some cases, reducing the number of estimated parameters is worth considering, and one way of doing that is backcasting.

In case of **backcasting** we do not need to estimate initials of ETS, ARIMA, and regression. What the model does in this case is goes through the series from $t = 1$ to $t = T$, fitting to the data, and then reverses and goes back from $t = T$ to $t = 1$ based on the following state space model:

$$\begin{aligned} y_t &= w(\mathbf{v}_{t+l}) + r(\mathbf{v}_{t+l})\epsilon_t \\ \mathbf{v}_t &= f(\mathbf{v}_{t+l}) + g(\mathbf{v}_{t+l})\epsilon_t \end{aligned} . \tag{11.21}$$

The new values of $\mathbf{v}_t$ for $t < 1$ are then used to fit the model to the data again. The procedure can be repeated several times for the initial states to converge to more reasonable values. `adam()` does that only two times.

The backcasting procedure implies the extended fitting process for the model, removing the need to estimate all the initials. It works exceptionally well on large samples of data (thousands of observations) and with models with several seasonal components. The bigger your model is, the more time the optimisation will take, and the more likely backcasting would do better. On the other hand, you might also prefer backcasting to optimisation on small samples when you do not have more than two seasons of data – estimation of initial seasonal components might become challenging and can lead to overfitting.

When discussing specific models, ADAM ARIMA works better (faster and more accurate) with backcasting than with optimisation because it does not need to estimate as many parameters as in the latter case. On the other hand, ADAM ETS typically works quite well in case of optimisation, when there is enough data to train the model on. Last but not least, if you introduce explanatory variables, the mechanism implemented in `adam()` will optimise their parameters instead of backcasting. In the case of a dynamic ETSX/ARIMAX (Section 10.3) this will mean that the initial values of the part of the state vector for explanatory variables will be estimated as well. If you want them not to be optimised you can use `initial="complete"` to do full backcasting of all elements of the state vector.

It is also important to note that **the information criteria of models with backcasting are typically lower than in the case of the optimised initials**. This is because the difference in the number of estimated parameters is substantial in these two cases, and the models are initialised differently. So, it is advised not to mix the model selection between the two initialisation techniques, although there is no theoretical ground for forbidding it.

Nonetheless, no matter what initialisation method is used, we need to start the fitting process from $t = 1$. This cannot be done unless we provide some starting (pre-initialised) values of parameters to the optimiser. The better we guess the starting values, the faster the optimiser will converge to the optimum. `adam()` uses several heuristics at this stage, explained in more detail in the following subsections.

11.4.2 Starting optimisation of parameters

Remark. In this subsection, we discuss how the values of smoothing parameters, damping parameters, and coefficients of ARIMA are preset before the initialisation. All the things discussed here are heuristics, developed based on my experience and many experiments with ADAM.

Depending on the model type, the vector of estimated parameters will have different lengths. We start with smoothing parameters of ETS:

1. For the unsafe mixed models ETS(A,A,M), ETS(A,M,A), ETS(M,A,A),

and ETS(M,A,M): $\hat{\alpha} = 0.01$, $\hat{\beta} = 0$ and $\hat{\gamma} = 0$. This is needed because the models listed above are susceptible to the changes in smoothing parameters and might fail for time series with actual values close to zero;

2. For one of the most complicated and sensitive models, ETS(M,M,A), $\hat{\alpha} = \hat{\beta} = \hat{\gamma} = 0$. The combination of additive seasonality and the multiplicative trend is the most difficult one. The multiplicative error makes estimation even more challenging in cases of low-level data. So starting from the deterministic model, that will work for sure is a safe option;
3. ETS(M,A,N) is slightly easier to estimate than ETS(M,A,M) and ETS(M,A,A), so $\hat{\alpha} = 0.2$, $\hat{\beta} = 0.01$. The low value for the trend is needed to avoid the difficult situations with low level data, when the fitted values become negative;
4. ETS(M,M,N) and ETS(M,M,M) have $\hat{\alpha} = 0.1$, $\hat{\beta} = 0.05$, and $\hat{\gamma} = 0.01$, making the trend and seasonal components a bit more conservative. The high values are typically not needed in this model as they might lead to explosive trends;
5. Other models with multiplicative components (ETS(M,N,N), ETS(M,N,A), ETS(M,N,M), ETS(A,N,M), ETS(A,M,N), and ETS(A,M,M)) are slightly easier to estimate and harder to break, so their parameters are set to $\hat{\alpha} = 0.1$, $\hat{\beta} = 0.05$, and $\hat{\gamma} = 0.05$;
6. Finally, pure additive models are initialised with $\hat{\alpha} = 0.1$, $\hat{\beta} = 0.05$, and $\hat{\gamma} = 0.11$. Their parameter space is the widest, and the models do not break on any data.

The smoothing parameter for the explanatory variables (Section 10.3) is set to $\hat{\delta} = 0.01$ in case of additive error and $\hat{\delta} = 0$ in case of the multiplicative one. The latter is done because the model might break if some ETS components are additive.

If a dampening parameter is needed in the model, then its starting value is $\hat{\phi} = 0.95$.

In the case of ARIMA, the parameters are pre-initialised based on ACF and PACF (Sections 8.3.2 and 8.3.3). First, the in-sample actual values are differenced, according to the selected order D_j for all $j = 0, \dots, n$, after which the ACF and PACF are calculated. Then the initials for AR parameters are taken from the PACF, while the initials for MA parameters are taken from the ACF, making sure that the sum of parameters is not greater than one in both cases. If it is, then the parameters are renormalised to satisfy the condition. This mechanism aims to get a potentially correct direction towards the optimal parameters of the model and makes sure that the initial values meet the basic stationarity and invertibility conditions. In cases when it is not possible to calculate ACF and PACF for the specified lags and orders, AR parameters are set to -0.1, while the MA parameters are set to 0.1, making sure that the conditions mentioned above hold.

In case of Generalised Normal distribution, the shape parameter is set to 2 (if it is estimated), making the optimiser start from the conventional Normal distribution.

The pre-initialisations described above guarantee that the model is estimable for a wide variety of time series and that the optimiser will reach the optimum in a limited time. If it does not work for a specific case, a user can provide their vector of pre-initialised parameters via the parameter `B` in the ellipsis of the model. Furthermore, the typical bounds for the parameters can be tuned as well. For example, the bounds for smoothing parameters in ADAM ETS are (-5, 5), and they are needed only to simplify the optimisation procedure. The function will check the violation of either `usual` or `admissible` bounds inside the optimiser, but having some ideas of where to search for optimal parameters, helps. A user can provide their vector for the lower bound via `lb` and the upper one via `ub`.

11.4.3 Starting optimisation of ETS states

After defining the pre-initial parameters, we need to provide similar values for the initial state vector $\mathbf{v}_t$. The steps explained below are based on my experience and typically lead to a robust model. The pre-initialisation of the states of ADAM ETS differs depending on whether the model is seasonal or not. If it is **seasonal**, then the multiple seasonal decomposition is done using the `msdecompose()` function from the `smooth` package with the seasonality set to "multiplicative" if either the error or seasonal component of ETS is multiplicative. After that:

- Initial level is set to be equal to the first initial value from the function (which is the back forecasted de-seasonalised series);
- The value is corrected if regressors are included to remove their impact on the value (either by subtracting the fitted of the regression part or by dividing by them – depending on the error type);
- If the trend is additive and seasonality is multiplicative, then the trend component is obtained by multiplying the initial level and trend from the decomposition (remember, the assumed model is multiplicative in this case) and then subtracting the previous level from the resulting value;
- If the trend is multiplicative and seasonality is additive, then the initials are added and then divided by the previous level to get the initial multiplicative trend component;
- If there is no seasonality and the trend is multiplicative, then the initial trend is set to 1. This is done to avoid the potentially explosive behaviour in the model;

- If the trend is multiplicative and the level is negative, then the level is substituted by the first actual value. This might happen in some weird cases of time series with low values;
- When it comes to seasonal components, if we have a pure additive or a pure multiplicative ETS model or ETS(A,Z,M), we use the seasonal indices obtained from the `msdecompose()` function (discussed in Subsection 3.2.3), making sure that they are normalised. The type of seasonality in `msdecompose()` corresponds to the seasonal component of ETS in this case, and nothing additional needs to be done;
- The situation is more challenging with ETS(M,Z,A), for which the decomposition would return the multiplicative seasonal components. To convert them to the additive ones, we take their logarithm and multiply them by the minimum value of the actual time series. This way, we guarantee that the seasonal components are closer to the optimal ones.

In the case of the **non-seasonal** model, the algorithm is more straightforward:

- The initial level is equal to either arithmetic or geometric mean (depending on the type of trend component) of the first $\max(m_1, \dots, m_n)$ observations, where m_j is the model lag (e.g. in case of ARIMA(1,1,2), the components will have lags of 1 and 2). If the length of this mean is smaller than 20% of the sample, then the arithmetic mean of the first 20% of actual values is used;
- If regressors are included, then the value is modified, similarly to how it is done in the seasonal case discussed above;
- If the model has an additive trend, then its initial value is equal to the mean difference between first $\max(m_1, \dots, m_n)$ observations;
- In the case of multiplicative trend, the initial value is equal to the geometric mean of ratios between first $\max(m_1, \dots, m_n)$ observations.

In cases of the small samples (less than two seasonal periods), the procedure is similar to the one above. However, the seasonal indices are obtained by taking the actual values and either subtracting the arithmetic mean from them or dividing them by the geometric one of the first m_j observations, depending on the seasonality type, normalising them afterwards.

Finally, to ensure that the safe initials were provided, for the ETS(M,Z,Z) models, if the initial level contains a negative value, it is substituted by the Global Mean of the series.

The pre-initialisation described here is not simple, but it guarantees that any ETS model can be constructed and estimated to almost any data. Yes, there might still be some issues with mixed ETS models, but the mechanism used in ADAM is quite robust.

11.4.4 Starting optimisation of ARIMA states

ADAM ARIMA models have as many states as the number of polynomials K (see Subsection 9.1.2). Each state $v_{i,t}$ needs to be initialised with i values (e.g. 1 for the first state, 2 for the second, etc). This leads in general to more initial values for states than the SSARIMA from Svetunkov and Boylan (2020): $\frac{K(K+1)}{2}$ instead of K. However, we can reduce the number of initial seeds to estimate either by using a different initialisation procedure (e.g. backcasting) or using the following trick. First, we take the conditional expectations of all ARIMA states, which leads to:

$$\mathrm{E}(v_{i,t}|t) = \eta_i y_t \text{ for } t = \{-K+1, -K+2, \dots, 0\}, \tag{11.22}$$

and then we use these expectations for the initialisation of ARIMA states. This still implies calculating a lot of initials, but we can further reduce their number. We can express the actual value in terms of the state and error from (9.4) for the last state K:

$$y_t = \frac{v_{K,t} - \theta_K \epsilon_t}{\eta_K}. \tag{11.23}$$

We select the last state K because it has the highest number of initials to estimate among all states. We can then insert the value (11.23) in each formula (11.22) for each state for $i = \{1, 2, \dots, K-1\}$ and take their expectations:

$$\mathrm{E}(v_{i,t}|t) = \frac{\eta_i}{\eta_K} \mathrm{E}(v_{K,t}|t) \text{ for } t = \{-i+1, -i+2, \dots, 0\}. \tag{11.24}$$

This formula shows how the expectation of each state i depends on the expectation of the state K. We can use it to propagate the values of the last state to the previous ones. However, this strategy will only work for the states corresponding to the ARI elements of the model. In the case of MA, using the same principle of initialisation via the conditional expectation, we can set the initial MA states to zero and estimate only ARI states. This is a crude but relatively simple way to pre-initialise ADAM ARIMA.

Having said all that, we need to point out that it is advised to use backcasting in the ADAM ARIMA model – this is a more reliable and faster procedure for initialisation of ARIMA than the optimisation.

11.4.5 Starting optimisation of regressor states and constant

When it comes to the initials for the regressors, they are obtained from the parameters of the `alm()` model based on the rules below:

- The model with the logarithm of the response variable is constructed, if the **error term is multiplicative** and one of the following distributions has been selected: Normal, Laplace, S, or Generalised Normal;

- Otherwise, the model is constructed based on the provided formula and selected distribution;
- In any case, the global trend is added to the formula to make sure that its effect on the values of parameters is reduced;
- If the data contains categorical variables (aka "factors" in R), then they are expanded to a set of dummy variables, adding the baseline value to the mix (i.e. not dropping any categories). While the classical multiple regression would not be estimable in this situation, dynamic models like ETSX and ARIMAX can work with the complete set of levels of categorical variables (see discussion in Section 14.9). To get the missing level, the intercept is added to the parameters of dummy variables, after which the obtained vector is normalised. This way, we can get, for example, all seasonal components if we want to model seasonality via X part of the model, not merging one of the components with level (see discussion in Section 10.5).

Finally, the initialisation of constant (if needed in the model) depends on the selected model. In case of ARIMA with all $D_j = 0$, the mean of the data is used. In all other cases, the arithmetic mean of difference or the geometric mean of ratios of all actual values is used depending on the error type. This is because the constant acts as a drift in the model in the case of non-zero differences. In case of ETS, the impact of the constant is removed from the level in ETS and the states of ARIMA by either subtraction or division, again depending on the error term type.

11.4.6 Example in R

All the details discussed above allow us to tell ADAM what values to start from if we want to help it in optimisation. For demonstration purposes, we consider the ETS(A,A,N)+ARIMA(2,0,0) applied to Box-Jenkins data:

```
adamETSBJ <- adam(BJsales, "AAN", orders=c(2,0,0))
```

The function will return the vector of parameters in the form it was used by the optimiser:

```
adamETSBJ$B
```

```
##        alpha         beta      phi1[1]      phi2[1]        level
    trend
##    0.9370660    0.2454796    0.1000000    0.1000000 194.9945583
    0.1622094
## ARIMAState1 ARIMAState2
##   13.1291317    9.8841468
```

It can be amended to help the optimiser if we have an idea of what values we should have or just reused again to get to better set of values:

```
adamETSBJ <- adam(BJsales, "AAN", orders=c(2,0,0),
                  B=adamETSBJ$B)
```

If we are dissatisfied with the result, we can print the solution found by the optimiser to understand why it stopped (we do not provide the output here):

```
adamETSBJ <- adam(BJsales, "AAN", orders=c(2,0,0),
                  B=adamETSBJ$B, print_level=41)
```

But hopefully after several iterations, we will get a better estimates of parameters of the model:

```
adamETSBJ
```

```
## Time elapsed: 0.13 seconds
## Model estimated using adam() function: ETS(AAN)+ARIMA(2,0,0)
## Distribution assumed in the model: Normal
## Loss function type: likelihood; Loss function value: 259.1489
## Persistence vector g:
##  alpha   beta
## 0.9531 0.2398
##
## ARMA parameters of the model:
## AR:
## phi1[1] phi2[1]
##  0.0908  0.0733
##
## Sample size: 150
## Number of estimated parameters: 9
## Number of degrees of freedom: 141
## Information criteria:
##      AIC     AICc      BIC     BICc
## 536.2977 537.5834 563.3934 566.6146
```

Given that we are trying the ETS+ARIMA model, we can use backcasting, which should help the optimiser by reducing the number of estimated parameters:

```
adam(BJsales, "AAN", orders=c(2,0,0),
     initial="backcasting")
```

```
## Time elapsed: 0.08 seconds
## Model estimated using adam() function: ETS(AAN)+ARIMA(2,0,0)
## Distribution assumed in the model: Normal
## Loss function type: likelihood; Loss function value: 258.6827
## Persistence vector g:
##  alpha   beta
## 0.9692 0.2518
##
## ARMA parameters of the model:
## AR:
## phi1[1] phi2[1]
##  0.0287  0.0281
##
## Sample size: 150
## Number of estimated parameters: 5
## Number of degrees of freedom: 145
## Information criteria:
##      AIC     AICc      BIC     BICc
## 527.3654 527.7820 542.4185 543.4624
```

Hopefully this gives an idea of how the estimation of parameters in `adam()` can be fine-tuned.

12

Multiple frequencies in ADAM

When we work with weekly, monthly, or quarterly data, we do not have more than one seasonal cycle. In this case, one and the same pattern can repeat itself only once a year. For example, we might see an increase in ski equipment sales over winter, so the seasonal component for December will typically be higher than the same component in August. However, we might see several seasonal patterns when moving to the data with higher granularity. For example, daily sales of the product will have a time of year seasonal pattern and a day of week one. If we move to hourly data, then the number of seasonal elements might increase to three: the hour of the day, the day of the week, and the time of year. Note that from the modelling point of view, these seasonal patterns should be called either "periodicities" or "frequencies" as the hour of the day cannot be considered a proper "season". But it is customary to refer to them as "seasonality" in forecasting literature.

To correctly capture such a complicated structure in the data, we need to have a model that includes these multiple frequencies. In this chapter, we discuss how this can be done in the ADAM framework for both ETS and ARIMA. In addition, when we move to modelling high granularity data, there appear several fundamental issues related to how the calendar works and how human beings make their lives more complicated by introducing daylight saving-related time changes over the year. Finally, we will discuss a simpler modelling approach, relying on the explanatory variables (mentioned in Chapter 10).

Among the papers related to the topic, we should start with Taylor (2003a), who proposed an Exponential Smoothing model with double seasonality and applied it to energy data. Since then, the topic was developed by Gould et al. (2008), Taylor (2008), Taylor (2010), De Livera (2010), and De Livera, Hyndman, and Snyder (2011). In this chapter, we will discuss some of the proposed models, how they relate to the ADAM framework and can be implemented.

12.1 Model formulation

Multiple Seasonal ARIMA has already been discussed in Subsections 8.2.3 and 9.1. Therefore, here we focus the discussion on ETS.

Roughly, the idea of a model with multiple seasonalities is in introducing additional seasonal components. For the general framework this means that the state vector (for example, in a model with trend and seasonality) becomes:

$$\mathbf{v}'_t = \begin{pmatrix} l_t & b_t & s_{1,t} & s_{2,t} & \dots & s_{n,t} \end{pmatrix}, \tag{12.1}$$

where n is the number of seasonal components (e.g. hour of day, hour of week, and hour of year components). The lag matrix in this case is:

$$\boldsymbol{l}' = \begin{pmatrix} 1 & 1 & m_1 & m_2 & \dots & m_n \end{pmatrix}, \tag{12.2}$$

where m_i is the i-th seasonal periodicity. While, in theory, there can be combinations between additive and multiplicative seasonal components, I argue that such a mixture does not make sense, and the components should align with each other. This means that in the case of ETS(M,N,M), all seasonal components should be multiplicative, while in ETS(A,A,A), they should be additive. This results fundamentally in two types of models:

1. Additive seasonality:

$$\begin{aligned} & y_t = \breve{y}_t + s_{1,t-m_1} + \dots + s_{n,t-m_n} \epsilon_t \\ & \vdots \\ & s_{1,t} = s_{1,t-m_1} + \gamma_1 \epsilon_t \\ & \vdots \\ & s_{n,t} = s_{n,t-m_n} + \gamma_n \epsilon_t \end{aligned}, \tag{12.3}$$

   where $\hat{y}_t$ is the point value based on all non-seasonal components (e.g. $\hat{y}_t = l_{t-1} + \phi b_{t-1}$ in case of damped trend model) and γ_i is the i-th seasonal smoothing parameter;

2. Multiplicative seasonality:

$$\begin{aligned} & y_t = \breve{y}_t \times s_{1,t-m_1} \times \dots \times s_{n,t-m_n} \times (1 + \epsilon_t) \\ & \vdots \\ & s_{1,t} = s_{1,t-m_1} (1 + \gamma_1 \epsilon_t) \\ & \vdots \\ & s_{n,t} = s_{n,t-m_n} (1 + \gamma_n \epsilon_t) \end{aligned}. \tag{12.4}$$

Depending on a specific model, the number of seasonal components can be 1, 2, 3, or more (although more than three might not make much sense from the modelling point of view). But there are two issues with this model:

1. Estimation;
2. Fractional seasonality.

The first one is discussed in Section 12.2. As for the second one, it appears if we think that a year contains 365.25 days, not 365 (because of the leap year). There are several solutions to this. One of those is discussed in Section 12.4. Another one was proposed by De Livera (2010) by introducing new types of components, based on Fourier terms, updated over time via smoothing parameters. This feature is not yet fully supported in `adam()`, but it is possible to substitute some of the seasonal components (especially those that have fractional periodicity) with Fourier terms via explanatory variables and update them over time. The explanatory variables idea was discussed in Chapter 10 and will also be addressed in Section 12.3.

Another issue with models (12.3) and (12.4) is that they have intersecting components, competing for the same space: if we deal with hourly data, we might have hour of day, hour of week, and hour of year seasonality. However it might be more logical not to have such intersections and introduce hour of day, day of week, and week of year instead. This is discussed to some extent in Section 12.3.

12.2 Estimation of multiple seasonal model

While the main principles of model estimation discussed in Chapter 11 can be widely used for the multiple seasonal models, there are some specific aspects that require additional attention. They mainly apply to ADAM ETS and to ADAM ARIMA.

12.2.1 ADAM ETS issues

Estimating a multiple seasonal ETS model is challenging because it implies a large optimisation task. The number of parameters related to seasonal components is equal in general to $\sum_{j=1}^{n} m_j + n$: $\sum_{j=1}^{n} m_j$ initial values and n smoothing parameters. For example, in the case of hourly data, a triple seasonal model for hours of day, hours of week, and hours of year will have: $m_1 = 24$, $m_2 = 24 \times 7 = 168$, and $m_3 = 24 \times 365 = 8760$, resulting overall in $24 + 168 + 8760 + 3 = 8955$ parameters related to seasonal components to estimate. This is not a trivial task and would take hours to converge to optimum unless the pre-initials (Section 11.4) are already close to optimum. So, if you want to construct multiple seasonal ADAM ETS model, it makes sense to use a different initialisation (see discussion in Section 11.4), reducing the number of estimated parameters. A possible solution in this case is backcasting (Subsection 11.4.1). The number of parameters in our example would reduce from 8955 to 3 (smoothing parameters), substantially speeding up the model estimation process.

Another consideration is a fitting model to the data. In the conventional ETS, the size of the transition matrix is equal to the number of initial states, which makes it too slow to be practical on high-frequency data (multiplication of a 8952×8952 matrix by a vector is a challenging task even for modern computers). But due to the lagged structure of the ADAM (discussed in Section 5), the construction of multiple seasonal models does not take as much time for ADAM ETS because we end up multiplying a matrix of 3×3 by a vector with three rows (skipping level and trend, which would add two more elements). So, in ADAM, the main computational burden comes from recursive relation in the state space model's transition equation because this operation needs to be repeated at least T times, whatever the sample size T is. This is still a computationally expensive task, so you would want to get to the optimum with as few iterations as possible. This gives another motivation for reducing the number of parameters to estimate, and thus for using backcasting.

Another potential simplification would be to use deterministic seasonality for some seasonal frequencies. The possible solution, in this case, is to use explanatory variables (Section 10) for the higher frequency states (see discussion in Section 12.3) or use multiple seasonal ETS, setting some of the smoothing parameters to zero.

Finally, given that we deal with large samples, some states of ETS might become more reactive than needed, having higher than required smoothing parameters. One of the possible ways to overcome this limitation is by using the multistep loss functions (Section 11.3). For example, Kourentzes and Trapero (2018) showed that using such loss functions as TMSE (from Subsection 11.3.2) in the estimation of ETS models on high-frequency data leads to improvements in accuracy due to the shrinkage of parameters towards zero, mitigating the potential overfitting issue. The only problem with this approach is that it is more computationally expensive than the conventional likelihood and thus would take more time than the conventional estimation procedures (at least h times more, where h is the length of the forecast horizon).

12.2.2 ADAM ARIMA issues

It is also possible to fit Multiple Seasonal ARIMA (discussed partially in Subsection 8.2.3) to the high-frequency data, and, for example, Taylor (2010) used triple seasonal ARIMA on the example of two time series and demonstrated that it produced more accurate forecasts than other ARIMAs under consideration, even slightly outperforming ETS. The main issue with ARIMA arises in the order selection stage. While in the case of ETS, one can decide what model to use based on judgment (e.g. there is no apparent trend, and the amplitude increases with the increase of level so we will fit the ETS(M,N,M) model), ARIMA requires more careful consideration of possible orders of the model. Selecting appropriate orders of ARIMA is not a trivial task on its own, but choosing the orders on high-frequency data (where correlations might appear significant just because of the sample size) becomes an even more challenging task than usual. Furthermore, while on monthly data, we typically set maximum AR and MA orders to 3 or 5, this does not have any merit in the case of high-frequency data. If the first seasonal component has a lag of 24, then, in theory, anything up until 24 might be helpful for the model. Long story short, be prepared for the lengthy investigation of appropriate ARIMA orders. While ADAM ARIMA implements an efficient order selection mechanism (see Section 15.2), it does not guarantee that the most appropriate model will be applied to the data. Inevitably, you would need to analyse the residuals of the applied model, add higher ARIMA orders, and see if there is an improvement in the model's performance.

The related issue to this in the context of ADAM ARIMA (Section 9.1) is the dimensionality problem. The more orders you introduce in the model, the bigger the transition matrix becomes. This leads to the same issues as in the ADAM ETS, discussed in the previous subsection. There is no unique recipe in this challenging situation, but backcasting (Section 11.4.1) addresses some of these issues. You might also want to fine-tune the optimiser to get a balance between speed and accuracy in the estimation of parameters (see discussion in Subsection 11.4.1).

12.3 Using explanatory variables for multiple seasonalities

The conventional way of introducing several seasonal components in ETS discussed in Section 12.1 has several issues:

1. It only works with the data with fixed periodicity (the problem sometimes referred to as "fractional frequency"): if m_i is not fixed and changes from period to period, the model becomes misaligned. An example of such a problem is fitting ETS on daily data with $m = 365$, while there are leap years that contain 366 days;
2. If the model fits high-frequency data, the parameter estimation problem becomes non-trivial. Indeed, on daily data with $m = 365$, we need to estimate 364 initial seasonal indices together with other parameters (as discussed in Section 12.2);
3. Different seasonal indices would "compete" with each other for each observation, thus making the model overfit the data. An example is daily data with $m_1 = 7$ and $m_2 = 365$, where both seasonal components are updated on each observation based on the same error but with different smoothing parameters. In this situation, the model implies that the day of year seasonality should be updated together with the day of week one, and this mixture might not lead to the correct split of

the dynamic effects, i.e. one of seasonalities being updated more often than needed.

The situation becomes even more complicated when the model has more than two seasonal components. But there are at least two ways of resolving these issues in the ADAM framework.

The first is based on the idea of De Livera (2010) and the dynamic ETSX (discussed in Section 10.3). In this case, we generate Fourier series and use them as explanatory variables in the model, turning on the mechanism of adaptation. For example, for the pure additive model, in this case, we will have:

$$\begin{aligned} y_t &= \breve{y}_t + \sum_{i=1}^{p} a_{i,t-1} x_{i,t} + \epsilon_t \\ &\vdots \\ a_{i,t} &= a_{i,t-1} + \delta_i \frac{\epsilon_t}{x_{i,t}} \text{ for each } i \in \{1, \dots, p\} \end{aligned} , \tag{12.5}$$

where $x_{i,t}$ is the i-th harmonic and p is the number of Fourier harmonics. In this case, we can introduce the conventional seasonal part of the model for the fixed periodicity (e.g. days of the week) in $\hat{y}_t$ and use the updated harmonics for the non-fixed one. This approach is not the same as De Livera (2010) but might lead to similar results. The only issue here is selecting the number of harmonics. This can be done judgmentally or via the variables selection mechanism (which will be discussed in Section 15.3), but would inevitably increase computational time.

The second option is based on the idea of a dynamic model with categorical variables (from Section 10.5). Instead of trying to fix the problem with days of the year, we first introduce the categorical variables for days of week and then for weeks of year (or months of year if we can assume that the effects of months are more appropriate than the weekly ones). After that, we can introduce both categorical variables in the model, using a similar adaptation mechanism to (12.5). If some variables have fixed periodicity, we can substitute them with the conventional seasonal components. So, for example, ETSX(M,N,M)$_7${D} could be written as:

$$\begin{aligned} y_t &= l_{t-1} s_{t-7} \times \prod_{i=1}^{q} \exp(a_{i,t-1} x_{i,t}) (1 + \epsilon_t) \\ l_t &= l_{t-1}(1 + \alpha \epsilon_t) \\ s_t &= s_{t-7}(1 + \gamma \epsilon_t) \\ a_{i,t} &= a_{i,t-1} + \left\{ \begin{aligned} &\delta \log(1 + \epsilon_t) \text{ for each } i \in \{1, \dots, q\}, \text{ if } x_{i,t} = 1 \\ &0 \text{ otherwise} \end{aligned} \right. \end{aligned} , \tag{12.6}$$

where q is the number of levels in the categorical variable (for weeks of year, this should be 53). The number of parameters to estimate in this situation might be greater than the number of harmonics in the first case, but this type of model resolves all three issues as well and does not have the dilemma of the number of harmonics selection.

Remark. A multiplicative model might make more sense in this context, because the seasonal effect captured by categorical variables will be multiplied by the baseline value, which might help in modelling a potentially more complicated seasonal pattern.

12.4 Dealing with daylight saving and leap years

One of the problems that arises in the case of data with high frequency is the change of local time due to **daylight saving** (DST). This happens in some countries two times a year: in spring, the time is moved one hour forward (typically at 1 a.m. to 2 a.m.), while in the autumn, it is moved back one hour. The implications of this are terrifying from a forecasting point of view because one day of the year has 23 hours, while the other has 25 hours. This leads to modelling difficulties because all the business processes are typically aligned with the local time. This means that if the conventional seasonal ETS model with $m = 24$ fits the data, it will only work correctly in half of the year. If the smoothing parameter γ is high enough then after the DST change, the model will start updating the states and eventually will adapt to the new patterns, but this implies that γ will be higher than needed, introducing unnecessary reactivity in the model and thus wider prediction intervals.

There are two solutions to this problem:

1. Shift the periodicity for one day, when the time changes from 24 to either 23, or 25, depending on the time of year;
2. Introduce categorical variables for factors, which will mark specific hours of the day.

The first option is more challenging to formalise mathematically and implement in software, but does not require estimation of additional parameters – we only need to change the seasonality lag from 24 to either 23 or 25 for a specific day depending on the specific time change. This approach for seasonal ETS is implemented in `adam()` if the data has appropriate timestamps and is framed as a `zoo` object or something similar. The second option relies on the already discussed mechanism of ETSX{D} with categorical variables (Section 10.5) and is in general simpler. Given the connection between seasonality in the conventional ETS model and the ETSX{D} with categorical variables for seasonality, both approaches should be equivalent in terms of final forecasts.

The second problem in the high frequency data is the **leap years**. It can also be solved shifting the periodicity from $m = 365$ to $m = 366$ on 29th February in the spirit of option (1) or using the categorical variables approach (2). There is a difference, however: the latter assumes the estimation of an additional parameter, while the former would be suitable for the data with only one leap year in the data, where the estimation of the seasonal index for 29th February might be difficult. However, given the discussion in Section 12.3, maybe we should not bother with $m = 365$ in the first place and rethink the problem, if possible. Having 52/53 weeks in a year has similar difficulties but at least does not involve the estimation of so many initial seasonal states.

Alternatively, De Livera (2010) proposed to tackle the problem of leap years, introducing the fractional seasonality via Fourier series. The model that implements this is called TBATS (it is an Exponential Smoothing state space model with Box-Cox transformation, ARMA errors, Trend, and Seasonal components, De Livera et al., 2011). While this resolves the aforementioned problem with leap years, the approach introduces an additional complexity, because now we need to select the number of harmonics to use, which in general is not straightforward.

Summarising, when trying to resolve the problem with DST and leap years, there are several possible solutions, each one of them having advantages and disadvantages. In order to decide which to use in the end, it makes sense to try out several of them and select the one that works better (e.g. produces lower forecast errors).

12.5 Examples of application

12.5.1 ADAM ETS

We will use the `taylor` series from the `forecast` package to see how ADAM can be applied to high-frequency data. This is half-hourly electricity demand in England and Wales from Monday 5th June 2000 to Sunday 27th August 2000, used in Taylor (2003b).

```
library(zoo)
y <- zoo(forecast::taylor,
         order.by=as.POSIXct("2000/06/05")+
           (c(1:length(forecast::taylor))-1)*60*30)
plot(y)
```

The series in Figure 12.1 does not exhibit an apparent trend but has two seasonal cycles: a half-hour of the day and a day of the week. Seasonality

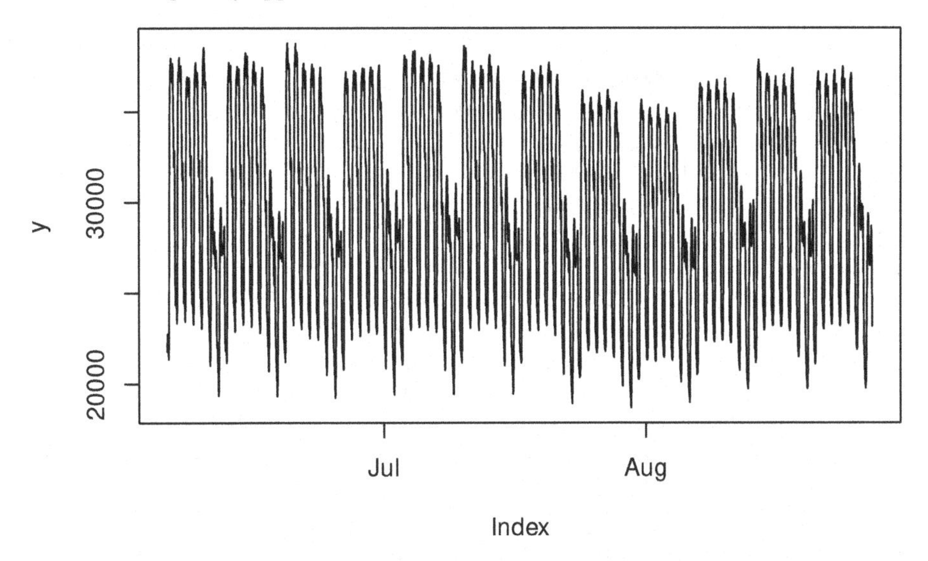

FIGURE 12.1 Half-hourly electricity demand in England and Wales.

seems to be multiplicative because, with the reduction of the level of series, the amplitude of seasonality also reduces. We will try several different models and see how they compare. In all the cases below, we will use backcasting to initialise the model. We will use the last 336 observations (48×7) as the holdout to see whether models perform adequately or not.

Remark. When we have data with DST or leap years (as discussed in Section 12.4), `adam()` will automatically correct the seasonal lags as long as your data contains specific dates (as `zoo` objects have, for example).

First, it is ADAM ETS(M,N,M) with `lags=c(48,7*48)`:

```
adamETSMNM <- adam(y, "MNM", lags=c(1,48,336),
                   h=336, holdout=TRUE,
                   initial="back")
adamETSMNM
```

```
## Time elapsed: 1.14 seconds
## Model estimated using adam() function: ETS(MNM)[48, 336]
## Distribution assumed in the model: Gamma
## Loss function type: likelihood; Loss function value: 25682.88
## Persistence vector g:
##  alpha gamma1 gamma2
## 0.1357 0.2813 0.2335
##
```

```
## Sample size: 3696
## Number of estimated parameters: 4
## Number of degrees of freedom: 3692
## Information criteria:
##      AIC     AICc      BIC     BICc
## 51373.76 51373.77 51398.62 51398.66
##
## Forecast errors:
## ME: 625.221; MAE: 716.941; RMSE: 817.796
## sCE: 709.966%; Asymmetry: 90.4%; sMAE: 2.423%; sMSE: 0.076%
## MASE: 1.103; RMSSE: 0.867; rMAE: 0.107; rRMSE: 0.1
```

As we see from the output above, the model was constructed in 1.14 seconds. Notice that the seasonal smoothing parameters are relatively high in this model. For example, the second γ is equal to 0.2335, which means that the model adapts the seasonal profile to the data substantially (takes 23.35% of the error from the previous observation in it). Furthermore, the smoothing parameter α is equal to 0.1357, which is also potentially too high, given that we have well-behaved data and that we deal with a multiplicative model. This might indicate that the model overfits the data. To see if this is the case, we can produce the plot of components over time (Figure 12.2).

```
plot(adamETSMNM, which=12)
```

As the plot in Figure 12.2 shows, the level of series repeats the seasonal pattern in the original data, although in a diminished way. In addition, the second seasonal component repeats the intra-day seasonality in it, although it is also reduced. Ideally, we want to have a smooth level component and for the second seasonal component not to have those spikes for half-hour of day seasonality.

Next, we can plot the fitted values and forecasts to see how the model performs overall (Figure 12.3).

```
plot(adamETSMNM, which=7)
```

As we see from Figure 12.3, the model fits the data well and produces reasonable forecasts. Given that it only took one second for the model estimation and construction, this model can be considered a good starting point. If we want to improve upon it, we can try one of the multistep estimators, for example, GTMSE (Subsection 11.3.3):

```
adamETSMNMGTMSE <- adam(y, "MNM", lags=c(1,48,336),
                        h=336, holdout=TRUE,
                        initial="back", loss="GTMSE")
```

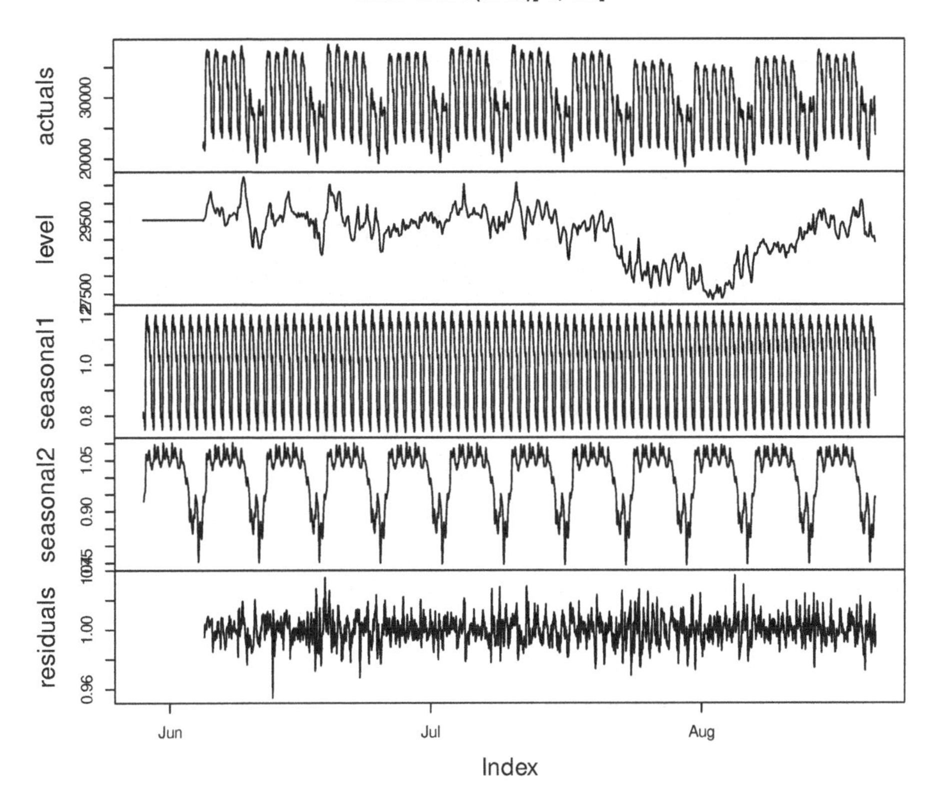

FIGURE 12.2 Half-hourly electricity demand data decomposition according to ETS(M,N,M)[48,336].

The function will take more time due to complexity in the loss function calculation, but hopefully, it will produce more accurate forecasts due to the shrinkage of smoothing parameters:

```
adamETSMNMGTMSE
```

```
## Time elapsed: 23.06 seconds
## Model estimated using adam() function: ETS(MNM)[48, 336]
## Distribution assumed in the model: Normal
## Loss function type: GTMSE; Loss function value: -2648.698
## Persistence vector g:
##  alpha gamma1 gamma2
## 0.0314 0.2604 0.1414
##
## Sample size: 3696
```

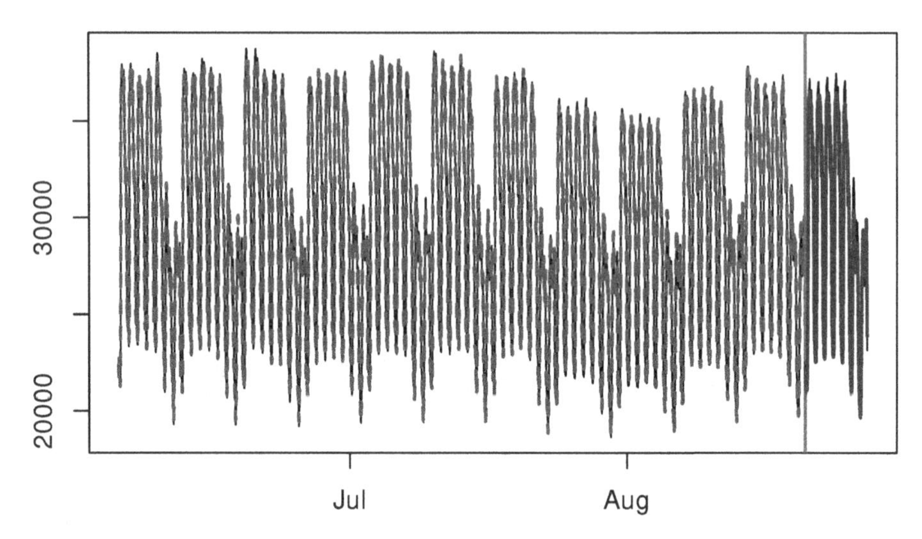

FIGURE 12.3 The fit and the forecast of the ETS(M,N,M)[48,336] model on half-hourly electricity demand data.

```
## Number of estimated parameters: 3
## Number of degrees of freedom: 3693
## Information criteria are unavailable for the chosen loss &
    distribution.
##
## Forecast errors:
## ME: 216.71; MAE: 376.291; RMSE: 505.375
## sCE: 246.084%; Asymmetry: 63%; sMAE: 1.272%; sMSE: 0.029%
## MASE: 0.579; RMSSE: 0.535; rMAE: 0.056; rRMSE: 0.062
```

The smoothing parameters of the second model are closer to zero than in the first one, which might mean that it does not overfit the data as much. We can analyse the components of the second model by plotting them over time, similarly to how we did it for the previous model (Figure 12.4):

```
plot(adamETSMNMGTMSE, which=12)
```

The components on the plot in Figure 12.4 are still not ideal, but at least the level does not seem to contain the seasonality anymore. The seasonal components could still be improved if, for example, the initial seasonal indices were smoother (this applies especially to the seasonal component 2).

Comparing the accuracy of the two models, for example, using RMSSE, we can

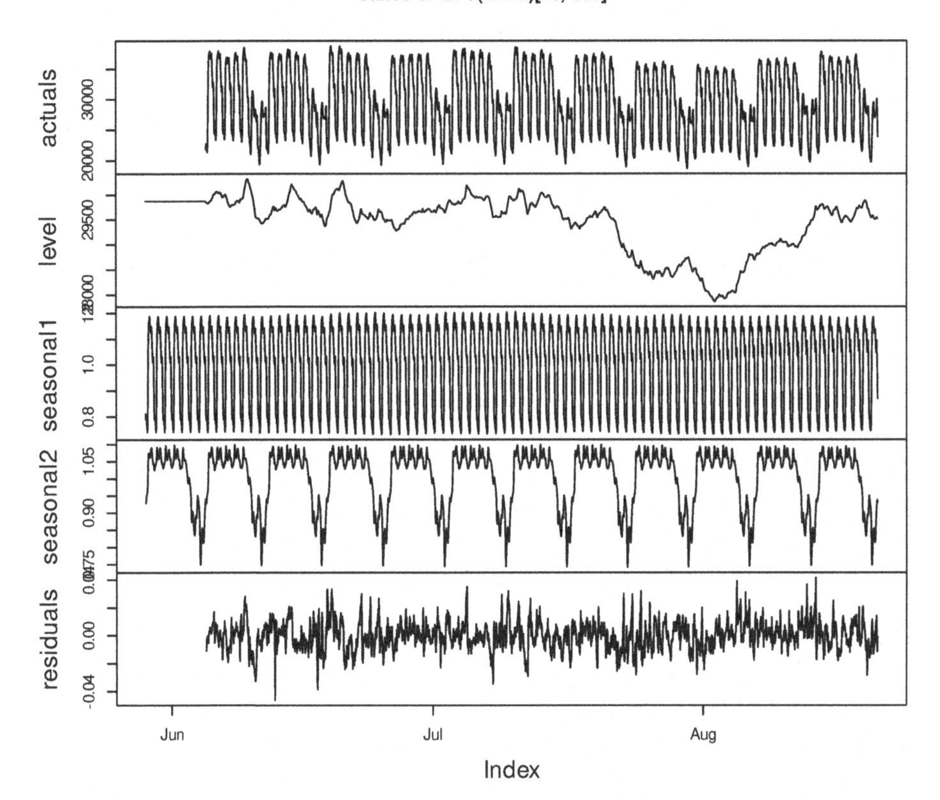

FIGURE 12.4 Half-hourly electricity demand data decomposition according to ETS(M,N,M)[48,336] estimated with GTMSE.

conclude that the one with GTMSE was more accurate than the one estimated using the conventional likelihood.

Another potential way of improvement for the model is the inclusion of an AR(1) term, as for example done by Taylor (2010). This might take more time than the first model, but could also lead to some improvements in the accuracy:

```
adamETSMNMAR <- adam(y, "MNM", lags=c(1,48,336),
                     initial="back", orders=c(1,0,0),
                     h=336, holdout=TRUE, maxeval=1000)
```

Estimating the ETS+ARIMA model is a complicated task because of the increase of dimensionality of the matrices in the transition equation. Still, by default, the number of iterations would be restricted by 160, which might not be enough to get to the minimum of the loss. This is why I increased the number of iterations in the example above to 1000 via `maxeval=1000`. If you want to get more feedback on how the optimisation has been carried out, you can ask the function to print details via `print_level=41`. This is how the output of the model looks:

```
adamETSMNMAR
```

```
## Time elapsed: 2.07 seconds
## Model estimated using adam() function: ETS(MNM)[48, 336]+ARIMA
    (1,0,0)
## Distribution assumed in the model: Gamma
## Loss function type: likelihood; Loss function value: 24108.2
## Persistence vector g:
##  alpha gamma1 gamma2
## 0.1129 0.2342 0.3180
##
## ARMA parameters of the model:
## AR:
## phi1[1]
##  0.6923
##
## Sample size: 3696
## Number of estimated parameters: 5
## Number of degrees of freedom: 3691
## Information criteria:
##      AIC     AICc      BIC     BICc
## 48226.39 48226.41 48257.47 48257.54
##
## Forecast errors:
## ME: 257.38; MAE: 435.476; RMSE: 561.237
## sCE: 292.266%; Asymmetry: 67.2%; sMAE: 1.472%; sMSE: 0.036%
## MASE: 0.67; RMSSE: 0.595; rMAE: 0.065; rRMSE: 0.069
```

In this specific example, we see that the ADAM ETS(M,N,M)+AR(1) leads to a slight improvement in accuracy in comparison with the ADAM ETS(M,N,M) estimated using the conventional loss function (it also has a lower AICc), but cannot beat the one estimated using GTMSE. We could try other options in `adam()` to get further improvements in accuracy, but we do not aim to get the best model in this section. An interested reader is encouraged to do that on their own.

12.5.2 ADAM ETSX

Another option of dealing with multiple seasonalities, as discussed in Section 12.3, is the introduction of explanatory variables. We start with a static model that captures half-hours of the day via its seasonal component and days of week frequency via an explanatory variable. We will use the `temporaldummy()` function from the `greybox` package to create respective categorical variables. This function works better when the data contains proper time stamps and, for example, is of class `zoo` or `xts`. It becomes especially useful when dealing with DST and leap years (see Section 12.4) because it will encode the dummy variables based on dates, allowing us to sidestep the issue with changing frequency in the data.

```
x1 <- temporaldummy(y,type="day",of="week",factors=TRUE)
x2 <- temporaldummy(y,type="hour",of="day",factors=TRUE)
taylorData <- data.frame(y=y,x1=x1,x2=x2)
```

Now that we created the data with categorical variables, we can fit the ADAM ETSX model with dummy variables for days of the week and use complete backcasting for initialisation:

```
adamETSXMNN <- adam(taylorData, "MNN", h=336, holdout=TRUE,
                    initial="complete")
```

In the code above, the initialisation method leads to potentially biased estimates of parameters at the cost of a reduced computational time (see discussion in Section 11.4.1). Here is what we get as a result:

```
adamETSXMNN
```

```
## Time elapsed: 0.61 seconds
## Model estimated using adam() function: ETSX(MNN)
## Distribution assumed in the model: Gamma
## Loss function type: likelihood; Loss function value: 30155.54
## Persistence vector g (excluding xreg):
##  alpha
## 0.6182
##
## Sample size: 3696
## Number of estimated parameters: 2
## Number of degrees of freedom: 3694
## Information criteria:
##      AIC     AICc      BIC     BICc
## 60315.08 60315.09 60327.51 60327.53
##
## Forecast errors:
## ME: -1664.294; MAE: 1781.472; RMSE: 2070.321
## sCE: -1889.878%; Asymmetry: -92.3%; sMAE: 6.021%; sMSE: 0.49%
```

```
## MASE: 2.74; RMSSE: 2.194; rMAE: 0.266; rRMSE: 0.253
```

The resulting model produces biased forecasts (they are consistently higher than needed). This is mainly because the smoothing parameter α is too high, and the model frequently changes the level. We can see that in the plot of the state (Figure 12.5):

```
plot(adamETSXMNN$states[,1], ylab="Level")
```

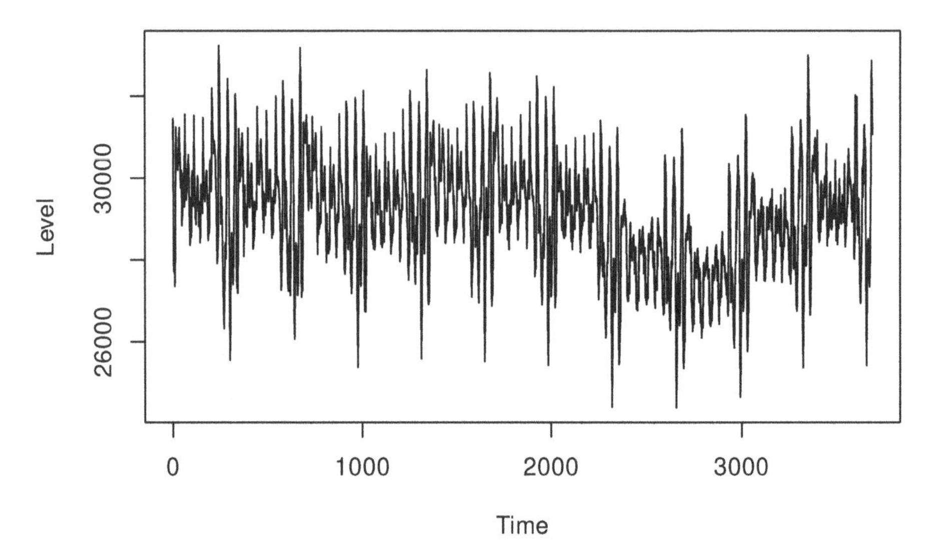

FIGURE 12.5 Plot of the level of the ETSX model.

As we see from Figure 12.5, the level component absorbs seasonality, which causes forecasting accuracy issues. However, the obtained value did not happen due to randomness – this is what the model does when seasonality is fixed and is not allowed to evolve. To reduce the model's sensitivity, we can shrink the smoothing parameter using a multistep estimator (discussed in Section 11.3). But as discussed earlier, these estimators are typically slower than the conventional ones, so that they might take more computational time:

```
adamETSXMNNGTMSE <- adam(taylorData, "MNN",
                         h=336, holdout=TRUE,
                         initial="complete", loss="GTMSE")
adamETSXMNNGTMSE
```

```
## Time elapsed: 39.26 seconds
## Model estimated using adam() function: ETSX(MNN)
## Distribution assumed in the model: Normal
```

```
## Loss function type: GTMSE; Loss function value: -2044.705
## Persistence vector g (excluding xreg):
##  alpha
## 0.0153
##
## Sample size: 3696
## Number of estimated parameters: 1
## Number of degrees of freedom: 3695
## Information criteria are unavailable for the chosen loss &
    distribution.
##
## Forecast errors:
## ME: 105.462; MAE: 921.897; RMSE: 1204.967
## sCE: 119.757%; Asymmetry: 18%; sMAE: 3.116%; sMSE: 0.166%
## MASE: 1.418; RMSSE: 1.277; rMAE: 0.138; rRMSE: 0.147
```

While the model's performance with GTMSE has improved due to the shrinkage of α to zero, the seasonal states are still deterministic and do not adapt to the changes in data. We could adapt them via `regressors="adapt"`, but then we would be constructing the ETS(M,N,M)[48,336] model but in a less efficient way. Alternatively, we could assume that one of the seasonal states is deterministic and, for example, construct the ETSX(M,N,M) model:

```
adamETSXMNMGTMSE <- adam(taylorData, "MNM", lags=48,
                         h=336, holdout=TRUE,
                         initial="complete", loss="GTMSE",
                         formula=y~x1)
adamETSXMNMGTMSE
```

```
## Time elapsed: 33.56 seconds
## Model estimated using adam() function: ETSX(MNM)
## Distribution assumed in the model: Normal
## Loss function type: GTMSE; Loss function value: -2082.17
## Persistence vector g (excluding xreg):
##  alpha  gamma
## 0.0135 0.0769
##
## Sample size: 3696
## Number of estimated parameters: 2
## Number of degrees of freedom: 3694
## Information criteria are unavailable for the chosen loss &
    distribution.
##
## Forecast errors:
## ME: 146.436; MAE: 830.332; RMSE: 1055.372
```

```
## sCE: 166.284%; Asymmetry: 27.1%; sMAE: 2.806%; sMSE: 0.127%
## MASE: 1.277; RMSSE: 1.118; rMAE: 0.124; rRMSE: 0.129
```

We can see an improvement compared to the previous model, so the seasonal states do change over time, which means that the deterministic seasonality is not appropriate in our example. However, it might be more suitable in some other cases, producing more accurate forecasts than the models assuming stochastic seasonality.

12.5.3 ADAM ARIMA

Another model we can try on this data is Multiple Seasonal ARIMA. We have not yet discussed the order selection mechanism for ARIMA, so I will construct a model based on my judgment. Keeping in mind that ETS(A,N,N) is equivalent to ARIMA(0,1,1) and that the changing seasonality in the ARIMA context can be modelled with seasonal differences, I will construct SARIMA$(0,1,1)(0,1,1)_{336}$, skipping the frequencies for a half-hour of the day. Hopefully, this will suffice to model: (a) changing level of data; (b) changing seasonal amplitude. Here is how we can construct this model using `adam()`:

```
adamARIMA <- adam(y, "NNN", lags=c(1,336), initial="back",
                  orders=list(i=c(1,1),ma=c(1,1)),
                  h=336, holdout=TRUE)
adamARIMA
```

```
## Time elapsed: 0.4 seconds
## Model estimated using adam() function: SARIMA(0,1,1)[1](0,1,1)[336]
## Distribution assumed in the model: Normal
## Loss function type: likelihood; Loss function value: 26098.77
## ARMA parameters of the model:
## MA:
##   theta1[1] theta1[336]
##      0.5086     -0.1969
##
## Sample size: 3696
## Number of estimated parameters: 3
## Number of degrees of freedom: 3693
## Information criteria:
##      AIC     AICc      BIC     BICc
## 52203.53 52203.54 52222.18 52222.20
##
## Forecast errors:
## ME: 49.076; MAE: 373.297; RMSE: 499.539
## sCE: 55.728%; Asymmetry: 18.5%; sMAE: 1.262%; sMSE: 0.029%
## MASE: 0.574; RMSSE: 0.529; rMAE: 0.056; rRMSE: 0.061
```

As we see from the output above, this model has the lowest RMSSE value among all models we tried. Furthermore, this model is directly comparable with ADAM ETS via information criteria, and as we can see, it is worse than ADAM ETS(M,N,M)+AR(1) and multiple seasonal ETS(M,N,M) in terms of AICc. Figure 12.6 shows the fit and forecast from this model.

```
plot(adamARIMA, which=7)
```

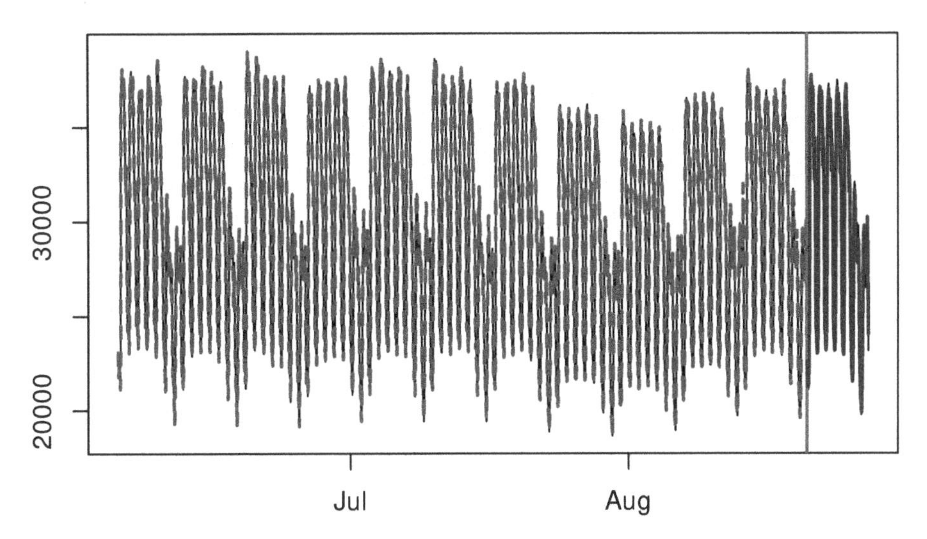

FIGURE 12.6 The fit and the forecast of the ARIMA(0,1,1)(0,1,1)$_{336}$ model on half-hourly electricity demand data.

We could analyse the residuals of this model and iteratively test whether the addition of AR terms and a half-hour of day seasonality improves the model's accuracy. We could also try ARIMA models with different distributions, compare them and select the most appropriate one or estimate the model using multistep losses. All of this can be done in `adam()`. The reader is encouraged to do this on their own.

13

Intermittent State Space Model

So far, we have discussed data that has regular occurrence (i.e. happening every observation). For example, daily sales of bread in a supermarket would have this regularity. However, there are time series where non-zero values do not happen on every observation. In the context of demand forecasting, this is called "**intermittent demand**". The conventional example of such demand is monthly sales of jet engines: they will contain a lot of zeroes, when nobody buys the product and then all of a sudden several units when an aeroplane company decides to buy it, again followed by zeroes. However, this problem does not only apply to such expensive exotic products – the retailers face this all the time for various products, mainly when sales are recorded daily or on an even higher frequency.

One of the most straightforward definitions of intermittent demand is that *it is the demand that happens at irregular frequency*. This means that we cannot tell with 100% certainty, when specifically customers will buy our product. In fact, if we zoom in on the sales of any product and observe it at a higher frequency, we will most probably see intermittent demand. For example, if we work with hourly sales of products in a supermarket, we will most probably observe intermittent demand because there will be some hours of day, when supermarket will have few customers and thus will not sell some products at all. However, if we aggregate this data to a weekly level, the intermittence will typically disappear, making the demand look regular. So, intermittent demand is related to the regular one, and in many cases arises on the lower aggregation levels.

You might also meet the term "count data" (or "integer-valued data") in a similar context, but there is a fundamental difference between the count and intermittent data. The former implies that demand can take integer values only and can be typically modelled via Poisson, Binomial, or Negative Binomial distributions. It does not necessarily contain zeroes and does not explicitly allow demand to happen at random. In this case, if there are zeroes, then it is assumed that they are just one of the possible values of a distribution. On the other hand, in the case of intermittent demand, we explicitly acknowledge that demand might not happen, but if it happens, then the value will be greater than zero. Furthermore, intermittent demand does not necessarily need to be integer-valued. For example, daily energy consumption for charging electric vehicles would typically be intermittent (because the vehicle owners do not charge them every day), but the non-zero consumption will not be an integer. Still, count distributions can be used in some cases of intermittent demand, but they do not necessarily always provide a good approximation of complex reality.

Before we move towards the proper discussion of the topic in the context of ADAM, we should acknowledge that at the heart of what follows, there lies the following model (Croston, 1972):

$$y_t = o_t z_t, \tag{13.1}$$

where o_t is the demand occurrence variable, which can be either zero or one and has some probability of occurrence p_t, z_t is the demand sizes captured by a model (for example, ETS), and y_t is the final observed demand. In the context of intermittent demand, this model was originally proposed by Croston (1972), but similar models (e.g. Hurdle and Zero Inflated Poisson) exist in other, non-forecasting related literature.

In this chapter, we will discuss the intermittent state space model (13.1), both parts of which can be modelled via ADAM, and we will see how they can be used, what they imply and how they connect to the conventional regular demand. If an ETS model is used for z_t then (13.1) will be called iETS. So, the iETS(M,N,N) model refers to the intermittent state space model, where demand sizes are modelled via ETS(M,N,N). ETS can also be used for the occurrence part of the model, so if the discussion is focused on the demand occurrence part of the model, o_t (as in Section 13.1), we will use "oETS" instead. Similarly, we can use the terms iARIMA and oARIMA, referring either to the whole model or just to its occurrence part. Note, however, that while ARIMA can be used in theory, it is not yet implemented for the occurrence part of the model. So we will focus the discussion in this chapter on the ADAM ETS. Furthermore, depending on how the occurrence part is modelled, the notations above can be expanded to include references to specific parts of the occurrence part of the model. This is discussed in detail in Section 13.1.

This chapter is based on Svetunkov and Boylan (2023). If you want to know more about intermittent demand forecasting, Boylan and Syntetos (2021) is an excellent textbook on the topic, covering all the main aspects in appropriate detail.

13.1 Occurrence part of the model

The general model (13.1) assumes that demand occurs randomly and that the variable o_t can be either zero (no demand) or one (there is some demand). While this process can be modelled using different distributions, Svetunkov and Boylan (2023) proposed using Bernoulli with a time varying probability:

$$o_t \sim \text{Bernoulli}(p_t), \tag{13.2}$$

where p_t is the probability of occurrence. The higher it is, the more frequently the demand will happen. If $p_t = 1$, then the demand becomes regular, while if $p_t = 0$, nobody buys the product at all. This section will discuss different types of models for the probability of occurrence. For each of them, there are different mechanisms of the model construction, estimation, error calculation, update of the states, and the generation of forecasts. To estimate any of these models, I recommend using likelihood, which can be calculated based on the Probability Mass Function (PMF) of Bernoulli distribution:

$$f_o(o_t, p_t) = p_t^{o_t}(1-p_t)^{1-o_t}. \tag{13.3}$$

The parameters of the occurrence part of the model can be then estimated via the maximisation of the log-likelihood function, which comes directly from (13.3), and in the most general case is:

$$\ell(\boldsymbol{\theta}_o|\mathbf{o}) = \sum_{o_t=1} \log(\hat{p}_t) + \sum_{o_t=0} \log(1-\hat{p}_t), \tag{13.4}$$

where $\hat{p}_t$ is the in-sample conditional one step ahead expectation of the probability on observation t, given the information on observation $t-1$, which depends on the vector of estimated parameters for the occurrence part of the model $\boldsymbol{\theta}_o$.

In order to demonstrate the difference between specific types of oETS models, we will use the following artificial data:

```
y <- ts(c(rpois(20,0.25), rpois(20,0.5), rpois(20,1),
          rpois(20,2), rpois(20,3), rpois(20,5)))
```

Figure 13.1 shows how the data looks:

```
plot(y, ylab="Sales")
```

The probability of occurrence in this example increases together with the demand sizes. This example corresponds to the situation of intermittent demand for a product evolving to the regular one.

13.1.1 Fixed probability model, oETS$_F$

We start with the simplest case of fixed probability of occurrence, the oETS$_F$ model:

$$o_t \sim \text{Bernoulli}(p), \tag{13.5}$$

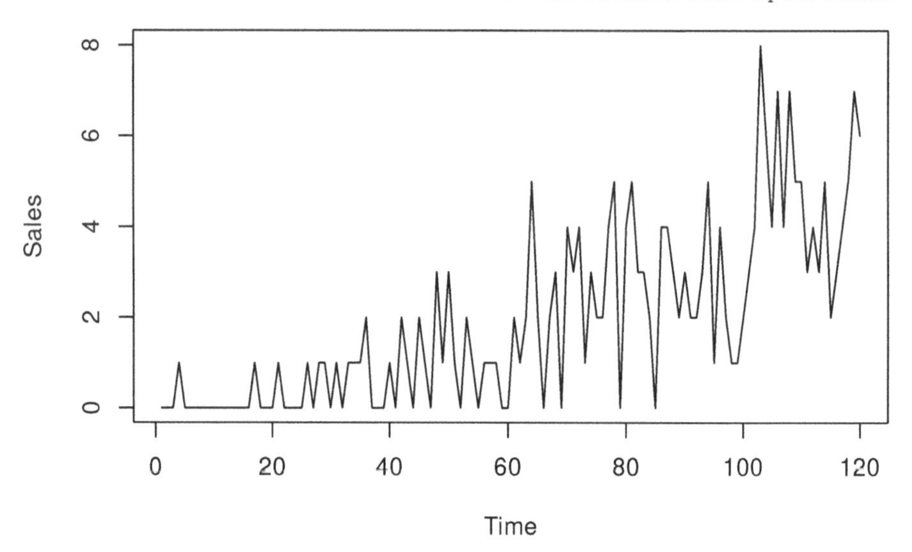

FIGURE 13.1 Example of intermittent demand data.

This model assumes that demand happens with the same probability no matter what. This might sound exotic because, in practice, there might be many factors influencing customers' desire to purchase, and the impact of these factors might change over time. But this is a basic model, which can be used as a benchmark on intermittent demand data. Furthermore, it might be suitable for modelling demand on expensive high-tech products, such as jet engines, which is "very slow" in its nature and typically does not evolve much over time.

When estimated via maximisation of the likelihood function (13.4), the probability of occurrence in this model is equal to:

$$\hat{p} = \frac{T_1}{T}, \tag{13.6}$$

where T_1 is the number of non-zero observations and T is the number of all the available observations in-sample.

The occurrence part of the model, oETS$_F$ can be constructed using the `oes()` function from the `smooth` package:

```
oETSFModel <- oes(y, h=10, holdout=TRUE,
                  occurrence="fixed")
oETSFModel
```

```
## Occurrence state space model estimated: Fixed probability
## Underlying ETS model: oETS[F](MNN)
```

```
## Vector of initials:
##  level
## 0.6455
##
## Error standard deviation: 1.0909
## Sample size: 110
## Number of estimated parameters: 1
## Number of degrees of freedom: 109
## Information criteria:
##      AIC     AICc      BIC     BICc
## 145.0473 145.0844 147.7478 147.8349
```

The oETS$_F$ model produces the straight line for the probability of 0.65, ignoring that in our example, the probability of occurrence has increased over time.

```
plot(oETSFModel)
```

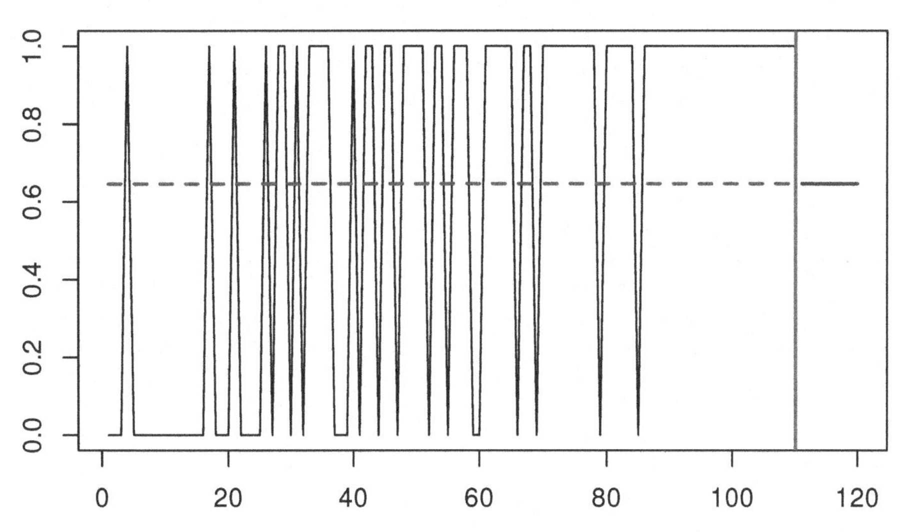

FIGURE 13.2 Demand occurrence and probability of occurrence in the oETS$_F$ model.

The plot in Figure 13.2 demonstrates the dynamics of the occurrence variable o_t and the fitted and predicted probabilities. The solid line shows when zeroes and ones happen, depicting the variable o_t. The dashed line corresponds to the fixed probability of occurrence $\hat{p}$ in the sample.

13.1.2 Odds ratio model, oETS$_O$

In this model, it is assumed that the update of the probability is driven by the previously observed occurrence of a variable. It is more complicated than the previous model, as the probability now changes over time and can be modelled, for example, with an ETS(M,N,N) model:

$$\begin{aligned} o_t &\sim \text{Bernoulli}\left(p_t\right) \\ p_t &= \frac{\mu_{a,t}}{\mu_{a,t}+1} \\ a_t &= l_{a,t-1}\left(1+\epsilon_{a,t}\right) \\ l_{a,t} &= l_{a,t-1}(1+\alpha_a \epsilon_{a,t}) \\ \mu_{a,t} &= l_{a,t-1} \end{aligned}, \tag{13.7}$$

where $l_{a,t}$ is the unobserved level component, α_a is the smoothing parameter, $1+\epsilon_{a,t}$ is the error term , which is positive, has the mean of one, and follows an unknown distribution, and $\mu_{a,t}$is the conditional expectation for the unobservable shape variable a_t. The measurement and transition equations in (13.7) can be substituted by any other ETS, ARIMA, or regression model if it is reasonable to assume that the probability dynamic has some additional components, such as trend, seasonality, or exogenous variables. This model is called the "odds ratio" because the probability of occurrence in (13.7) is calculated using the classical logistic transform. This also means that $\mu_{a,t}$ equals to:

$$\mu_{a,t} = \frac{p_t}{1-p_t}. \tag{13.8}$$

When $\mu_{a,t}$ increases in the oETS$_\text{O}$ model, the odds ratio increases as well, meaning that the probability of occurrence goes up. Svetunkov and Boylan (2023) explain that this model is, in theory, appropriate for the demand for products becoming obsolescent, because the multiplicative error ETS models asymptotically almost surely converge to zero. Still, given the updating mechanism, it should also work fine on other types of intermittent data.

When it comes to the application of the model to the data, its construction is done via the following set of equations (example with oETS$_O$(M,N,N)):

$$\begin{aligned} \hat{p}_t &= \frac{\hat{\mu}_{a,t}}{\hat{\mu}_{a,t}+1} \\ \hat{\mu}_{a,t} &= \hat{l}_{a,t-1} \\ \hat{l}_{a,t} &= \hat{l}_{a,t-1}(1+\hat{\alpha}_a e_{a,t}), \\ 1+e_{a,t} &= \frac{u_t}{1-u_t} \\ u_t &= \frac{1+o_t-\hat{p}_t}{2} \end{aligned} \tag{13.9}$$

where $e_{a,t}$ is the proxy for the unobservable error term $\epsilon_{a,t}$ and $\hat{\mu}_t$ is the estimate of $\mu_{a,t}$. If a multiple steps ahead forecast for the probability is needed from this model, then the formulae discussed in Section 4.2 can be used to get $\hat{\mu}_{a,t+h}$, which then can be inserted in the first equation of (13.9) to get the final conditional multiple steps ahead probability of occurrence.

Finally, to estimate the parameters of the model (13.9), the likelihood (13.4) can be used.

The occurrence model oETS$_O$ is constructed using the very same `oes()` function, but also allows specifying the ETS model to use. For example, here is the oETS$_O$(M,M,N) model:

```
oETSOModel <- oes(y, model="MMN", h=10, holdout=TRUE,
                  occurrence="odds-ratio")
oETSOModel
```

```
## Occurrence state space model estimated: Odds ratio
## Underlying ETS model: oETS[O](MMN)
## Smoothing parameters:
##  level  trend
## 0.0209 0.0000
## Vector of initials:
##  level  trend
## 0.0986 1.0448
##
## Error standard deviation: 2.0989
## Sample size: 110
## Number of estimated parameters: 4
## Number of degrees of freedom: 106
## Information criteria:
##      AIC     AICc      BIC     BICc
## 103.5230 103.9040 114.3250 115.2203
```

In this example, we introduce the multiplicative trend in the model, which is supposed to reflect the idea of demand building up over time.

Figure 13.3 shows that the model captures the changing probability of occurrence well, reflecting that it increases over time. Also notice how the model reacts more to the zeroes, in the beginning of the zeroes rather than to ones at the end. This is the main distinguishing characteristic of the model.

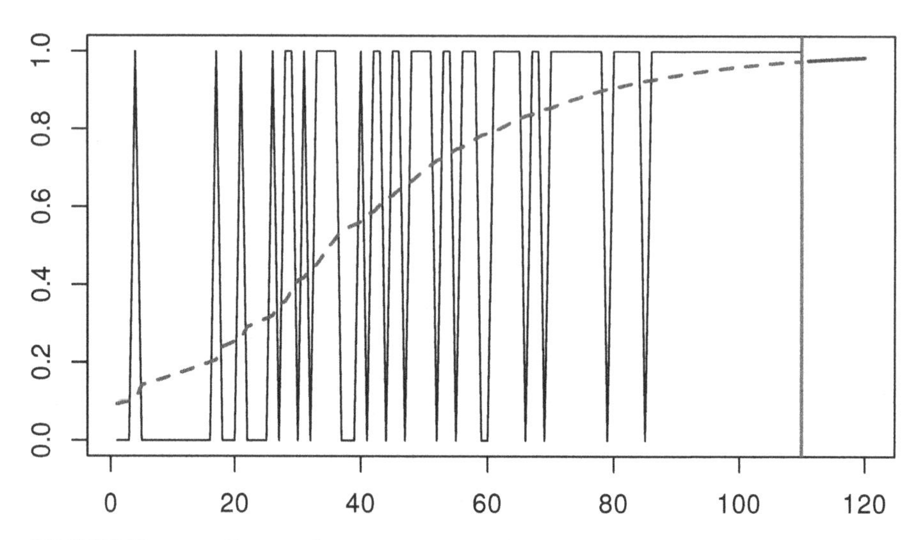

FIGURE 13.3 Demand occurrence and probability of occurrence in the oETS(M,M,N)$_O$ model.

13.1.3 Inverse odds ratio model, oETS$_I$

Using a similar approach to the oETS$_O$, we can formulate the "inverse odds ration" model oETS$_I$(M,N,N):

$$\begin{aligned} o_t &\sim \text{Bernoulli}\left(p_t\right) \\ p_t &= \frac{1}{1+\mu_{b,t}} \\ b_t &= l_{b,t-1}\left(1+\epsilon_{b,t}\right) \\ l_{b,t} &= l_{b,t-1}(1+\alpha_b\epsilon_{b,t}) \\ \mu_{b,t} &= l_{b,t-1} \end{aligned}, \tag{13.10}$$

where similarly to (13.9), $l_{b,t}$ is the unobserved level component, α_b is the smoothing parameter, $1+\epsilon_{b,t}$ is the positive error term with a mean of one, and $\mu_{b,t}$ is the one step ahead conditional expectation for the unobservable shape parameters b_t. The main difference between this model and the previous one is in the mechanism of probability calculation, which relies on the probability of "inoccurrence", i.e. on zeroes of data rather than on ones. This type of model should be more appropriate for cases of demand building up (Svetunkov & Boylan, 2023). The probability calculation mechanism in (13.10) implies that $\mu_{b,t}$ can be expressed as:

$$\mu_{b,t} = \frac{1-p_t}{p_t}. \tag{13.11}$$

The construction of the model (13.10) is similar to (13.9):

$$\begin{aligned} \hat{p}_t &= \frac{1}{1+\hat{\mu}_{b,t}} \\ \hat{\mu}_{b,t} &= \hat{l}_{b,t-1} \\ \hat{l}_{b,t} &= l_{b,t-1}(1+\hat{\alpha}_b e_{b,t}), \\ 1+e_{b,t} &= \frac{1-u_t}{u_t} \\ u_t &= \frac{1+o_t-\hat{p}_t}{2} \end{aligned} \tag{13.12}$$

where $e_{b,t}$ is the proxy for the unobservable error term $\epsilon_{b,t}$ and $\hat{\mu}_{b,t}$ is the estimate of $\mu_{b,t}$. Once again, we refer an interested reader to Subsection 4.2 for the discussion of the multiple steps ahead conditional expectations from the ETS(M,N,N) model.

Svetunkov and Boylan (2023) show that the oETS$_I$(M,N,N) model, in addition to (13.12), can be estimated using Croston's method, as long as we can assume that the probability does not change over time substantially. In this case the demand intervals (the number of zeroes between demand sizes) can be used instead of $\hat{\mu}_{b,t}$ in (13.12). So the iETS(M,N,N)$_I$(M,N,N) can be considered as a model underlying Croston's method.

The function `oes()` implements the oETS$_I$ model as well. For example, here is the oETS$_I$(M,M,N) model:

```
oETSIModel <- oes(y, model="MMN", h=10, holdout=TRUE,
                  occurrence="inverse-odds-ratio")
oETSIModel
```

```
## Occurrence state space model estimated: Inverse odds ratio
## Underlying ETS model: oETS[I](MMN)
## Smoothing parameters:
##  level  trend
## 0.0419 0.0000
## Vector of initials:
##   level   trend
## 25.2308  0.8982
##
## Error standard deviation: 4.1508
## Sample size: 110
## Number of estimated parameters: 4
## Number of degrees of freedom: 106
## Information criteria:
##      AIC     AICc      BIC     BICc
## 105.3722 105.7531 116.1741 117.0694
```

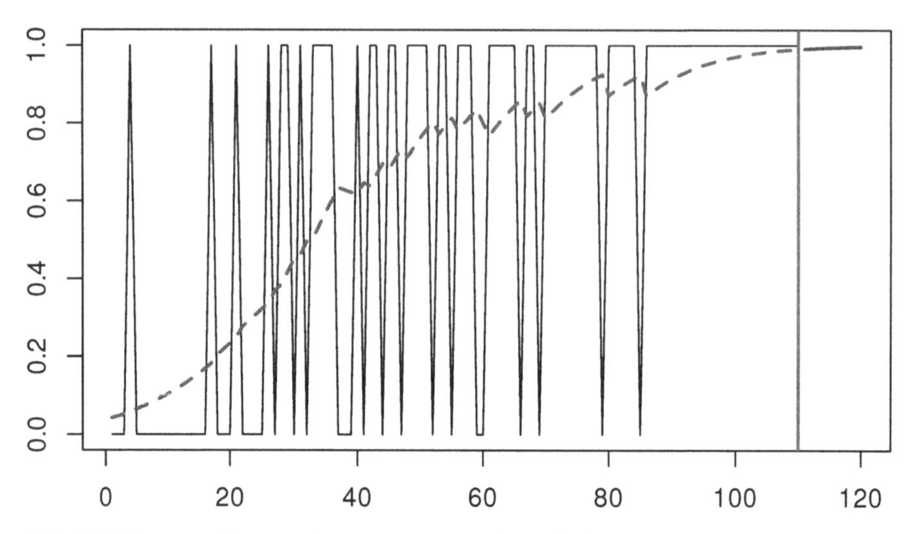

FIGURE 13.4 Demand occurrence and probability of occurrence in the oETS(M,M,N)$_I$ model.

Figure 13.4 shows that similarly to the oETS$_O$, the model captures the trend in the probability of occurrence but has a higher smoothing parameter α_b. Also notice how the update of states happens more often towards the second part of the data, when demand is building up and more ones appear in the data. The model is much more inert when it has many zeroes. This behaviour can be contrasted to the behaviour of oETS$_O$, which updated the states more frequently in the beginning of the series in our example.

13.1.4 General oETS model, oETS$_G$

Uniting the oETS$_O$ with oETS$_I$, we can obtain the "general" model, which in the most general case can be summarised in the following set of state space equations:

$$\begin{aligned} p_t &= f_p(\mu_{a,t}, \mu_{b,t}) \\ a_t &= w_a(\mathbf{v}_{a,t-\boldsymbol{l}}) + r_a(\mathbf{v}_{a,t-\boldsymbol{l}})\epsilon_{a,t} \\ \mathbf{v}_{a,t} &= f_a(\mathbf{v}_{a,t-\boldsymbol{l}}) + g_a(\mathbf{v}_{a,t-\boldsymbol{l}})\epsilon_{a,t}, \\ b_t &= w_b(\mathbf{v}_{b,t-\boldsymbol{l}}) + r_b(\mathbf{v}_{b,t-\boldsymbol{l}})\epsilon_{b,t} \\ \mathbf{v}_{b,t} &= f_b(\mathbf{v}_{b,t-\boldsymbol{l}}) + g_b(\mathbf{v}_{b,t-\boldsymbol{l}})\epsilon_{b,t} \end{aligned} \tag{13.13}$$

where $\epsilon_{a,t}$, $\epsilon_{b,t}$, $\mu_{a,t}$, and $\mu_{b,t}$ have been defined in previous subsections, and the other elements correspond to the ADAM discussed in Chapter 5. Note that in this case two state space models similar to the one discussed in Section 7.1

are used for the modelling of a_t and b_t. These can be either ETS, or ARIMA, or regression, or any combination of those.

The general formula for the probability in the case of the multiplicative error model is:

$$p_t = \frac{\mu_{a,t}}{\mu_{a,t} + \mu_{b,t}}, \tag{13.14}$$

while for the additive one, it is:

$$p_t = \frac{\exp(\mu_{a,t})}{\exp(\mu_{a,t}) + \exp(\mu_{b,t})}. \tag{13.15}$$

This is because both $\mu_{a,t}$ and $\mu_{b,t}$ need to be strictly positive, while the additive error models support negative and positive values and zero. The canonical oETS model assumes that the pure multiplicative model is used for a_t and b_t. This model type is positively defined for any values of error, trend, and seasonality, which is essential for the values of a_t and b_t and their expectations. If a combination of additive and multiplicative error models is used, then the additive part should be exponentiated before using the formulae to calculate the probability. So, the $f_p(\cdot)$ function from (13.13) maps the expectations from models A and B to the probability of occurrence, depending on the error type of the respective models:

$$p_t = f_p(\mu_{a,t}, \mu_{b,t}) = \begin{cases} \frac{\mu_{a,t}}{\mu_{a,t} + \mu_{b,t}} & \text{when both have multiplicative errors} \\ \frac{\mu_{a,t}}{\mu_{a,t} + \exp(\mu_{b,t})} & \text{when model B has additive error} \\ \frac{\exp(\mu_{a,t})}{\exp(\mu_{a,t}) + \mu_{b,t}} & \text{when model A has additive error} \\ \frac{\exp(\mu_{a,t})}{\exp(\mu_{a,t}) + \exp(\mu_{b,t})} & \text{when both have additive errors.} \end{cases} \tag{13.16}$$

An example of the oETS model is the one based on two local level models (see discussion in Subsection 4.3), oETS$_G$(M,N,N)(M,N,N):

$$\begin{aligned} & o_t \sim \text{Bernoulli}(p_t) \\ & p_t = \frac{\mu_{a,t}}{\mu_{a,t} + \mu_{b,t}} \\ & \\ & a_t = l_{a,t-1}(1 + \epsilon_{a,t}) \\ & l_{a,t} = l_{a,t-1}(1 + \alpha_a \epsilon_{a,t}), \\ & \mu_{a,t} = l_{a,t-1} \\ & \\ & b_t = l_{b,t-1}(1 + \epsilon_{b,t}) \\ & l_{b,t} = l_{b,t-1}(1 + \alpha_b \epsilon_{b,t}) \\ & \mu_{b,t} = l_{b,t-1} \end{aligned} \tag{13.17}$$

where all the parameters have already been defined in Subsections 13.1.2 and 13.1.3. More advanced models can be constructed for a_t and b_t by specifying the ETS models for each part and/or adding explanatory variables.

The construction of the model (13.17) is done via the following set of recursive equations:

$$\begin{aligned} e_{a,t} &= \frac{u_t}{1-u_t} - 1 \\ \hat{a}_t &= \hat{l}_{a,t-1} \\ \hat{l}_{a,t} &= \hat{l}_{a,t-1}(1 + \alpha_a e_{a,t}) \\ e_{b,t} &= \frac{1-u_t}{u_t} - 1 \\ \hat{b}_t &= \hat{l}_{b,t-1} \\ \hat{l}_{b,t} &= \hat{l}_{b,t-1}(1 + \alpha_b e_{b,t}) \end{aligned} . \tag{13.18}$$

In R, there is a separate function for the oETS$_G$ model, called `oesg()`. It has twice as many parameters as `oes()`, because it allows fine tuning of the models for the both parts a_t and b_t. This gives an additional flexibility. For example, here is how we can use ETS(M,N,N) for the a_t and ETS(A,A,N) for the b_t, resulting in oETS$_G$(M,N,N)(A,A,N):

```
oETSGModel <- oesg(y, modelA="MNN", modelB="AAN",
                   h=10, holdout=TRUE)
oETSGModel
```

```
## Occurrence state space model estimated: General
## Underlying ETS model: oETS[G](MNN)(AAN)
##
## Sample size: 110
## Number of estimated parameters: 6
## Number of degrees of freedom: 104
## Information criteria:
##      AIC     AICc      BIC     BICc
## 107.1447 107.9603 123.3476 125.2643
```

We can also analyse models separately for a_t and b_t from the saved variable. Here is, for example, Model A:

```
oETSGModel$modelA
```

```
## Occurrence state space model estimated: Odds ratio
## Underlying ETS model: oETS[G](MNN)_A
## Smoothing parameters:
##  level
```

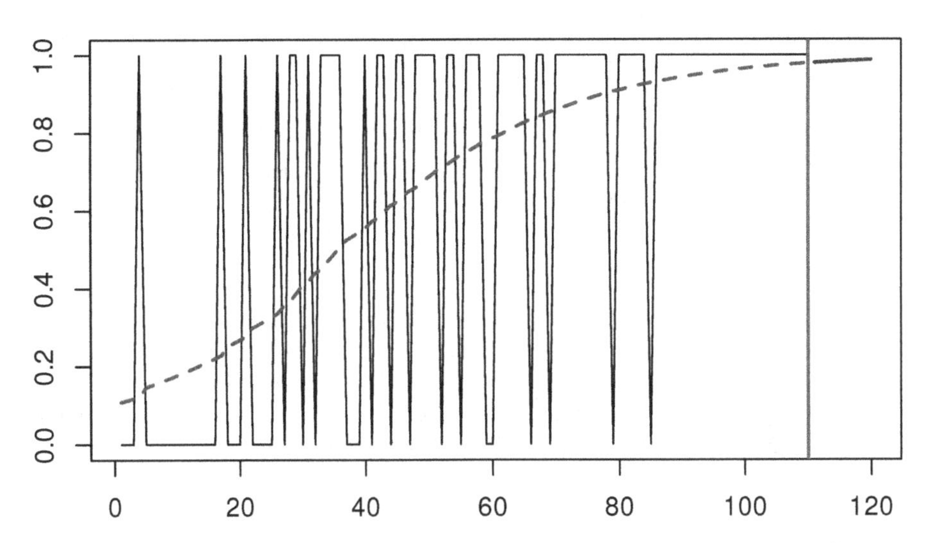

FIGURE 13.5 Demand occurrence and probability of occurrence in the oETS(M,N,N)(A,A,N)$_G$ model.

```
## 0.0118
## Vector of initials:
##  level
## 1.2764
##
## Error standard deviation: 1.8449
## Sample size: 110
## Number of estimated parameters: 2
## Number of degrees of freedom: 108
## Information criteria:
##      AIC     AICc      BIC     BICc
##  99.1447  99.2569 104.5457 104.8093
```

The experiments that I have done so far show that oETS$_G$ very seldom brings improvements in comparison with oETS$_O$ or oETS$_I$ in terms of forecasting accuracy. Besides, selecting models for each of the parts is a challenging task. So, this model is theoretically attractive, being more general than the other oETS models, but is not very practical. Still it is useful because we can introduce different oETS models by restricting a_t and b_t. For example, we can get:

1. oETS$_F$, when $\mu_{a,t} = \text{const}$, $\mu_{b,t} = \text{const}$ for all t;
2. oETS$_O$, when $\mu_{b,t} = 1$ for all t;
3. oETS$_I$, when $\mu_{a,t} = 1$ for all t;

4. oETS_D, when $\mu_{a,t} + \mu_{b,t} = 1$, $\mu_{a,t} \in [0,1]$ for all t (discussed in Subsection 13.1.5);
5. oETS_G, when there are no restrictions.

13.1.5 Direct probability model, oETS_D

The last model in the family of oETS is the "Direct probability". It appears, when the following restriction is imposed on the oETS_G:

$$\mu_{a,t} + \mu_{b,t} = 1, \mu_{a,t} \in [0,1]. \tag{13.19}$$

This restriction is inspired by the mechanism for the probability update proposed by Teunter, Syntetos, and Babai (2011) (TSB method). Their paper uses SES (discussed in Section 3.4) to model the time-varying probability of occurrence. Based on this idea and the restriction (13.19), we can formulate an oETS_D(M,N,N) model, which will underly the occurrence part of the TSB method:

$$\begin{aligned} & o_t \sim \text{Bernoulli}\left(\mu_{a,t}\right) \\ & a_t = l_{a,t-1}\left(1 + \epsilon_{a,t}\right) \\ & l_{a,t} = l_{a,t-1}(1 + \alpha_a \epsilon_{a,t}) \\ & \mu_{a,t} = \min(l_{a,t-1}, 1) \end{aligned} . \tag{13.20}$$

There is also an option with the additive error for the occurrence part (also underlying TSB), which has a different, more complicated form:

$$\begin{aligned} & o_t \sim \text{Bernoulli}\left(\mu_{a,t}\right) \\ & a_t = l_{a,t-1} + \epsilon_{a,t} \\ & l_{a,t} = l_{a,t-1} + \alpha_a \epsilon_{a,t} \\ & \mu_{a,t} = \max\left(\min(l_{a,t-1}, 1), 0\right) \end{aligned} . \tag{13.21}$$

The estimation of the oETS_D(M,N,M) model can be done using the following set of equations:

$$\begin{aligned} & \hat{\mu}_{a,t} = \hat{l}_{a,t-1} \\ & \hat{l}_{a,t} = \hat{l}_{a,t-1}(1 + \hat{\alpha}_a e_{a,t}) \end{aligned} , \tag{13.22}$$

where

$$e_{a,t} = \frac{o_t(1 - 2\kappa) + \kappa - \hat{\mu}_{a,t}}{\hat{\mu}_{a,t}}, \tag{13.23}$$

and κ is a very small number (for example, $\kappa = 10^{-10}$), needed only in order to make the model estimable. The estimate of the error term in case of the additive model is much simpler and does not need any specific tricks to work:

$$e_{a,t} = o_t - \hat{\mu}_{a,t}, \tag{13.24}$$

which is directly related to the TSB method. Note that equation (13.22) does not contain the minimum function, because the estimated error (13.23) will always guarantee that the level will lie between 0 and 1 as long as the smoothing parameter lies in the [0, 1] region (which is the conventional assumption for both th ETS(A,N,N) and ETS(M,N,N) models). This also applies for the $oETS_D$(A,N,N) model, where the maximum and minimum functions can be dropped as long as the smoothing parameter lies in [0,1].

An important feature of this model is that it allows probability to become either 0 or 1, thus implying either that there is no demand on the product or that the demand for the product has become regular. No other oETS model permits that – they assume that probability might become very close to bounds but can never reach them.

Here's an example of the application of the $oETS_D$(M,M,N) to the same artificial data:

```
oETSDModel <- oes(y, model="MMN", h=10, holdout=TRUE,
                  occurrence="direct")
oETSDModel
```

```
## Occurrence state space model estimated: Direct probability
## Underlying ETS model: oETS[D](MMN)
## Smoothing parameters:
##  level  trend
## 0.0439 0.0000
## Vector of initials:
##  level  trend
## 0.2351 1.0111
##
## Error standard deviation: 1.2292
## Sample size: 110
## Number of estimated parameters: 4
## Number of degrees of freedom: 106
## Information criteria:
##      AIC     AICc      BIC     BICc
## 109.9171 110.2981 120.7191 121.6144
```

From Figure 13.6, we can see that the probability of occurrence increases rapidly and reaches the bound of one around the 90th observation.

Practically speaking, using $oETS_D$ makes sense, when you expect the demand on your product either to disappear completely or to become regular. In other cases, $oETS_I$ and $oETS_O$ could be used efficiently. On M5 data, I found that the latter two models to performed very well in the majority of cases (Svetunkov & Boylan, 2023).

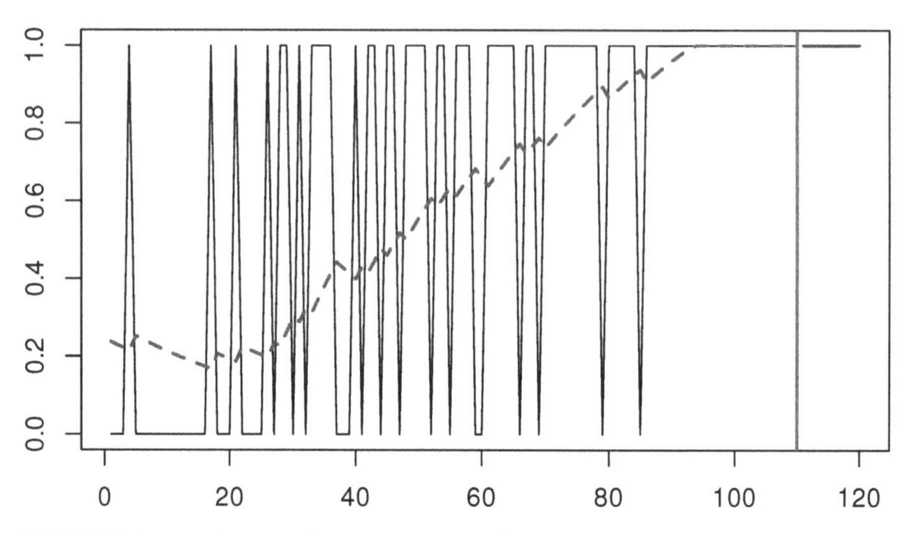

FIGURE 13.6 Demand occurrence and probability of occurrence in the oETS(M,M,N)$_D$ model.

13.1.6 Model selection in oETS

There are two dimensions for the model selection in the oETS model:

1. Selection of the type of occurrence;
2. Selection of the model type for the occurrence part.

The solution in this situation is simple. Given the formula (13.4), we can try each of the models, calculate log-likelihoods, the number of all estimated parameters, and then select the one that has the lowest information criterion. The demand occurrence models discussed in this section will have:

1. oETS$_F$: 1 parameter for the probability of occurrence;
2. oETS$_O$, oETS$_I$, and oETS$_D$: initial values, smoothing and dampening parameters;
3. oETS$_G$: initial values, smoothing, and dampening parameters for models A and B.

For example, if the oETS(M,N,N) model is constructed, the overall number of estimated parameters for the models will be:

1. oETS(M,N,N)$_F$ – 1 parameter: the probability of occurrence $\hat{p}$;
2. oETS(M,N,N)$_O$, oETS(M,N,N)$_I$ and oETS(M,N,N)$_D$ – 2 each: the initial value of level and the smoothing parameter;
3. oETS(M,N,N)$_G$ – 4: the initial values of $\hat{l}_{a,0}$ and $\hat{l}_{b,0}$ and the smoothing parameters $\hat{\alpha}_a$ and $\hat{\alpha}_b$.

This implies that the selection between models in (2) will come to the best fit to the demand occurrence data, while oETS(M,N,N)$_G$ will only be selected if it provides a much better fit to the data. Given that intermittent demand typically does not have many observations, the selection of oETS(M,N,N)$_G$ becomes highly improbable.

When it comes to selecting the most appropriate demand occurrence model (e.g. selecting ETS components or ARIMA orders), the approach would be similar: estimate the pool of models via likelihood, calculate numbers of parameters, select the model with the lowest information criterion. Given that we assume that the demand occurrence is independent of the demand sizes, the selection of models in the occurrence part can be done based on the likelihood (13.4) independently of the demand sizes part of the model, which significantly simplifies the estimation and model selection process for the whole iETS model.

13.2 Demand sizes part of the model

So far, we have discussed the occurrence part of the model o_t and how to capture the probability of demand occurrence p_t. But this is only half of the intermittent state space model. The second one is the model for the demand sizes z_t, which focuses on how many units of product will be sold if our customers decide to buy in a specific period of time. This can be modelled with any ADAM (ETS/ARIMA/regression), but it would need to be amended slightly to take intermittent demand features into account.

We start the discussion with analysis of an iETS(M,N,N)$_F$ model, which can be formulated as:

$$\begin{aligned} y_t &= o_t z_t \\ z_t &= l_{z,t-1}(1 + \epsilon_{z,t}) \\ l_{z,t} &= l_{z,t-1}(1 + \alpha_z \epsilon_{z,t}) , \\ o_t &\sim \text{Bernoulli}(p) \end{aligned} \tag{13.25}$$

where the subscript z refers to the components and parameters of demand sizes. This model assumes that there is always a potential demand on the product which evolves over time (even when $o_t = 0$), but we do not always observe it. This model's main properties have already been discussed in Section 6.1. The main challenge appears when this model needs to be constructed and estimated because z_t is not observable when $o_t = 0$. In these instances, the error term cannot be estimated, but according to the model, it still exists, thus impacting the level of demand $l_{z,t}$. To construct the model in the cases of no demand, we propose taking the conditional expectation for these periods, given the last available non-zero observation. This means that iETS(M,N,N)$_F$

can be constructed using the following set of equations:

$$\begin{aligned}
e_{z,t} &= \frac{z_t - \hat{\mu}_{z,t}}{\hat{\mu}_{z,t}}, \text{ when } o_t = 1 \\
\hat{\mu}_{z,t} &= \hat{l}_{z,t-1} \\
\hat{l}_{z,t} &= \begin{cases} \hat{l}_{z,t-1}(1 + \hat{\alpha}_z e_t), & \text{when } o_t = 1 \\ \hat{l}_{z,t-1}, & \text{when } o_t = 0 \end{cases}
\end{aligned} . \tag{13.26}$$

This is only possible if $\mathrm{E}(1 + \epsilon_{z,t}) = 1$, which is an important assumption for multiplicative error models, discussed in Section 6.5. If this is violated, then the formula for the calculation of the level in (13.26) will become more complicated, involving the expectation of products of random variables.

In a similar way, we can construct more complicated models for the demand sizes. In a more general case (Section 5) this can be written as:

$$\begin{aligned}
e_{z,t} &= \frac{z_t - \hat{\mu}_{z,t}}{\hat{\mu}_{z,t}}, \text{ when } o_t = 1 \\
\hat{\mathbf{v}}_t &= \begin{cases} f(\hat{\mathbf{v}}_{t-\mathbf{l}}) + g(\hat{\mathbf{v}}_{t-\mathbf{l}}) e_t, & \text{when } o_t = 1 \\ f(\hat{\mathbf{v}}_{t-\mathbf{l}}), & \text{when } o_t = 0 \end{cases}
\end{aligned} , \tag{13.27}$$

where all the functions and vectors have been defined for the original ADAM (7.1) in Section 7.1.

13.2.1 Additive vs multiplicative ETS for demand sizes

The approach above supports any type of ADAM, including pure additive ETS (Section 5.1), pure multiplicative ETS (Section 6.1) or mixed ETS (Section 7.2). While selection of the appropriate model can be automated, I argue that the better approach is to do it based on the understanding of the problem. In demand forecasting, typically we expect the values to be non-negative: people want to buy our product, and usually, the business does not want to buy products back from customers (unless we are dealing with a circular supply chain, but this is a different topic). This means that the pure multiplicative models should be preferred to the additive ones, as they will always produce meaningful results, as long as the assumption of positivity of $(1 + \epsilon_{z,t})$ holds. This assumption is important because the intermittent demand would typically have low volume, and the model might generate unreasonable (negative) point and interval forecasts if a non-positive distribution is used for the error term (e.g. Normal). Thus, it is important to use Inverse Gaussian, or Gamma, or Log-Normal distribution (see discussion in Section 6.5) for the error term of the demand sizes part of the model when the volume of data is low, and you expect the non-zero values to be strictly positive.

The main difficulty with pure multiplicative models arises from the construction point of view. As discussed in Section 6.3, the point forecasts of such models, in general, do not correspond to the conditional h steps ahead expectations (the only exclusion is the ETS(M,N,N) model). At the same time, the construction of the model for demand sizes assumes that the conditional expectations are equal to point forecasts when demand is not observed. If this is violated, then (13.27) is no longer the correct way to construct the model. This problem becomes especially important for the models with the multiplicative trend, where the conditional expectation might differ from point forecasts substantially (Svetunkov & Boylan, 2022). Still, point forecasts can be considered proxies for the conditional expectations, especially when smoothing parameters are close to zero. For example, the conditional expectation coincides with the point forecast in the boundary case with $\alpha = 0$ and $\beta = 0$ in ETS(M,M,N). The higher the smoothing parameters are, the more significant the discrepancy will be, implying that the model for the demand sizes is constructed incorrectly.

The pure additive models do not have the issue with the conditional expectation and thus can be constructed easily in the case of intermittent demand. But as discussed earlier, they might violate the non-negativity assumption of the model. So, in practice, they should be used with care.

13.2.2 Using ARIMA for demand sizes

ADAM ARIMA can also be used for demand sizes, resulting in the iARIMA model. All the discussions in the previous subsection would apply to ARIMA as well, keeping in mind that ADAM ARIMA can be either pure additive (Section 9.1.2) or pure multiplicative (Section 9.1.4). Given that the multiplicative ARIMA is formulated via logarithms and still has the error term with the expectation of one, any ARIMA model can be used for the variable z_t and can be constructed via (13.27). This can also be used for the cases when a pure multiplicative model with the trend is needed, and there are difficulties with the construction of ETS(M,M,N) (i.e. smoothing parameters are not close to zero). The relation between ARIMA and ETS (discussed in Section 8.4) might be useful in this case: instead of constructing ETS(M,M,N) we can construct logARIMA(0,2,2) (see Section 9.1.4), sidestepping the aforementioned problem.

13.2.3 Rounding up forecasts

Finally, when it comes to using an intermittent state space model on count data, there is a temptation to round up the resulting forecasts. If this is done for the point forecasts (conditional expectations), then this should be avoided, because the values show what happens on average and thus are allowed to take any values, not only the integers. However, when it comes to predictive quantiles, Svetunkov and Boylan (2023) show that rounding them up is equivalent to generating quantiles from a model with discretised distribution (see discussion on discretised distributions in Chakraborty, 2015) and improves both the forecasting and inventory performance of the model. So, the simple approach of generating a prediction interval (see Section 18.3) and then rounding it up has a theoretical rationale behind it and works well in practice.

13.3 The complete ADAM

Uniting demand occurrence (from Section 13.1) with the demand sizes (Section 13.2) parts of the model, we can now discuss the complete ADAM model, which in the most general form can be represented as:

$$
\begin{aligned}
& y_t = o_t z_t, \\
& z_t = w_z(\mathbf{v}_{z,t-\boldsymbol{l}}) + r_z(\mathbf{v}_{z,t-\boldsymbol{l}})\epsilon_{z,t} \\
& \mathbf{v}_{z,t} = f_z(\mathbf{v}_{z,t-\boldsymbol{l}}) + g_z(\mathbf{v}_{z,t-\boldsymbol{l}})\epsilon_{z,t} \\
& \\
& o_t \sim \text{Bernoulli}(p_t), \\
& p_t = f_p(\mu_{a,t}, \mu_{b,t}) \\
& a_t = w_a(\mathbf{v}_{a,t-\boldsymbol{l}}) + r_a(\mathbf{v}_{a,t-\boldsymbol{l}})\epsilon_{a,t} \\
& \mathbf{v}_{a,t} = f_a(\mathbf{v}_{a,t-\boldsymbol{l}}) + g_a(\mathbf{v}_{a,t-\boldsymbol{l}})\epsilon_{a,t} \\
& b_t = w_b(\mathbf{v}_{b,t-\boldsymbol{l}}) + r_b(\mathbf{v}_{b,t-\boldsymbol{l}})\epsilon_{b,t} \\
& \mathbf{v}_{b,t} = f_b(\mathbf{v}_{b,t-\boldsymbol{l}}) + g_b(\mathbf{v}_{b,t-\boldsymbol{l}})\epsilon_{b,t}
\end{aligned}
, \qquad (13.28)
$$

where the elements of the demand size and demand occurrence parts have been discussed in Sections 13.2 and 13.1, respectively. The model (13.28) can also be considered a more general one than the conventional ADAM ETS and ARIMA models: if the probability of occurrence p_t is equal to one for all observations, then the model reverts to them.

Summarising the discussions in previous sections of this chapter, the complete ADAM has the following assumptions:

1. The demand sizes variable z_t is continuous. This is a reasonable assumption for many contexts, including, for example, energy forecasting. But even when we deal with integer values, Svetunkov and Boylan (2023) showed that such a model does not perform worse than count data models. And if the integer values are needed, Svetunkov and Boylan (2023) demonstrated that rounding up quantiles from such a model is a reasonable and efficient approach that performs well in terms of forecasting accuracy and inventory performance;
2. Potential demand size may change over time even when $o_t = 0$. This means that the states evolve even when demand is not observed;
3. Demand sizes and demand occurrence are independent. This simplifies many derivations and makes the model estimable. If the assumption is

violated, a different model with different properties would need to be constructed. My understanding of the problem tells me that the model (13.28) will work well even if this is violated, but I have not done any experiments in this direction.

Depending on the specific model for each part and restrictions on $\mu_{a,t}$ and $\mu_{b,t}$, we might have different types of iETS models. To distinguish one model from another, we introduce the notation of iETS models of the form "iETS(demand sizes model)$_{\text{type of occurrence}}$(model A type)(model B type)". For example, in the iETS(M,N,N)$_G$(A,N,N)(M,M,N), the first brackets say that ETS(M,N,N) was applied to the demand sizes, and the underscored letter points out that this is the "general probability" model (Subsection 13.1.4), which has ETS(A,N,N) for the model A and ETS(M,M,N) for the model B. If only one demand occurrence part is used (either a_t or b_t), then the redundant brackets are dropped, and the notation is simplified. For example, iETS(M,N,N)$_O$(M,M,N) has only one part for demand occurrence, which is captured using the ETS(M,M,N) model. If the same type of model is used for both demand sizes and demand occurrence, then the second brackets can be dropped as well, simplifying this further to iETS(M,N,N)$_O$ (odds ratio model with ETS(M,N,N) for both parts, Section 13.1.2). All these models are implemented in the `adam()` function for the `smooth` package in R.

Similar notations and principles can be used for models based on ARIMA. Note that oARIMA is not yet implemented in `smooth`, but in theory, a model like iARIMA(0,1,1)$_O$(1,1,2) could be constructed in the ADAM framework.

Furthermore, in some cases, we might have explanatory variables, such as promotions, prices, weather, etc. They would impact both demand occurrence and demand sizes. In ADAM, they are implemented in respective `oes()` and `adam()` functions. Remember that when you include explanatory variables in the occurrence part, you are modelling the probability of occurrence, not the occurrence itself. So, for example, a promotional effect in this situation would mean a higher chance of having sales. In some other situations, we might not need dynamic models, such as ETS and ARIMA, and can focus on static regression. While `adam()` supports this, the `alm()` function from the `greybox` might be more suitable in this situation. It supports similar parameters, but its `occurrence` parameter accepts either the type of transform (`plogis` for the logit model and `pnorm` for the probit one) or a previously estimated occurrence model (either from `alm()` or from `oes()`).

In addition, there might be some cases, when the demand itself happens at random, but the demand sizes are at the same time not random. This means that when someone buys a product, they buy a fixed amount of z. Typically, in these situations $z = 1$, but there might be some cases with other values. An example of such a demand process is shown in Figure 13.7.

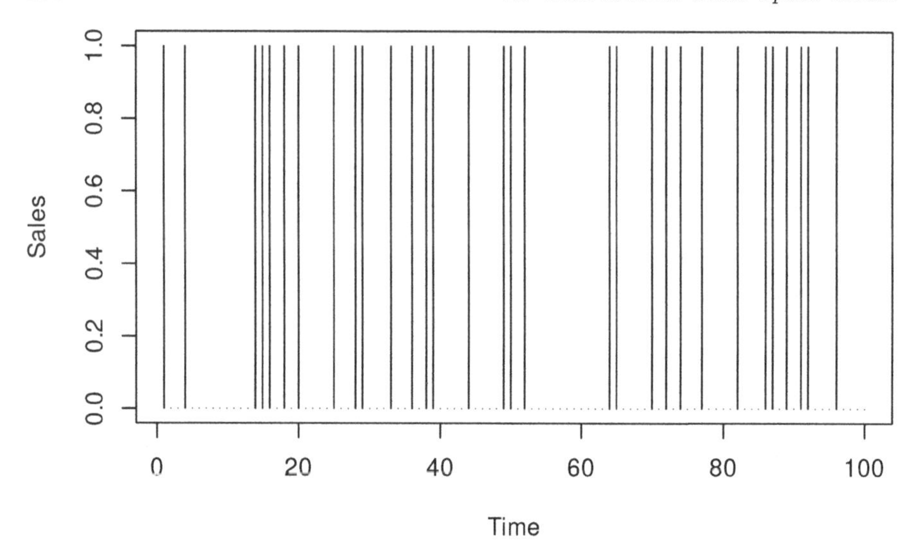

FIGURE 13.7 Example of intermittent time series with fixed demand sizes.

In this case, the model (13.28) simplifies to:

$$\begin{aligned} y_t &= o_t z, \\ o_t &\sim \text{Bernoulli}(p_t) \end{aligned}, \tag{13.29}$$

where p_t can be captured via one of the demand occurrence models discussed in previous sections. Given that z is not random anymore, it does not require estimation, and the likelihood simplifies to the one for the demand occurrence (from Section 13.1). The forecast from this model can be generated by producing conditional expectation for the probability and then multiplying it by the value of z.

Finally, in some situations the occurrence might be not random, i.e. we deal with not an intermittent demand, but with a demand, where sales sometimes do not happen (i.e. $o_t = 0$), but we can predict perfectly when they happen. If we know when the demand occurs, we can use the respective values of zeroes and ones in the variable o_t, simplifying the model (13.28) to:

$$\begin{aligned} y_t &= o_t z_t, \\ z_t &= w_z(\mathbf{v}_{z,t-\boldsymbol{l}}) + r_z(\mathbf{v}_{z,t-\boldsymbol{l}})\epsilon_{z,t} \\ \mathbf{v}_{z,t} &= f_z(\mathbf{v}_{z,t-\boldsymbol{l}}) + g_z(\mathbf{v}_{z,t-\boldsymbol{l}})\epsilon_{z,t} \end{aligned}, \tag{13.30}$$

where the values of o_t are known in advance. An example of such a situation is the demand on watermelons, which reaches zero in specific periods of time in winter (at least in the UK). This special type of model is implemented in both the `adam()` and `alm()` functions, where the user needs to provide the vector of zeroes and ones in the `occurrence` parameter.

13.3.1 Maximum Likelihood Estimation

While there are different ways of estimating the parameters of the ADAM (13.28), it is worth focusing on likelihood estimation (Section 11.1) for consistency with other ADAMs. The likelihood of the model will consist of several parts:

1. The probability density function (PDF) of demand sizes when demand occurs;
2. The probability of occurrence;
3. The probability of inoccurrence.

When demand occurs the likelihood is:

$$\mathcal{L}(\boldsymbol{\theta}|y_t, o_t = 1) = p_t f_z(z_t|\mathbf{v}_{z,t-\boldsymbol{l}}), \tag{13.31}$$

while in the opposite case it is:

$$\mathcal{L}(\boldsymbol{\theta}|y_t, o_t = 0) = (1 - p_t) f_z(z_t|\mathbf{v}_{z,t-\boldsymbol{l}}), \tag{13.32}$$

where $\boldsymbol{\theta}$ includes all the estimated parameters of the model (for demand sizes and demand occurrence, including parameters of distribution, such as scale). Note that because the model assumes that the demand evolves over time even when it is not observed ($o_t = 0$), we have a probability density function of demand sizes, $f_z(z_t|\mathbf{v}_{z,t-\boldsymbol{l}})$ in (13.32). Based on equations (13.31) and (13.32), we can summarise the likelihood for the whole sample of T observations:

$$\mathcal{L}(\boldsymbol{\theta}|\mathbf{y}) = \prod_{o_t=1} p_t \prod_{o_t=0} (1 - p_t) \prod_{t=1}^{T} f_z(z_t|\mathbf{v}_{z,t-\boldsymbol{l}}), \tag{13.33}$$

or in logarithms:

$$\ell(\boldsymbol{\theta}|\mathbf{y}) = \sum_{o_t=1} \log(p_t) + \sum_{o_t=0} \log(1 - p_t) + \sum_{t=1}^{T} f_z(z_t|\mathbf{v}_{z,t-\boldsymbol{l}}), \tag{13.34}$$

where $\mathbf{y}$ is the vector of all actual values and $f_z(z_t|\mathbf{v}_{z,t-\boldsymbol{l}})$ can be substituted by a PDF of the assumed distribution from the list of candidates in Section 11.1. The main issue in calculating the likelihood (13.34) is that the demand sizes are not observable when $o_t = 0$. This means that we cannot calculate the likelihood using the conventional approach. We need to use something else. Svetunkov and Boylan (2023) proposed using the Expectation Maximisation (EM) algorithm for this purpose, which is typically done in the following stages:

1. Take *Expectation* of the likelihood;
2. *Maximise* it with the obtained parameters;
3. Go to (1) with the new set of parameters if the likelihood has not converged to the maximum.

TABLE 13.1 Differential entropies for different distributions. $\Gamma(\cdot)$ is the Gamma function, while $\psi(\cdot)$ is the digamma function.

	Assumption	Differential Entropy
Normal	$\epsilon_t \sim \mathcal{N}(0, \sigma^2)$	$\frac{1}{2}\left(\log(2\pi\sigma^2) + 1\right)$
Laplace	$\epsilon_t \sim \mathcal{L}(0, s)$	$1 + \log(2s)$
S	$\epsilon_t \sim \mathcal{S}(0, s)$	$2 + 2\log(2s)$
Generalised Normal	$\epsilon_t \sim \mathcal{GN}(0, s, \beta)$	$\beta^{-1} - \log\left(\frac{\beta}{2s\Gamma(\beta^{-1})}\right)$
Inverse Gaussian	$1 + \epsilon_t \sim \mathcal{IG}(1, \sigma^2)$	$\frac{1}{2}\left(\log \pi e \sigma^2 - \log(2)\right)$
Gamma	$1 + \epsilon_t \sim -(\sigma^{-2}, \sigma^2)$	$\sigma^{-2} + \log \Gamma\left(\sigma^{-2}\right) + \left(1 - \sigma^{-2}\right)\psi\left(\sigma^{-2}\right)$
Log-Normal	$1 + \epsilon_t \sim \log\mathcal{N}\left(-\frac{\sigma^2}{2}, \sigma^2\right)$	$\frac{1}{2}\left(\log(2\pi\sigma^2) + 1\right) - \frac{\sigma^2}{2}$

A classic example with EM is when several samples have different parameters, and we need to split them (e.g. do clustering). In that case, we do not know what cluster specific observations belong to and what the probability that each observation belongs to one of the groups is. In our context, the problem is slightly different: we know probabilities, but we do not observe some of the values. As a result the application of EM gives a different result. If we take the expectation of (13.34) with respect to the unobserved demand sizes, we will get:

$$\begin{aligned} \mathrm{E}\left(\ell(\boldsymbol{\theta}|\mathbf{y})\right) = & \sum_{o_t=1} \log f_z\left(z_t|\mathbf{v}_{z,t-\boldsymbol{l}}\right) + \sum_{o_t=0} \mathrm{E}\left(\log f_z\left(z_t|\mathbf{v}_{z,t-\boldsymbol{l}}\right)\right) \\ & + \sum_{o_t=1} \log(p_t) + \sum_{o_t=0} \log(1-p_t) \end{aligned} . \tag{13.35}$$

The first term in (13.35) is known because the z_t are observed when $o_t = 1$. The expectation in the second term of (13.35) is known in statistics as "Differential Entropy" (in the formula above, we have the negative differential entropy). It will differ from one distribution to another. Table 13.1 summarises differential entropies for the distributions used in ADAM.

The majority of formulae for differential entropy in Table 13.1 are taken from Lazo and Rathie (1978) with the exclusion of the one for $\mathcal{IG}$, which was derived by Mudholkar and Tian (2002). These values can be inserted instead of the $\mathrm{E}\left(\log f_z\left(z_t|\mathbf{v}_{z,t-\boldsymbol{l}}\right)\right)$ in the formula (13.35), leading to the expected likelihood for respective distributions, which can then be maximised. For example, for Inverse Gaussian distribution (using the PDF from the Table 11.2 and the entropy from Table 13.1), we get:

$$\begin{aligned} \mathrm{E}\left(\ell(\boldsymbol{\theta}|\mathbf{y})\right) = & -\frac{T_1}{2} \log\left(2\pi\sigma^2\right) - \frac{1}{2}\sum_{o_t=1} \left(1+\epsilon_t\right)^3 \\ & - \frac{3}{2}\sum_{o_t=1} \log y_t - \frac{1}{2\sigma^2}\sum_{o_t=1} \frac{\epsilon_t^2}{1+\epsilon_t} \\ & - \frac{T_0}{2}\left(\log \pi e \sigma^2 - \log(2)\right) \\ & + \sum_{o_t=1} \log(p_t) + \sum_{o_t=0} \log(1-p_t) \end{aligned} , \tag{13.36}$$

where T_0 is the number of zeroes in the data and T_1 is the number of non-zero values. Luckily, the EM process in our specific situation does not need to be iterative – the obtained likelihood can then be maximised directly by changing the values of parameters $\boldsymbol{\theta}$. It is also possible to derive analytical formulae for parameters of some of distributions based on (13.35) and the values from Table 13.1. For example, in the case of the Inverse Gaussian distribution the estimate of scale parameter is:

$$\hat{\sigma}^2 = \frac{1}{T}\sum_{o_t=1} \frac{e_t^2}{1+e_t}. \tag{13.37}$$

This is obtained by taking the derivative of (13.36) with respect to σ^2 and equating it to zero. It can be shown that the likelihood estimates of scales for different distributions correspond to the conventional formulae from Section 11.1, but with the summation over $o_t = 1$ instead of all the observations. Note, however, that the division in (13.37) is done by the whole sample T. This implies that the scale estimate will be biased, similarly to the classical bias of the sample variance. Svetunkov and Boylan (2023) show that in the full ADAM, the estimate of scale is biased not only in-sample but also asymptotically, implying that with the increase of the sample size, it will be consistently lower than needed. This is because the summation is done over the non-zero values, while the division is done over the whole sample. This proportion of non-zeroes impacts the scale in (13.37), deflating its value. The only situation when the bias will be reduced is when the probability of occurrence reaches 1 (demand becomes regular). Still, the value (13.37) will maximise the expected likelihood (13.35) and is useful for inference. However, if one needs to construct prediction intervals, this bias needs to be addressed, which can be done using the conventional correction:

$$\hat{\sigma}^{2\prime} = \frac{T}{T_1 - k}\hat{\sigma}^2, \tag{13.38}$$

where k is the number of all estimated parameters.

Finally, for the two exotic cases with known demand occurrence and known demand sizes, the likelihood (13.35) simplifies to the first two and the last two terms in the formula, respectively.

13.3.2 Conditional expectation and variance

Now that we have discussed how the model is formulated and how it can be estimated, we can move to the discussion of conditional expectation and variance from it. The former is needed to produce point forecasts, while the latter might be required for different inventory decisions.

The conditional h steps ahead expectation of the model can be obtained easily based on the assumption of independence of demand occurrence and demand sizes discussed earlier in Section 13.3:

$$\mathrm{E}(y_{t+h}|t) = \mu_{y,t+h|t} = \mathrm{E}(o_{t+h}|t)\mathrm{E}(z_{t+h}|t) = \mu_{o,t+h|t}\mu_{z,t+h|t}, \tag{13.39}$$

where $\mu_{o,t+h|t}$ is the conditional expectation of the occurrence variable (the conditional h steps ahead probability of occurrence) and $\mu_{z,t+h|t}$ is the conditional expectation of the demand sizes variable z_t. So, the forecast from the complete ADAM relies on the probability of occurrence of the variable and will reflect an average demand per period of time. As a result, it might be less than one in some cases, implying that the product is not sold every day. Consequentially, Kourentzes (2014) argued that the term "demand rate" should be used in this context instead of the conditional expectation. However, any forecasting model produces "demand per period" forecasts. They just typically assume that the probability of occurrence is equal to one ($p_t = 1$) for all observations. So, there is no conceptual difference between the point forecasts produced by regular and intermittent demand models, and I do not see the point in using the "demand rate" term.

As for the conditional variance, it is slightly trickier than the conditional expectation, because the variance of a product involves not only variances, but expectations as well (assuming that two variables are independent):

$$\mathrm{V}(y_{t+h}|t) = \mathrm{V}(o_{t+h}|t)\mathrm{V}(z_{t+h|t}) + \mathrm{E}(o_{t+h}|t)^2\mathrm{V}(z_{t+h}|t) + \mathrm{V}(o_{t+h}|t)\mathrm{E}(z_{t+h}|t)^2. \tag{13.40}$$

Given that we use Bernoulli distribution for the variable o_t, its variance is equal to $\mu_{o,t+h|t}(1-\mu_{o,t+h|t})$. In our context this implies that the conditional h steps ahead variance for the complete ADAM is:

$$\sigma^2_h = \mu_{o,t+h|t}(1-\mu_{o,t+h|t})\sigma^2_{z,h} + \mu^2_{o,t+h|t}\sigma^2_{z,h} + \mu_{o,t+h|t}(1-\mu_{o,t+h|t})\mu^2_{z,t+h|t}, \tag{13.41}$$

or after some manipulations:

$$\sigma^2_h = \mu_{o,t+h|t}\left(\sigma^2_{z,h} + (1-\mu_{o,t+h|t})\mu^2_{z,t+h|t}\right). \tag{13.42}$$

All the elements of the formula (13.42) are available and have been discussed in Sections 5.3, 6.3, and 13.1.

When it comes to the two exotic cases, discussed in the beginning of this section, we have the following conditional expectation and variance:

1. When demand sizes are fixed and known, i.e. $z_t = z$:

$$\begin{aligned} \mu_{y,t+h|t} &= \mu_{o,t+h|t} z, \\ \sigma^2_h &= \mu_{o,t+h|t}(1 - \mu_{o,t+h|t}) z^2; \end{aligned} \tag{13.43}$$

2. When demand occurrence is known:

$$\begin{aligned} \mu_{y,t+h|t} &= o_{t+h} \mu_{z,t+h|t}, \\ \sigma^2_h &= o_{t+h} \left(\sigma^2_{z,h} + (1 - o_{t+h}) \mu^2_{z,t+h|t} \right). \end{aligned} \tag{13.44}$$

These two cases might not arise often, but it is important to understand how they change the moments of the model.

13.4 Examples of application

We consider the same example from Section 13.1. Just as a reminder, in that example, both demand occurrence and demand sizes increase over time (Figure 13.1), meaning that we can try the model with the trend for both parts. This can be done using the `adam()` function from the `smooth` package, defining the type of occurrence to use. We will try several options and select the one that has the lowest AICc:

```
adamiETSy <- vector("list",4)
adamiETSy[[1]] <- adam(y, "MMdN", h=10, holdout=TRUE,
                       occurrence="odds-ratio")
adamiETSy[[2]] <- adam(y, "MMdN", h=10, holdout=TRUE,
                       occurrence="inverse-odds-ratio")
adamiETSy[[3]] <- adam(y, "MMdN", h=10, holdout=TRUE,
                       occurrence="direct")
adamiETSy[[4]] <- adam(y, "MMdN", h=10, holdout=TRUE,
                       occurrence="general")
adamiETSyAICcs <-
    setNames(sapply(adamiETSy,AICc),
             c("odds-ratio", "inverse-odds-ratio",
               "direct", "general"))
adamiETSyAICcs
```

```
##          odds-ratio inverse-odds-ratio             direct
              general
##            359.5494           360.5086           354.4810
    371.6503
```

Based on this, we can see that the model with direct probability has the lowest AICc. We can show how the model approximates the data and produces forecasts for the holdout:

```
i <- which.min(adamiETSyAICcs)
par(mfcol=c(2,1), mar=c(2,1,2,1))
plot(adamiETSy[[i]],7)
plot(adamiETSy[[i]]$occurrence,7)
```

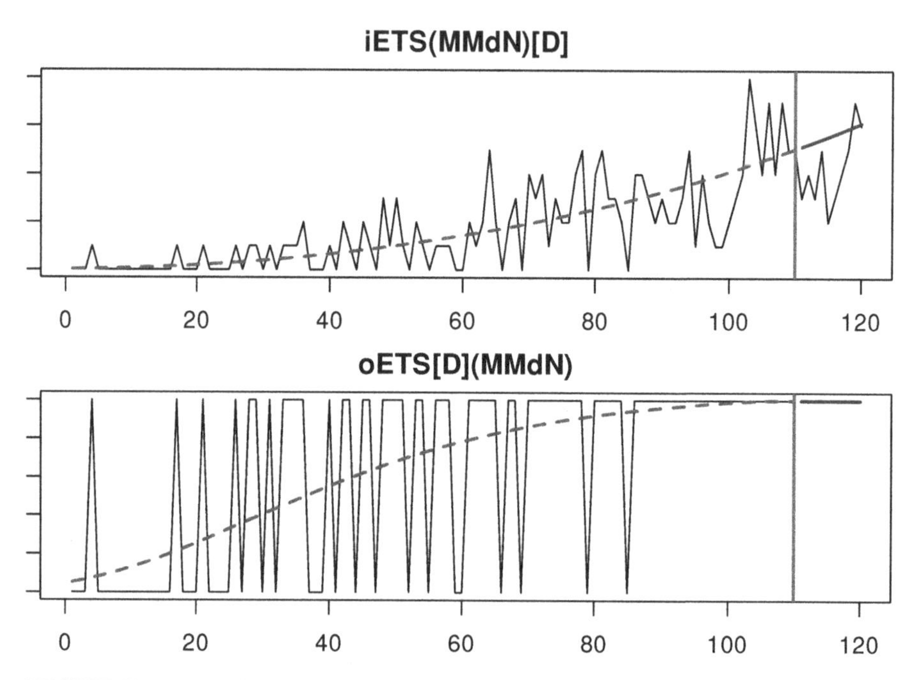

FIGURE 13.8 The fit of the best model to the intermittent data: final demand and the demand occurrence parts.

Figure 13.8 shows that the model captured the trend well for both the demand occurrence and demand sizes parts. It forecasts that the mean demand will increase for the holdout period. We can also explore the demand occurrence part of this model by typing:

```
adamiETSy[[i]]$occurrence
```

```
## Occurrence state space model estimated: Direct probability
## Underlying ETS model: oETS[D](MMdN)
## Smoothing parameters:
## level trend
```

```
##      0      0
## Vector of initials:
##  level  trend
## 0.0463 1.1343
##
## Error standard deviation: 1.4805
## Sample size: 110
## Number of estimated parameters: 5
## Number of degrees of freedom: 105
## Information criteria:
##      AIC     AICc      BIC     BICc
## 103.5314 104.1083 117.0338 118.3897
```

In our example, the smoothing parameters are equal to zero for the demand occurrence part, which makes sense because the selected model is the damped multiplicative trend one, which should capture the increasing probability of occurrence well.

Depending on the generated data, there might be issues in the ETS(M,Md,N) model for demand sizes, if the smoothing parameters are too big, especially for the trend component. So, we can also try out the logARIMA(1,1,2) to see how it compares with this model. Given that ARIMA is not yet implemented for the occurrence part of the model, we need to construct the oETS separately and then use in `adam()`:

```
oETSModel <- oes(y, "MMdN", h=10, holdout=TRUE,
                 occurrence=names(adamiETSyAICcs)[i])
adamiARIMA <- adam(y, "NNN", h=10, holdout=TRUE,
                   occurrence=oETSModel,
                   orders=c(1,1,2),
                   distribution="dlnorm")
adamiARIMA
```

```
## Time elapsed: 0.18 seconds
## Model estimated using adam() function: iARIMA(1,1,2)[D]
## Occurrence model type: Direct
## Distribution assumed in the model: Mixture of Bernoulli and Log-
   Normal
## Loss function type: likelihood; Loss function value: 128.0496
## ARMA parameters of the model:
## AR:
## phi1[1]
##  -0.209
## MA:
## theta1[1] theta2[1]
```

```
##   -0.4984    0.0283
##
## Sample size: 110
## Number of estimated parameters: 6
## Number of degrees of freedom: 104
## Information criteria:
##      AIC      AICc       BIC      BICc
## 361.6306 362.4461 377.8335 379.7502
##
## Forecast errors:
## Asymmetry: -70.393%; sMSE: 46.286%; rRMSE: 1.071; sPIS: 3196.961%;
    sCE: -386.317%
```

Comparing the iARIMA model with the previous iETS based on AIC would not be fair because as soon as the occurrence model is provided to the `adam()`, it does not count the parameters estimated in that part towards the overall number of estimated parameters. To make the comparison fair, we need to make ADAM iETS comparable by estimating it in a similar way:

```
adamiETSy[[i]] <- adam(y, "MMdN", h=10, holdout=TRUE,
                       occurrence=oETSModel)
adamiETSy[[i]]
```

```
## Time elapsed: 0.1 seconds
## Model estimated using adam() function: iETS(MMdN)[D]
## Occurrence model type: Direct
## Distribution assumed in the model: Mixture of Bernoulli and Gamma
## Loss function type: likelihood; Loss function value: 119.067
## Persistence vector g:
## alpha  beta
##     0     0
## Damping parameter: 0.9985
## Sample size: 110
## Number of estimated parameters: 6
## Number of degrees of freedom: 104
## Information criteria:
##      AIC      AICc       BIC      BICc
## 343.6655 344.4810 359.8684 361.7851
##
## Forecast errors:
## Asymmetry: -88.492%; sMSE: 53.022%; rRMSE: 1.146; sPIS: 3703.305%;
    sCE: -542.075%
```

Comparing information criteria, the iETS model is more appropriate for this data. But this might be due to different distributional assumptions and difficulties estimating the ARIMA model. If you want to experiment more with iARIMA, you might try fine tuning its parameters (see Section 11.4.1) for the data either by increasing the `maxeval` or changing the initialisation, for example:

```
adamiARIMA <- adam(y, "NNN", h=10, holdout=TRUE,
                   occurrence=oETSModel, orders=c(1,1,2),
                   distribution="dgamma", initial="back")
```

Finally, we can produce point and interval forecasts from either of the models via the `forecast()` method. Here is an example:

```
forecast(adamiETSy[[i]], h=10,
         interval="prediction", nsim=10000) |>
    plot()
```

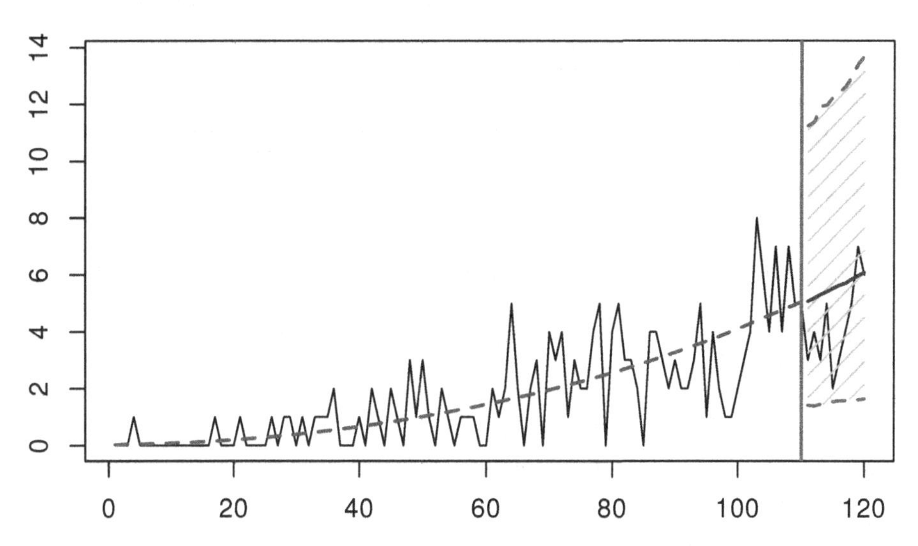

FIGURE 13.9 Point forecasts and prediction interval from the iETS(M,Md,N)$_D$ model.

In Figure 13.9, the interval is expanding, reflecting the captured tendency of growth in the data. The prediction interval produced from multiplicative ETS models will typically be simulated if `interval="prediction"`, so to make them smoother, you might need to increase the `nsim` parameter, for example, to `nsim=100000`.

13.5 Intermittent demand challenges

Intermittent demand is complicated and is difficult to work with. As a result, several challenges are related to the ADAM specifically and to the intermittent demand at large are worth discussing.

First, given the presence of zeroes, the decomposition (Section 3.2) of intermittent time series does not make sense. The classical time series model assumes that the demand happens on every observation, while the intermittent demand happens irregularly. This makes all the conventional models inapplicable to the problem, although some of them might still work well in some cases (for example, SES from Section 3.4 in case of mildly intermittent data).

The second follows directly from the first point. While, in theory, it is possible to use any ETS/ARIMA model for both demand occurrence and demand sizes of the ADAM, some of the specific model types are either impossible or very difficult to estimate. For example, seasonality on intermittent data is not very well pronounced, so estimating the initial values of components of seasonal models (such as ETS(M,N,M)) is not a trivial task. In some cases, if we have several products in a group that exhibit similar seasonal patterns, we can aggregate them to the group level to get a better estimate of seasonal indices, and then use them on the lower level. The `adam()` function allows doing that via `initial=list(seasonal=seasonalIndices)`. But in all the other cases, the estimation of seasonal models might fail.

Third, in some cases, you might know when specifically demand will happen (for example, kiwis stop growing in New Zealand from May till September, so the crop will go down around that time). In this case, you do not need a proper intermittent demand model, you just need to deal with the demand sizes via ADAM ETS/ARIMA and provide zeroes and ones in the demand occurrence part for the variable o_t. We have discussed this type of model in Section 13.3. Practically speaking, this can be done in `adam()` via `occurrence=ot`, where `ot` would contain zeroes and ones for the sample. This can also be done for the holdout sample in the `forecast()` function in a similar manner:

```
forecast(ourModel, occurrence=otFuture, h=h)
```

where `otFuture` should contain the values of the occurrence variable in the future.

Fourth, more specialised models, such as iETS, will produce positively biased estimates of the smoothing parameters, whatever the estimator is used (see explanation in Svetunkov & Boylan, 2023). This is caused by the assumption that the potential demand might change between the observed sales. In this situation, the components would evolve slowly, while we would only see their values before the set of zeroes and afterwards, which will make the applied model catch up to the data, thus inflating the estimates of smoothing parameters. This also implies that such forecasting methods as Croston (Croston, 1972) and TSB (Teunter et al., 2011) would also result in positively biased estimates of parameters if we assume that demand might change between the non-zero observations. Practically speaking, this means that the smoothing parameters will be higher than needed, implying more rapid changes in components and thus higher uncertainty in final forecasts. There is currently no solution to this problem.

Finally, summarising this chapter, intermittent demand forecasting is a complex problem. Differences between various forecasting models and methods on such data might be insignificant, and it would be challenging to select the appropriate one. Furthermore, point forecasts on intermittent demand are difficult to grasp and make actionable (unless you are interested in lead time forecasts, to get an idea about the expected demand over a period of time). All of this means that intermittent demand should be avoided if possible. Yes, you can have fancy models for it, but do you need to? For example, do you need to look at daily demand on products if decisions are made on a weekly basis (e.g. how many units of pasta should a supermarket order for the next week)? In many cases thinking about the problem carefully would allow avoiding intermittent demand, making the life of the analyst easier. But if it is not possible, then ADAM iETS and iARIMA models can be considered potential solutions in some situations.

14

Model diagnostics

In this chapter, we investigate how ADAM can be diagnosed and improved. Most topics will build upon the typical model assumptions discussed in Subsection 1.4.1 and in Chapter 15 of Svetunkov (2022). Some of the assumptions cannot be diagnosed properly, but there are well-established instruments for others. All the assumptions about statistical models can be summarised as follows:

1. Model is correctly specified:
    a. No omitted variables;
    b. No redundant variables;
    c. The necessary transformations of the variables are applied;
    d. No outliers in the residuals of the model.
2. Residuals are i.i.d.:
    a. They are not autocorrelated;
    b. They are homoscedastic;
    c. The expectation of residuals is zero, no matter what;
    d. The residuals follow the specified distribution;
    e. The distribution of residuals does not change over time.
3. The explanatory variables are not correlated with anything but the response variable:
    a. No multicollinearity;
    b. No endogeneity (not discussed in the context of ADAM).

Technically speaking, (3) is not an assumption about the model, it is just a requirement for the estimation to work correctly. In regression context, the satisfaction of these assumptions implies that the estimates of parameters are efficient and unbiased (respectively for (3a) and (3b)).

In general, all model diagnostics are aimed at spotting patterns in residuals. If there are patterns, then some assumption is violated and something is probably missing in the model. In this chapter, we will discuss which instruments can be used to diagnose different types of violations of assumptions.

Remark. The analysis carried out in this chapter is based mainly on visual inspection of various plots. While there are statistical tests for some assumptions, we do not discuss them here. This is because in many cases human judgment is at least as good as automated procedures (Petropoulos, Kourentzes, Nikolopoulos, & Siemsen, 2018), and people tend to misuse the latter (Wasserstein & Lazar, 2016). So, if you can spend time on improving the model for a specific data, the visual inspection will typically suffice.

To make this more actionable, we will consider a conventional regression model on `Seatbelts` data, discussed in Section 10.6. We start with a pure regression model, which can be estimated equally well with the `adam()` function from the `smooth` package or the `alm()` from the `greybox` in R. In general, I recommend using `alm()` when no dynamic elements are present in the model (or only AR(p) and/or I(d) are needed). Otherwise, you should use `adam()` in the following way:

```
adamSeat01 <- adam(Seatbelts, "NNN",
                   formula=drivers~PetrolPrice+kms)
plot(adamSeat01, 7, main="")
```

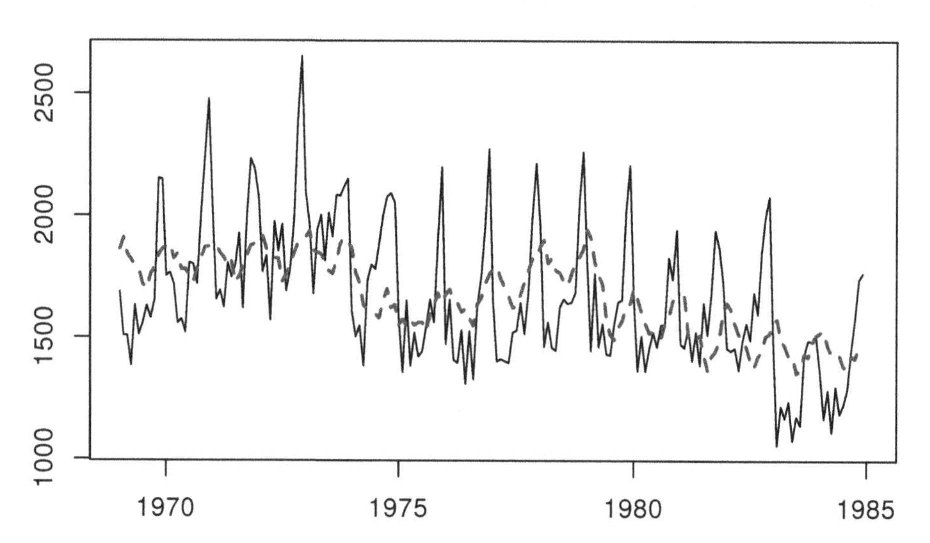

FIGURE 14.1 Basic regression model for the data on road casualties in Great Britain 1969–1984.

This model has several issues, and in this chapter, we will discuss how to diagnose and fix them.

14.1 Model specification: Omitted variables

We start with one of the most critical assumptions for models: the model does not omit important variables. If it does then the point forecasts might not be as accurate as we would expect and in some serious cases exhibit substantial bias. For example, if the model omits a seasonal component, it will not be able to produce accurate forecasts for specific observations (e.g. months of year).

In general, this issue is difficult to diagnose because it is typically challenging to identify what is missing if we do not have it in front of us. The best thing one can do is a mental experiment, trying to comprise a list of all theoretically possible components and variables that would impact the variable of interest. If you manage to come up with such a list and realise that some of them are missing, the next step would be to collect the variables themselves or use their proxies. The proxies are variables that are expected to be correlated with the missing variables and can partially substitute them. We would need to add the missing information in the model one way or another.

In case of ETS components, the diagnostics of this issue is possible, because the pool of components is restricted and we know what the data with different components should look like (see Section 4.1). In the case of ARIMA, the diagnostics comes to analysing ACF/PACF (Section 14.5). But in the case of regression, it might be difficult to say what specifically is missing.

However, in some cases, we might be able to diagnose this even for the regression model. For example, with the model that we estimated in the previous section, we have a set of variables not included in it. A simple thing to do is to see if the residuals of our model are correlated with any of the omitted variables. We can either produce scatterplots or calculate measures of association (see Chapter 9 of Svetunkov, 2022) to see if there are relations in the residuals. I will use `assoc()` and `spread()` functions from `greybox` for this:

```
# Create a new matrix, removing the variables that are already
# in the model
SeatbeltsWithResiduals <-
  cbind(as.data.frame(residuals(adamSeat01)),
        Seatbelts[,-c(2,5,6)])
colnames(SeatbeltsWithResiduals)[1] <- "residuals"
# Spread plot
greybox::spread(SeatbeltsWithResiduals)
```

The `spread()` function automatically detects the type of variable and based on that produces scatterplot/`boxplot()`/`tableplot()` between them, making the final plot more readable. The plot in Figure 14.2 tells us that residuals are

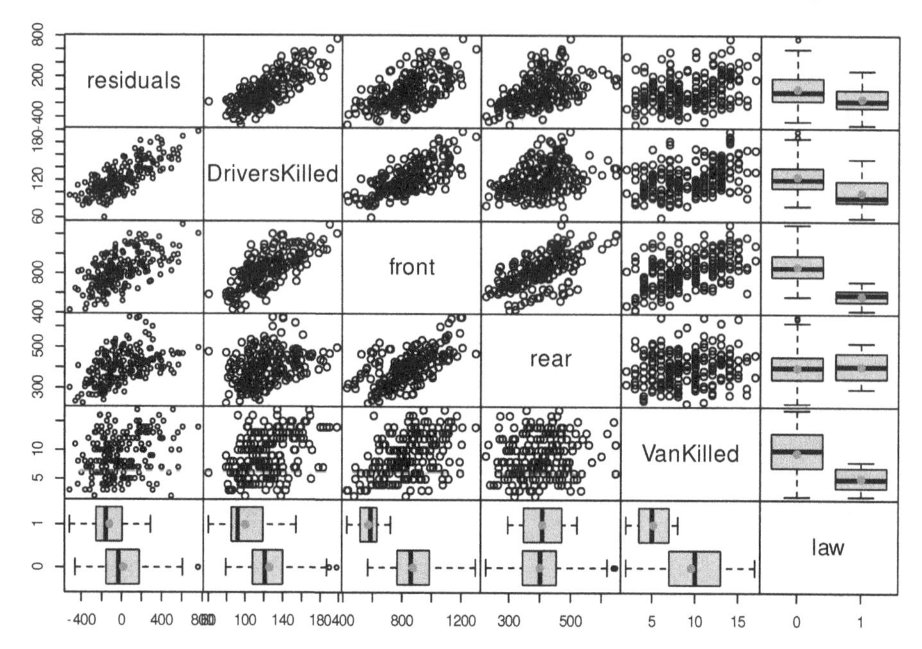

FIGURE 14.2 Spread plot for the residuals of model vs omitted variables.

correlated with `DriversKilled`, `front`, `rear`, and `law`, so some of these variables can be added to the model to improve it. However, not all of them make sense in our model. For example, `VanKilled` might have a weak relation with `drivers`, but judging by the description should not be included in the model (this is a part of the `drivers` variable). Also, I would not add `DriversKilled`, as it seems not to drive the number of deaths and injuries (based on our understanding of the problem), but is just correlated with it for obvious reasons (`DriversKilled` is included in `drivers`). The variables `front` and `rear` should not be included in the model either, because they do not explain injuries and deaths of drivers, they are impacted by similar factors, and can be considered as output variables. So, only `law` can be safely added to the model, because it makes sense: the introduction of law should hopefully impact the number of casualties amongst drivers.

We can also calculate measures of association between these variables to get an idea of the strength of linear relations bewteen them:

```
greybox::assoc(SeatbeltsWithResiduals)
```

```
## Associations:
## values:
##                 residuals DriversKilled  front   rear VanKilled
    law
```

```
## residuals         1.0000          0.7826 0.6121 0.4811     0.2751
    0.1892
## DriversKilled     0.7826          1.0000 0.7068 0.3534     0.4070
    0.3285
## front             0.6121          0.7068 1.0000 0.6202     0.4724
    0.5624
## rear              0.4811          0.3534 0.6202 1.0000     0.1218
    0.0291
## VanKilled         0.2751          0.4070 0.4724 0.1218     1.0000
    0.3949
## law               0.1892          0.3285 0.5624 0.0291     0.3949
    1.0000
##
## p-values:
##               residuals DriversKilled front   rear VanKilled    law
## residuals        0.0000             0     0 0.0000    0.0001 0.0086
## DriversKilled    0.0000             0     0 0.0000    0.0000 0.0000
## front            0.0000             0     0 0.0000    0.0000 0.0000
## rear             0.0000             0     0 0.0000    0.0925 0.6890
## VanKilled        0.0001             0     0 0.0925    0.0000 0.0000
## law              0.0086             0     0 0.6890    0.0000 0.0000
##
## types:
##               residuals DriversKilled front     rear      VanKilled
     law
## residuals     "none"    "pearson"     "pearson" "pearson" "pearson"
     "mcor"
## DriversKilled "pearson" "none"        "pearson" "pearson" "pearson"
     "mcor"
## front         "pearson" "pearson"     "none"    "pearson" "pearson"
     "mcor"
## rear          "pearson" "pearson"     "pearson" "none"    "pearson"
     "mcor"
## VanKilled     "pearson" "pearson"     "pearson" "pearson" "none"
       "mcor"
## law           "mcor"    "mcor"        "mcor"    "mcor"    "mcor"
       "none"
```

Technically speaking, the output of this function tells us that all variables are correlated with residuals and can be considered in the model. This is because p-values are lower than my favourite significance level of 1%, so we can reject the null hypothesis for each of the tests (which is that the respective parameter is equal to zero in the population). I would still prefer not to add `DriversKilled`, `VanKilled`, `front`, and `rear` variables in the model for the reasons explained earlier, but I would add `law`:

```
adamSeat02 <- adam(Seatbelts, "NNN",
                   formula=drivers~PetrolPrice+kms+law)
```

The model now fits the data differently (Figure 14.3):

```
plot(adamSeat02, 7, main="")
```

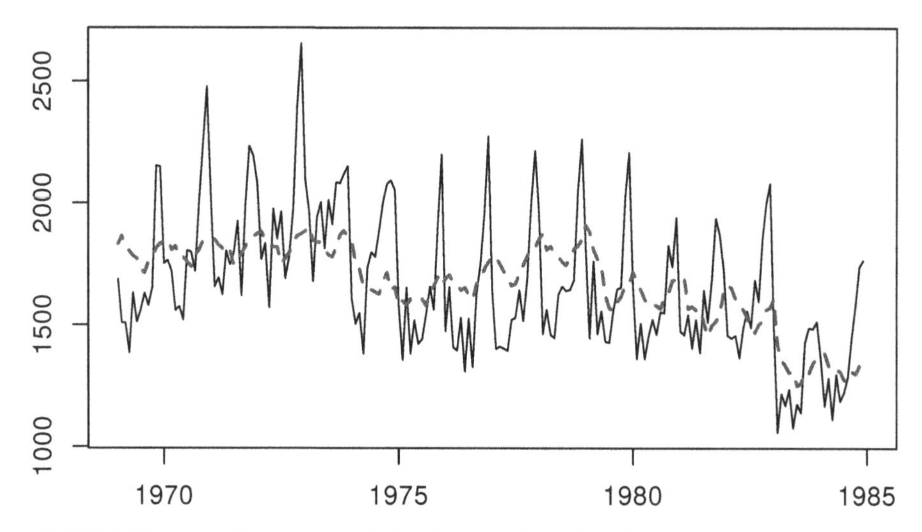

FIGURE 14.3 The data and the fitted values for the second model.

How can we know that we have not omitted any important variables in our new model? Unfortunately, there is no good way of knowing that. In general, we should use judgment to decide whether anything else is needed or not. But given that we deal with time series, we can analyse residuals over time and see if there is any structure left (Figure 14.4):

```
plot(adamSeat02, 8, main="")
```

Plot in Figure 14.4 shows that the model has not captured seasonality correctly and that there is still some structure left in the residuals. In order to address this, we will add an ETS(A,N,A) element to the model, estimating ETSX instead of just regression:

```
adamSeat03 <- adam(Seatbelts, "ANA",
                   formula=drivers~PetrolPrice+kms+law)
```

We can then produce additional plots to do model diagnostics (Figure 14.5):

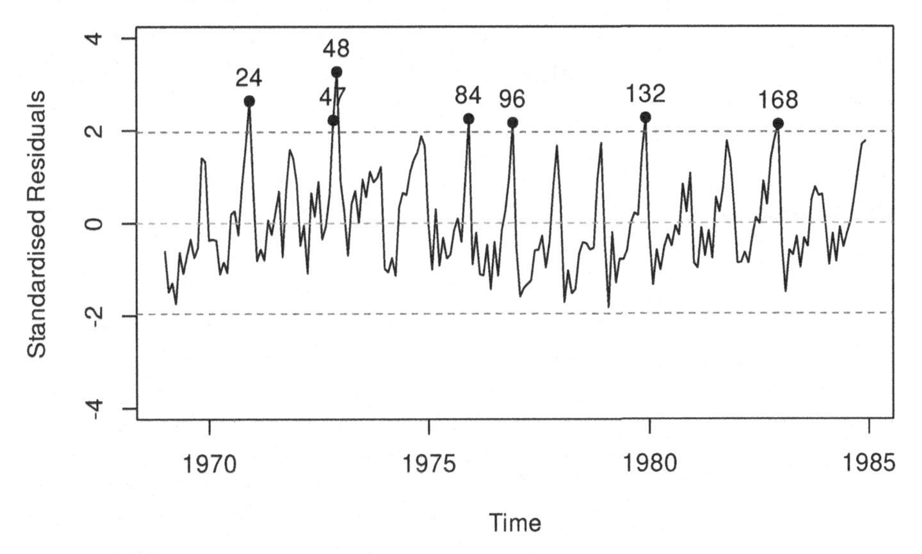

FIGURE 14.4 Standardised residuals vs time plot.

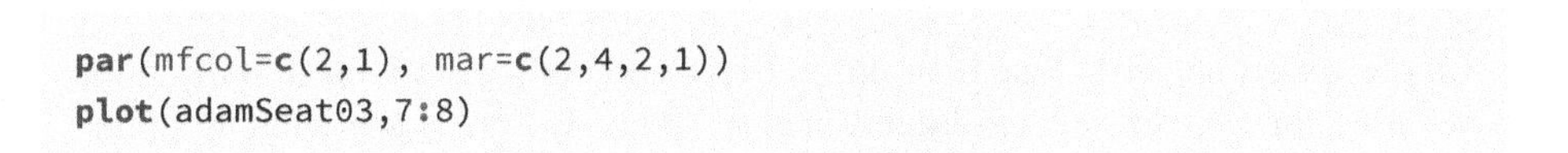

```
par(mfcol=c(2,1), mar=c(2,4,2,1))
plot(adamSeat03,7:8)
```

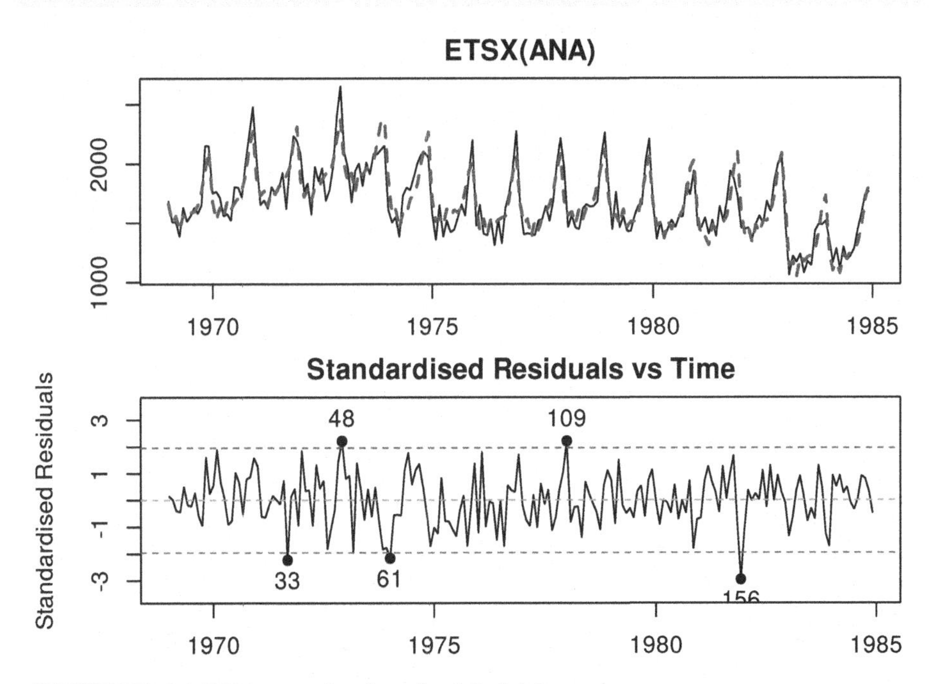

FIGURE 14.5 Diagnostic plots for Model 3.

In Figure 14.5, we do not see any apparent missing structure in the data and we do not have any additional variables to add. So, we can now move to the next step of diagnostics.

14.2 Model specification: Redundant variables

While there are some ways of testing for omitted variables, the redundant ones are sometimes even more challenging to diagnose. Yes, we could look at the significance of variables (Section 7.1 of Svetunkov, 2022) or compare models with and without some variables based on information criteria (Section 16.4 of Svetunkov, 2022): this will show which variables contribute towards the model fit, and which do not. However, even if our approaches say that a variable is not significant, this does not mean that it is not needed in the model. There can be many reasons why a test would fail to reject H_0: $a_j = 0$, and AIC would prefer a model without the variable under consideration. So, in the end, it comes to using judgment, trying to figure out whether a variable is needed in the model or not.

In the example with Seatbelt data, `DriversKilled` would be a redundant variable for the reasons explained in Section 14.1. Let us see what happens with the model if we include it:

```
adamSeat04 <- adam(Seatbelts, "NNN",
                   formula=drivers~PetrolPrice+kms+
                     law+DriversKilled)
par(mfcol=c(2,1), mar=c(4,4,2,1))
plot(adamSeat04,7:8)
```

The residuals from this model look adequate, and it is not apparent that there is an issue in the model. The summary of this model is:

```
summary(adamSeat04)
```

```
##
## Model estimated using alm() function: Regression
## Response variable: drivers
## Distribution used in the estimation: Normal
## Loss function type: likelihood; Loss function value: 1189.274
## Coefficients:
##                 Estimate Std. Error Lower 2.5% Upper 97.5%
## (Intercept)     905.6559   115.0935   678.6073   1132.6294 *
## PetrolPrice   -1603.7772   827.8145 -3236.8326     28.7384
```

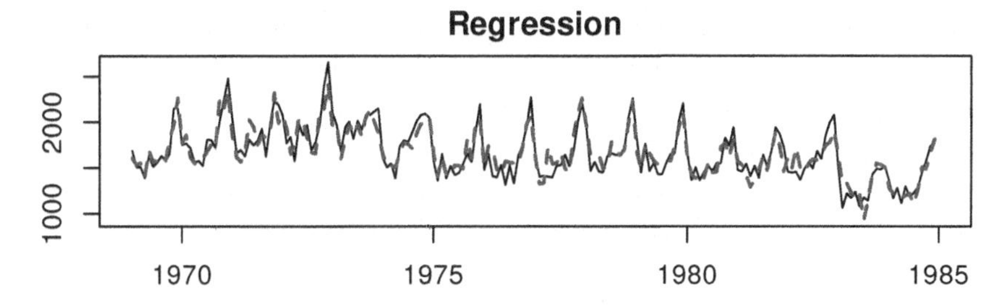

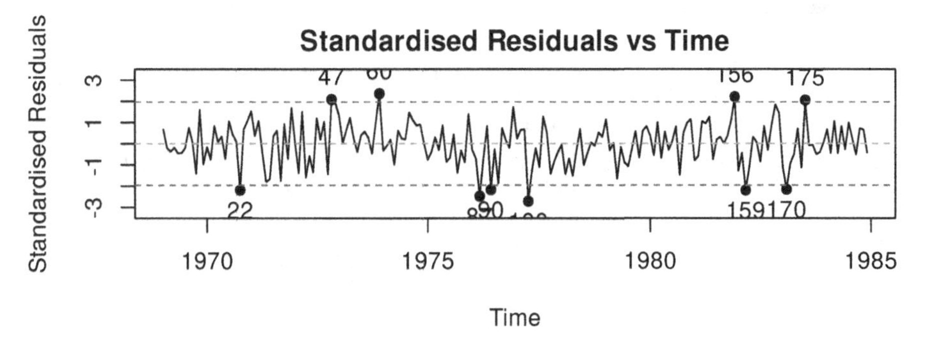

FIGURE 14.6 Diagnostic plots for Model 4.

```
## kms                  -0.0112     0.0035    -0.0182      -0.0043 *
## law                 -91.2672    31.9765  -154.3483     -28.2070 *
## DriversKilled         9.0423     0.3831     8.2866       9.7978 *
##
## Error standard deviation: 120.1081
## Sample size: 192
## Number of estimated parameters: 5
## Number of degrees of freedom: 187
## Information criteria:
##      AIC     AICc      BIC     BICc
## 2388.549 2388.871 2404.836 2405.684
```

The uncertainty around the parameter `DriversKilled` is narrow, showing that the variable positively impacts the `drivers`. If we used automated techniques for variables selection (based on AIC or statistical tests), we would conclude that the variable is important and is needed in the model. However, the issue here is not statistical but rather fundamental: we have included the variable that is a part of our response variable. It does not explain why drivers get injured and killed, it just reflects a part of the variable itself. So it approximates some proportion of the variance, which should have been explained by other variables (e.g. `PetrolPrice`), making them statistically not significant. So, based on the technical analysis, we would be inclined to keep the variable, but based on our understanding of the problem, we should not.

When it comes to the impact of this issue on forecasting, if the model contains redundant variables then it will overfit the data, which could lead to narrower prediction intervals and biased point forecasts. The parameters of such models are typically unbiased but inefficient (Section 6.3 of Svetunkov, 2022).

14.3 Model specification: Transformations

The question of appropriate transformations for variables in the model is challenging, because it is difficult to decide, what sort of transformation is needed, if needed at all. In many cases, this comes to selecting between an additive linear model and a multiplicative one. This implies that we compare the model:

$$y_t = a_0 + a_1 x_{1,t} + \dots + a_n x_{n,t} + \epsilon_t, \tag{14.1}$$

with

$$y_t = \exp\left(a_0 + a_1 x_{1,t} + \dots + a_n x_{n,t} + \epsilon_t\right). \tag{14.2}$$

(14.2) is equivalent to the so called "log-linear" model, but can also include logarithms of explanatory variables instead of the variables themselves to become a "log-log" model. Fundamentally, the transformations of variables should be done based on the understanding of the problem rather than on technicalities.

There are different ways of diagnosing the problem with wrong transformations. The first one is the actuals vs fitted plot (Figure 14.7):

```
plot(adamSeat03, which=1, main="")
```

The grey dashed line on the plot in Figure 14.7 corresponds to the situation when actuals and fitted coincide (100% fit). The red line on the plot is the LOWESS line (Cleveland, 1979), produced by the `LOWESS()` function in R, smoothing the scatterplot to reflect the potential tendencies in the data. This red line should coincide with the grey line in the ideal situation. In addition, the variability around the line should not change with the increase of fitted values. In our case, there is a slight U-shape in the red line and a slight rise in variability around the middle of the data. This could either be due to pure randomness and thus should be ignored or indicate a slight non-linearity in the data. After all, we have constructed a pure additive model on the data that exhibits seasonality with multiplicative characteristics, which becomes especially apparent at the end of the series, where the drop in level is accompanied by the decrease of the variability of the data (Figure 14.8):

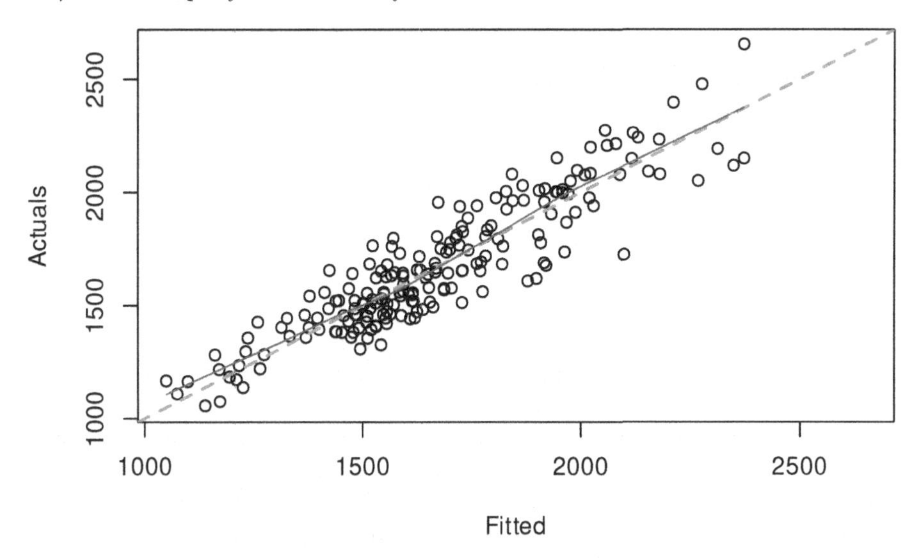

FIGURE 14.7 Actuals vs fitted for Model 3.

```
plot(adamSeat03, which=7, main="")
```

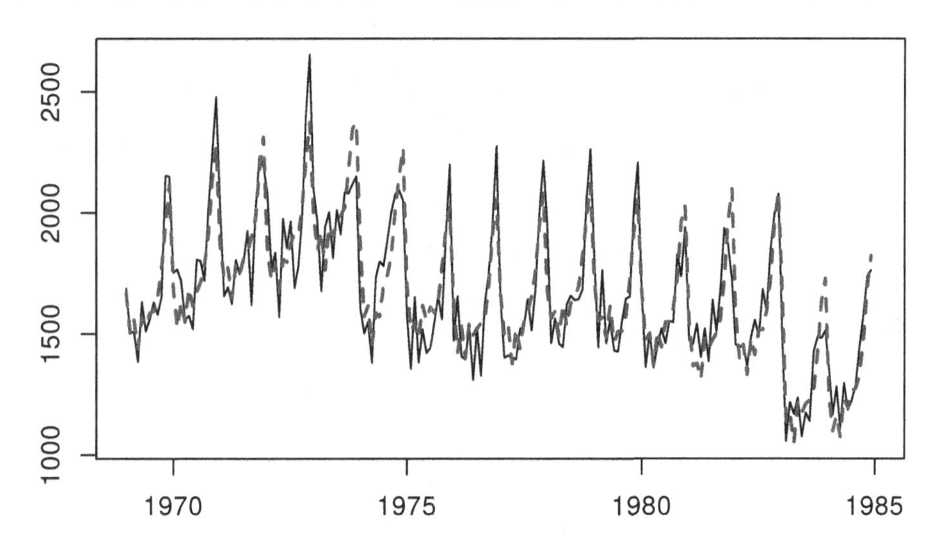

FIGURE 14.8 Actuals and fitted values for Model 3.

To diagnose this properly, we might use other instruments. One of these is the analysis of standardised residuals. The formula for the standardised residuals u_t will differ depending on the assumed distribution. For some of them it comes to the value inside the "exp" part of the Probability Density Function:

1. Normal, $\epsilon_t \sim \mathcal{N}(0, \sigma^2)$: $u_t = \frac{e_t - \bar{e}}{\hat{\sigma}}$;
2. Laplace, $\epsilon_t \sim \mathcal{L}aplace(0, s)$: $u_t = \frac{e_t - \bar{e}}{\hat{s}}$;
3. S, $\epsilon_t \sim \mathcal{S}(0, s)$: $u_t = \frac{e_t - \bar{e}}{\hat{s}^2}$;
4. Generalised Normal, $\epsilon_t \sim \mathcal{GN}(0, s, \beta)$: $u_t = \frac{e_t - \bar{e}}{\hat{s}^{\frac{1}{\beta}}}$;
5. Inverse Gaussian, $1 + \epsilon_t \sim \mathcal{IG}(1, \sigma^2)$: $u_t = \frac{1+e_t}{\bar{e}}$;
6. Gamma, $1 + \epsilon_t \sim -(\sigma^{-2}, \sigma^2)$: $u_t = \frac{1+e_t}{\bar{e}}$;
7. Log Normal, $1 + \epsilon_t \sim \log\mathcal{N}\left(-\frac{\sigma^2}{2}, \sigma^2\right)$: $u_t = \frac{e_t - \bar{e} + \frac{\hat{\sigma}^2}{2}}{\hat{\sigma}}$.

Here $\bar{e}$ is the mean of residuals, which is typically assumed to be zero, and u_t is the value of standardised residuals. Note that the scales in the formulae above should be calculated via the formula with the bias correction, i.e. with the division by degrees of freedom, not the number of observations, otherwise the bias of scale might impact the standardised residuals. Also, note that in the cases of the Inverse Gaussian, Gamma, and Log-Normal distributions and the additive error, the formulae for the standardised residuals will be the same as shown above.

The standardised residuals can then be plotted against something else to do diagnostics of the model. Here is an example of a plot of fitted vs standardised residuals in R (Figure 14.9):

```
plot(adamSeat03, which=2, main="")
```

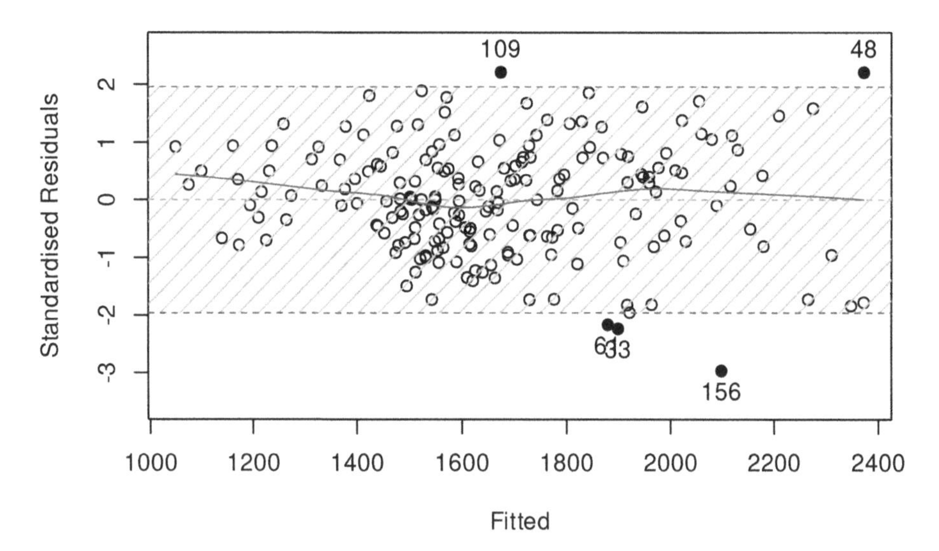

FIGURE 14.9 Standardised residuals vs fitted for a pure additive ETSX model.

Given that the scale of the original variable is now removed in the standardised residuals, it might be easier to spot the non-linearity. In our case, in Figure 14.9, it is still not apparent, but there is a slight curvature in the LOWESS line and a slight change in the variance: the variability in the beginning of the plot seems to be lower than the variability in the middle. Another plot that might be helpful (we have already used it before) is standardised residuals over time (Figure 14.10):

```
plot(adamSeat03, which=8, main="")
```

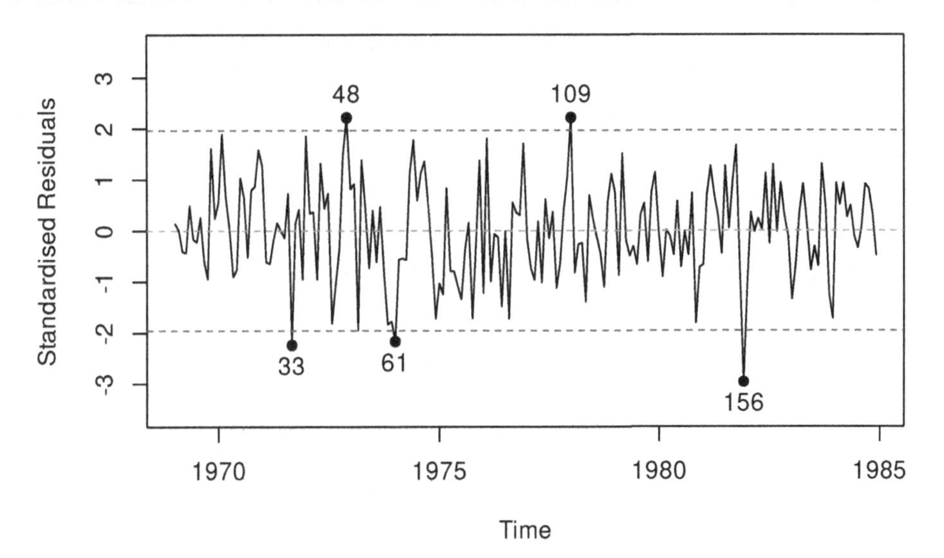

FIGURE 14.10 Standardised residuals vs time for a pure additive ETSX model.

The plot in Figure 14.10 does not show any apparent non-linearity in the residuals, so it is not clear whether any transformations are needed or not.

However, based on my judgment and understanding of the problem, I would expect the number of injuries and deaths to change proportionally to the change of the level of the data. If, after some external interventions, the overall level of injuries and deaths would decrease, then with a change of already existing variables in the model, we would expect a percentage decline, not a unit decline. This is why I will try a multiplicative model next (transforming explanatory variables as well):

```
adamSeat05 <- adam(Seatbelts, "MNM",
                   formula=drivers~log(PetrolPrice)+log(kms)+law)
plot(adamSeat05, 2, main="")
```

The plot in Figure 14.11 shows that the variability is now slightly more uniform

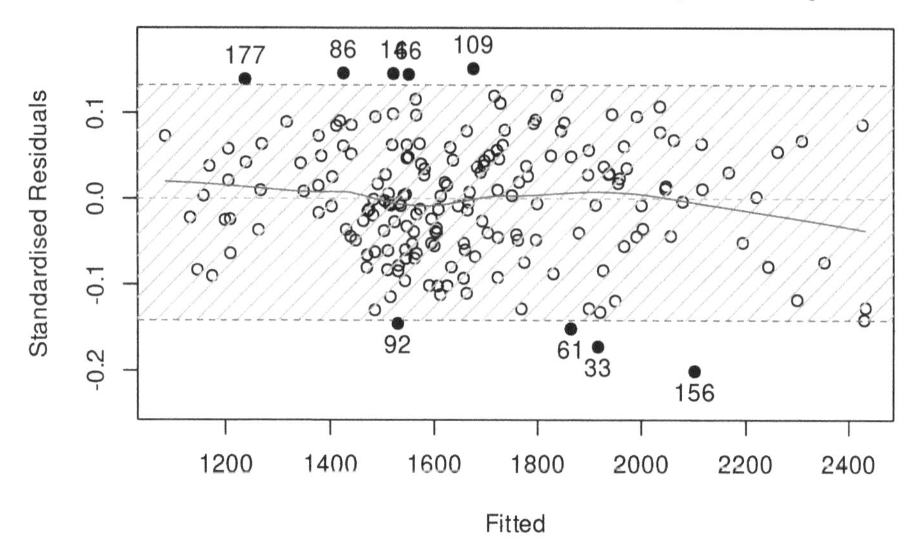

FIGURE 14.11 Standardised residuals vs fitted for a pure multiplicative ETSX model.

across all fitted values, but the difference between Figures 14.9 and 14.11 is not very prominent. One of the potential ways of deciding what to choose in this situation is to compare the models using information criteria:

```
setNames(c(AICc(adamSeat03), AICc(adamSeat05)),
         c("Additive model", "Multiplicative model"))
```

```
##       Additive model Multiplicative model
##             2424.123             2406.366
```

Based on this, we would be inclined to select the multiplicative model. My judgment in this specific case agrees with the information criterion.

We could also investigate the need for transformations of explanatory variables, but the interested reader is encouraged to do this analysis on their own.

Finally, the non-linear transformations are not limited with logarithm only. There are more of them, some of which are discussed in Chapter 14 of Svetunkov (2022).

14.4 Model specification: Outliers

In statistics, it is assumed that if the model is correctly specified then it does not have influential outliers. This means that there are no observations that cannot be explained by the model and the variability around it. If they happen then this might mean that:

1. We missed some important information (e.g. promotion) and did not include a respective variable in the model;
2. There was an error in recordings of the data, e.g. a value of 2000 was recorded as 200;
3. We did not miss anything predictable, we just face a distribution with fat tails (i.e. we assumed a wrong distribution).

In any of these cases, outliers might impact estimates of parameters of our model. With ETS, this might lead to higher than needed smoothing parameters, which in turn results in wider than needed prediction intervals and potentially biased forecasts. In the case of ARIMA, the mechanism is more complicated, still usually leading to widened intervals and biased forecasts. Finally, in regression, they might lead to biased estimates of parameters. So, it is important to identify outliers and deal with them.

14.4.1 Outliers detection

While it is possible to do preliminary analysis of the data and analyse distribution of the variable of interest, this does not allow identifying the outliers. This is because the simple univariate analysis (e.g. using boxplots) assumes that the variable has a fixed mean and a fixed distribution. This is violated by any model which has either a dynamic element or explanatory variables in it. So, the detection should be done based on residuals of an applied model.

One of the simplest ways of identifying outliers relies on distributional assumptions about the residuals and/or the response variable. For example, if we assume that our data follows the Normal distribution, we would expect 95% of observations to lie inside the bounds with approximately $\pm 1.96\sigma$ and 99.8% of them to lie inside the $\pm 3.09\sigma$. Sometimes these values are substituted by heuristics "values lying inside 2/3 sigmas", which is not precise and works only for Normal distribution. Still, based on this, we could flag the values outside these bounds and investigate them to see if any of them are indeed outliers.

Remark. If some observations lie outside these bounds, they are not necessarily outliers: building a 95% confidence interval always implies that approximately 5% of observations will lie outside the bounds.

Given that the ADAM supports different distributions, the heuristics mentioned above in general is inappropriate. We need to get proper quantiles for each of the assumed distributions. Luckily, this is not difficult because the quantile functions for all the distributions supported by ADAM either have analytical forms or can be obtained numerically.

Here is an example in R with the same multiplicative ETSX model and the standardised residuals vs fitted values with the 95% bounds (Figure 14.12):

```
plot(adamSeat05, which=2, level=0.95, main="")
```

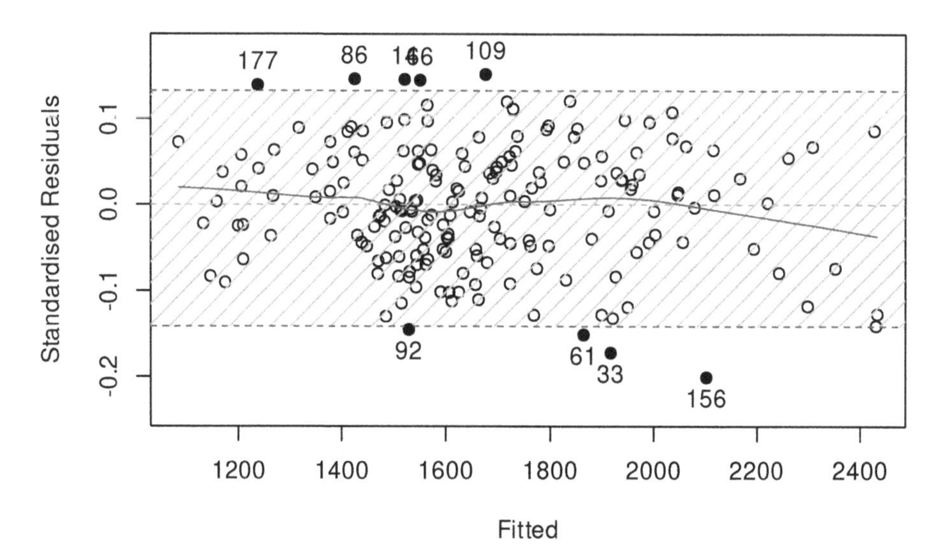

FIGURE 14.12 Standardised residuals vs fitted for the pure multiplicative ETSX model.

Remark. In the case of Inverse Gaussian, Gamma, and Log-Normal distributions, the function will produce the residuals in logarithms to make the plot more readable.

The plot in Figure 14.12 demonstrates that there are outliers, some of which are further away from the bounds. Although the amount of outliers is not large (it is roughly 5% of our sample), it would make sense to investigate why they happened.

Given that we deal with time series, plotting residuals vs time is also sometimes helpful (Figure 14.13):

```
plot(adamSeat05, which=8, main="")
```

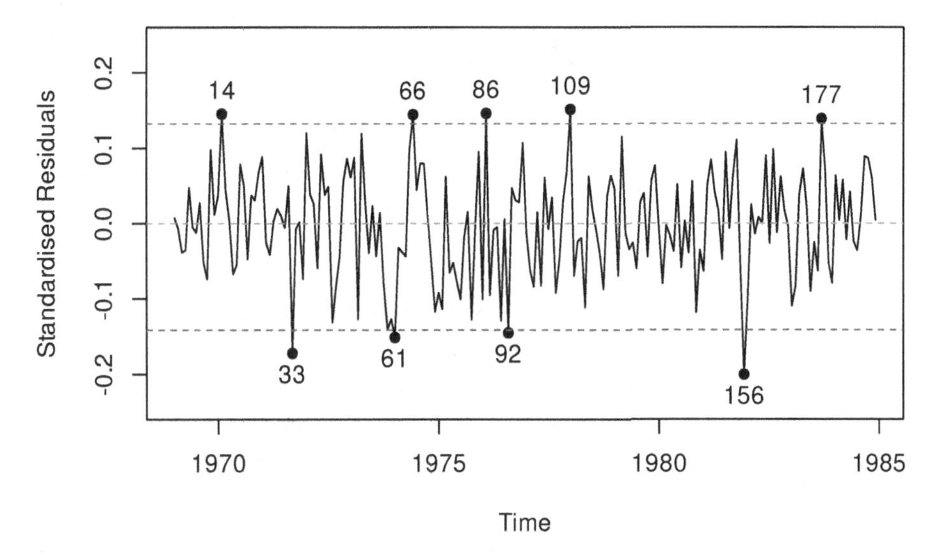

FIGURE 14.13 Standardised residuals vs time for the pure multiplicative ETSX model.

We see in Figure 14.13 that there is no specific pattern in the outliers, they happen randomly, so they appear not because of the omitted variables or wrong transformations. We have nine observations lying outside the bounds, which means that the 95% interval contains $\frac{192-9}{192} \times 100\% \approx 95.3\%$ of observations (192 is the sample size). This is close to the nominal value.

In some cases, the outliers might impact the scale of distribution and will lead to wrong standardised residuals, distorting the picture. This is where **studentised residuals** come into play. They are calculated similarly to the standardised ones, but the scale of distribution is recalculated for each observation by considering errors on all but the current observation. So, in a general case, this is an iterative procedure that involves looking through $t = \{1, \dots, T\}$ and that should, in theory, guarantee that the real outliers do not impact the scale of distribution. This procedure is simplified for the Normal distribution and has an analytical solution. We do not discuss it in the context of ADAM. Here is how they can be analysed in R:

```
par(mfcol=c(2,1), mar=c(4,4,2,0))
plot(adamSeat05, which=c(3,9))
```

In many cases (ours included), the standardised and studentised residuals will

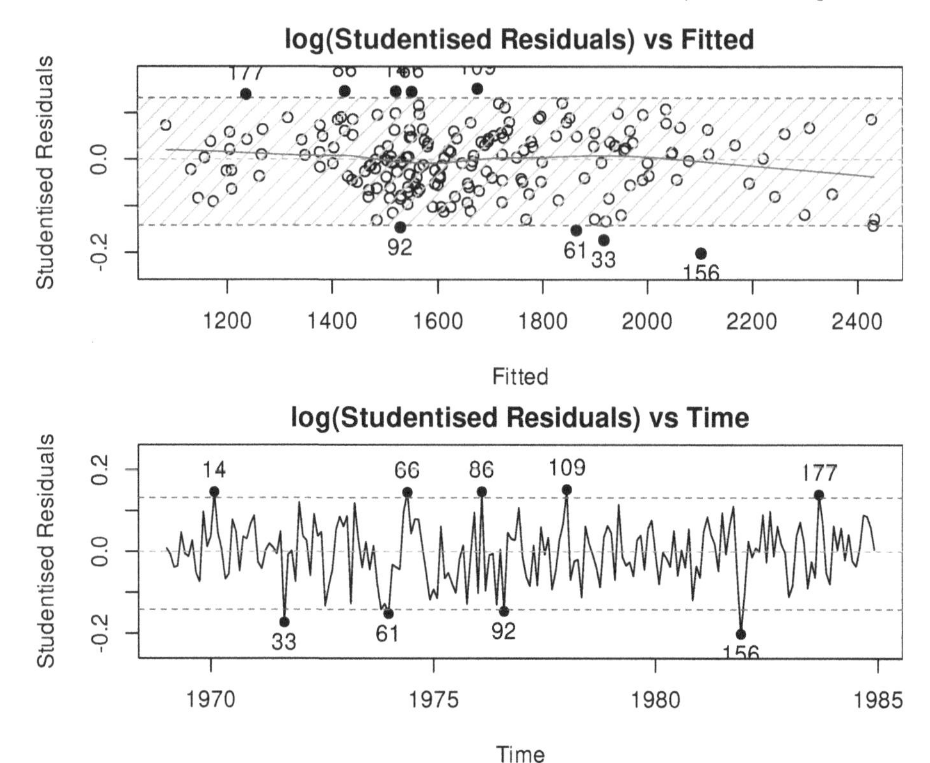

FIGURE 14.14 Studentised residuals analysis for the pure multiplicative ETSX model.

look very similar. But in some cases of extreme outliers, they might differ, and the latter might show outliers better than the former.

Given the situation with outliers in our case, we could investigate when they happen in the original data to understand better whether they need to be taken care of. But instead of manually recording which of the observations lie beyond the bounds, we can get their ids via the `outlierdummy()` method from the `greybox` package, which extracts either standardised or studentised residuals (depending on what we want) and flags those observations that lie outside the constructed interval, automatically creating dummy variables for these observations. Here is how it works:

```
adamSeat05Outliers <-
  outlierdummy(adamSeat05,
               level=0.95, type="rstandard")
```

The method returns several objects (see the documentation for details), including the ids of outliers:

```
adamSeat05Outliers$id
```

```
## [1]  14  33  61  66  86  92 109 156 177
```

These ids can be used to produce additional plots. For example:

```
# Plot actuals
plot(actuals(adamSeat05), main="",
    ylab="Drivers injured")
# Add fitted
lines(fitted(adamSeat05),col="grey")
# Add points for the outliers
points(time(Seatbelts)[adamSeat05Outliers$id],
       Seatbelts[adamSeat05Outliers$id,"drivers"],
       col="red", pch=16)
# Add the text with ids of outliers
text(time(Seatbelts)[adamSeat05Outliers$id],
     Seatbelts[adamSeat05Outliers$id,"drivers"],
     adamSeat05Outliers$id, col="red", pos=2)
```

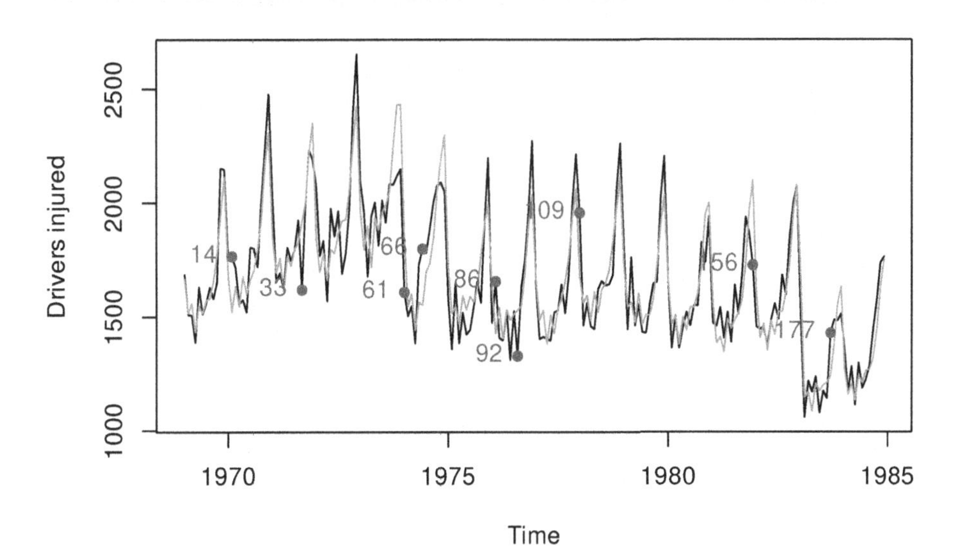

FIGURE 14.15 Actuals over time with points corresponding to outliers of the pure multiplicative ETSX model.

We cannot see any peculiarities in the appearance of outliers in Figure 14.15.

They seem to happen at random. There might be some external factors, leading to those unexpected events (for example, the number of injuries being much lower than expected on observation 156, in November 1981), but investigation of these events is outside of the scope of this demonstration.

Remark. As a side note, in R, there are several methods for extracting residuals:

- `resid()` or `residuals()` will extract either e_t or $1 + e_t$, depending on the error type in the model;
- `rstandard()` will extract the standardised residuals u_t;
- `rstudent()` will do the same for the studentised ones.

The `smooth` package also introduces the `rmultistep` function, which extracts multiple steps ahead in-sample forecast errors. We do not discuss this method here, but we will return to it in Subsection 14.7.3.

14.4.2 Dealing with outliers

While in the case of cross-sectional data some of the observations with outliers can be removed without damaging the model, in the case of time series, this is usually not possible to do. If we remove an observation from time series then we break its structure, and the applied model will not work as intended. So, we need to do something different to either interpolate the outliers or to tell the model that they happen.

Based on the output of the `outlierdummy()` method from the previous example, we can construct a model with explanatory variables to diminish the impact of outliers on the model:

```
# Add outliers to the data
SeatbeltsWithOutliers <-
    cbind(as.data.frame(Seatbelts[,-c(1,3,4,7)]),
          adamSeat05Outliers$outliers)
# Transform the drivers variable into time series object
SeatbeltsWithOutliers$drivers <- ts(SeatbeltsWithOutliers$drivers,
                                    start=start(Seatbelts),
                                    frequency=frequency(Seatbelts))
# Run the model with all the explanatory variables in the data
adamSeat06 <- adam(SeatbeltsWithOutliers, "MNM", lags=12,
                   formula=drivers~.)
```

In order to decide, whether the dummy variables help or not, we can use information criteria, comparing the two models:

```
setNames(c(AICc(adamSeat05), AICc(adamSeat06)),
         c("ETSX", "ETSXOutliers"))
```

```
##         ETSX ETSXOutliers
##     2406.366     2377.058
```

Comparing the two values above, we would conclude that adding dummies improves the model. However, this could be a mistake, given that we do not know the reasons behind most of them. In general, we should not include dummy variables for the outliers unless we know why they happened. If we do that, we might overfit the data. Still, if we have good reasons for this, we could add explanatory variables for outliers in the function to remove their impact on the response variable.

14.4.3 An automatic mechanism

An automated mechanism based on what we have done manually in the previous subsection, is implemented in the `adam()` function, which has the `outliers` parameter, defining what to do with them if there are any with the following three options:

1. `"ignore"` – do nothing;
2. `"use"` – create the model with explanatory variables including all of them, as shown in the previous subsection, and see if it is better than the model without the variables in terms of an information criterion;
3. `"select"` – create lags and leads of dummies from `outlierdummy()` and then select the dummies based on the explanatory variables selection mechanism (discussed in Section 15.3). Lags and leads are needed for cases when the effect of outliers is carried over to neighbouring observations.

Here is how this works in our case:

```
adamSeat08 <- adam(Seatbelts, "MNM", lags=12,
                   formula=drivers~PetrolPrice+kms+law,
                   outliers="select", level=0.95)
AICc(adamSeat08)
```

```
## [1] 2401.044
```

This model has selected some dummies for outliers. We can see them by looking at the coefficients of the model:

```
coef(adamSeat08)
```

```
##         alpha         gamma         level    seasonal_1
    seasonal_2
##  4.384319e-01  8.840989e-03  1.656792e+03  1.016679e+00  8.782549e
    -01
##    seasonal_3    seasonal_4    seasonal_5    seasonal_6
    seasonal_7
##  9.327608e-01  8.632158e-01  9.455688e-01  9.065027e-01  9.582421e
    -01
##    seasonal_8    seasonal_9   seasonal_10   seasonal_11
    outlier5
##  9.807905e-01  1.009272e+00  1.088397e+00  1.205244e+00 -1.457101e
    -02
##      outlier8      outlier2      outlier1      outlier7
    outlier9
##  1.534476e-02  1.304409e-02 -1.085061e-02 -1.270869e-02 -1.004567e
    -02
##      outlier4      outlier3      outlier6 outlier4Lead1
    outlier3Lead1
## -7.970148e-03  9.859827e-03  1.374919e-02 -3.385482e-03  8.286090e
    -03
## outlier5Lead1
##  7.793977e-03
```

Given that this is an automated approach, it is prone to potential mistakes. It needs to be treated with care as it might select unnecessary dummy variables and lead to overfitting. I would recommend exploring the outliers manually when possible and not relying too much on the automated procedures.

14.4.4 Final remarks

Koehler et al. (2012) explored the impact of outliers on ETS performance in terms of forecasting accuracy. They found that if outliers happen at the end of the time series, it is important to take them into account in a model. If they happen much earlier, their impact on the final forecast will be negligible. Unfortunately, the authors did not explore the impact of outliers on the prediction intervals. Based on my experience, I can tell that the outliers typically impact the width of the interval rather than the point forecasts. So, it is important to take care of them when they happen.

14.5 Residuals are i.i.d.: Autocorrelation

One of the typical characteristics of time series models is the dynamic relation between variables. Even if fundamentally, the sales of ice cream on Monday do not impact sales of the same ice cream on Tuesday, they might impact sales of a competing product on Tuesday, Wednesday, or next week. Missing this structure might lead to the autocorrelation of residuals, influencing the estimates of parameters and final forecasts. This can influence the prediction intervals (causing miscalibration) and in serious cases can lead to biased forecasts. Autocorrelations might also arise due to wrong transformations of variables, where the model would systematically underforecast the actuals, producing autocorrelated residuals. In this section, we will see one of the potential ways for the diagnostics of this issue and try to improve the model in a stepwise manner, adding different orders of the ARIMA model (Section 9).

As an example, we continue with the same seatbelts data, dropping the dynamic part to see what would happen in this case:

```
adamSeat09 <- adam(Seatbelts, "NNN", lags=12,
                   formula=drivers~log(PetrolPrice)+log(kms)+law)
AICc(adamSeat09)
```

```
## [1] 2651.191
```

There are different ways to diagnose the autocorrelation in this model. We start with a basic plot of residuals over time (Figure 14.16):

```
plot(adamSeat09, which=8, main="")
```

We see in Figure 14.16 that on one hand the residuals still contain seasonality and on the other, they do not look stationary. We could conduct ADF and/or KPSS tests to get a formal answer to the stationarity question:

```
adamSeat09 |> resid() |> tseries::kpss.test()
```

```
## Warning in tseries::kpss.test(resid(adamSeat09)): p-value greater
    than printed
## p-value

##
##  KPSS Test for Level Stationarity
##
```

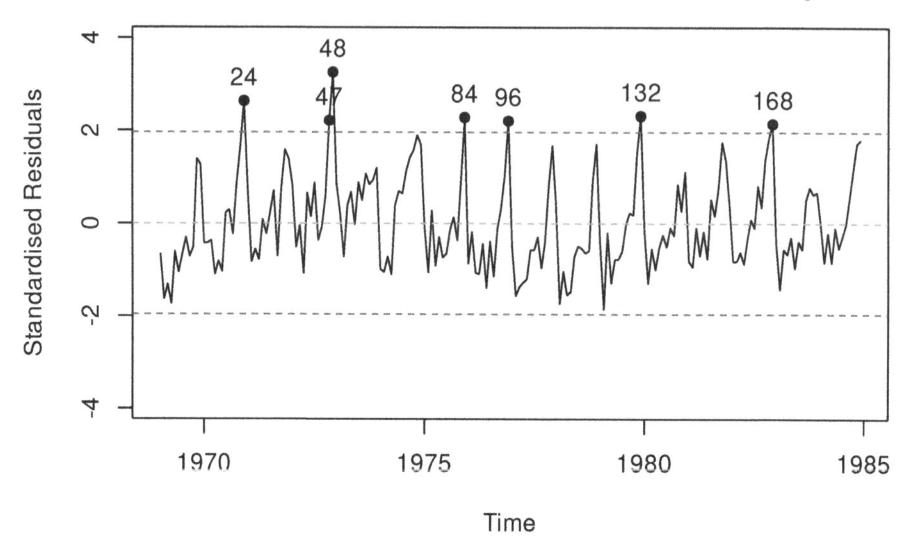

FIGURE 14.16 Standardised residuals vs time for Model 9.

```
## data:  resid(adamSeat09)
## KPSS Level = 0.12844, Truncation lag parameter = 4, p-value = 0.1
```

```
adamSeat09 |> resid() |> tseries::adf.test()
```

```
## Warning in tseries::adf.test(resid(adamSeat09)): p-value smaller
    than printed
## p-value
```

```
##
##  Augmented Dickey-Fuller Test
##
## data:  resid(adamSeat09)
## Dickey-Fuller = -6.8851, Lag order = 5, p-value = 0.01
## alternative hypothesis: stationary
```

The tests have opposite null hypotheses, and in our case, we would fail to reject H_0 on a 1% significance level for the KPSS test and reject H_0 on the same level for the ADF, which means that the residuals look stationary. The main problem in the residuals is the seasonality, which formally makes the residuals non-stationary (their mean changes from month to month), but which cannot be detected by these tests. Yes, there is a Canova-Hansen test (implemented in the `ch.test` function of the `uroot` package in R), which tests the seasonal unit root, but instead of trying it out and coming to conclusions, I will try the model in seasonal differences and see if it is better than the one without it:

```
# SARIMAX(0,0,0)(0,1,0)_12
adamSeat10 <- adam(Seatbelts,"NNN",lags=12,
                   formula=drivers~log(PetrolPrice)+log(kms)+law,
                   orders=list(i=c(0,1)))
AICc(adamSeat10)
```

```
## [1] 2547.274
```

Remark. While in general models in differences are not comparable with the models applied to the original data, `adam()` allows such comparison, because the ARIMA model implemented in it is initialised before the start of the sample and does not loose any observations.

This leads to an improvement in AICc in comparison with the previous model. The residuals of the model are now also better behaved (Figure 14.17):

```
plot(adamSeat10, which=8, main="")
```

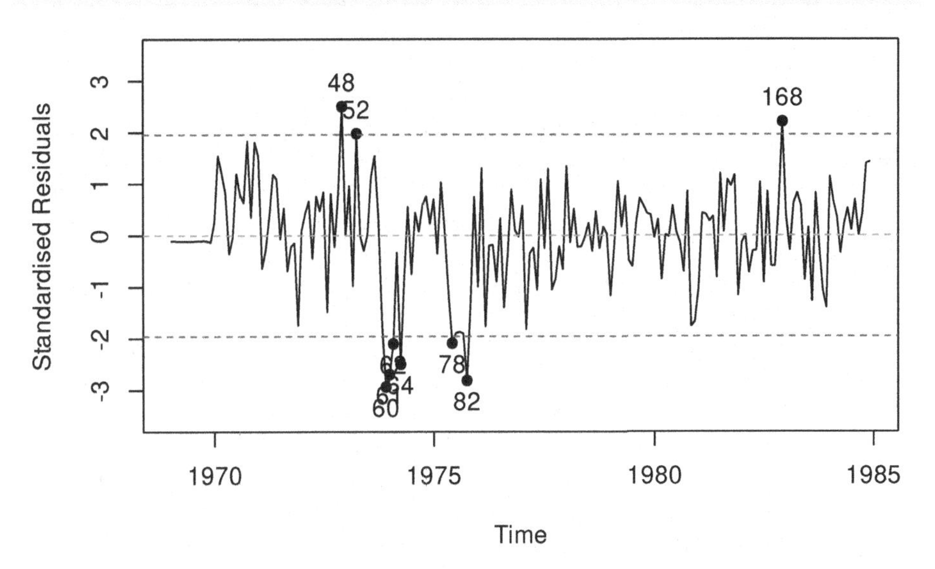

FIGURE 14.17 Standardised residuals vs time for Model 10.

In order to see whether there are any other dynamic elements left, we plot the ACF and PACF (discussed in Subsections 8.3.2 and 8.3.3) of residuals (Figure 14.18):

```r
par(mfcol=c(1,2), mar=c(4,4,1,1))
plot(adamSeat10, which=10:11, level=0.99, main="")
```

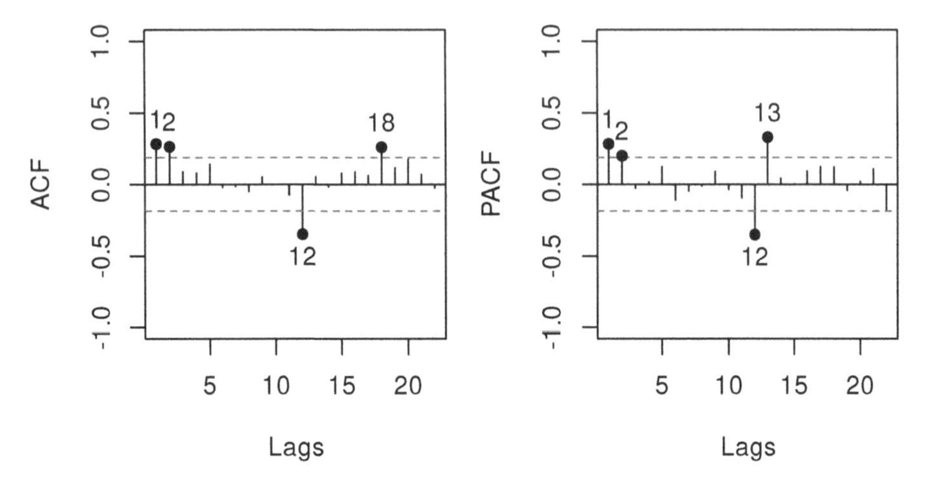

FIGURE 14.18 ACF and PACF of residuals of Model 10.

In the case of the `adam` objects, these plots, by default, will always have the range for the y-axis from -1 to 1 and will start from lag 1 on the x-axis. The red horizontal lines represent the "non-rejection" region. If a point lies inside the region, it is not statistically different from zero on the selected confidence `level` (the uncertainty around it is so high that it covers zero). The points with numbers are those that are statistically significantly different from zero. So, the ACF/PACF analysis might show the statistically significant lags on the selected level (the default one is 0.95, but I use `level=0.99` in this section). Given that this is a statistical instrument, we expect that approximately (1-level)% (e.g. 1%) of lags will lie outside these bounds just due to randomness, even if the null hypothesis is correct. So it is okay if we do not see all points lying inside them. However, we should not see any patterns there, and we might need to investigate the suspicious lags to improve the model (low orders of up to 3 – 5 and the seasonal lags if they appear). In our example in Figure 14.18, we see that there are spikes in lag 12 for both ACF and PACF, which means that we have missed the seasonal element in the data. There is also a suspicious lag 1 on PACF and lags 1 and 2 on ACF, which could potentially indicate that MA(1) and/or AR(1), AR(2) elements are needed in the model. While it is not clear what specifically is needed here, we can try out several models and see which one is better to determine the appropriate order of ARIMA. We should start with the seasonal part of the model, as it is easier to deal with in the first step.

```
# SARIMAX(0,0,0)(1,1,0)_12
adamSeat11 <- adam(Seatbelts,"NNN",lags=12,
                   formula=drivers~log(PetrolPrice)+log(kms)+law,
                   orders=list(ar=c(0,1),i=c(0,1)))
AICc(adamSeat11)
```

```
## [1] 2598.068
```

```
# SARIMAX(0,0,0)(0,1,1)_12
adamSeat12 <- adam(Seatbelts,"NNN",lags=12,
                   formula=drivers~log(PetrolPrice)+log(kms)+law,
                   orders=list(i=c(0,1),ma=c(0,1)))
AICc(adamSeat12)
```

```
## [1] 2474.985
```

```
# SARIMAX(0,0,0)(1,1,1)_12
adamSeat13 <- adam(Seatbelts,"NNN",lags=12,
                   formula=drivers~log(PetrolPrice)+log(kms)+law,
                   orders=list(ar=c(0,1),i=c(0,1),ma=c(0,1)))
AICc(adamSeat13)
```

```
## [1] 2506.93
```

Based on this analysis, we would be inclined to include the seasonal MA(1) only in the model. The next step in our iterative process – another ACF/PACF plot of the residuals to investigate whether there is anything else left:

In this case, there is a spike on PACF for lag 1 and a textbook exponential decrease in ACF starting from lag 1, which might mean that we need to include AR(1) component in the model:

```
# ARIMAX(1,0,0)(0,1,1)_12
adamSeat14 <- adam(Seatbelts,"NNN",lags=12,
                   formula=drivers~log(PetrolPrice)+log(kms)+law,
                   orders=list(ar=c(1,0),i=c(0,1),ma=c(0,1)))
AICc(adamSeat14)
```

```
## [1] 2463.552
```

Choosing between the new model and the old one, we should give preference to the former, which has a lower AICc than the previous one. So, adding AR(1) to the model leads to further improvement in AICc. Using this iterative procedure,

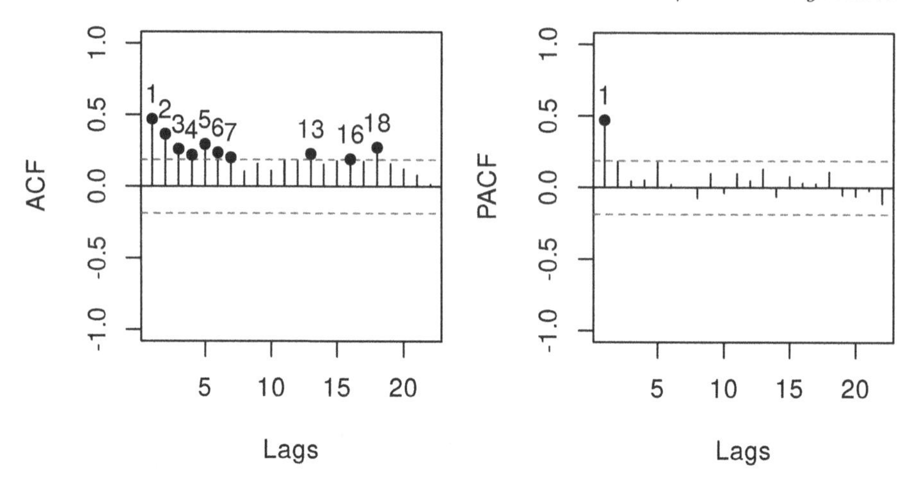

FIGURE 14.19 ACF and PACF of residuals of the model 12.

we could continue our investigation to find the most suitable SARIMAX model for the data. We stop our discussion here, because this example should suffice in providing a general idea of how the diagnostics and the fix of autocorrelation issue can be done in ADAM. As an alternative, we could simplify the process and use the automated ARIMA selection algorithm (see discussion in Section 15.2). This is built on the principles discussed above (testing sequentially the potential orders based on ACF/PACF and AIC):

```
adamSeat15 <-
  adam(Seatbelts,"NNN",lags=12,
       formula=drivers~log(PetrolPrice)+log(kms)+law,
       orders=list(ar=c(3,2),i=c(2,1),ma=c(3,2),select=TRUE))
adamSeat15$model
```

```
## [1] "SARIMAX(0,1,1)[1](0,1,1)[12]"
```

```
AICc(adamSeat15)
```

```
## [1] 2426.486
```

This newly constructed model is SARIMAX$(0,1,1)(0,1,1)_{12}$, which differs from the model that we achieved manually. Its residuals are better behaved than the ones of Model 5 (although we might need to analyse the residuals for the potential outliers, as discussed in Subsection 14.4):

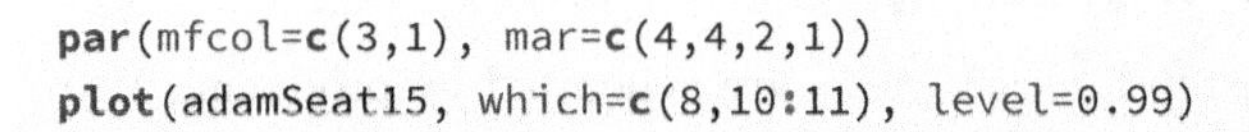

```
par(mfcol=c(3,1), mar=c(4,4,2,1))
plot(adamSeat15, which=c(8,10:11), level=0.99)
```

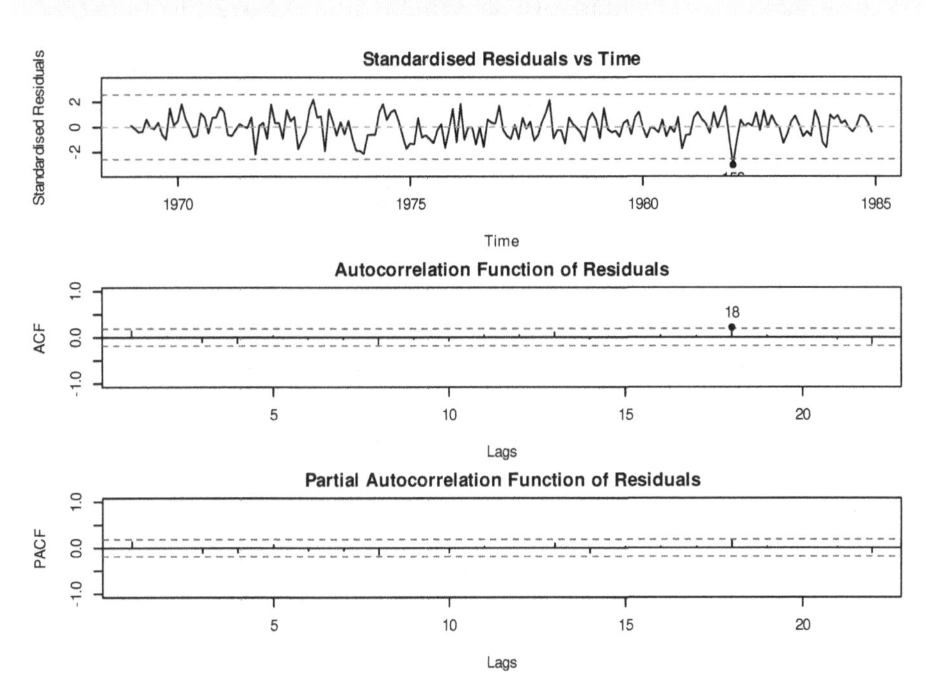

FIGURE 14.20 Residuals of Model 16.

As a final word for this section, we have focused our discussion on the visual analysis of time series, ignoring the statistical tests (we only used ADF and KPSS). Yes, there is the Durbin-Watson (Durbin & Watson, 1950) test for AR(1) in residuals, and yes, there are Ljung-Box (Ljung & Box, 1978), Box-Pierce (Box & Pierce, 1970) and Breusch-Godfrey (Breusch, 1978; Godfrey, 1978) tests for multiple AR elements. But visual inspection of time series is not less powerful than hypothesis testing. It makes you think and analyse the model and its assumptions, while the tests are the lazy way out that might lead to wrong conclusions because they have the standard limitations of any hypotheses tests (as discussed in Section 7.1 of Svetunkov, 2022). After all, if you fail to reject H_0, it does not mean that the effect does not exist. Having said that, the statistical tests become extremely useful when you need to process many time series simultaneously and cannot inspect them manually. So, if you are in that situation, I would recommend reading more about them (even Wikipedia has good articles about them).

14.5.1 Correlations of multistep forecast errors

As a reminder, the multistep forecast errors are generated by producing 1 to h steps ahead forecasts from each observation in-sample starting from $t = 1$ and ending with $t = T - h$ and subtracting it from the respective actual values. We have already discussed them and loss functions based on them in Section 11.3. They can be a useful diagnostic instrument, but they have one important property, which is sometimes ignored by researchers. In contrast with the residual e_t, which is expected not to be autocorrelated, the forecast errors e_{t+i} and e_{t+j} for $i \neq j$ should always be correlated if the persistence matrix of the model contains non-zero values. This is discussed by Svetunkov, Kourentzes, and Killick (2023) who showed that the only case when the forecast errors are uncorrelated is when the model is deterministic and does not have autocorrelated residuals. The correlation between the forecast errors will be stronger the closer i and j are to each other (e.g. it will be stronger between e_{t+3} and e_{t+4} than between e_{t+3} and e_{t+6}).

In R, the multistep forecast errors can be extracted via `rmultistep()` function from the `smooth` package. For demonstration purposes we use model 5:

```
# Extract multistep errors
adamSeat05ResidMulti <- rmultistep(adamSeat05, h=10)
# Give adequate names to the columns
colnames(adamSeat05ResidMulti) <- c(1:10)
# Calculate correlation matrix for forecast errors
cor(adamSeat05ResidMulti) |>
    round(3)
```

```
##        1     2     3     4     5     6     7     8     9    10
## 1  1.000 0.307 0.275 0.163 0.096 0.206 0.138 0.115 0.006 0.093
## 2  0.307 1.000 0.358 0.314 0.196 0.150 0.247 0.181 0.142 0.052
## 3  0.275 0.358 1.000 0.387 0.336 0.238 0.182 0.272 0.188 0.169
## 4  0.163 0.314 0.387 1.000 0.392 0.356 0.250 0.195 0.268 0.198
## 5  0.096 0.196 0.336 0.392 1.000 0.404 0.359 0.249 0.176 0.259
## 6  0.206 0.150 0.238 0.356 0.404 1.000 0.421 0.367 0.242 0.188
## 7  0.138 0.247 0.182 0.250 0.359 0.421 1.000 0.427 0.357 0.243
## 8  0.115 0.181 0.272 0.195 0.249 0.367 0.427 1.000 0.425 0.362
## 9  0.006 0.142 0.188 0.268 0.176 0.242 0.357 0.425 1.000 0.419
## 10 0.093 0.052 0.169 0.198 0.259 0.188 0.243 0.362 0.419 1.000
```

As we can see from the output above, the correlations between the neighbouring forecast errors is higher than between the ones with larger distance. We cannot use this property for diagnostics purposes, it is just one of the features of dynamic models that we should keep in mind.

14.6 Residuals are i.i.d.: Heteroscedasticity

Another important assumption for conventional models is that the residuals are homoscedastic, meaning that their variance stays the same (no matter what). If it does not then the parameters of the model might be inefficient (thus change substantially with the change of the sample size) and prediction intervals from the model might be miscalibrated (either narrower or wider than needed, depending on the circumstances). This section will show how the issue can be diagnosed and resolved.

14.6.1 Detecting heteroscedasticity

Building upon our previous example, we will use the ETSX(A,N,A) model, which has some issues, as we remember from Section 14.3. One of those is the wrong type of model – additive instead of multiplicative. This is also related to the variance of residuals, because the multiplicative error model takes care of one of the types of heteroscedasticity. To see if the residuals of the ETSX(A,N,A) model are homoscedastic, we can plot them against the fitted values (Figure 14.21):

```
par(mfcol=c(2,1), mar=c(4,4,2,1))
plot(adamSeat03, which=4:5)
```

The two plots in Figure 14.21 allow detecting a specific type of heteroscedasticity when the residuals' variability changes with the increase of fitted values. The plot of absolute residuals vs fitted is more appropriate for models, where the scale parameter is calculated based on absolute values of residuals (e.g. the model with Laplace distribution) and relates to MAE (Subsection 11.2.1), while the squared residuals vs fitted shows whether the variance of residuals is stable or not (thus making it more suitable for models with Normal and related distributions). However, the squared residuals plot might be challenging to read due to outliers, so the first one might help detect the heteroscedasticity even when the scale is supposed to rely on squared errors. What we want to see on these plots is for all the points to lie in the same corridor for lower and for the higher fitted values and for the red LOWESS line to be parallel to the x-axis. In our case, there is a slight increase in the line. Furthermore, the variability of residuals around 1000 is lower than the one around 2000, indicating that we have heteroscedasticity in residuals. In our case, this is caused by the wrong transformations in the model (see Section 14.3), so to fix the issue, we should switch to a multiplicative model. In fact, switching to a multiplicative model (aka model in logarithms) fixes the heteroscedasticity issue in many cases in practice.

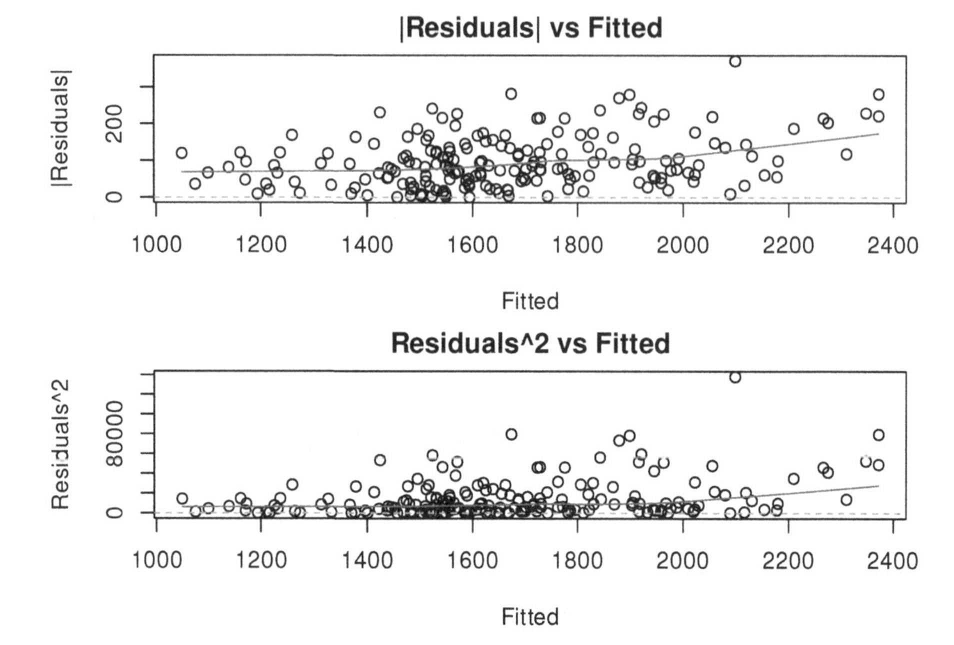

FIGURE 14.21 Absolute and squared residuals vs fitted of Model 3.

Another diagnostics tool that might become useful in some situations is the plot of absolute and squared *standardised residuals* versus fitted values. They have a similar idea to the previous plots, but they might change slightly because of the standardisation (mean is equal to 0 and scale is equal to 1). These plots become especially useful if the changing variance is modelled explicitly (e.g. via a regression model or a GARCH-type of model, I will discuss this in Chapter 17):

```
par(mfcol=c(2,1), mar=c(4,4,2,0))
plot(adamSeat03, which=13:14)
```

In our case, the plots in Figure 14.22 do not give an additional message. We already know that there is a slight heteroscedasticity and that we need to transform the response variable.

If we suspect that there are some specific variables that might cause heteroscedasticity, we can plot absolute or squared residuals vs these variables to see if they indeed cause it. For example, here is how we can produce a basic plot of absolute residuals vs all explanatory variables included in the model:

```
cbind(as.data.frame(abs(resid(adamSeat03))),
```

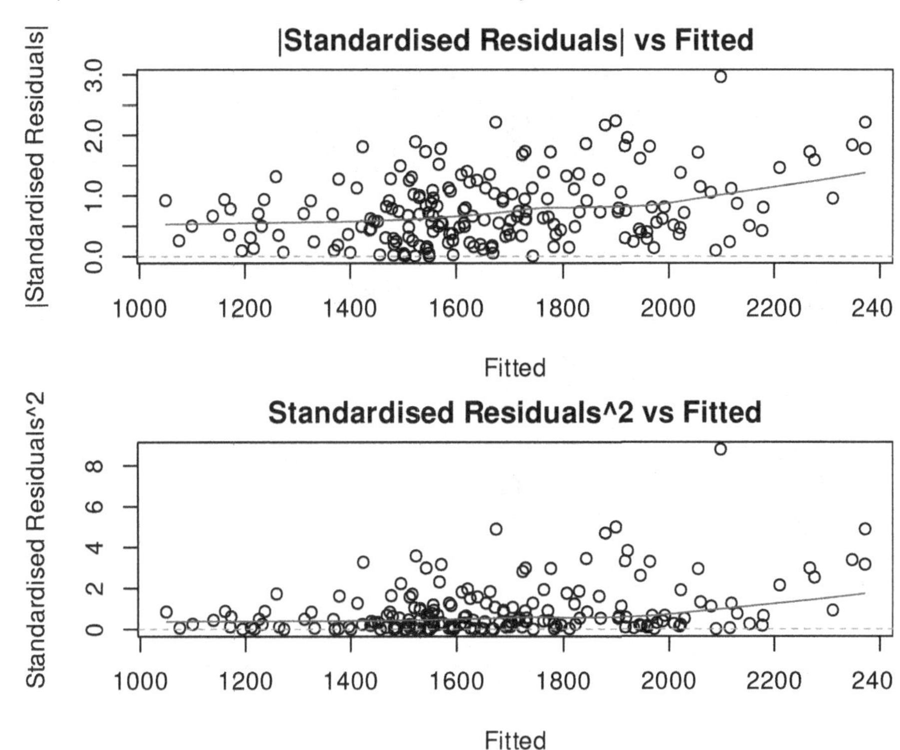

FIGURE 14.22 Absolute and squared standardised residuals vs fitted of Model 3.

```
adamSeat03$data[,all.vars(formula(adamSeat03))[-1]]) |>
  spread(LOWESS=TRUE)
```

The plot in Figure 14.23 can be read similarly to the plots discussed above: if we notice a change in variability of residuals or a change (increase or decrease) in the LOWESS lines with the change of a variable, then this might indicate that the respective variable causes heteroscedasticity. In our example, it looks like the variable `law` causes the most significant issue – all the other variables do not cause as substantial a change in the variance as this one. If we want to fix this specific issue then we might need to consider a scale model, modelling the change of scale based on variables like `law` directly (this will be discussed in Chapter 17).

We already know that we need to use a multiplicative model instead of the additive one in our example, so we will see how the residuals look for the correctly specified model in Figure 14.24.

The plots in Figure 14.24 do not demonstrate any substantial issues: the

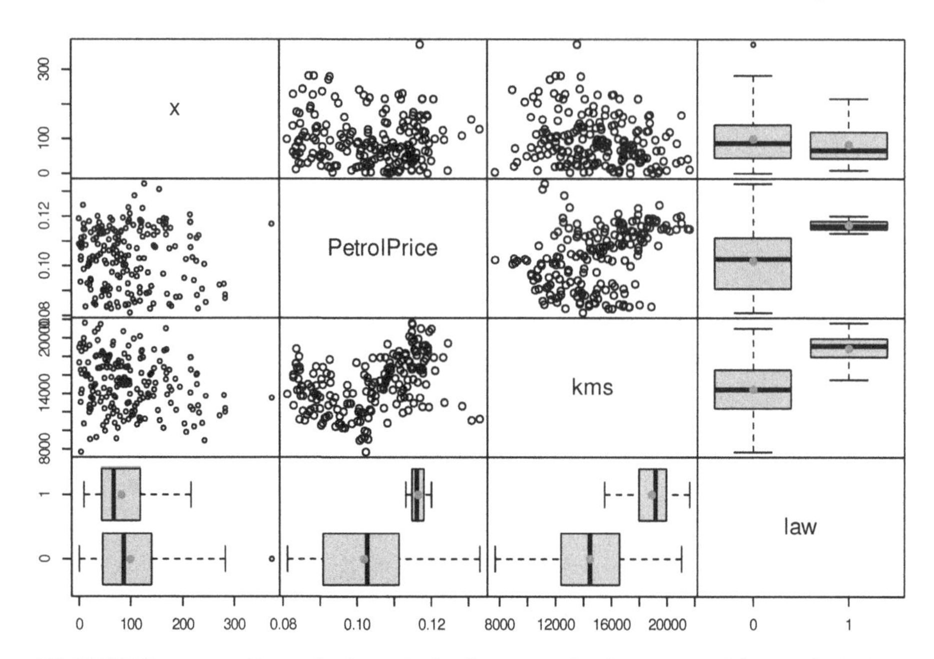

FIGURE 14.23 Spread plot of absolute residuals vs variables included in Model 3.

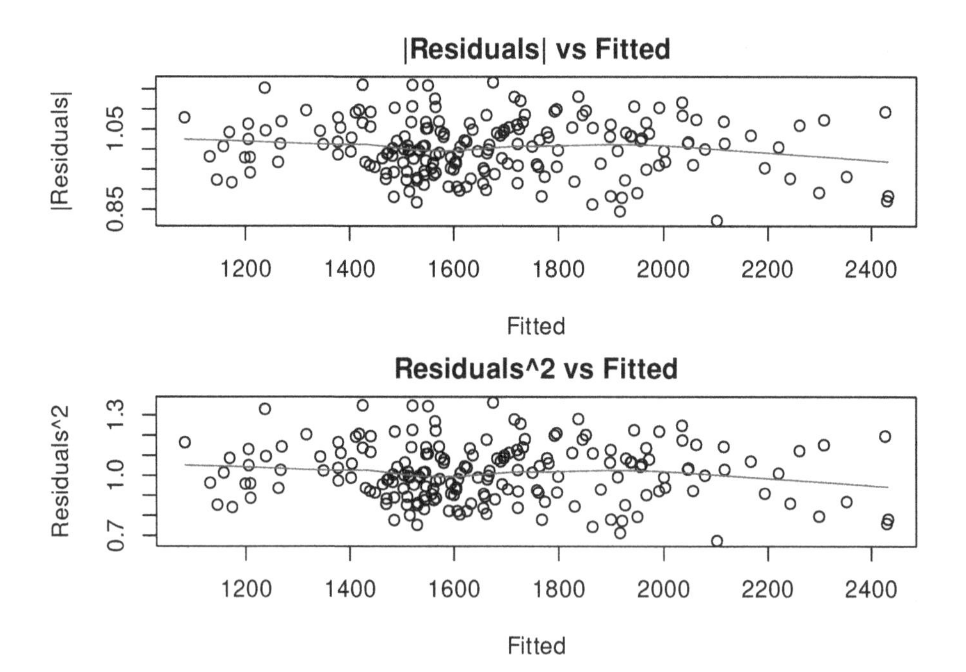

FIGURE 14.24 Absolute and Squared residuals vs Fitted of Model 5.

residuals look homoscedastic, and given the scale of residuals, the change of LOWESS line does not reflect significant changes in the scale of the residuals. An additional plot of absolute residuals vs explanatory variables does not show any severe issues either (Figure 14.25). So, we can conclude that the multiplicative model resolves the issue with heteroscedasticity.

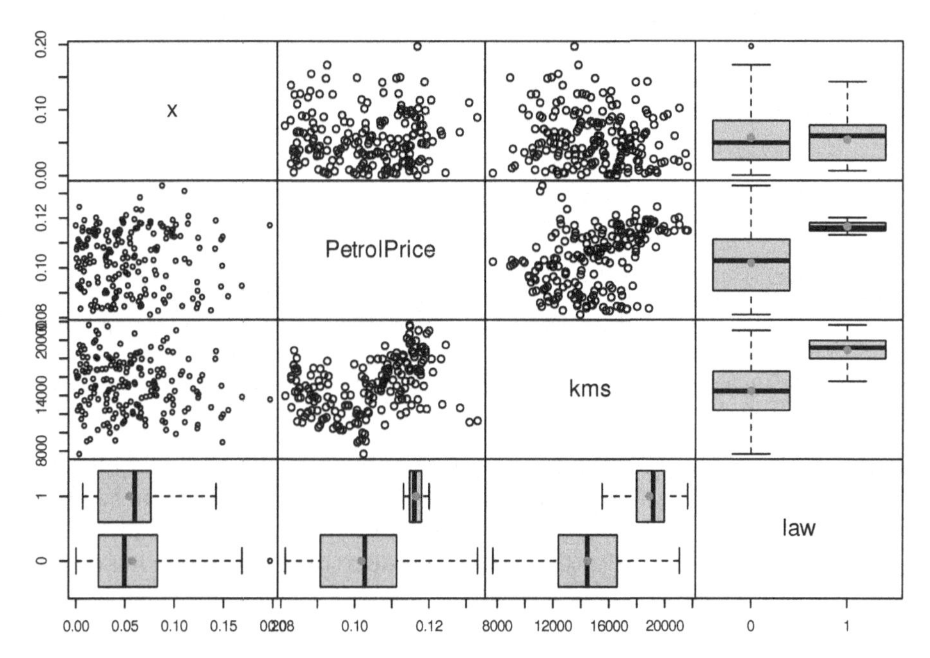

FIGURE 14.25 Spread plot of absolute residuals vs variables included in the Model 5.

Concluding this section, we have focused the analysis on the visual diagnostics. But there are formal statistical tests for heteroscedasticity, such as White (White, 1980), Breusch-Pagan (Breusch & Pagan, 1979), and Bartlett's (Bartlett, 1937) tests and others. They all test different types of heteroscedasticity and can be used when the visual diagnostics are not possible. We do not discuss them here for a reason outlined in Section 14.5.

14.7 Residuals are i.i.d.: Zero expectation

14.7.1 Unconditional expectation of residuals

This assumption only applies for the additive error models (Section 5.1). In the case of the multiplicative error models, it is changed to "expectation of the error term is equal to one" (Section 6.5). It does not make sense to check this assumption unconditionally because it does not mean anything in-sample: it will hold automatically for the residuals of a model in the case of OLS estimation. The observed mean of the residuals might not be equal to zero in other cases, but this does not give any helpful information. In fact, when we work with dynamic models (ETS, ARIMA), the in-sample residuals being equal to zero might imply for some of them that the final values of components are identical to the initial ones. For example, in the case of ETS(A,N,N) (from Section 4.3), we can use the transition equation from (4.3) to express the final value of level via the previous values up until $t = 0$:

$$\begin{aligned} \hat{l}_t = & \hat{l}_{t-1} + \hat{\alpha} e_t = \hat{l}_{t-2} + \hat{\alpha} e_{t-1} + \hat{\alpha} e_t = \\ & \hat{l}_0 + \hat{\alpha} \sum_{j=1}^{t} e_{t-j}. \end{aligned} \tag{14.3}$$

If the mean of the residuals in-sample is indeed equal to zero, then the equation (14.3) reduces to $\hat{l}_t = \hat{l}_0$. So, this assumption does not make sense in-sample and cannot be properly checked, meaning that it is all about the true model and the asymptotic behaviour rather than the model applied to data.

On the other hand, if for some reason the mean of residuals is not equal to zero in the population, then the model will change. For example, if we have an ETS(A,N,N) model with the non-zero mean of residuals μ_ϵ, then the residuals can be represented in the form $\epsilon_t = \mu_\epsilon + \xi_t$, where $\mathrm{E}(\xi_t) = 0$ which leads to a different model than ETS(A,N,N):

$$\begin{aligned} & y_t = l_{t-1} + \mu_\epsilon + \xi_t \\ & l_t = l_{t-1} + \alpha \mu_\epsilon + \alpha \xi_t \end{aligned}. \tag{14.4}$$

If we apply an ETS(A,N,N) model to the data instead of (14.4), we will omit an important element and thus the estimated smoothing parameter will be higher than needed. The same logic applies to the multiplicative error models: the mean of residuals $1 + \epsilon_t$ should be equal to one for them, otherwise the model would change.

This phenomenon arises in dynamic models because of the "pull-to-centre" effect, where due to the presence of residuals in the transition equations, the model updates the states so that they become closer to the conditional mean of data.

Summarising this subsection, the expectation of residuals of ADAM should be equal to zero asymptotically, but it cannot be tested in-sample.

14.7.2 Conditional expectation of residuals

The more valuable part of this assumption that can be checked is whether the expectation of the residuals *conditional on some variables* (or time) equals to zero (or one in the case of a multiplicative error model). In a way, this comes to ensuring that there are no patterns in the residuals and thus there are no parts of data where residuals have systematically non-zero expectation.

There are different ways to diagnose this. First, we could use the already discussed plot of standardised (or studentised) residuals vs fitted values from Section 14.3. If the LOWESS line differs substantially from zero, we might suspect that the conditional expectation of residuals is not zero. Second, we could use the plot of residuals over time, which we have already discussed in Section 14.5. The logic here is similar to the previous one with the main difference being in spotting patterns, implying non-zero residuals over time. Furthermore, we can also plot residuals vs some of the variables to see if they cause the change in mean. But in a way, all of these methods might also mean that the residuals are autocorrelated and/or some transformations of variables are needed.

Related to the conditional expectation of the residuals, is the effect called "endogeneity" (discussed briefly in Section 15.3 of Svetunkov, 2022). According to the econometrics literature (see for example, Hanck, Arnold, Gerber, & Schmelzer, 2020), it implies that the residuals are correlated with some variables. This becomes equivalent to the situation when the expectation of residuals changes with the change of a variable. The most prominent cause of this effect is the omission of important variables (discussed in Section 14.1), which can sometimes be diagnosed by looking at correlations between the residuals and the omitted variables if the latter are available. While econometricians propose using other estimation methods (such as Instrumental Variables) to diminish the effect of endogeneity on regression models, the forecasters cannot do that because we need to fix the problem to get more reasonable forecasts rather than better estimates of parameters. Unfortunately, there is no universal recipe for the solution to this problem, but in some cases transforming variables, adding the omitted ones, or substituting them by proxies (variables that act similarly to the omitted ones) might resolve the issue to some extent.

14.7.3 Multistep forecast errors have zero mean

This follows from the previous assumption if the model is correctly specified and its residuals are i.i.d (also discussed in Subsection 14.5.1). In that situation, we would expect the multiple steps ahead forecast errors to have zero mean. In practice, this might be violated if some structural changes or level shifts are not taken into account by the model. The only thing to note is that the multistep forecast errors imply defining the forecast horizon h. This should typically come from the task itself and the decisions made.

Practically speaking, the diagnostics of this assumption can be done using the `rmultistep()` method for `adam()`. This method would apply the estimated model and produce multiple steps ahead forecasts from each in-sample observation to the horizon h, stacking the forecast errors in rows. Whether we use an additive or multiplicative error model, the method will produce the residual e_t.

Here is an example of the code for the extraction and plotting of multistep forecast errors for the multiplicative model 5 from the previous sections:

```
# Extract multistep errors
adamSeat05ResidMulti <- rmultistep(adamSeat05, h=12)
# Give adequate names to the columns
colnames(adamSeat05ResidMulti) <- c(1:12)
# Produce boxplots
boxplot(adamSeat05ResidMulti, xlab="horizon")
# Add the zero line
abline(h=0, col="red")
# Add mean values
apply(adamSeat05ResidMulti,2,mean) |>
    points(col="red", pch=16)
```

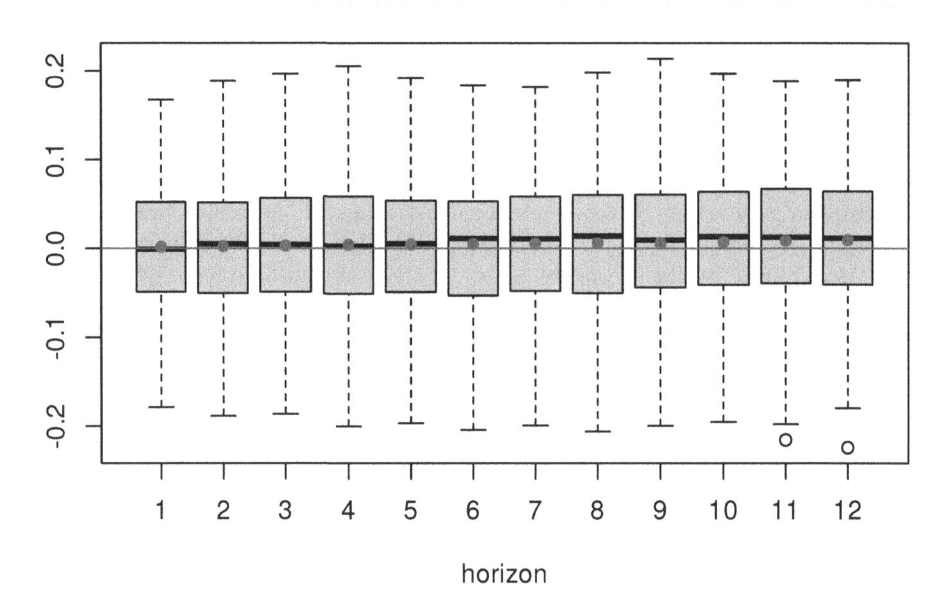

FIGURE 14.26 Boxplot of multistep forecast errors extracted from Model 5.

As the plot in Figure 14.26 demonstrates, the mean of the residuals does not increase substantially with the increase of the forecast horizon, which indicates that the model has captured the main structure in the data correctly.

14.8 Residuals are i.i.d.: Distributional assumptions

Finally, we come to the distributional assumptions of ADAM. If we use a wrong distribution then we might get incorrect estimates of parameters and we might end up with miscallibrated prediction intervals (i.e. quantiles from the model will differ from the theoretical ones substantially).

As discussed earlier (for example, in Section 11.1), the ADAM framework supports several distributions. The specific parts of assumptions will change depending on the type of error term in the model. Given that, it is relatively straightforward to see if the residuals of the model follow the assumed distribution or not, and there exist several tools for that.

The simplest one is called a Quantile-Quantile (QQ) plot. It produces a figure with theoretical vs actual quantiles and shows whether they are close to each other or not. Here is, for example, how the QQ plot will look for one of the previous models, assuming Normal distribution (Figure 14.27):

```
plot(adamSeat03, which=6)
```

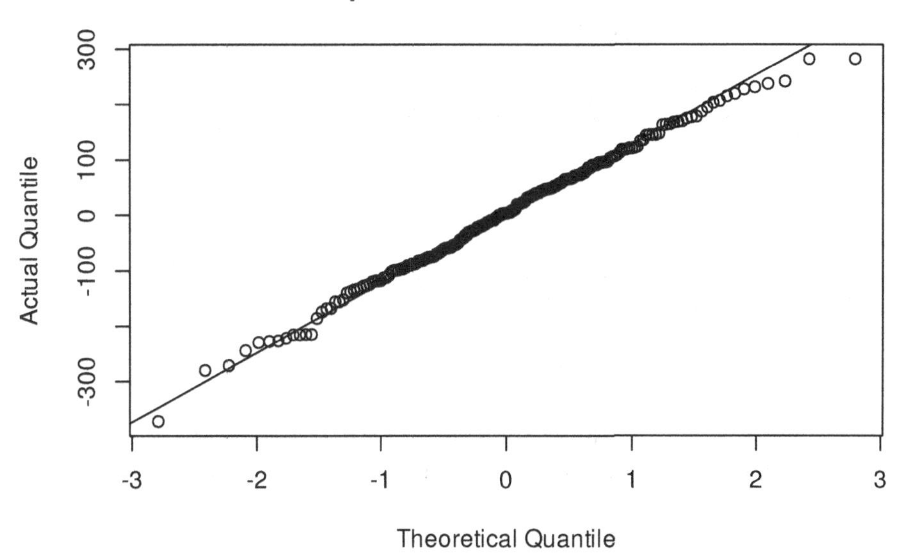

FIGURE 14.27 QQ plot of residuals extracted from model 3.

If the residuals do not contradict the assumed distribution, all the points should lie either very close to or on the line. In our case, in Figure 14.27, most points are close to the line, but the tails (especially the right one) are slightly off. This might mean that we should either use a different error type or a different distribution. Just for the sake of argument, we can try an ETSX(M,N,M) model, with the same set of explanatory variables as in the model `adamSeat03`, and with the same Normal distribution:

```
adamSeat16 <- adam(Seatbelts, "MNM",
                   formula=drivers~log(PetrolPrice)+log(kms)+law,
                   distribution="dnorm")
plot(adamSeat16, which=6)
```

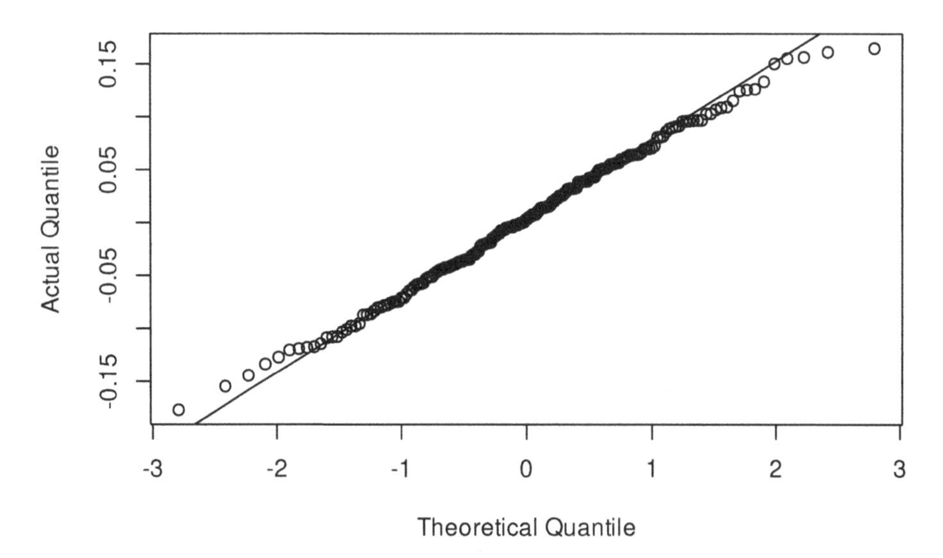

FIGURE 14.28 QQ plot of residuals extracted from the multiplicative model with Normal distribution.

According to the QQ plot in Figure 14.28, the residuals of the new model are still not very close to the theoretical ones. The tails have a slight deviation from normality: both of them are slightly shorter than expected. If our aim is to capture the distribution correctly then this can be addressed by using a Generalised Normal distribution with a higher `shape` parameter, which will have lighter tails. Hopefully, ADAM can estimate the shape parameters correctly in our case:

```
adamSeat17 <- adam(Seatbelts, "MNM",
                   formula=drivers~log(PetrolPrice)+log(kms)+law,
```

```
                    distribution="dgnorm")
plot(adamSeat17, which=6)
```

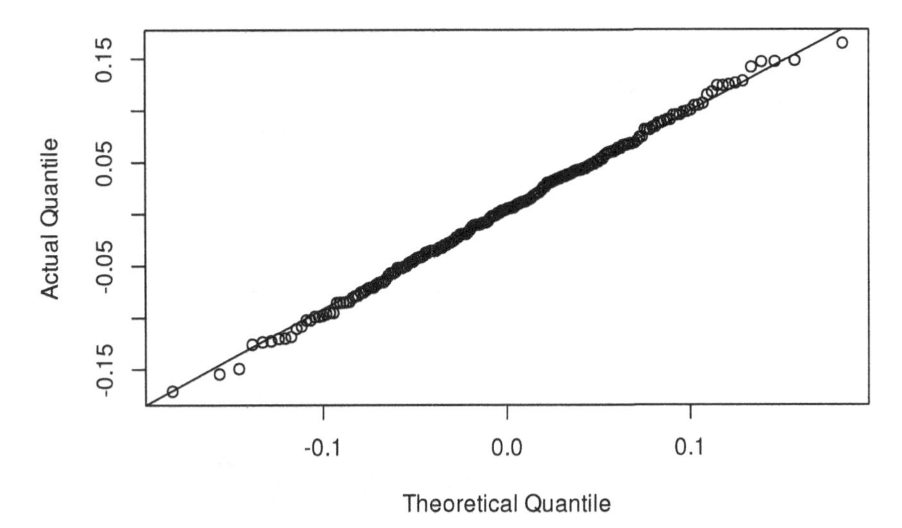

FIGURE 14.29 QQ plot of residuals extracted from the multiplicative model with Gamma distribution.

The QQ plot in Figure 14.29 shows that the residuals of Model 17 are closer to the parametric distribution than in the cases of the two previous models. We could use AICc to select between the two models if we are not sure, which of them to prefer:

```
AICc(adamSeat16)
```

```
## [1] 2405.49
```

```
AICc(adamSeat17)
```

```
## [1] 2404.881
```

Based on these results, we can conclude that the model with the Generalised Normal distribution is more suitable for this situation than the one assuming Normality.

Another way to analyse the distribution of residuals is to plot a histogram together with the theoretical probability density function (PDF). Here is an example for Model 3:

```
# Plot histogram of residuals
hist(residuals(adamSeat03), probability=TRUE,
     xlab="Residuals", main="", ylim=c(0,0.0035))
# Add density line of the theoretical distribution
lines(seq(-400,400,1),
      dnorm(seq(-400,400,1),
            mean(residuals(adamSeat03)),
            adamSeat03$scale),
      col="red")
```

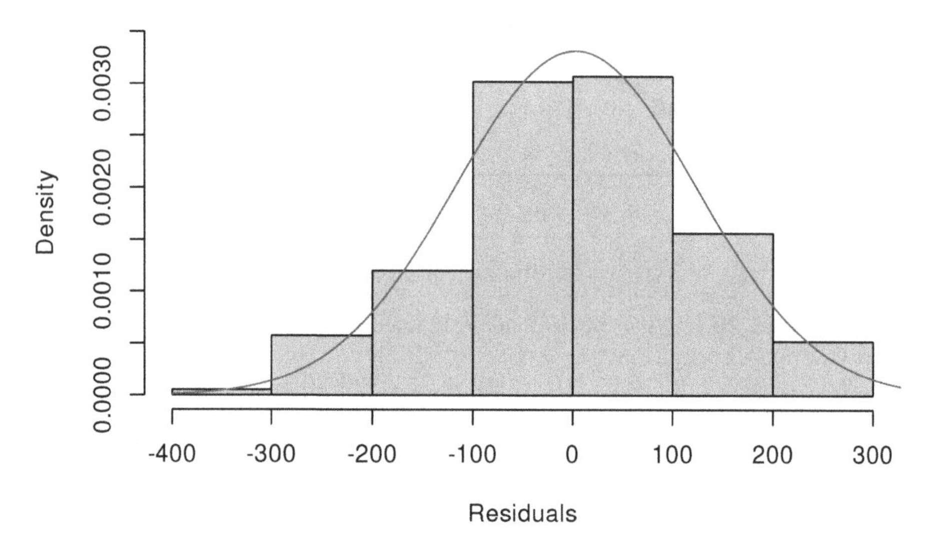

FIGURE 14.30 Histogram and density line for the residuals from Model 3 (assumed to follow Normal distribution).

However, the plot in Figure 14.30 is arguably more challenging to analyse than the QQ plot – it is not clear whether the distribution is close to the theoretical one or not. For example, Figure 14.31 shows how the histogram and the PDF curve would look for Model 17 which had the best distributional fit (assuming Generalised Normal distribution).

```
# Plot histogram of residuals
hist(residuals(adamSeat17), probability=TRUE,
     xlab="Residuals", main="")
# Add density line of the theoretical distribution
lines(seq(-0.2,0.2,0.01),
```

```
dgnorm(seq(-0.2,0.2,0.01), mean(residuals(adamSeat17)),
       adamSeat17$scale, adamSeat17$other$shape),
col="red")
```

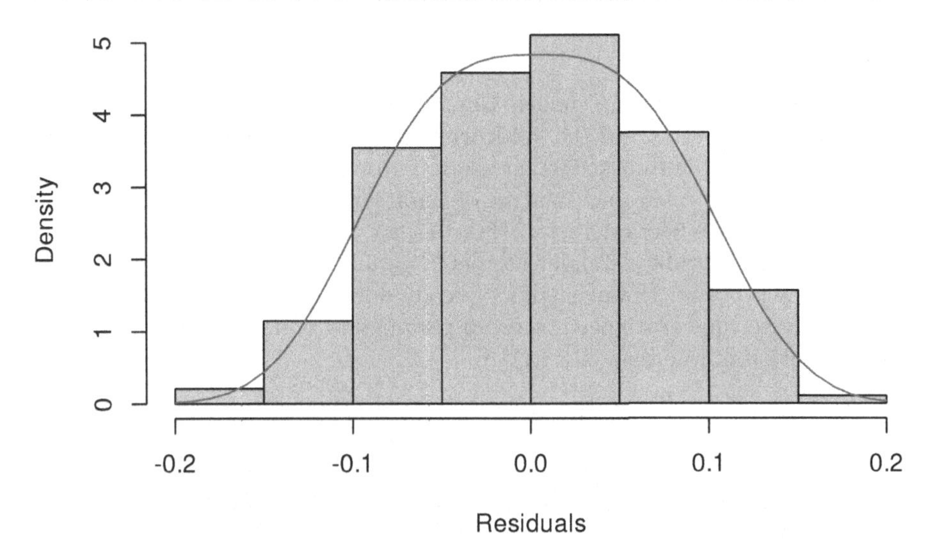

FIGURE 14.31 Histogram and density line for the residuals from model 17 (assumed to follow Generalised Normal distribution).

Comparing the plots in Figures 14.30 and 14.31 is a challenging task. This is why in general, I would recommend using QQ plots instead of histograms.

There are also formal tests for the distribution of residuals, such as Shapiro-Wilk (Shapiro & Wilk, 1965), Anderson-Darling (Anderson & Darling, 1952), and others. However, I prefer to use visual inspection when possible instead of these tests because, as discussed in Section 7.1 of Svetunkov (2022), the null hypothesis is always wrong whatever the test you use. In practice, it will inevitably be rejected with the increase of the sample size, which does not mean that it is either correct or wrong. Besides, if you fail to reject H_0, it does not mean that your variable follows the assumed distribution. It only means that you have not found enough evidence to reject the null hypothesis.

14.9 Multicollinearity

While this is not an assumption about a model and can be considered as a natural phenomenon, the issue with multicollinearity is considered one of the common issues in regression analysis. An extensive discussion on this topic in that setting is provided in Section 15.3 of the Svetunkov (2022) textbook.

When it comes to dynamic models, the implications of a model with multicollinearity might differ from the ones in regression context. So, in this section we will focus our discussion on several aspects that might not be relevant to regression.

In the conventional ARIMA model (Chapter 8), multicollinearity is inevitable by construction because of the autocorrelations between actual values. This is why sometimes heteroskedasticity- and autocorrelation-consistent (HAC) estimators of the covariance matrix of parameters are used instead of the standard ones (see Section 15.4 of Hanck et al., 2020). They are designed to fix the issue and produce standard errors of parameters that are close to those without the problem. However, this typically does not impact the forecasting itself, because the covariance matrix of parameters typically plays no role in the generation of forecasts of ARIMA.

Furthermore, multicollinearity can be considered a serious issue in static models that are estimated using conventional estimators, such as OLS. However, when it comes to state space models, and specifically to ETS, multicollinearity might not cause as severe issues as in the case of regression. For example, it is possible to use all the values of a categorical variable (Section 10.5) and still avoid the trap of dummy variables. The values of a categorical variable, in this case, are considered as changes relative to the baseline. The classic example of this is the seasonal model, for example, ETS(A,A,A), where the seasonal components can be considered as a set of parameters for dummy variables, expanded from the seasonal categorical variable (e.g. months of year variable). If we set $\gamma = 0$, thus making the seasonality deterministic, the ETS can still be estimated even though all variable levels are used, in which case the conventional regression would be inestimable. This becomes apparent with the conventional ETS model, for example, from the `forecast` package for R:

```
etsModel <- forecast::ets(AirPassengers, "AAA")
# Calculate determination coefficients for seasonal states
# These correspond to squares of multiple correlation coefficients
determ(etsModel$states[,-c(1:2)])
```

```
##        s1        s2        s3        s4        s5        s6
    s7        s8
## 0.9999992 0.9999992 0.9999991 0.9999991 0.9999992 0.9999992
    0.9999992 0.9999991
##        s9       s10       s11       s12
## 0.9999991 0.9999991 0.9999992 0.9999992
```

As we see, the states of the model are almost perfectly correlated, but still, the model works and does not have the issue that the classical linear regression would have. This is because the state-space models are constructed and estimated differently than the conventional regression (see discussion in Section 10).

Note however that this does not mean that ADAM will work easily in cases of strong multicollinearity of explanatory variables. The example above demonstrates that the issue is not necessarily as severe as in the regression context. Still, in some cases techniques for dimensionality reduction (such as Principle Components Analysis) might be required for ADAMX to work properly.

15

Model selection and combinations in ADAM

So far, we have managed to avoid discussing the topic of model selection and combinations. However, it is important to understand how the most appropriate model can be selected and how to capture the uncertainty around the model form (this comes to one of the fundamental sources of uncertainty discussed by Chatfield, 1996). There are several ways to decide which model to use, and there are several dimensions in which a decision needs to be made:

1. Which of the base models to use: ETS/ARIMA/ETS+ARIMA/Regression/ETSX/ARIMAX/ETSX+ARIMA?
2. What components of the ETS model to select?
3. What order of ARIMA model to select?
4. Which of the explanatory variables to use?
5. What distribution to use?
6. Should we select the best model or combine forecasts from different ones?
7. Do we need all models in the pool?
8. What about the demand occurrence part of the model? (Luckily, this question has already been answered in Subsection 13.1.6.)

In this chapter, we discuss these questions. We start with principles based on information criteria (see discussion in Chapter 16 of Svetunkov, 2022) for ETS and ARIMA. We then move to selecting explanatory variables and finish with topics related to the combination of models.

Before we do that, we need to recall the distributional assumptions in ADAM, which play an essential role in estimation and selection if the maximum likelihood is used (Section 11.1). In that case, an information criterion can be calculated and used for the selection of the most appropriate model across the eight dimensions mentioned above. Typically, this is done by fitting all the candidate models and then selecting the one that has the lowest information criterion. For example, when a best-fitting distribution needs to be selected, we could fit ADAMs with all the supported distributions and then select the one that gives the lowest AIC. Here is the list of the supported distributions in ADAM:

- Normal;
- Laplace;
- S;

- Generalised Normal;
- Log-Normal;
- Inverse Gaussian;
- Gamma.

The function `auto.adam()` implements this automatic selection of distribution based on an information criterion for the provided vector of `distribution` by a user. This selection procedure can be combined with other selection techniques for different elements of the ADAM discussed in the following sections of this chapter. Here is an example of selection of distribution for a specific model, ETS(M,M,N) on Box-Jenkins data using `auto.adam()`:

```
auto.adam(BJsales, model="MMN", h=10, holdout=TRUE)
```

```
## Time elapsed: 3 seconds
## Model estimated using auto.adam() function: ETS(MMN)
## Distribution assumed in the model: Log-Normal
## Loss function type: likelihood; Loss function value: 245.3716
## Persistence vector g:
##  alpha   beta
## 1.0000 0.2406
##
## Sample size: 140
## Number of estimated parameters: 5
## Number of degrees of freedom: 135
## Information criteria:
##      AIC     AICc      BIC     BICc
## 500.7432 501.1909 515.4514 516.5577
##
## Forecast errors:
## ME: 3.219; MAE: 3.332; RMSE: 3.786
## sCE: 14.133%; Asymmetry: 91.6%; sMAE: 1.463%; sMSE: 0.028%
## MASE: 2.819; RMSSE: 2.484; rMAE: 0.925; rRMSE: 0.922
```

In this case, the function has applied one and the same model but with different distributions, estimated each one of them using likelihood, and selected the one that has the lowest AICc value. It looks like Log-Normal is the most appropriate distribution for ETS(M,M,N) on this data.

15.1 ETS components selection

Remark. The model selection mechanism explained in this section is also used in the `es()` function from the `smooth` package. So, all the options for ADAM discussed here can be used in the case of `es()` as well.

Having 30 ETS models to choose from, selecting the most appropriate one becomes challenging. Petropoulos, Kourentzes, et al. (2018) showed that human experts can do this task successfully if they need to decide which components to include in the time series. However, when you face the problem of fitting ETS to thousands of time series, the judgmental selection becomes infeasible. Using some sort of automation becomes critically important.

The components selection in ETS is based on information criteria (Section 16.4 of Svetunkov, 2022). The general procedure consists of the following three main steps (in the ETS context, this approach was first proposed by Hyndman et al., 2002):

1. Define a pool of models;
2. Fit all models in the pool;
3. Select the one that has the lowest information criterion.

Depending on what is included in step (1), we will get different results. So, the pool needs to be selected carefully based on our understanding of the problem. The `adam()` function in the `smooth` package supports the following options:

1. Pool of all 30 models (Section 4.1), `model="FFF"`;
2. `model="ZZZ"`, which triggers the selection among all possible models based on a branch-and-bound algorithm (see below);
3. Pool of pure additive models (Section 5.1), `model="XXX"`. As an option, "X" can also be used to tell function to only try additive components on the selected place. e.g. `model="MXM"` will tell function to only test ETS(M,N,M), ETS(M,A,M), and ETS(M,Ad,M) models. Branch-and-bound is used in this case as well;
4. Pool of pure multiplicative models (Section 6.1), `model="YYY"`. Similarly to (3), we can tell `adam()` to only consider multiplicative components in a specific place. e.g. `model="YNY"` will consider only ETS(M,N,N) and ETS(M,N,M). Similarly to (2) and (3), in this case `adam()` will use a branch-and-bound algorithm in the components selection;
5. Pool of pure models only, `model="PPP"` – this is a shortcut for doing (2) and (3) and then selecting the best between the two pools;
6. Manual pool of models, which can be provided as a vector of models, for example: `model=c("ANN","MNN","ANA","AAN")`.

There is a trade-off when deciding which pool to use: if you provide the large one, it will take more time to find the appropriate model, and there is a risk of overfitting the data; if you provide the small pool, then the optimal model might be outside of it, giving you the sub-optimal one.

Furthermore, in some situations, you might not need to go through all models in the pool because, for example, the seasonal component is not required for the data. Trying out all the models would be just a waste of time. So, to address this issue, I have developed a branch-and-bound algorithm for the selection of the most appropriate ETS model, which is triggered via `model="ZZZ"`. The idea of the algorithm is to drop the components that do not improve the model. It allows forming a much smaller pool of models after identifying what components improve the fit. Here is how it works:

1. Apply ETS(A,N,N) to the data, calculate an information criterion;
2. Apply ETS(A,N,A) to the data, calculate information criterion. If it is lower than (1), then this means that there is some seasonal component in the data, move to step (3). Otherwise, go to (4);
3. Apply ETS(M,N,M) model and calculate information criterion. If it is lower than the previous one, then the data exhibits multiplicative seasonality. Go to (4);
4. Fit the model with the additive trend component and the seasonal component selected from the previous steps, which can be either "N", "A", or "M". Calculate information criterion for the new model and compare it with the best information criterion so far. If it is lower than any of criteria before, there is some trend component in the data. If it is not, then the trend component is not needed.

Remark. In case of multiple seasonal ETS (e.g. day of week and day of year seasonality), all the seasonal components should have the same type (see discussion in Section 12.1), so there is no need to test mixed ones (e.g. one seasonal component being additive, while the other one is multiplicative). Also, while it is possible to test whether each of the seasonal components is needed (e.g. day of week is needed, while day of year is unnecessary), this is not yet implemented in `adam()`. The simple solution here would be fitting models with and without some of the seasonal components and then selecting the one that has the lowest information criterion.

Based on these four steps, we can kick off the unnecessary components and reduce the pool of models to check. For example, if the algorithm shows that seasonality is not needed, but there is a trend, then we only have ten models to check overall instead of 30: ETS(A,N,N), ETS(A,A,N), ETS(A,Ad,N), ETS(M,N,N), ETS(M,M,N), ETS(M,Md,N), ETS(A,M,N), ETS(A,Md,N), ETS(M,A,N), and ETS(M,Ad,N). In steps (2) and (3), if there is a trend in the seasonal data, the model will have a higher than needed smoothing parameter α. While it will not approximate the data perfectly, the seasonality will play a more important role than the trend in reducing the value of the information criterion, which will help in correctly selecting the component. This is why the algorithm is, in general, efficient. It might not guarantee that the optimal model is selected all the time, but it substantially reduces the computational time.

The branch-and-bound algorithm can be combined with different types of model pools and is also supported in `model="XXX"` and `model="YYY"`, where the pool of models for steps (1) – (4) is restricted by the pure ones only. This would also work in the combinations of the style `model="XYZ"`, where the function would form the pool of the following models: ETS(A,N,N), ETS(A,M,N), ETS(A,Md,N), ETS(A,N,A), ETS(A,M,A), ETS(A,Md,A), ETS(A,N,M), ETS(A,M,M), and ETS(A,Md,M).

Finally, while the branch-and-bound algorithm is efficient, it might end up providing a mixed model, which might not be suitable for your data. So, it is recommended to think of the possible pool of models before applying it to the data. For example, in some cases, you might realise that additive seasonality is unnecessary and that the data can be either non-seasonal or with multiplicative seasonality. In this case, you can explore the `model="YZY"` option, aligning the error term with the seasonal component.

Here is an example with an automatically selected ETS model using the branch-and-bound algorithm described above:

```
adam(AirPassengers, model="ZZZ", h=12, holdout=TRUE)
```

```
## Time elapsed: 1.99 seconds
## Model estimated using adam() function: ETS(MAM)
## Distribution assumed in the model: Gamma
## Loss function type: likelihood; Loss function value: 467.2981
## Persistence vector g:
##  alpha   beta  gamma
## 0.7691 0.0053 0.0000
##
## Sample size: 132
## Number of estimated parameters: 17
## Number of degrees of freedom: 115
## Information criteria:
##       AIC      AICc       BIC      BICc
##  968.5961  973.9646 1017.6038 1030.7102
##
## Forecast errors:
## ME: 9.537; MAE: 20.784; RMSE: 26.106
## sCE: 43.598%; Asymmetry: 64.8%; sMAE: 7.918%; sMSE: 0.989%
## MASE: 0.863; RMSSE: 0.833; rMAE: 0.273; rRMSE: 0.254
```

In this specific example, the optimal model will coincide with the one selected via `model="FFF"` and `model="ZXZ"` (the reader is encouraged to try these pools on their own), although this does not necessarily hold universally.

15.2 ARIMA order selection

While ETS has 30 models to choose from, ARIMA has thousands if not more. For example, selecting the non-seasonal ARIMA with/without constant restricting the orders with $p \leq 3$, $d \leq 2$, and $q \leq 3$ leads to the combination of $3 \times 2 \times 3 \times 2 = 36$ possible models. If we increase the possible orders to 5 or even more, we will need to go through hundreds of models. Adding the seasonal part increases this number by order of magnitude. Having several seasonal cycles, increases it further. This means that we cannot just test all possible ARIMA models and select the most appropriate one. We need to be smart in the selection process.

Hyndman and Khandakar (2008) developed an efficient mechanism of ARIMA order selection based on statistical tests (for stationarity and seasonality), reducing the number of models to test to a reasonable amount. Svetunkov and Boylan (2020) developed an alternative mechanism, relying purely on information criteria, which works well on seasonal data, but potentially may lead to models overfitting the data (this is implemented in the `auto.ssarima()` and `auto.msarima()` functions in the `smooth` package). We also have the Box-Jenkins approach discussed in Section 8.3 for ARIMA orders selection, which relies on the analysis of ACF (Subsection 8.3.2) and PACF (Subsection 8.3.3). Still, we should not forget the limitations of that approach (Subsection 8.3.4). Finally, Sagaert and Svetunkov (2022) proposed the stepwise trace forward approach (discussed briefly in Section 15.3), which relies on partial correlations and uses the information criteria to test the model on each iteration. Building upon all of that, I have developed the following algorithm for order selection of ADAM ARIMA:

1. Determine the order of differences by fitting all possible combinations of ARIMA models with $P_j = 0$ and $Q_j = 0$ for all lags j. This includes trying the models with and without the constant term. The order D_j is then determined via the model with the lowest information criterion;
2. Then iteratively, starting from the highest seasonal lag and moving to the lag of 1 do for every lag m_j:
    a. Calculate the ACF of residuals of the model;
    b. Find the highest value of autocorrelation coefficient that corresponds to the multiple of the respective seasonal lag m_j;
    c. Define what should be the order of MA based on the lag of the

autocorrelation coefficient on the previous step and include it in the ARIMA model;
   d. Estimate the model and calculate an information criterion. If it is lower than for the previous best model, keep the new MA order;
   e. Repeat (a) – (d) while there is an improvement in the information criterion;
   f. Do steps (a) – (e) for AR order, substituting ACF with PACF of the residuals of the best model;
   g. Move to the next seasonal lag and go to step (a);
3. Try out several restricted ARIMA models of the order $q = d$ (this is based on (1) and the restrictions provided by the user). The motivation for this comes from the idea of the relation between ARIMA and ETS (Section 8.4);
4. Select the model with the lowest information criterion.

As you can see, this algorithm relies on the Box-Jenkins methodology but takes it with a pinch of salt, checking every time whether the proposed order is improving the model or not. The motivation for doing MA orders before AR is based on understanding what the AR model implies for forecasting (Section 8.1.1). In a way, it is safer to have an ARIMA(0,d,q) model than ARIMA(p,d,0) because the former is less prone to overfitting than the latter. Finally, the proposed algorithm is faster than the algorithm of Svetunkov and Boylan (2020) and is more modest in the number of selected orders of the model.

In R, in order to start the algorithm, you would need to provide the parameter `select=TRUE` in the `orders`. Here is an example with Box-Jenkins sales data:

```
adamARIMAModel <- adam(BJsales, model="NNN",
                       orders=list(ar=3,i=2,ma=3,select=TRUE),
                       h=10, holdout=TRUE)
```

In this example, "`orders=list(ar=3,i=2,ma=3,select=TRUE)`" tells function that the maximum orders to check are $p \leq 3$, $d \leq 2$ $q \leq 3$. The resulting model is ARIMA(0,2,2), which has the fit shown in Figure 15.1.

```
plot(adamARIMAModel, which=7)
```

The resulting model will be parsimonious when optimal initials are used. If we want to have a more flexible model, we can use a different initialisation (e.g. backcasting as discussed in Section 11.4), and in some cases, the algorithm will select a model with higher orders of AR, I, and MA.

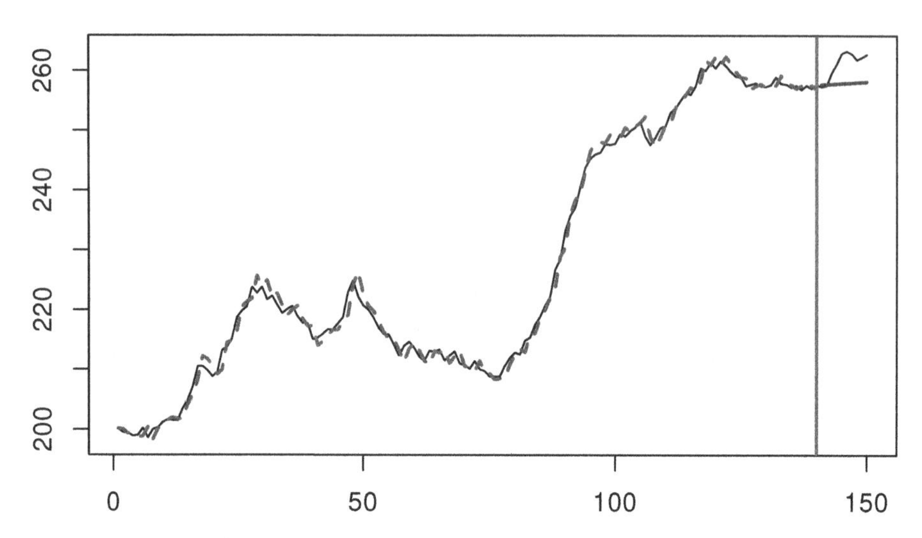

FIGURE 15.1 Actuals, fitted, and forecast for the Box-Jenkins sales data.

15.2.1 ETS + ARIMA restrictions

Based on the relation between ARIMA and ETS (see Section 8.4), we do not need to test some of the combinations of models when selecting ARIMA orders. For example, if we already consider ETS(A,N,N), we do not need to check the ARIMA(0,1,1) model. The recommendations for what to skip in different circumstances have been discussed in Section 9.4. Still, there are various ways to construct an ETS + ARIMA model, with different sequences between ETS/ARIMA selection. We suggest starting with ETS components, then moving to the selection of ARIMA orders. This way, we are building upon the robust forecasting model and seeing if it can be improved further by introducing elements that are not there. Note that given the complexity of the task of estimating all parameters for ETS and ARIMA, it is advised to use backcasting (see Section 11.4.1) for the initialisation of such model. Here is an example in R:

```
adam(AirPassengers, model="PPP",
     orders=list(ar=c(3,3),i=c(2,1),ma=c(3,3),select=TRUE),
     h=10, holdout=TRUE, initial="back")
```

```
## Time elapsed: 2.41 seconds
## Model estimated using auto.adam() function: ETS(MMM)+SARIMA(3,0,0)
    [1](1,0,0)[12]
## Distribution assumed in the model: Gamma
```

```
## Loss function type: likelihood; Loss function value: 468.391
## Persistence vector g:
##  alpha   beta  gamma
## 0.5109 0.0046 0.0000
##
## ARMA parameters of the model:
## AR:
##  phi1[1]  phi2[1]  phi3[1] phi1[12]
##   0.2154   0.2296  -0.0402   0.2084
##
## Sample size: 134
## Number of estimated parameters: 8
## Number of degrees of freedom: 126
## Information criteria:
##      AIC     AICc      BIC     BICc
## 952.7819 953.9339 975.9647 978.7858
##
## Forecast errors:
## ME: 3.416; MAE: 16.347; RMSE: 20.154
## sCE: 12.908%; Asymmetry: 31.9%; sMAE: 6.178%; sMSE: 0.58%
## MASE: 0.681; RMSSE: 0.646; rMAE: 0.164; rRMSE: 0.163
```

The resulting model is ETS(M,M,M) with AR elements: three non-seasonal and one seasonal AR, which improve the fit of the model and hopefully result in more accurate forecasts.

15.3 Explanatory variables selection

There are different approaches for automatic variable selection, but not all of them are efficient in the context of dynamic models. For example, conventional stepwise approaches might be either not feasible in the case of small samples or may take too much time to converge to an optimal solution (it has polynomial computational time). This well known problem in regression context is magnified in the context of dynamic models, because each model fit takes much more time than in the case of regression. This is because the ADAMX needs to be refitted and re-estimated repeatedly using recursive relations based on the state space model (10.4). So, there need to be some simplifications, which will make variables selection in ADAMX doable in a reasonable time.

To make the mechanism efficient in a limited time, I propose using the Sagaert and Svetunkov (2022) approach of stepwise trace forward selection of variables. It uses partial correlations between variables to identify which of them to include in each iteration. While it has linear computational time instead of the polynomial, doing that in the proper ADAMX would still take a lot of time, because of the fitting of the dynamic model. So one of the possible solutions is to do variables selection in ADAMX based on models residuals, in the following steps:

1. Estimate and fit the ADAM;
2. Extract the residuals of the ADAM;
3. Select the most suitable variables, explaining the residuals, based on the trace forward stepwise approach and the selected information criterion;
4. Re-estimate the ADAMX with the selected explanatory variables.

The residuals in step (2) might vary from model to model, depending on the type of the error term and the selected distribution:

- Normal, Laplace, S, Generalised Normal or Asymmetric Laplace: e_t;
- Additive error and Log-Normal, Inverse Gaussian or Gamma: $\left(1 + \frac{e_t}{\hat{y}_t}\right)$;
- Multiplicative error and Log-Normal, Inverse Gaussian or Gamma: $1 + e_t$.

So, the extracted residuals should be aligned with the distributional assumptions of each model.

In R, step (3) is done using the `stepwise()` function from the `greybox` package, supporting all the distributions implemented in ADAM. The only thing that needs to be modified is the number of degrees of freedom: the function should consider all estimated parameters (including the number of parameters of the dynamic part). This is done internally via the `df` parameter in `stepwise()`.

While the suggested approach has obvious limitations (e.g. smoothing parameters can be higher than needed, explaining the variability otherwise explained by variables), it is efficient in terms of computational time.

To see how it works, we use SeatBelt data:

```
SeatbeltsData <- Seatbelts[,c("drivers","kms","PetrolPrice","law")]
```

We have already had a look at this data earlier in Section 10.6, so we can move directly to the selection part:

```
adamETSXMNMSelectSeat <- adam(SeatbeltsData, "MNM",
                              h=12, holdout=TRUE,
                              regressors="select")
summary(adamETSXMNMSelectSeat)
```

```
## Warning: Observed Fisher Information is not positive semi-definite,
     which means
## that the likelihood was not maximised properly. Consider
     reestimating the
```

```
## model, tuning the optimiser or using bootstrap via bootstrap=TRUE.

##
## Model estimated using adam() function: ETSX(MNM)
## Response variable: drivers
## Distribution used in the estimation: Gamma
## Loss function type: likelihood; Loss function value: 1117.189
## Coefficients:
##               Estimate Std. Error Lower 2.5% Upper 97.5%
## alpha           0.2877     0.0856     0.1186      0.4565 *
## gamma           0.0000     0.0414     0.0000      0.0816
## level        1655.9759    97.3924  1463.6713   1848.0473 *
## seasonal_1      1.0099     0.0155     0.9808      1.0459 *
## seasonal_2      0.9053     0.0153     0.8762      0.9413 *
## seasonal_3      0.9352     0.0156     0.9061      0.9712 *
## seasonal_4      0.8696     0.0147     0.8405      0.9056 *
## seasonal_5      0.9465     0.0162     0.9174      0.9825 *
## seasonal_6      0.9152     0.0155     0.8861      0.9513 *
## seasonal_7      0.9623     0.0160     0.9332      0.9983 *
## seasonal_8      0.9706     0.0159     0.9416      1.0067 *
## seasonal_9      1.0026     0.0169     0.9735      1.0386 *
## seasonal_10     1.0824     0.0178     1.0533      1.1184 *
## seasonal_11     1.2012     0.0183     1.1721      1.2372 *
## law             0.0200     0.1050    -0.1873      0.2271
##
## Error standard deviation: 0.0752
## Sample size: 180
## Number of estimated parameters: 16
## Number of degrees of freedom: 164
## Information criteria:
##      AIC     AICc      BIC     BICc
## 2266.378 2269.715 2317.465 2326.131
```

Remark. The `summary()` method might complain about the observed Fisher Information. This only means that the estimated variances of parameters might be lower than they should be in reality. This is discussed in Section 16.2.

Based on the summary from the model, we can see that neither `kms` nor `PetrolPrice` improve the model in terms of AICc (they were not included in the model). We could check them manually to see if the selection worked out well in our case (construct sink regression as a benchmark):

```
adamETSXMNMSinkSeat <- adam(SeatbeltsData, "MNM",
```

```
                                h=12, holdout=TRUE)
summary(adamETSXMNMSinkSeat)
```

```
## Warning: Observed Fisher Information is not positive semi-definite,
     which means
## that the likelihood was not maximised properly. Consider
    reestimating the
## model, tuning the optimiser or using bootstrap via bootstrap=TRUE.

##
## Model estimated using adam() function: ETSX(MNM)
## Response variable: drivers
## Distribution used in the estimation: Gamma
## Loss function type: likelihood; Loss function value: 1234.278
## Coefficients:
##             Estimate Std. Error Lower 2.5% Upper 97.5%
## alpha         0.9508     2.5767     0.0000      1.0000
## gamma         0.0000     0.0131     0.0000      0.0259
## level        23.2952     1.2746    20.7782     25.8087 *
## seasonal_1    1.1340     0.0621     1.0115      4.2731 *
## seasonal_2    0.9924     0.9516     0.8698      4.1314 *
## seasonal_3    0.9248     0.8549     0.8023      4.0639 *
## seasonal_4    0.8342     0.7888     0.7117      3.9733 *
## seasonal_5    0.9068     0.8962     0.7843      4.0459 *
## seasonal_6    0.8625     0.9153     0.7400      4.0016 *
## seasonal_7    0.8370     0.8126     0.7144      3.9761 *
## seasonal_8    0.8477     0.7795     0.7252      3.9868 *
## seasonal_9    0.9798     0.9959     0.8573      4.1189 *
## seasonal_10   1.1417     1.2918     1.0192      4.2808 *
## seasonal_11   1.3273     1.5918     1.2047      4.4663 *
## kms           0.0000     0.0000    -0.0001      0.0001
## PetrolPrice  -2.7216     3.1292    -8.9009      3.4494
## law           0.0181     5.2249   -10.2996     10.3216
##
## Error standard deviation: 0.1413
## Sample size: 180
## Number of estimated parameters: 18
## Number of degrees of freedom: 162
## Information criteria:
##      AIC     AICc      BIC     BICc
## 2504.556 2508.804 2562.029 2573.060
```

We can see that the sink regression model has a higher AICc value than the model with the selected variables, which means that the latter is closer to the "true model". While `adamETSXMNMSelectSeat` might not be the best possible model in terms of information criteria, it is still a reasonable one and can be used for further inference.

15.4 Forecasts combinations

When it comes to achieving the most accurate forecasts possible in practice, the most robust (in terms of not failing) approach is producing combined forecasts. The primary motivation for combining is that there is no one best forecasting method for everything – methods might perform very well in some conditions and fail in others. It is typically not possible to say which of the cases you face in practice. Furthermore, the model selected on one sample might differ from the model chosen for the same sample but with one more observation. Thus there is a model uncertainty (as defined by Chatfield, 1996), which can be mitigated by producing forecasts from several models and then combining them to get the final forecast. This way, the potential damage from an inaccurate forecast is typically reduced.

There are many different techniques for combining forecasts. The non-exhaustive list includes:

1. **Simple average**, which works fine as long as you do not have exceptionally poorly performing methods;
2. **Median**, which produces good combinations when the pool of models is relatively small and might contain those that produce very different forecasts from the others (e.g. explosive trajectories). However, when a big pool of models is considered, the median might ignore vital information and decrease accuracy, as noted by Jose and Winkler (2008). Stock and Watson (2004) conducted an experiment on macroeconomic data, and medians performed poorer than the other approaches (probably because of the high number of forecasting methods), while a median-based combination worked well for Petropoulos and Svetunkov (2020), who considered only four forecasting approaches;
3. **Trimmed and/or Winsorized mean**, which drop extreme forecasts, when calculating the mean and, as was shown by Jose and Winkler (2008), work well in cases of big pools of models, outperforming medians and simple average;
4. **Weighted mean**, which assigns weights to each forecast and produces a combined forecast based on them. While this approach sounds more reasonable than the others, there is no guarantee that it will work better because the weights need to be estimated and might change with the change of sample size or a pool of models. Claeskens, Magnus, Vasnev, and Wang (2016) explain why the simple average approach outperforms

weighted averages in many cases: it does not require estimation of weights and thus does not introduce as much uncertainty. However, when done smartly, combinations can be beneficial in terms of accuracy, as shown, for example, by Kolassa (2011) and Kourentzes, Barrow, and Petropoulos (2019).

The forecast combination approach implemented in ADAM is the weighted mean, based on Kolassa (2011), who used AIC weights as proposed by Burnham and Anderson (2004). This approach aims to estimate all models in the pool, calculate information criteria for each of them (see discussion in Section 16.4 in Svetunkov, 2022) and then calculate weights for each model. Those models with lower ICs will have higher weights, while the poorly performing ones will have the lower ones. The only requirement of the approach is for the parameters of models to be estimated via likelihood maximisation (see Section 11.1). Furthermore, it is not important what model is used or what distribution is assumed, as long as the models are initialised (see discussion in Section 11.4) and constructed in the same way and the likelihood is used in the estimation.

When it comes to prediction intervals, the correct way of calculating them for the combination is to consider the joint distribution of all forecasting models in the pool and take quantiles based on that. However, Lichtendahl, Grushka-Cockayne, and Winkler (2013) showed that a simpler approach of averaging the quantiles works well in practice. This approach implies producing prediction intervals for all the models in the pool and then averaging the obtained values. It is fast and efficient in terms of getting prediction intervals from a combined model.

In R, the `adam()` function supports the combination of ETS models via `model="CCC"` or any other combination of letters, as long as the model contains "C" in its name. For example, the function will combine all non-seasonal models if `model="CCN"` is provided. Consider the following example on Box-Jenkins sales series:

```
adamETSCCN <- adam(BJsales, "CCN", h=10, holdout=TRUE, ic="AICc")
```

In the code above, the function will estimate all non-seasonal models, extract AICc for each of them and then calculate weights, which we can be extracted for further analysis:

```
round(adamETSCCN$ICw, 3)
```

```
##   ANN   MAN  AAdN   MMN  AMdN   MNN   AAN  MAdN   AMN  MMdN
## 0.000 0.014 0.252 0.010 0.511 0.000 0.073 0.031 0.050 0.059
```

As can be seen from the output of weights, the level models ETS(A,N,N) and

ETS(M,N,N) were further away from the best model and, as a result, got weights very close to zero.

In the ADAM combination, the fitted values are combined from all models, while the residuals are calculated as $e_t = y_t - \hat{y}_t$, where $\hat{y}_t$ is the combined value. The final forecast together with the prediction interval can be generated via the `forecast()` function (Figure 15.2):

```
adamETSCCN |>
    forecast(h=10, interval="prediction") |>
    plot()
```

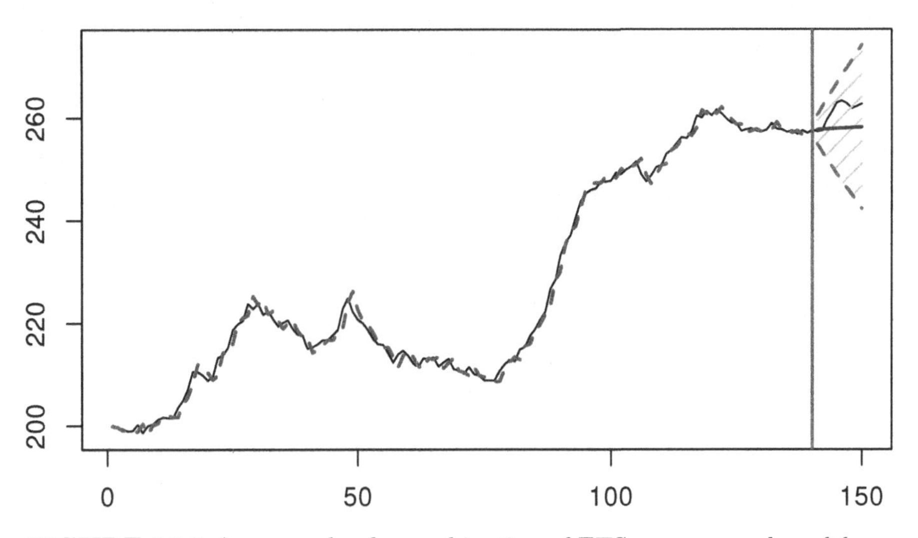

FIGURE 15.2 An example of a combination of ETS non-seasonal models on Box-Jenkins sale time series.

What the function above does, in this case, is produces forecasts and prediction intervals from each model and then uses original weights to combine them. Each model can be extracted and used separately if needed. Here is an example with the ETS(A,Ad,N) model from the estimated pool:

```
adamETSCCN$models$AAdN |>
    forecast(h=10,interval="prediction") |>
    plot()
```

As can be seen from the plots in Figures 15.2 and 15.3, due to the highest

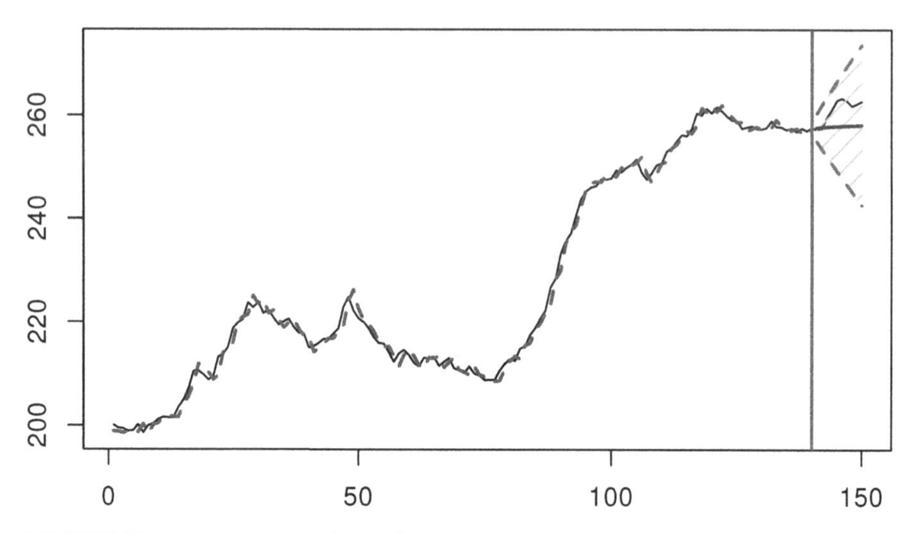

FIGURE 15.3 An example of the combination of ETS non-seasonal models on Box-Jenkins sale time series.

weight, ETS(A,Ad,N) and ETS(C,C,N) models have produced very similar point forecasts and prediction intervals.

Alternatively, if we do not need to consider all ETS models, we can provide the pool of models, including a model with "C" in its name. Here is an example of how pure additive non-seasonal models can be combined:

```
adamETSCCNPureAdditive <- adam(BJsales,
                               c("CCN","ANN","AAN","AAdN"),
                               h=10, holdout=TRUE,
                               ic="AICc")
```

The main issue with the combined ETS approach is that it is computationally expensive due to the estimation of all models in the pool and can also result in high memory usage (because it saves all the estimated models). As a result, it is recommended to be smart in deciding which models to include in the pool.

15.4.1 Other combination approaches

While `adam()` supports information criterion weights combination of ETS models only, it is also possible to combine ARIMA, regression models, and models with different distributions in the framework. Given that all models are initialised in the same way and that the likelihoods are calculated using similar principles, the weights can be calculated manually using the formula from Burnham and Anderson (2004):

$$w_i = \frac{\exp\left(-\frac{1}{2}\Delta_i\right)}{\sum_{j=1}^{n} \exp\left(-\frac{1}{2}\Delta_j\right)}, \tag{15.1}$$

where $\Delta_i = \mathrm{IC}_i - \min_{i=1}^{n}(\mathrm{IC}_i)$ is the information criteria distance from the best performing model, IC_i is the value of an information criterion of model i, and n is the number of models in the pool. For example, here how we can combine the best ETS with the best ARIMA and the ETSX(M,M,N) model in the ADAM framework, based on BICc:

```
# Prepare data with explanatory variables
BJsalesData <- cbind(as.data.frame(BJsales),
                     xregExpander(BJsales.lead,c(-5:5)))

# Apply models
adamPoolBJ <- vector("list",3)
adamPoolBJ[[1]] <- adam(BJsales, "ZZN",
                        h=10, holdout=TRUE,
                        ic="BICc")
adamPoolBJ[[2]] <- adam(BJsales, "NNN",
                        orders=list(ar=3,i=2,ma=3,select=TRUE),
                        h=10, holdout=TRUE,
                        ic="BICc")
adamPoolBJ[[3]] <- adam(BJsalesData, "MMN",
                        h=10, holdout=TRUE,
                        ic="BICc",
                        regressors="select")

# Extract BICc values
adamsICs <- sapply(adamPoolBJ, BICc)

# Calculate weights
adamsICWeights <- adamsICs - min(adamsICs)
adamsICWeights[] <- exp(-0.5*adamsICWeights) /
                    sum(exp(-0.5*adamsICWeights))
names(adamsICWeights) <- c("ETS","ARIMA","ETSX")
round(adamsICWeights, 3)
```

```
##   ETS ARIMA  ETSX
## 0.524 0.424 0.052
```

These weights can then be used for the combination of the fitted values, forecasts, and prediction intervals:

```
# Produce forecasts from the three models
adamPoolBJForecasts <- lapply(adamPoolBJ, forecast,
                              h=10, interval="pred")

# Produce combined conditional means and prediction intervals
finalForecast <- cbind(sapply(adamPoolBJForecasts,
                              "[[","mean") %*% adamsICWeights,
                       sapply(adamPoolBJForecasts,
                              "[[","lower") %*% adamsICWeights,
                       sapply(adamPoolBJForecasts,
                              "[[","upper") %*% adamsICWeights)
# Give the appropriate names
colnames(finalForecast) <- c("Mean",
                             "Lower bound (2.5%)",
                             "Upper bound (97.5%)")
# Transform the table in the ts format (for convenience)
finalForecast <- ts(finalForecast,
                    start=start(adamPoolBJForecasts[[1]]$mean))
finalForecast
```

```
## Time Series:
## Start = 141
## End = 150
## Frequency = 1
##         Mean Lower bound (2.5%) Upper bound (97.5%)
## 141 257.6481           254.8609            260.4284
## 142 257.7053           253.2997            262.0983
## 143 257.7465           251.8092            263.6947
## 144 257.7930           250.2730            265.2801
## 145 257.8512           248.7825            266.9468
## 146 257.8915           247.2128            268.6438
## 147 257.9279           245.5952            270.1407
## 148 257.9769           244.0815            271.8091
## 149 258.0166           242.4309            273.6080
## 150 258.0684           240.7363            275.3822
```

In order to see how the forecast looks, we can plot it via `graphmaker()` function from `greybox`:

```
graphmaker(BJsales, finalForecast[,1],
           lower=finalForecast[,2], upper=finalForecast[,3],
           level=0.95)
```

Figure 15.4 demonstrates the slightly increasing trajectory with an expand-

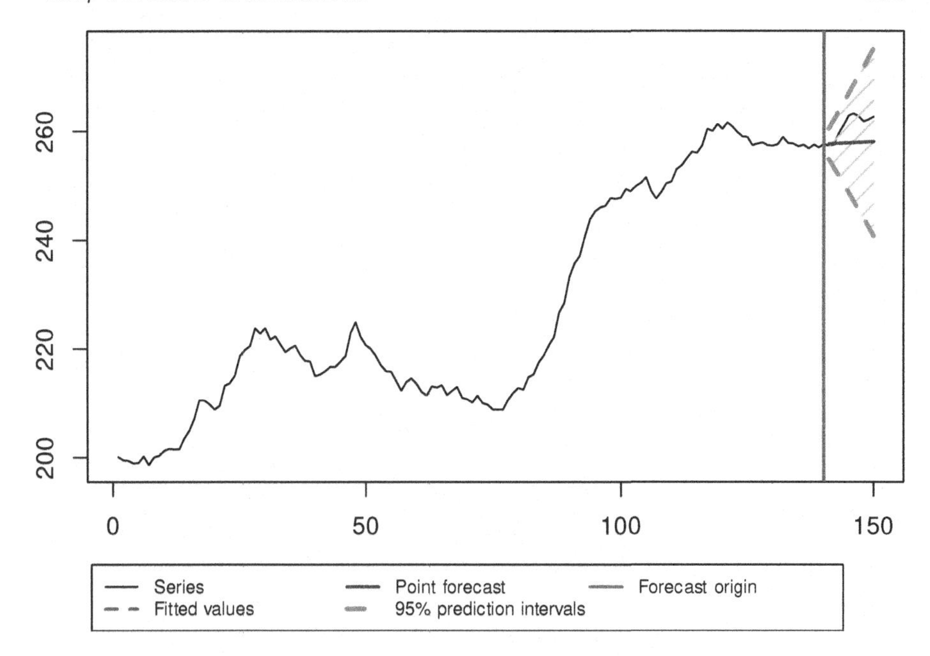

FIGURE 15.4 Final forecast from the combination of ETS, ARIMA, and ETSX models.

ing prediction interval. The point forecast (conditional mean) continues the trajectory observed on the last several observations, just before the forecast origin. The future values are inside the prediction interval, so overall, this can be considered a reasonable forecast.

16

Handling uncertainty in ADAM

So far, when we discussed forecasts from ADAM, we have assumed that the smoothing parameters and initial values are known, even though we have acknowledged in Chapter 11 that they are estimated. This is the conventional assumption of ETS models from Hyndman et al. (2008), which also applies to ARIMA models. However, in reality, the parameters are never known and are always estimated in-sample. This means that the estimates of parameters will inevitably change with the change of sample size. This uncertainty will impact the model fit, the point forecasts, and prediction intervals. To overcome this issue, Bergmeir, Hyndman, and Benítez (2016) proposed bagging ETS – the procedure that decomposes time series using STL (Cleveland, Cleveland, McRae, & Terpenning, 1990) then recreates many time series by bootstrapping the remainder then fits best ETS model to each of the newly created time series and combines the forecasts from the models. This way (as was explained by Petropoulos, Hyndman, & Bergmeir, 2018), the parameters of the models will differ from one generated time series to another. Thus, the final forecasts will handle the uncertainty about the parameters. In addition, this approach also mitigates to some extent the model uncertainty, which was discussed in Section 15.4, because models are selected automatically on each bootstrapped series. The main issue with the approach is that it is computationally expensive and assumes that STL decomposition is appropriate for time series. Furthermore, it assumes that the residuals from this decomposition do not contain any information and are independent.

In this chapter, we focus on a discussion of uncertainty in ADAM, specifically about the estimates of parameters. We start with a discussion of how the data can be simulated from an estimated ADAM, then move to how to deal with confidence intervals for the parameters and after that – how the parameters' uncertainty can be propagated to the states and fitted values of the model. Some parts of this chapter are based on Svetunkov and Pritularga (2023a).

16.1 Simulating data from ADAM

Before we move to the discussion of parameters' uncertainty and how it propagates to the states, fitted values, and final forecasts, it makes sense to understand how data can be generated based on an ADAM with some parameters. The data generation in this case is done in the following steps:

1. Decide what structure will be used for data generation. This depends on the type and order of the specific ADAM. For example, if it a pure additive ETS(A,N,A)+ARIMA(2,0,0) model then the set of equations (9.30) will be used with matrices and vectors defined by (9.31);
2. Generate an error term for the values of $t = 1 \dots T$, where T is the desired sample size. The error term can be generated using any known distribution as long as its mean equals to zero in the case of additive error ($\mathrm{E}(\epsilon_t) = 0$) or equals to one in the case of the multiplicative one ($\mathrm{E}(1 + \epsilon_t) = 1$);
3. Set the values of the persistence vector $\mathbf{g}$ and the matrices $\mathbf{w}$ and $\mathbf{F}$. In the case of ARIMA, the values are defined based on the order of the model and the values of its parameters (see discussion in Section 9.1). The specific values will change depending on the type of model and the elements it has;
4. Define the initial value of the state vector $\mathbf{v}_t$ for $t < 1$;
5. Apply recursively the formula of the state space model (7.1) (discussed in Section 7.1), collecting the values of the state vector $\mathbf{v}_t$ and the generated actual values y_t for all $t = 1, \dots, T$.

Note that the simulation process allows using distributions that are not officially supported by ADAM on the step (2). In fact, in some instances the error term can be provided by a user, so it would not be random.

In the `smooth` package for R, there are several functions that implement this simulation procedure:

- `sim.es()` allows generating data from an arbitrary ETS model;
- `sim.ssarima()` supports data generation from a state space ARIMA model with parameters and initial states provided by the user;
- `sim.sma()` generates data from Simple Moving Average, discussed in Subsection 3.3.3;
- `simulate()` method for `adam` and `smooth` classes allows generating the data using the parameters of an estimated model, thus creating time series similar to the original one.

As an example, consider estimating ADAM ETS(M,M,M) on `AirPassengers` data and then generating several time series from it (see Figure 16.1) using the following R code:

```
adam(AirPassengers, "MMM") |>
    simulate() |>
    plot(main="")
```

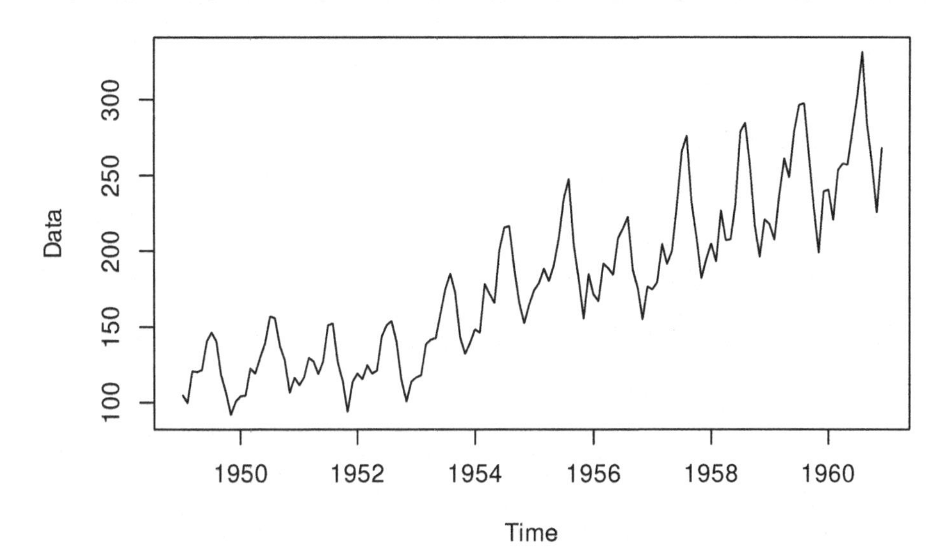

FIGURE 16.1 Data generated by ADAM using the parameters of the estimated ETS model.

You might notice that the trend in Figure 16.1 differs from the one in the original data. This is because the ETS model captured some changes in the trend in the data, which was then reflected in the generated series. The resulting simulated data can be used for the experiments in a controlled environment.

In some cases, some specific data might be required with specific parameters and values for the error term. In those cases, the `sim.es()`, `sim.ssarima()`, and other similar functions from the `smooth` package can be used. Here is an example of a code for `sim.es()` with an arbitrary function for the generation of the error term:

```
# A function to generate error term
randomizer <- function(n, mu=0, s=1){
    return(mu + s * rlogis(n, 0, 1))
}
# Generation of data from ETS(A,N,N)
x <- sim.es("ANN", obs=120, nsim=10,
            persistence=0.1, initial=1000,
            randomizer="randomizer", mu=0, s=1)
```

```
# Plot a randomly selected series from the 10 generated ones
plot(x)
```

16.2 Covariance matrix of parameters

One of the basic conventional statistical ways of capturing uncertainty about estimates of parameters is via the calculation of the covariance matrix of parameters. The covariance matrix that is typically calculated in regression context can be based either on OLS estimates of parameters, or on MLE assuming that the residuals of the model follow Normal distribution. In the latter case, it relies on the "Fisher Information", which in turn is calculated as a negative expectation of Hessian of parameters (the matrix of second derivatives of the likelihood function with respect to all the estimates of parameters). The idea of Hessian is to capture the curvatures of the likelihood function in its optimal point to understand what impact each of the parameters has on the likelihood. If we calculate the Hessian matrix and have the Fisher Information, then using the Cramer-Rao bound (Rao, 1945), the true variance of the parameters will be greater than or equal to the inverse of the Fisher Information:

$$V(\hat{\theta}_j) \geq \frac{1}{\mathrm{FI}(\hat{\theta}_j)}, \tag{16.1}$$

where θ_j is the parameter under consideration. The property (16.1) can then be used for the calculation of the covariance matrix of parameters. While in the case of the linear regression this calculation has an analytical solution, in cases of ETS and ARIMA, this can only be done via numeric methods, because the models rely on recursive relations and there is no closed analytical solution for the parameters of these models.

In R, an efficient calculation of Hessian can be done via the `hessian()` function from the `pracma` package (Borchers, 2022). In `smooth` there is a method `vcov()` that does all the calculations, estimating the negative Hessian inside the `adam()` and then inverting the result. Here is an example of how this works:

```
adamETSBJ <- adam(BJsales, h=10, holdout=TRUE)
adamETSBJVcov <- vcov(adamETSBJ)
round(adamETSBJVcov, 3)
```

```
##          alpha   beta    phi  level  trend
## alpha  0.013 -0.010  0.007 -0.206  0.167
## beta  -0.010  0.019 -0.014  0.470 -0.376
## phi    0.007 -0.014  0.016 -0.482  0.389
```

```
## level -0.206  0.470 -0.482  22.287 -16.158
## trend  0.167 -0.376  0.389 -16.158  12.704
```

The precision of the estimate will depend on the closeness of the likelihood function to its maximum in the estimated parameters. If the likelihood was not properly maximised and the function stopped prematurely, then the covariance matrix might be incorrect and contain errors. In that case, `vcov()` will produce a warning, saying that the resulting matrix might not be estimated correctly. The covariance matrix itself shows the variability of each of the parameters and whether they have any relations between them or not. In order to get a clearer picture about the latter, we could calculate the correlation matrix of the estimated parameters:

```
cov2cor(adamETSBJVcov) |>
  round(3)
```

```
##        alpha   beta    phi  level  trend
## alpha  1.000 -0.657  0.467 -0.381  0.410
## beta  -0.657  1.000 -0.803  0.714 -0.757
## phi    0.467 -0.803  1.000 -0.813  0.870
## level -0.381  0.714 -0.813  1.000 -0.960
## trend  0.410 -0.757  0.870 -0.960  1.000
```

This matrix demonstrates that the estimates of the initial level and trend of the ETS(A,Ad,N) model applied to this data are strongly negatively correlated. This means that on average with the increase of the initial level, the initial trend tends to be lower, which makes sense as a starting point of a model. The values from the covariance matrix can also be used, for example, for calculation of confidence intervals of parameters (Section 16.3), construction of confidence intervals for the fitted values and point forecasts (Section 16.5), and for the construction of more adequate prediction intervals (i.e. taking the uncertainty of estimates of parameters into account, see Subsection 18.3.6).

In some cases, the `vcov()` method would complain that the Fisher Information cannot be inverted. This typically means that the `adam()` failed to reach the maximum of the likelihood function. Re-estimating the model with different initial values and/or optimiser settings might resolve the problem (see Section 11.4).

Remark. This method only works when `loss="likelihood"` or when the loss is aligned with the assumed distribution (e.g. `loss="MSE"` and `distribution="dnorm"`). In all the other cases, other approaches (such as bootstrap) would need to be used to estimate the covariance matrix of parameters.

16.2.1 Bootstrapped covariance matrix

An alternative way of constructing the matrix is via the bootstrap. The one implemented in `smooth` is based on the `coefbootstrap()` method from the `greybox` package, which implements the modified case resampling. It is less efficient than the Fisher Information method in terms of computational time and works only for larger samples. The algorithm implemented in the function creates continuous sub-samples of the original data, starting from the initial point $t = 1$ (if backcasting is used, as discussed in Section 11.4.1, then the starting point will be allowed to vary). These sub-samples are then used for re-estimation of `adam()` to get the empirical estimates of parameters. The procedure is repeated `nsim` times, which for `adam()` is by default equal to 100. This approach is far from ideal and will typically lead to the underestimated variance of initials because of the sample size restrictions. Still, it does not break the data structure and allows obtaining results relatively fast without imposing any additional assumptions on the model and the data. I personally recommend using it in the case of the initialisation via backcasting.

Here is an example of how the function works on the data above – it is possible to speed up the process by doing parallel calculations:

```
adamETSBJBoot <- coefbootstrap(adamETSBJ, parallel=TRUE)
adamETSBJBoot
```

```
## Bootstrap for the adam model with nsim=100 and size=70
## Time elapsed: 2.72 seconds
```

The `size` in the output above refers to the sub-sample size, which by default is 75% of the original data length. The covariance matrix can then be extracted from the result via `adamETSBJBoot$vcov`. The same procedure is used in the `vcov()` method if `bootstrap=TRUE`:

```
adamETSBJ |>
    vcov(bootstrap=TRUE, parallel=TRUE) |>
    round(3)
```

```
##        alpha   beta    phi  level  trend
## alpha  0.028  0.008  0.007  0.032 -0.036
## beta   0.008  0.013  0.000 -0.005  0.015
## phi    0.007  0.000  0.016  0.071 -0.075
## level  0.032 -0.005  0.071  0.537 -0.682
## trend -0.036  0.015 -0.075 -0.682  0.958
```

16.3 Confidence intervals for parameters

As it is well known in statistics (e.g. see Section 6.4 of Svetunkov, 2022), if several vital model assumptions (discussed in Section 14) are satisfied and CLT holds, then the distribution of estimates of parameters will follow the Normal one, which will allow us to construct confidence intervals to capture the uncertainty around the parameters. In cases of ETS and ARIMA models in the ADAM framework, the estimated parameters include smoothing, dampening, and ARMA parameters together with the initial states. In the case of explanatory variables, the pool of parameters is increased by the coefficients for those variables and their smoothing parameters (if the dynamic model from Section 10.3 is used). Furthermore, in the case of the intermittent state space model, the parameters will also include the elements of the occurrence part of the model. The CLT should hold for all of them if:

1. Estimates of parameters are consistent (e.g. MSE or likelihood is used in estimation, see Section 11);
2. The parameters do not lie near the bounds;
3. The model is correctly specified;
4. Moments of the distribution of error term are finite.

In case of ETS and ARIMA, some of the parameters are bounded (e.g. to satisfy stability condition from Section 5.4), and the estimates might lie near the bounds. This means that the distribution of estimates of parameters might not be Normal. However, given that the bounds of the parameters are typically fixed, and all estimates that exceed them are set to the boundary values in the optimisation routine, the estimates of parameters will follow Rectified Normal distribution (Socci, Lee, & Seung, 1997). This is important because knowing the distribution, we can derive the confidence intervals for the parameters. However, given that we estimate the standard errors of parameters in-sample, we need to use t-statistics to correctly capture the uncertainty. The confidence intervals will be constructed in a conventional way in this case, using the formula (see Section 6.4 of Svetunkov, 2022):

$$\theta_j \in (\hat{\theta}_j + t_{\alpha/2}(df) s_{\theta_j}, \hat{\theta}_j + t_{1-\alpha/2}(df) s_{\theta_j}), \tag{16.2}$$

where $t_{\alpha/2}(df)$ is Student's t-statistics for $df = T - k$ degrees of freedom (T is the sample size and k is the number of all estimated parameters) and α is the significance level. Then, after constructing the intervals, we can cut their values with the bounds of parameters, thus rectifying the distribution.

To construct the interval, we need to know the standard errors of parameters. Luckily, they can be calculated as square roots of the diagonal of the covariance matrix of parameters (discussed in Section 16.2):

```
diag(adamETSBJVcov) |>
  sqrt()
```

```
##     alpha      beta       phi     level     trend
## 0.1144386 0.1394695 0.1256504 4.7208625 3.5642472
```

Based on these values and the formula (16.2), we can produce confidence intervals for the parameters of any ADAM, which is done in R using the `confint()` method. For example, here are the intervals for the model estimated before with the significance level of 1% (confidence level of 99%):

```
confint(adamETSBJ, level=0.99)
```

```
##            S.E.        0.5%       99.5%
## alpha 0.1144386   0.6517003   1.0000000
## beta  0.1394695   0.0000000   0.6550654
## phi   0.1256504   0.5348598   1.0000000
## level 4.7208625 190.4171720 215.0739522
## trend 3.5642472 -11.7130838   6.9027646
```

In the output above, the distributions for α, β, and ϕ are rectified: α and ϕ are restricted with the region (0, 1) and thus are rectified from above, while $\beta \in (0, \alpha)$ and as a result is rectified from below.

Remark. We do not rectify the distribution of β from above, because $\hat{\alpha} \approx 0.9507$.

To have the bigger picture, we can produce the summary of the model, which will include the table above:

```
summary(adamETSBJ, level=0.99)
```

```
##
## Model estimated using adam() function: ETS(AAdN)
## Response variable: BJsales
## Distribution used in the estimation: Normal
## Loss function type: likelihood; Loss function value: 241.078
## Coefficients:
##       Estimate Std. Error Lower 0.5% Upper 99.5%
## alpha   0.9507     0.1144     0.6517      1.0000 *
## beta    0.2911     0.1395     0.0000      0.6551
## phi     0.8632     0.1257     0.5349      1.0000 *
## level 202.7529     4.7209   190.4172    215.0740 *
```

```
## trend  -2.3996    3.5642   -11.7131     6.9028
##
## Error standard deviation: 1.384
## Sample size: 140
## Number of estimated parameters: 6
## Number of degrees of freedom: 134
## Information criteria:
##      AIC     AICc      BIC     BICc
## 494.1560 494.7876 511.8058 513.3664
```

The output above shows the estimates of parameters and their 99% confidence intervals. Based on this output, for example, we can conclude that the uncertainty about the initial trend estimate is large, and in the "true model", it could be either positive or negative (or even close to zero). At the same time, the "true" parameter of the initial level will lie in 99% of the cases between 190.4172 and 215.0740. Just as a reminder, Figure 16.2 shows the model fit and point forecasts for the estimated ETS model on this data.

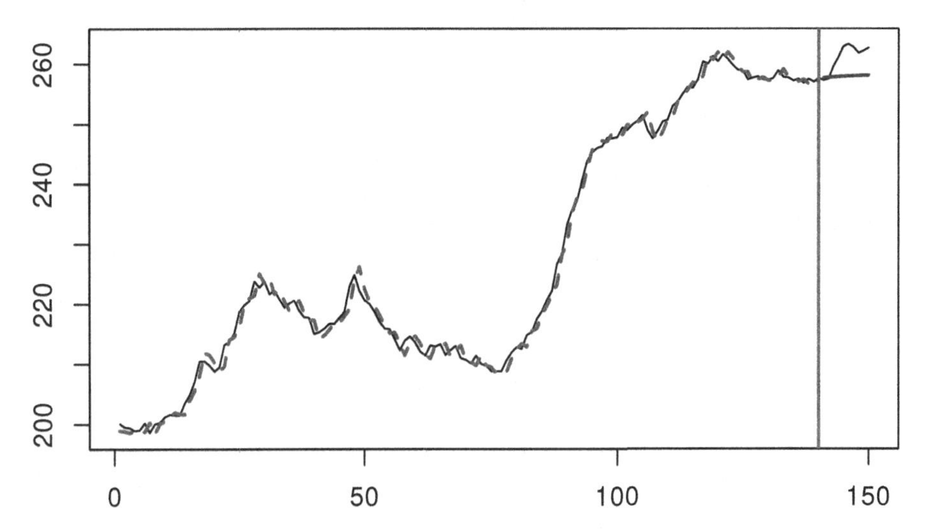

FIGURE 16.2 Model fit and point forecasts of ETS(A,Ad,N) on Box-Jenkins sales data.

As another example, we can have a similar summary for ARIMA models in ADAM:

```
adam(BJsales, "NNN", h=10, holdout=TRUE,
```

```
  order=list(ar=3,i=2,ma=3,select=TRUE)) |>
  summary()
```

```
##
## Model estimated using auto.adam() function: ARIMA(0,2,2)
## Response variable: BJsales
## Distribution used in the estimation: Normal
## Loss function type: likelihood; Loss function value: 243.2819
## Coefficients:
##              Estimate Std. Error Lower 2.5% Upper 97.5%
## theta1[1]     -0.7515     0.0830    -0.9156     -0.5875 *
## theta2[1]     -0.0109     0.0956    -0.1998      0.1780
## ARIMAState1 -200.1902     1.9521  -204.0508   -196.3321 *
## ARIMAState2 -200.2338     2.7554  -205.6830   -194.7879 *
##
## Error standard deviation: 1.4007
## Sample size: 140
## Number of estimated parameters: 5
## Number of degrees of freedom: 135
## Information criteria:
##      AIC     AICc      BIC     BICc
## 496.5638 497.0115 511.2720 512.3783
```

From the summary above, we can see that the parameter θ_2 is close to zero, and the interval around it is wide. So, we can expect that it might change the sign if the sample size increases or become even closer to zero. Given that the model above was estimated with the optimisation of initial states, we also see the values for the ARIMA states and their confidence intervals in the summary above. If we used `initial="backcasting"`, the summary would not include them.

Remark. If we faced difficulties estimating the covariance matrix of parameters using the standard Hessian-based approach, we could try bootstrap via `summary(adamETSBJ, bootstrap=TRUE)`.

This estimate of uncertainty via confidence intervals might also be helpful to see what can happen with the estimates of parameters if the sample size increases: will they change substantially or not? If they do, then the decisions made on Monday based on the available data might differ considerably from the decisions made on Tuesday. So, in the ideal world, we would want to have as narrow confidence intervals of parameters as possible.

16.4 Conditional variance with uncertain parameters

Now that we have discussed how the covariance matrix of parameters and confidence intervals for parameters can be generated in ADAM, we can move to the discussion of propagating the effect of uncertainty of parameters to the states, fitted values, and forecasts. I consider two special cases with pure additive state space models:

(1) When the values of the initial state vector are estimated;
(2) When the model parameters (e.g. smoothing or AR/MA parameters) are estimated.

I discuss analytical formulae for the conditional variance for these cases. This variance can then be used to construct the confidence interval of the fitted line and/or for the confidence/prediction interval for the holdout period. I do not cover the more realistic case when both initials and parameters are estimated because there is no closed analytical form for this due to potential correlations between the estimates of parameters. Furthermore, there are no closed forms for the conditional variance for the multiplicative and mixed models, which is why I focus my explanation on the pure additive ones only.

16.4.1 Estimated initial state

First, we need to recall the recursive relations discussed in Section 5.4, specifically formula (5.10). Just to simplify all the derivations in this section, we consider the non-seasonal case, in which all elements of $\boldsymbol{l}$ are equal to one. This can be ETS(A,N,N), ETS(A,A,N), ETS(A,Ad,N), or some ARIMA models.

Remark. The more general case is more complicated but is derivable using the same principles as discussed below.

The recursive relation from the first observation till some observation t can be written as:

$$\hat{\mathbf{v}}_t = \mathbf{D}^t \hat{\mathbf{v}}_0 + \sum_{j=0}^{t-1} \mathbf{D}^j \mathbf{g} y_{t-j}, \tag{16.3}$$

where $\mathbf{D} = \mathbf{F} - \mathbf{g}\mathbf{w}'$. The formula (16.3) shows that the most recent value of the state vector depends on the initial value $\hat{\mathbf{v}}_0$ and on the linear combination of actual values.

Remark. We assume in this part that the matrix $\mathbf{D}$ is known, i.e. the smoothing parameters are not estimated. Although this is an unrealistic assumption, it helps in showing how the variance of the initial state would influence the conditional variance of actual values at the end of the sample.

If we now take the variance of state vector conditional on the previous actual values y_{t-j} for all $j = \{0, \dots, t-1\}$, then we will have (due to the independence of two terms in (16.3)):

$$\mathrm{V}(\hat{\mathbf{v}}_t | y_1, y_2, \dots y_t) = \mathrm{V}\left(\mathbf{D}^t \hat{\mathbf{v}}_0\right) + \mathrm{V}\left(\sum_{j=0}^{t-1} \mathbf{D}^j y_{t-j} | y_1, y_2, \dots y_t \right). \tag{16.4}$$

We condition the variance on actual values in the formula above because they are given to us, and we want to see how different initial states would lead to the changes in the model fit given these values and thus how the uncertainty will propagate from $j = 1$ to $j = t$. In the formula (16.4), the right-hand side is equal to zero because all actual values are known, and $\mathbf{D}$ does not have any uncertainty due to the assumption above. This leads to the following covariance matrix of states on observation t:

$$\mathrm{V}(\hat{\mathbf{v}}_t | y_1, y_2, \dots y_t) = \mathbf{D}^t \mathrm{V}\left(\hat{\mathbf{v}}_0\right) \left(\mathbf{D}^t\right)'. \tag{16.5}$$

Inserting the values of matrix $\mathbf{D}$ in (16.5), we can then get the variance of the state vector on the observation t given the uncertainty of the initial state. For example, for ETS(A,N,N), the conditional variance of the level on observation t is:

$$\mathrm{V}(\hat{l}_t | y_1, y_2, \dots y_t) = (1-\alpha)^t \mathrm{V}\left(\hat{l}_0\right) (1-\alpha)^t. \tag{16.6}$$

As the formula above shows, if the smoothing parameter lies between zero and one, then the impact of the uncertainty of the initial level on the current one will be diminished with the increase of t. The closer α is to zero, the more impact the variance of the initial level will have on the variance of the current level. If we use admissible bounds (see Section 4.7), then the smoothing parameter might lie in the region (1, 2), and the impact of the variance of the initial state on the current one will be higher the closer α is to two.

Now that we have the variance of the state, we can also calculate the variance of the fitted values (or one step ahead in-sample forecast). In the pure additive model, the fitted values are calculated as:

$$\hat{y}_t = \mu_{y,t|t-1} = \mathbf{w}' \hat{\mathbf{v}}_{t-1}. \tag{16.7}$$

The variance of the fitted value conditional on all actual observations will then be:

$$\mathrm{V}(\hat{y}_t | y_1, y_2, \dots y_t) = \mathrm{V}\left(\mathbf{w}' \hat{\mathbf{v}}_{t-1}\right), \tag{16.8}$$

which after inserting (16.5) in (16.8) leads to:

$$\mathrm{V}(\hat{y}_t | y_1, y_2, \dots y_t) = \mathbf{w}' \mathbf{D}^{t-1} \mathrm{V}\left(\hat{\mathbf{v}}_0\right) \left(\mathbf{D}^{t-1}\right)' \mathbf{w}. \tag{16.9}$$

This variance can then be used to calculate the confidence interval for the fitted values, assuming that the estimates of the initial state follow a Normal distribution (due to CLT). In case of ETS(A,N,N), this equals to:

$$\mathrm{V}(\hat{y}_t|y_1, y_2, \dots y_t) = (1-\alpha)^{t-1}\mathrm{V}\left(\hat{l}_0\right)(1-\alpha)^{t-1}. \tag{16.10}$$

Finally, the variance of the initial states will also impact the conditional h steps ahead variance of the model. This can be seen from the recursion (5.20), which in the case of non-seasonal models simplifies to:

$$y_{t+h} = \mathbf{w}'\mathbf{F}^{h-1}\hat{\mathbf{v}}_t + \mathbf{w}'\sum_{j=1}^{h-1}\mathbf{F}^{j-1}\mathbf{g}e_{t+h-j} + e_{t+h}. \tag{16.11}$$

Taking the variance of y_{t+h} conditional on all the information until the observation t (all actual values) with $h > 1$ leads to:

$$\begin{aligned}\mathrm{V}(y_{t+h}|y_1, y_2, \dots y_t) =& \mathbf{w}'\mathbf{F}^{h-1}\mathbf{D}^{t-1}\mathrm{V}\left(\hat{\mathbf{v}}_0\right)\left(\mathbf{D}^{t-1}\right)'\left(\mathbf{F}'\right)^{h-1}\mathbf{w}+ \\ & \left(\left(\mathbf{w}'\sum_{j=1}^{h-1}\mathbf{F}^{j-1}\mathbf{g}\mathbf{g}'(\mathbf{F}')^{j-1}\mathbf{w}\right)+1\right)\sigma^2.\end{aligned} \tag{16.12}$$

This formula can then be used for the construction of prediction intervals of the model, for example using formula (5.13). The topic of construction of prediction intervals will be discussed later in Section 18.3. In the case of the ETS(A,N,N) model this simplifies to:

$$\begin{aligned}\mathrm{V}(y_{t+h}|y_1, y_2, \dots y_t) =& (1-\alpha)^{t-1}\mathrm{V}\left(\hat{l}_0\right)(1-\alpha)^{t-1}+ \\ & \left(1+(h-1)\alpha^2\right)\sigma^2.\end{aligned} \tag{16.13}$$

Remark. It is also possible to derive the variances for the seasonal models. The only thing that would change in comparison with the formulae above is that the matrices **F**, **w**, and **g** will need to be split into sub-matrices, similar to how it was done in Section 5.2.

16.4.2 Estimated parameters of ADAM

Now we discuss the case when the initial states are either known or not estimated directly. This, for example, corresponds to the situation with backcasted initials. Continuing our non-seasonal model example, we can use the following recursion (similar to (16.11)), keeping in mind that now the value of the initial state vector $\mathbf{v}_0$ is known:

$$\mathbf{v}_{t+h-1} = \hat{\mathbf{F}}^{h-1}\mathbf{v}_t + \sum_{j=1}^{h-1}\hat{\mathbf{F}}^{j-1}\hat{\mathbf{g}}e_{t+h-j}. \tag{16.14}$$

The conditional variance of the state, given the values on observation t in (16.14) in general does not have a closed-form because of the exponentiation of the transition matrix $\hat{\mathbf{F}}$. However, in a special case, when the matrix does not contain the parameters (e.g. non-damped trend ETS models or ARIMA without AR terms), there is an analytical solution to the variance. In this case, $\mathbf{F}$ is provided rather than being estimated, which simplifies the inference:

$$\mathrm{V}(\mathbf{v}_{t+h-1}|t) = \mathrm{V}\left(\sum_{j=1}^{h-1} \mathbf{F}^{j-1} \hat{\mathbf{g}} e_{t+h-j}\right) \tag{16.15}$$

The variance of the sum in (16.15) can be expanded as:

$$\begin{aligned} \mathrm{V}\left(\sum_{j=1}^{h-1} \mathbf{F}^{j-1} \hat{\mathbf{g}} e_{t+h-j}\right) = & \sum_{j=1}^{h-1} \mathrm{V}\left(\mathbf{F}^{j-1} \hat{\mathbf{g}} e_{t+h-j}\right) + \\ & 2 \sum_{j=2}^{h-1} \sum_{i=1}^{j-1} \mathrm{cov}(\mathbf{F}^{j-1} \hat{\mathbf{g}} e_{t+h-j}, \mathbf{F}^{i} \hat{\mathbf{g}} e_{t+h-i}). \end{aligned} \tag{16.16}$$

Each variance in the left-hand side of (16.16) can be expressed via:

$$\begin{aligned} \mathrm{V}\left(\mathbf{F}^{j-1} \hat{\mathbf{g}} e_{t+h-j}\right) = \mathbf{F}^{j-1} \big(& \mathrm{V}(\hat{\mathbf{g}}) \mathrm{V}(e_{t+h-j}) + \mathrm{V}(\hat{\mathbf{g}}) \mathrm{E}(e_{t+h-j})^2 + \\ & \mathrm{E}(\hat{\mathbf{g}}) \mathrm{E}(\hat{\mathbf{g}})' \mathrm{V}(e_{t+h-j}) \big) (\mathbf{F}^{j-1})'. \end{aligned} \tag{16.17}$$

Given that the expectation of the error term is assumed to be zero, and substituting $\mathrm{V}(e_{t+h-j}) = \sigma^2$ (assuming that the error term is homoscedastic), this simplifies to:

$$\mathrm{V}\left(\mathbf{F}^{j-1} \hat{\mathbf{g}} e_{t+h-j}\right) = \mathbf{F}^{j-1} \left(\mathrm{V}(\hat{\mathbf{g}}) + \mathrm{E}(\hat{\mathbf{g}}) \mathrm{E}(\hat{\mathbf{g}})'\right) (\mathbf{F}^{j})' \sigma^2. \tag{16.18}$$

As for the covariances in (16.16), after the expansion it can be shown that each of them is equal to:

$$\begin{aligned} \mathrm{cov}(\mathbf{F}^{j-1} \hat{\mathbf{g}} e_{t+h-j}, \mathbf{F}^{i} \hat{\mathbf{g}} e_{t+h-i}) = & \mathrm{V}(\mathbf{F}^{j-1} \hat{\mathbf{g}}) \mathrm{cov}(e_{t+h-i}, e_{t+h-j}) \\ & + \left(\mathbf{F}^{j-1} \hat{\mathbf{g}}\right)^2 \mathrm{cov}(e_{t+h-i}, e_{t+h-j}) \\ & + \mathrm{E}(e_{t+h-i}) \mathrm{E}(e_{t+h-j}) \mathrm{V}(\mathbf{F}^{j-1} \hat{\mathbf{g}}). \end{aligned} \tag{16.19}$$

Given the assumptions of the model, the autocovariances of error terms should all be equal to zero, and the expectation of the error term should be equal to zero as well, which means that all the value in (16.19) will be equal to zero as well. Based on this, the conditional variance of states equals to:

$$\mathrm{V}(\mathbf{v}_{t+h-1}|t) = \sum_{j=1}^{h-1} \mathbf{F}^{j-1} \left(\mathrm{V}(\hat{\mathbf{g}}) + \mathrm{E}(\hat{\mathbf{g}}) \mathrm{E}(\hat{\mathbf{g}})'\right) (\mathbf{F}^{j})' \sigma^2 \tag{16.20}$$

As discussed in Section 5.3, the conditional variance of the actual value h steps ahead is:

$$\mathrm{V}(y_{t+h}|t) = \mathbf{w}'\mathrm{V}(\mathbf{v}_{t+h-1}|t)\mathbf{w} + \sigma^2 \tag{16.21}$$

Inserting (16.20) in (16.21), we get the final conditional h steps ahead variance of the model:

$$\sigma^2_h = \mathrm{V}(y_{t+h}|t) = \left(\mathbf{w}' \sum_{j=1}^{h-1} \mathbf{F}^{j-1} \left(\mathrm{V}(\hat{\mathbf{g}}) + \mathrm{E}(\hat{\mathbf{g}})\mathrm{E}(\hat{\mathbf{g}})'\right) (\mathbf{F}^j)'\mathbf{w} + 1 \right) \sigma^2, \tag{16.22}$$

which looks similar to the formula (5.15) from Section 5.3, but now has the covariance of persistence vector in it. For a special case of ETS(A,N,N) this simplifies to:

$$\sigma^2_h = \mathrm{V}(y_{t+h}|t) = \left((h-1)\left(\mathrm{V}(\hat{\alpha}) + \hat{\alpha}^2\right) + 1\right) \sigma^2, \tag{16.23}$$

which as can be seen differs from the conventional variance by the value of the variance of the smoothing parameter $\mathrm{V}(\hat{\alpha})$. Similarly, the conditional variances for ETS(A,A,N), ETS(A,N,A), and ETS(A,A,A) can be produced using the formula (16.22).

Unfortunately, the conditional variances for the other models are more complicated due to the introduction of convolutions of parameters. Furthermore, the formula (16.22) only focuses on the conditional variance given the known $\mathbf{v}_t$ but does not take into account the uncertainty of it for the fitted values in-sample. Given the complexity of the problem, in the next section, we introduce a technique that allows correctly propagating the uncertainty of parameters and initial values to the forecasts of any ADAM.

16.5 Multi-scenarios for ADAM states

As discussed in Section 16.4, it is difficult to capture the impact of the uncertainty about the parameters on the states of the model and, as a result, difficult to take it into account on the forecasting stage. Furthermore, so far, we have only discussed pure additive models, for which it is at least possible to do some derivations. When it comes to models with multiplicative components, it becomes nearly impossible to demonstrate how the uncertainty propagates over time. To overcome these limitations, we developed (Svetunkov & Pritularga, 2023a) a simulation-based approach (similar to the one discussed in Section 16.1) that relies on the selected model form.

The idea of the approach is to get the covariance matrix of the parameters of the selected model (see Section 16.2) and then generate n sets of parameters randomly from a Rectified Multivariate Normal distribution using the matrix and the values of estimated parameters. After that, the model is applied to the data with each generated parameters combination to get the states, fitted values, and residuals (simulation is done using the principles from Section 16.1). This way, we propagate the uncertainty about the parameters from the first observation to the last one. The final states can then be used to produce point forecasts and prediction intervals based on each set of parameters. These scenarios allow creating more adequate prediction intervals from the model and/or confidence intervals for the fitted values, states, and conditional expectations. All of this is done without any additional model assumptions and without any additional modelling steps, relying entirely on the estimated ADAM. This approach is computationally expensive, as it requires fitting all the n models to the data, however no estimation is needed. Furthermore, if the uncertainty about the model needs to be taken into account, then the combination of models can be used, as described in Section 15.4, where the approach from this section would be applied for each of the models in the pool before combining the point forecasts and prediction intervals with information criteria weights.

The `smooth` package has the method `reapply()` that implements this approach for `adam()` models. This works with ADAM ETS, ARIMA, regression, and any combination of the three. Here is an example in R with $n = 1000$:

```
# Estimate the model
adamETSAir <- adam(AirPassengers, "MMM", h=10, holdout=TRUE,
                   maxeval=16*100)
# Produce the multiple scenarios
adamETSAirReapply <- reapply(adamETSAir, nsim=1000)
```

Remark. In the code above I have increased the number of iterations in the optimiser to $k \times 100$, because I noticed that the default value does not allow reaching the maximum of the likelihood, and as a result the variances of parameters become too large. At the moment, there is no automated solution to this problem and $k \times 100$ is a heuristic that can be used if a more precise optimum is required.

After producing the scenarios, we can plot them (see Figure 16.3).

```
plot(adamETSAirReapply)
```

Figure 16.3 demonstrates how the approach works on the example of `AirPassengers` data with an ETS(M,M,M) model. The grey areas around the fitted line show quantiles from the fitted values, forming confidence intervals of the width 95%, 80%, 60%, 40%, and 20%. They show how the fitted value would vary if the parameters would differ from the estimated ones. Notice

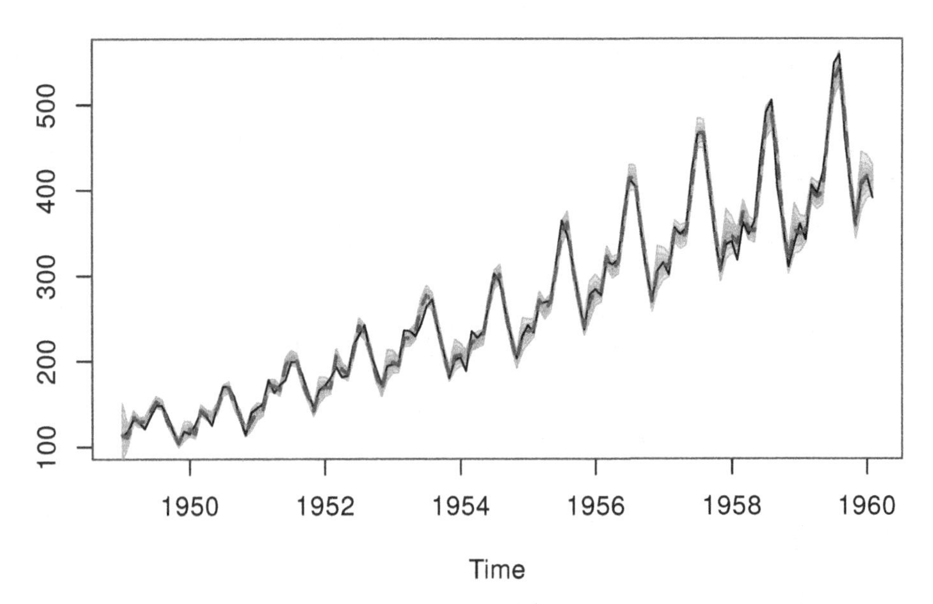

FIGURE 16.3 Refitted ADAM ETS(M,M,M) model on AirPassengers data.

that there was a warning about the covariance matrix of parameters, which typically appears if the optimal value of the loss function was not reached. If this happens, I would recommend tuning the optimiser (see Section 11.4).

The `adamETSAirReapply` object contains several variables, including:

- `adamETSAirReapply$states` – the array of states of dimensions $k \times (T + m) \times n$, where m is the maximum lag of the model, k is the number of components, and T is the sample size;
- `adamETSAirReapply$refitted` – fitted values produced from different parameters, dimensions $T \times n$;
- `adamETSAirReapply$transition` – the array of transition matrices of the size $k \times k \times n$;
- `adamETSAirReapply$measurement` – the array of measurement matrices of the size $(T + m) \times k \times n$;
- `adamETSAirReapply$persistence` – the persistence matrix of the size $k \times n$;

The last three will contain the random parameters (smoothing, dampening, and AR/MA parameters), which is why they are provided together with the other values.

As mentioned earlier, ADAM ARIMA also supports this approach. Here is an example on an artificial, non-seasonal data, generated from ARIMA(0,1,1) (see Figure 16.4):

```
# Generate the data
y <- sim.ssarima(orders=list(i=1,ma=1), obs=120, MA=-0.7)
# Apply ADAM, then refit it and plot
adam(y$data, "NNN", h=10, holdout=TRUE,
     orders=c(0,1,1)) |>
    reapply() |>
    plot()
```

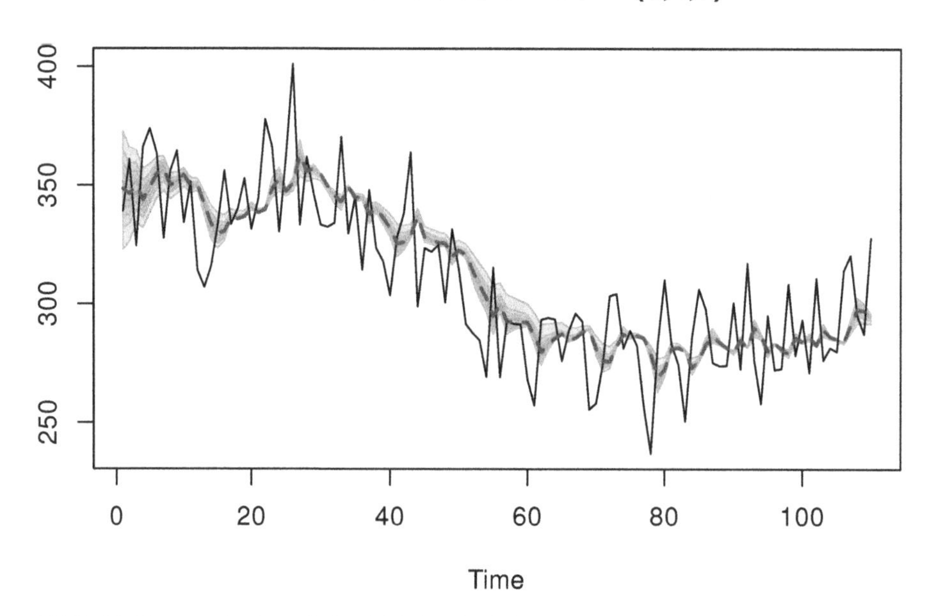

FIGURE 16.4 Refitted ADAM ARIMA(0,1,1) model on artificial data.

Note that the more complicated the fitted model is, the more difficult it is to optimise, and thus the more difficult it is to get accurate estimates of the covariance matrix of parameters. This might result in highly uncertain states and thus fitted values. The safer approach, in this case, is using bootstrap for the estimation of the covariance matrix, but this is more computationally expensive and would only work on longer time series. Here is how bootstrap can be used for the multi-scenarios in R (and Figure 16.5):

```
adam(y$data, "NNN", h=10, holdout=TRUE,
     orders=c(0,1,1)) |>
```

```
reapply(bootstrap=TRUE, parallel=TRUE) |>
  plot()
```

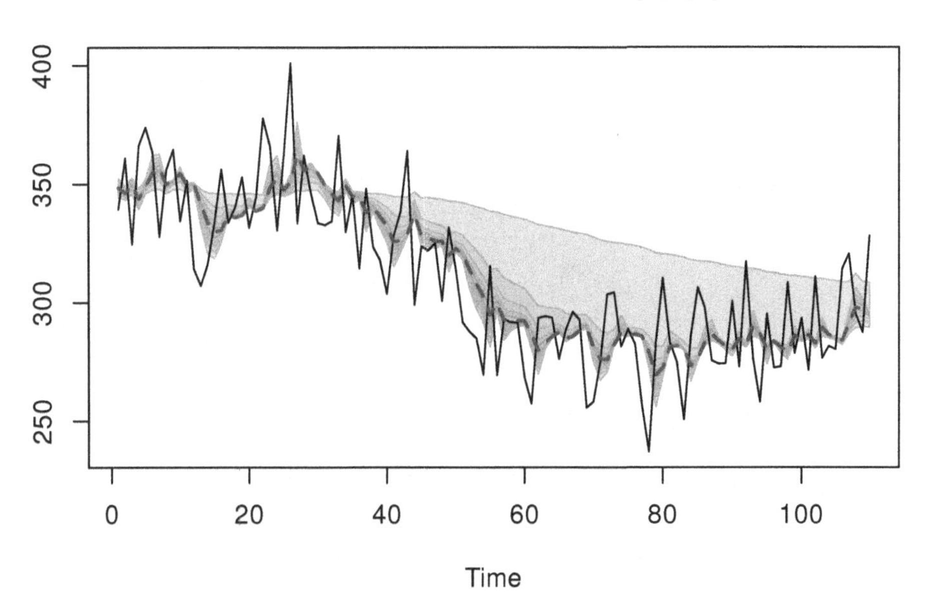

FIGURE 16.5 Refitted ADAM ARIMA(0,1,1) model on artificial data, bootstrapped covariance matrix.

The approach described in this section is still a work in progress. While it works in theory, there are computational difficulties with calculating the Hessian matrix in some situations. If the covariance matrix is not estimated accurately, it might contain high variances, leading to higher than needed uncertainty of the model. This might result in unreasonable confidence bounds and lead to extremely wide prediction intervals.

17

Scale model for ADAM

Up until this chapter, we have focused our discussion on modelling the location of a distribution (e.g. conditional mean, point forecasts), neglecting the fact that in some situations, the variance of a model might exhibit some time-varying patterns. In statistics, the effect of a non-constant variance is called "heteroscedasticity" (see discussion in Section 14.6). It implies that the variance of the residuals of a model changes either over time or under influence of some variables. In some cases, in order to capture this effect, we might use the multiplicative model – it takes care of heteroscedasticity caused by changing the level of data if the variance is proportional to its value. But there might be some situations where variance changes due to some external factors, not necessarily available to the analyst. In this situation, it should be captured separately using a different model. Hereafter the original ADAM that is used for producing of conditional mean will be called the **location model**, while the model for the variance will be called the **scale model**.

In this chapter, we discuss the scale model for ADAM ETS/ARIMA/Regression, the model that allows capturing time-varying variance and using it for forecasting. We discuss how this model is formulated, how it can be estimated, and then move to the discussion of its relation to such models as ARCH and GARCH. This chapter is inspired by the GAMLSS, which models the scale of distribution using functions of explanatory variables. We build upon that by introducing the dynamic element.

17.1 Model formulation

In order to understand better how the more general ADAM scale model works, we start the discussion with an example of a pure additive ETS model with Normal distribution for location and an ETS model for the scale.

17.1.1 An example with local level model with Normal distribution

Consider an ETS(A,N,N) model (which was discussed in Section 4.3), which has the following measurement equation:

$$y_t = l_{t-1} + \epsilon_t, \tag{17.1}$$

where the most commonly used assumption for the error term is:

$$\epsilon_t \sim \mathcal{N}(0, \sigma^2).$$

The same error term can be represented as a multiplication of the Standard Normal variable by the standard deviation:

$$\epsilon_t = \sigma \eta_t, \tag{17.2}$$

where $\eta_t \sim \mathcal{N}(0, 1)$. Now consider the situation when instead of the constant variance σ^2 we have one that changes over time either because of its own dynamics or because of the influence of explanatory variables. In that case we will add the subscript t to the standard deviation in (17.2) to have:

$$\epsilon_t = \sigma_t \eta_t. \tag{17.3}$$

The thing to keep in mind is that in the case of Normal distribution, the scale is equal to variance rather than to the standard deviation, so in the modelling, we should consider σ_t^2 rather than σ_t. The variance, in this case, can be modelled explicitly using any ADAM. However, the pure multiplicative ones make more sense because they guarantee that the variance will not become zero or even negative. For demonstration purposes, we use ETS(M,N,N) to model the variance:

$$\begin{aligned} & \epsilon_t^2 = \sigma_t^2 \eta_t^2 \\ & \sigma_t^2 = l_{\sigma,t-1} \\ & l_{\sigma,t} = l_{\sigma,t-1} \left(1 + \alpha_\sigma (\eta_t^2 - 1)\right) \end{aligned}, \tag{17.4}$$

Note that although the part $\left(1 + \alpha_\sigma (\eta_t^2 - 1)\right)$ looks slightly different than the respective part $(1 + \alpha \epsilon_t)$ in the conventional ETS(M,N,N), they are equivalent: if we substitute $\xi_t = \eta_t^2 - 1$ in (17.4), we will arrive at the conventional ETS(M,N,N) model. Furthermore, because η_t follows Standard Normal distribution, its square will follow Chi-squared distribution: $\eta_t^2 \sim \chi^2(1)$. This can be used for model diagnostics. Finally, we can use all the properties of the pure multiplicative models discussed in Chapter 6 to get the fitted values and forecasts from the model (17.4):

$$\begin{aligned} & \sigma_{t|t-1}^2 = l_{\sigma,t-1} \\ & \sigma_{t+h|t}^2 = l_{\sigma,t} \end{aligned}. \tag{17.5}$$

In order to construct this model, we need to collect the residuals e_t of the location model (17.1), square them, and use them in the following system of equations:

$$\begin{aligned} \hat{\sigma}_t^2 &= \hat{l}_{\sigma,t-1} \\ \hat{\eta}_t^2 &= \frac{e_t^2}{\hat{\sigma}_t^2} \\ \hat{l}_{\sigma,t} &= \hat{l}_{\sigma,t-1}(1 + \hat{\alpha}_\sigma(\hat{\eta}_t^2 - 1)) \end{aligned} . \tag{17.6}$$

In order for this to work, we need to estimate $\hat{l}_{\sigma,0}$ and $\hat{\alpha}_\sigma$, which can be done in the conventional way by maximising the log-likelihood function of the Normal distribution (see Section 11.1) – the only thing that will change in comparison with the conventional estimation of the location model is the fitted values of $\hat{\sigma}_t^2$ for the variance, which will need to be generated from (17.6):

$$\ell(\boldsymbol{\theta}, \sigma_t^2 | \mathbf{y}) = -\frac{T}{2} \log(2 \pi \sigma_t^2) - \frac{1}{2} \sum_{t=1}^T \frac{\epsilon_t^2}{\sigma_t^2} . \tag{17.7}$$

The thing to keep in mind is that the final ETS(A,N,N) model (17.1) with the scale model (17.4) will have four parameters instead of three (as it would in case of a simpler model): two for the location part of the model (the initial level l_0 and the smoothing parameter α) and two for the scale part of the model (the initial level $l_{\sigma,0}$ and the smoothing parameter α_σ).

Remark. The estimation of the scale model can be done after the estimation of the location one: given that the scale and location of distribution are assumed to be independent, the likelihood can be maximised first by estimating the parameters of the model (17.1) and then (17.4). This also means that identifying whether the scale model is needed at all and if yes, what components and variables it should have, can be done automatically via information criteria.

As can be seen from this example, constructing and estimating the scale model for ADAM is not a difficult task. Following similar principles, we can use any other pure multiplicative model for scale, including ETS(Y,Y,Y) (Chapter 6), log ARIMA (Section 9.1.4), or a multiplicative regression. Furthermore, the location model ETS(A,N,N) can be substituted by any other ETS/ARIMA/Regression model – the same principles as discussed in this Subsection can be applied equally effectively to them. The only restriction in all of this is that both parts of the model should be estimated via maximisation of likelihood – other methods typically do not explicitly estimate the scale of distribution.

17.1.2 General case with Normal distribution

More generally speaking, a scale model can be created for any ETS/ARIMA/Regression location model. Following from the discussion in Section 7.1, I remind here that the general location model can be written as:

$$\begin{aligned} y_t &= w(\mathbf{v}_{t-\boldsymbol{l}}) + r(\mathbf{v}_{t-\boldsymbol{l}})\epsilon_t \\ \mathbf{v}_t &= f(\mathbf{v}_{t-\boldsymbol{l}}) + g(\mathbf{v}_{t-\boldsymbol{l}})\epsilon_t \end{aligned}$$

If we assume that $\epsilon_t \sim \mathcal{N}(0, \sigma_t^2)$, then the general scale model with ETS+ARIMA+Regression elements can be formulated as (Section 6.1, Subsection 9.1.4 and Section 10.3):

$$\begin{aligned} \epsilon_t^2 &= \sigma_t^2 \eta_t^2 \\ \sigma_t^2 &= \exp\left(\mathbf{w}'_E \log(\mathbf{v}_{E,\sigma,t-\boldsymbol{l}_E}) + \mathbf{w}'_A \log(\mathbf{v}_{A,\sigma,t-\boldsymbol{l}_A}) + \mathbf{a}_t \mathbf{x}_t\right) \\ \log \mathbf{v}_{E,\sigma,t} &= \mathbf{F}_{E,\sigma} \log \mathbf{v}_{E,\sigma,t-\boldsymbol{l}_E} + \log(\mathbf{1}_k + \mathbf{g}_{E\sigma}(\eta_t^2 - 1)) \\ \log \mathbf{v}_{A,\sigma,t} &= \mathbf{F}_{A,\sigma} \log \mathbf{v}_{A,\sigma,t-\boldsymbol{l}_A} + \mathbf{g}_{A,\sigma} \log(\eta_t^2) \\ \mathbf{a}_t &= \mathbf{a}_{t-1} + \mathbf{z}_t \mathbf{g}_{R,\sigma} \log(\eta_t^2) \end{aligned} , \quad (17.8)$$

where subscripts E, A, and R denote components of ETS, ARIMA, and Regression respectively. The main thing that unites all the parts of the model is that they should be pure multiplicative in order to avoid potential issues with negative numbers. The construction and estimation of the scale model in this case becomes similar to the one discussed for the ETS(A,N,N) example above. When it comes to the forecasting of the conditional h steps ahead scale, given the limitations of the pure multiplicative model discussed in Section 6.3, it needs to be obtained via simulations – this way the forecast from the ADAM will coincide with the expectation, which in the case of (17.8) will give the conditional h steps ahead scale. All the principles discussed in Sections 6.1, 9.1.4 and 10.3 can be used directly for the scale model without any limitations, and the estimation of the model can be done via maximisation of likelihood as shown in the example above (equation (17.7)).

Finally, ADAM can be expanded even further by introducing the occurrence part of the model (i.e. dealing with the time-varying scale of distribution in case of intermittent demand, discussed in Chapter 13). This part will need to be introduced in the location model, while the scale model (17.8) can be used as-is in this case, applying it to the non-zero observations.

17.1.3 Other distributions

The examples above focused on the Normal distribution, but ADAM supports other distributions as well. Depending on the error term, these are:

1. Additive error term (Section 5.5):
    a. Normal: $\epsilon_t \sim \mathcal{N}(0, \sigma_t^2)$;
    b. Laplace: $\epsilon_t \sim \mathcal{Laplace}(0, s_t)$;
    c. S: $\epsilon_t \sim \mathcal{S}(0, s_t)$;
    d. Generalised Normal: $\epsilon_t \sim \mathcal{GN}(0, s_t, \beta)$;

   e. Log Normal: $\left(1 + \frac{\epsilon_t}{\mu_{y,t}}\right) \sim \log\mathcal{N}\left(-\frac{\sigma_t^2}{2}, \sigma_t^2\right)$;
   f. Inverse Gaussian: $\left(1 + \frac{\epsilon_t}{\mu_{y,t}}\right) \sim \mathcal{IG}(1, \sigma_t^2)$;
   g. Gamma: $\left(1 + \frac{\epsilon_t}{\mu_{y,t}}\right) \sim -(\sigma_t^{-2}, \sigma_t^2)$;
2. Multiplicative error term (Sections 6.5):
   a. Normal: $\epsilon_t \sim \mathcal{N}(0, \sigma_t^2)$;
   b. Laplace: $\epsilon_t \sim \mathcal{L}\dashv\sqrt{\updownarrow}\dashv\rfloor\rceil(0, s_t)$;
   c. S: $\epsilon_t \sim \mathcal{S}(0, s_t)$;
   d. Generalised Normal: $\epsilon_t \sim \mathcal{GN}(0, s_t, \beta)$;
   e. Log Normal: $(1 + \epsilon_t) \sim \log\mathcal{N}\left(-\frac{\sigma_t^2}{2}, \sigma_t^2\right)$;
   f. Inverse Gaussian: $(1 + \epsilon_t) \sim \mathcal{IG}(1, \sigma_t^2)$;
   g. Gamma: $(1 + \epsilon_t) \sim \Gamma(\sigma^{-2}, \sigma_t^2)$.

The error terms in these cases can also be presented in a form similar to (17.3) to get the first equation in (17.8) for the respective distributions:

1. Additive error term:
   a. Normal: $\epsilon_t^2 = \sigma_t^2 \eta_t^2$, where $\eta_t \sim \mathcal{N}(0, 1)$ or accidentally $\eta_t^2 \sim \chi^2(1)$;
   b. Laplace: $|\epsilon_t| = s_t|\eta_t|$, where $\eta_t \sim \mathcal{L}\dashv\sqrt{\updownarrow}\dashv\rfloor\rceil(0, 1)$;
   c. S: $0.5|\epsilon_t|^{0.5} = s_t|\eta_t|^{0.5}$, where $\eta_t \sim \mathcal{S}(0, 1)$;
   d. Generalised Normal: $\beta|\epsilon_t|^\beta = s_t|\eta_t|^\beta$, where $\eta_t \sim \mathcal{GN}(0, 1, \beta)$;
   e. Log Normal: $\log\left(1 + \frac{\epsilon_t}{\mu_{y,t}}\right) = \sigma_t\eta_t - \frac{\sigma_t^2}{2}$, where $\eta_t \sim \mathcal{N}(0, 1)$;
   f. Inverse Gaussian: $\frac{\left(\frac{\epsilon_t}{\mu_{y,t}}\right)^2}{\left(1+\frac{\epsilon_t}{\mu_{y,t}}\right)} = \sigma_t^2\eta_t^2$, where $\eta_t^2 \sim \chi^2(1)$;
   g. Gamma: $\left(1 + \frac{\epsilon_t}{\mu_{y,t}}\right) = \sigma_t^2\eta_t$, so that $\eta_t \sim -(\sigma_t^{-2}, 1)$;
2. Multiplicative error term (Sections 6.5):
   a. Normal: $\epsilon_t^2 = \sigma_t^2\eta_t^2$, where $\eta_t \sim \mathcal{N}(0, 1)$;
   b. Laplace: $|\epsilon_t| = s_t|\eta_t|$, where $\eta_t \sim \mathcal{L}\dashv\sqrt{\updownarrow}\dashv\rfloor\rceil(0, 1)$;
   c. S: $0.5|\epsilon_t|^{0.5} = s_t|\eta_t|^{0.5}$, where $\eta_t \sim \mathcal{S}(0, 1)$;
   d. Generalised Normal: $\beta|\epsilon_t|^\beta = s_t|\eta_t|^\beta$, where $\eta_t \sim \mathcal{GN}(0, 1, \beta)$;
   e. Log Normal: $\log(1 + \epsilon_t) = \sigma_t\eta_t - \frac{\sigma_t^2}{2}$, where $\eta_t \sim \mathcal{N}(0, 1)$;
   f. Inverse Gaussian: $\frac{\epsilon_t^2}{(1+\epsilon_t)} = \sigma_t^2\eta_t^2$, where $\eta_t^2 \sim \chi^2(1)$;
   g. Gamma: $(1 + \epsilon_t) = \sigma_t^2\eta_t$, so that $\eta_t \sim -(\sigma_t^{-2}, 1)$.

Remark.

- The relations between ϵ_t and η_t in $\mathcal{S}$ and $\mathcal{GN}$ introduce constants 0.5 and β, arising because of how the scales in those distributions are estimated (see Section 11.1);
- The relation between ϵ_t and η_t in Log-Normal distribution is complicated

because for the latter to be Standard Normal, the former needs to be transformed according to the formulae above;

- In case of Inverse Gaussian, transformations shown above are required to make the η_t independent of the scale parameter;
- Finally, in case of Gamma distribution, η_t cannot be made independent of the scale parameter, which makes it restrictive and less useful than other distributions.

The equations above can be used instead of the first equation in (17.8) to create and estimate the scale model for the chosen distribution: the other four equations in (17.8) will be exactly the same, substituting η_t^2 with the respective η_t, $|\eta_t|^{0.5}$ and η_t^β – depending on the distribution. The estimation of the respective models can be done via the maximisation of the respective likelihood functions (as discussed in Section 11.1).

The diagnostics of the scale model can be done in the same way as discussed in Chapter 14, keeping in mind the distributional assumptions about the η_t variable rather than ϵ_t. The `smooth` package already implements all the necessary diagnostic plots, similar to the ones discussed for the location model in Chapter 14.

Finally, the model selection for the scale part can be done using the same principles as discussed in Chapter 15. For example, one can select the most suitable ETS model similar to how it was discussed in Section 15.1, or the most suitable ARIMA, as in Section 15.2, or a set of explanatory variables, based on the approach in Section 15.3. All of this is available out of the box if the scale model is estimated via likelihood maximisation.

17.2 Connection with ARCH and GARCH

The scale model is not a new invention. The literature knows models that focus on modelling the dynamics of the scale of distribution or, more specifically, the second moment (variance). Engle (1982) proposed an Autoregressive Conditional Heteroscedasticity (ARCH) model to capture the time-varying variance using lagged values of the squared error term. Bollerslev (1986) expanded the idea by also including lagged values of variance, introducing a Generalised ARCH (GARCH), which can be formulated as:

$$\sigma_t^2 = a_0 + a_1 \sigma_{t-1}^2 + \dots + a_q \sigma_{t-q}^2 + b_1 \epsilon_{t-1}^2 + \dots + b_p \epsilon_{t-p}^2, \tag{17.9}$$

where $\epsilon_t \sim \mathcal{N}(0, \sigma_t^2)$ and a_j and b_j are the parameters of the model. Bollerslev (1986) argues that GARCH, being equivalent to ARMA, will be a stationary process if its parameters are restricted so that $\sum_{j=1}^p a_j + \sum_{j=1}^q b_j < 0$ and

$a_j, b_j \in [0, 1)$ for all j. The restriction on parameters guarantees that the resulting values of σ_t^2 are positive, but as Pantula (1986) noted such parameter space might be too restrictive. To make sure that the predicted variance is always positive, Geweke (1986) and Pantula (1986) suggested to build GARCH in logarithms, leading to the log-GARCH model:

$$\log \sigma_t^2 = a_0 + a_1 \log \sigma_{t-1}^2 + \dots + a_q \log \sigma_{t-q}^2 + b_1 \log \epsilon_{t-1}^2 + \dots + b_p \log \epsilon_{t-p}^2. \tag{17.10}$$

Pantula (1986) pointed out that the model (17.10) is equivalent to ARMA(p,q) applied to logarithms of squared error ϵ_t^2. In our notations it can be written as:

$$\log \epsilon_t^2 = a_0 + b_1' \log \epsilon_{t-1}^2 + \dots + b_p' \log \epsilon_{t-p}^2 + a_1' \log \eta_{t-1}^2 + \dots + a_q' \log \eta_{t-q}^2 + \log \eta_t^2, \tag{17.11}$$

which can be obtained by substituting $\log \sigma_{t-j}^2 = \log \epsilon_{t-j}^2 - \log \eta_{t-j}^2$ for all j. In this case $b_j' = b_j + a_j$ and $a_j' = -a_j$ for all j. Given the connection of log-ARMA with log-GARCH and the discussion in Subsection 9.1.4, the model (17.11) is just a special case of the scale model for ADAM, being a special case of (17.8). ADAM Scale Model not only supports the ARMA elements, but it also allows for explicit time series components modelling in the variance and the usage of explanatory variables.

Remark. GARCH and log-GARCH typically assume that the error term follows Normal distribution: $\epsilon_t \sim \mathcal{N}(0, \sigma_t^2)$ – and model the variance. At the same time, ADAM Scale Model works with several other distributions, discussed in Subsection 17.1.3 and is focused on modelling scale (variance is just a special case of it).

Furthermore, it can be shown that ETS(M,N,N) applied to the ϵ_t^2 can be considered as a special case of GARCH(1,1) with the restriction on parameter $a_1 = 1 - b_1$, $b_1 = \alpha$ and $a_0 = 0$, because the latter becomes equivalent to SES (Geweke, 1986):

$$\sigma_t^2 = (1 - \alpha)\sigma_{t-1}^2 + \alpha \epsilon_{t-1}^2. \tag{17.12}$$

The connection of SES and ETS(M,N,N) has been discussed in Subsection 4.3.2.

17.3 Examples in R

17.3.1 Example 1

To demonstrate how the scale model works, we will use the model that we constructed in Section 10.6 (we use Normal distribution for simplicity):

```
adamLocationSeat <- adam(SeatbeltsData, "MNM", h=12, holdout=TRUE,
                         formula=drivers~log(kms)+log(PetrolPrice)+law,
                         distribution="dnorm")
```

To see if the scale model is needed in this case, we will produce several diagnostics plots (Figure 17.1).

```
par(mfcol=c(2,1), mar=c(4,4,2,1))
plot(adamLocationSeat, which=c(4,8))
```

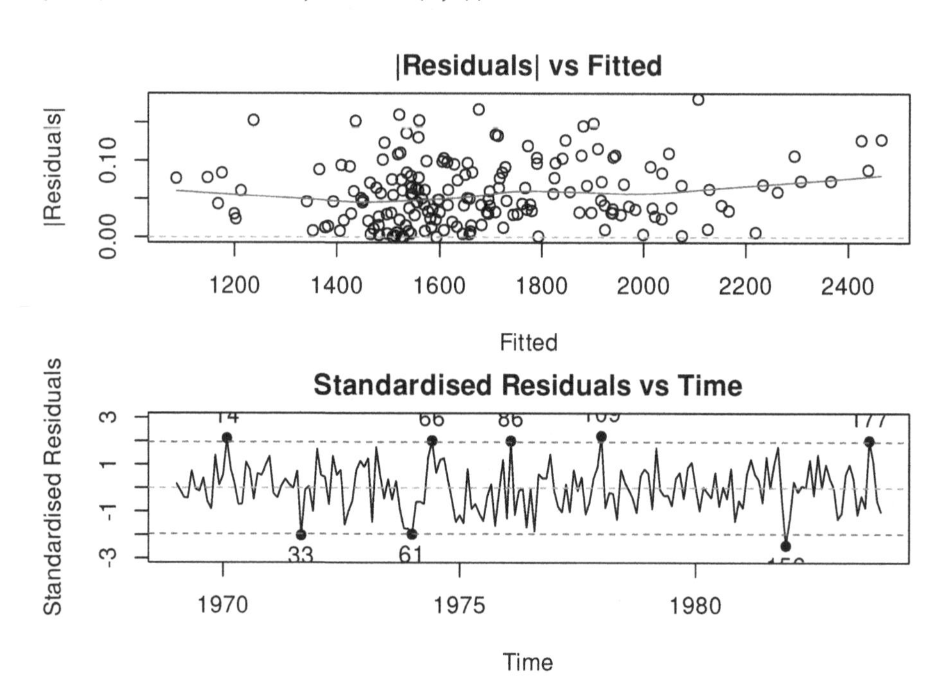

FIGURE 17.1 Diagnostic plots for the ETSX(M,N,M) model.

Figure 17.1 shows how the residuals change with: (1) change of fitted values, (2) change of time. The first plot shows that there might be a slight change in the variability of residuals, but it is not apparent. The second plot does not demonstrate any significant changes in variance over time. To check the latter point more formally, we can also produce ACF and PACF of the squared residuals, thus trying to see if the GARCH model is required:

```
par(mfcol=c(2,1), mar=c(4,4,2,1))
plot(adamLocationSeat, which=c(15,16))
```

In the plot of Figure 17.2, lag 8 is significant on both ACF and PACF and

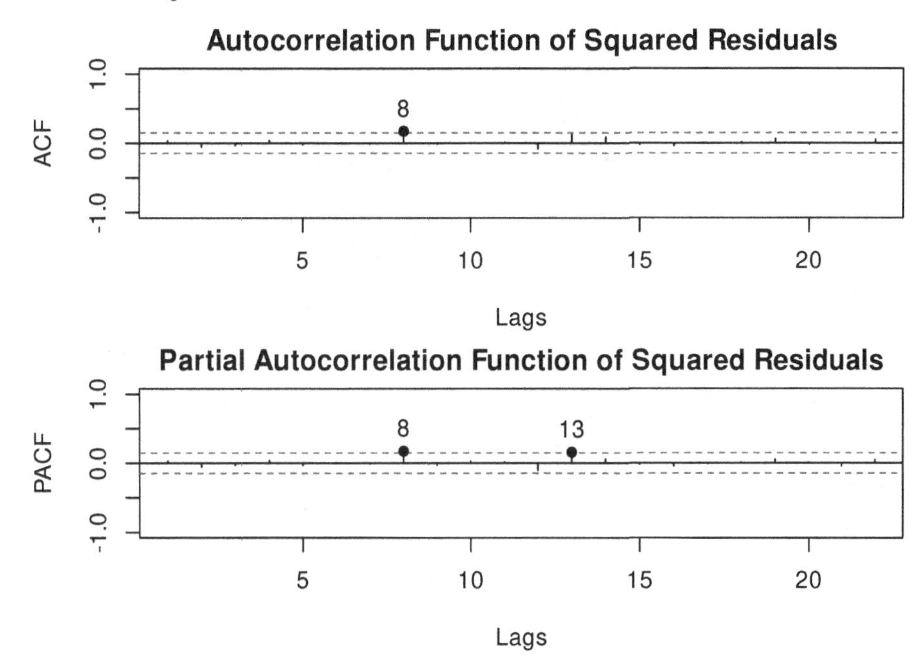

FIGURE 17.2 Diagnostic plots for the ETSX(M,N,M) model, continued.

lag 13 is significant on PACF. We could introduce the dynamic elements in the scale model to remove this autocorrelation, however the lags 8 and 13 do not have any meaning from the problem point of view: it is difficult to motivate how a variance 8 months ago might impact the variance this month. Furthermore, the bounds in the plot have a 95% confidence level, which means that these coefficients could have become significant by chance. So, based on this simple analysis, we can conclude that there might be some factors impacting the variance of residuals, but it does not exhibit any obvious and meaningful GARCH-style dynamics. To investigate this further, we can plot explanatory variables vs the squared residuals (Figure 17.3).

```
spread(cbind(residSq=as.vector(resid(adamLocationSeat)^2),
             as.data.frame(SeatbeltsData)[1:180,-1]),
       lowess=TRUE)
```

One of the potentially alarming features in the plot in Figure 17.3 is the slight change in variance for the variable `law`. The second one is a slight increase in the LOWESS line for squared residuals vs `PetrolPrice`. The `kms` variable does not demonstrate an apparent impact on the squared residuals. Based on this, we will investigate the potential effect of the law and petrol price on the scale of the model. We do that in an automated way, using the principles discussed in Section 15.3 via `regressors="select"` command:

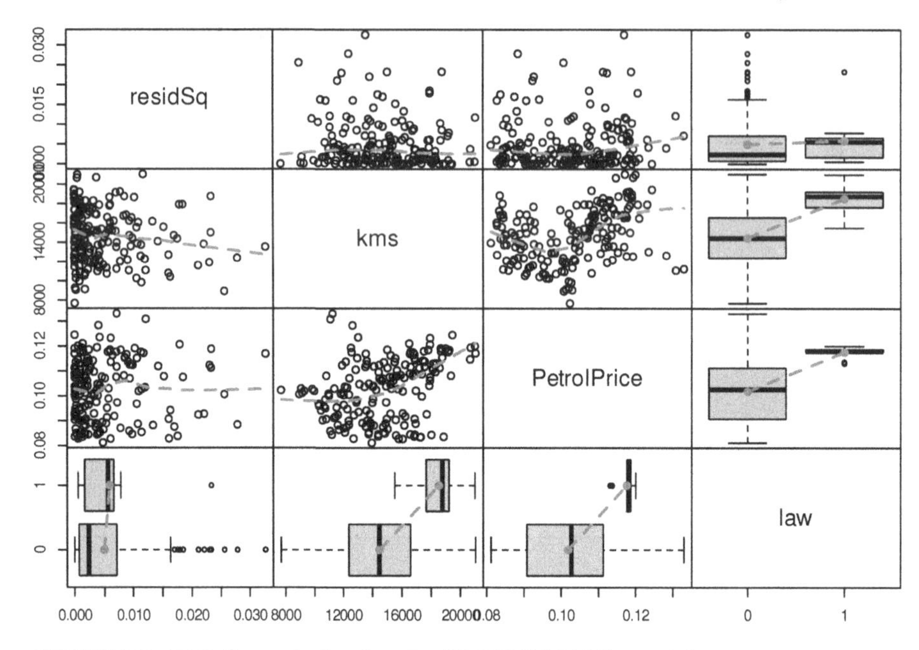

FIGURE 17.3 Spread plot for the ETSX(M,N,M) model.

```
adamScaleSeat <- sm(adamLocationSeat, model="NNN",
                    formula=drivers~law+PetrolPrice,
                    regressors="select")
```

```
## Warning: This type of model can only be applied to the data in
    logarithms.
## Amending the data
```

In the code above, we fit a scale model to the already estimated location model ETSX(M,N,M). We switch off the ETS element in the scale model and introduce the explanatory variables `law` and `PetrolPrice`, asking the function to select the most appropriate one based on AICc. The function provides a warning that it will create a model in logarithms, which is what we want anyway. But looking at the output of the scale model, we notice that none of the variables have been selected by the function:

```
summary(adamScaleSeat)
```

```
##
## Model estimated using sm.adam() function: Regression in logs
## Response variable: drivers
## Distribution used in the estimation: Normal
```

```
## Loss function type: likelihood; Loss function value: -418.9007
## Coefficients:
##             Estimate Std. Error Lower 2.5% Upper 97.5%
## (Intercept)  -6.5769     0.1416    -6.8563     -6.2974 *
##
## Error standard deviation: 1.9
## Sample size: 180
## Number of estimated parameters: 1
## Number of degrees of freedom: 179
## Information criteria:
##       AIC      AICc       BIC      BICc
## -835.8014 -835.7790 -832.6085 -832.5501
```

This means that the explanatory variables do not add value to the fit and should not be included in the scale model. As we can see, selection of variables in the scale model for ADAM can be done automatically without any additional external steps.

17.3.2 Example 2

The example in the previous subsection shows how the analysis can be done for the scale model when explanatory variables are considered. To see how the scale model can be used in forecasting, we will consider another example, now with dynamic elements. For this example, I will use `AirPassengers` data and an automatically selected ETS model with Inverse Gaussian distribution (for demonstration purposes):

```
adamLocationAir <- adam(AirPassengers, h=10, holdout=TRUE,
                        distribution="dinvgauss")
```

The diagnostics of the model indicate that a time-varying scale might be required in this situation (see Figure 17.4) because the variance of the residuals seems to decrease in the period from 1954 – 1958.

```
plot(adamLocationAir, which=8)
```

To capture this change, we will ask the function to select the best ETS model for the scale and will not use any other elements:

```
adamScaleAir <- sm(adamLocationAir, model="YYY")
```

The `sm` function accepts roughly the same set of parameters as `adam()` when applied to the `adam` class objects. So, you can experiment with other types

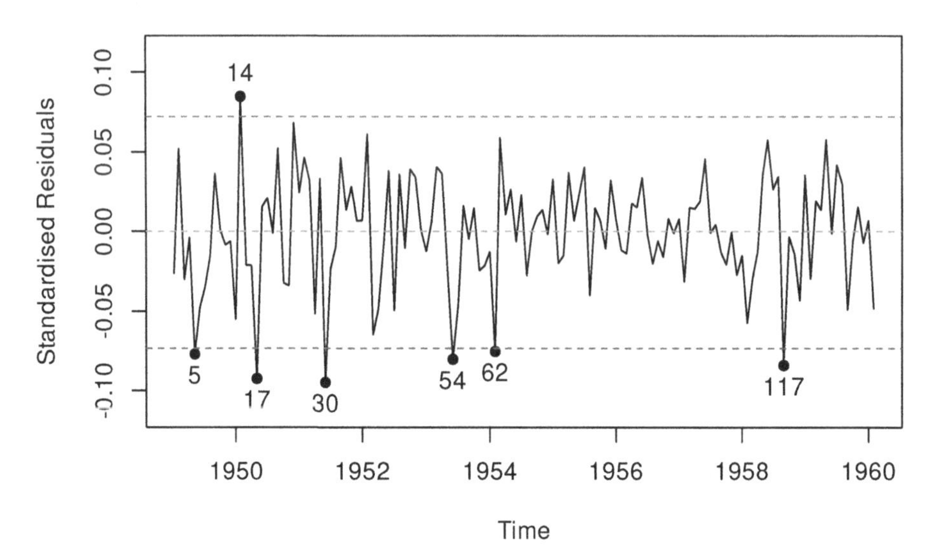

FIGURE 17.4 Diagnostic plot of the location model.

of ETS/ARIMA models if you want. Here is what the function has selected automatically for the scale model:

```
summary(adamScaleAir)
```

```
##
## Model estimated using sm.adam() function: ETS(MNN)
## Response variable: AirPassengers
## Distribution used in the estimation: Inverse Gaussian
## Loss function type: likelihood; Loss function value: 474.332
## Coefficients:
##        Estimate Std. Error Lower 2.5% Upper 97.5%
## alpha    0.0704     0.0569     0e+00      0.1828
## level    0.0021     0.0011    -1e-04      0.0043
##
## Error standard deviation: 1.41
## Sample size: 134
## Number of estimated parameters: 2
## Number of degrees of freedom: 132
## Information criteria:
##      AIC      AICc       BIC      BICc
## 952.6640 952.7556 958.4596 958.6840
```

Remark. Information criteria in the summary of the scale model do not include the number of estimated parameters in the location part of the model, so they are not directly comparable with the ones from the initial model. So, to understand whether there is an improvement in comparison with the model with a fixed scale, we can implant the scale model in the location one and compare the new model with the initial one:

```
adamMerged <- implant(adamLocationAir, adamScaleAir)
c(AICc(adamLocationAir), AICc(adamMerged)) |>
        setNames(c("Location", "Merged"))
```

```
## Location   Merged
## 991.4883 990.6118
```

The output above shows that introducing the ETS(M,N,N) model for the scale has improved the ADAM, reducing the AICc.

Judging by the output, we see that the most suitable model for the location is ETS(M,N,N) with $\alpha = 0.0704$, which, as we discussed in Subsection 17.2, is equivalent to GARCH(1,1) with restricted parameters:

$$\sigma_t^2 = 0.9296\sigma_{t-1}^2 + 0.0704\epsilon_{t-1}^2.$$

After implanting the scale in the model, we can access it via `adamMerged$scale` and extract the predicted scale via `extractScale(adamMerged)` command in R. The model fits the data in the following way (see Figure 17.5):

```
plot(adamMerged$scale, which=7)
```

As can be seen from Figure 17.5, the scale changes over time, indeed decreasing in the middle of the series and going slightly up at the end. The actual values in the holdout part can in general be ignored because they show the forecast errors several steps ahead for the whole holdout and are not directly comparable with the in-sample ones.

We can also do diagnostics of the merged model using the same principles as discussed in Chapter 14 – the function will automatically use the fitted values of scale where needed (see Figure 17.6).

```
par(mfcol=c(2,2), mar=c(4,4,2,1))
plot(adamMerged, which=c(2,6,8,7))
```

```
## Note that residuals diagnostics plots are produced for scale model
```

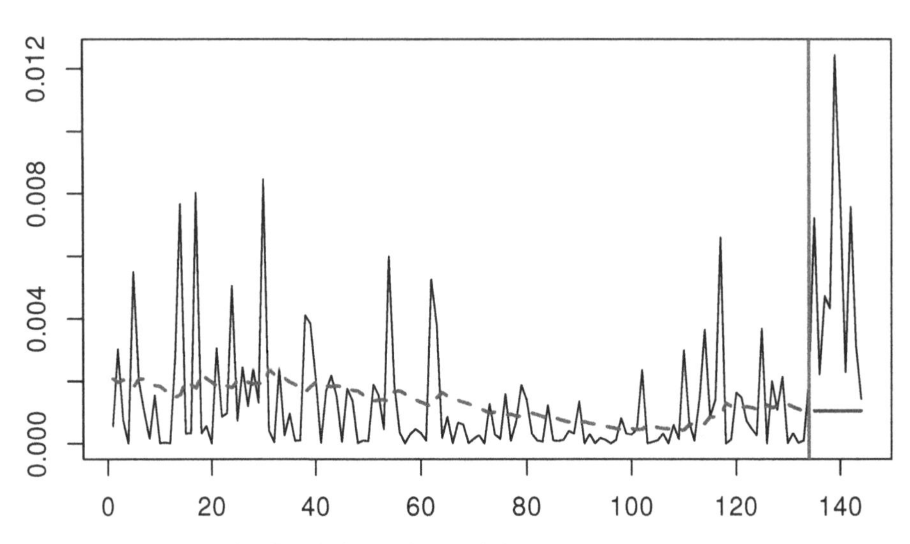

FIGURE 17.5 The fit of the scale model.

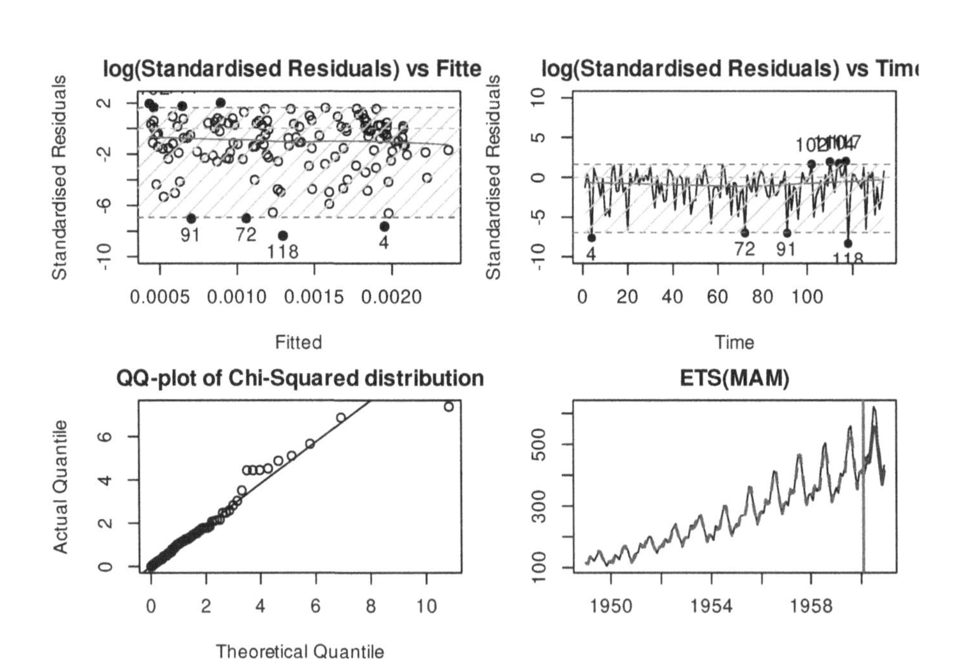

FIGURE 17.6 Diagnostics of the merged ADAM

Remark. In the case of a merged model, the plots that produce residuals will be done for the scale model rather than for the location (the plot function will produce a message on that), meaning that they plot η_t instead of ϵ_t (see discussion in Subsection 17.1.3 for the assumptions about the residual in this case).

Figure 17.6 demonstrates the diagnostics for the constructed model. Given that we used the Inverse Gaussian distribution, the $\eta_t^2 \sim \chi^2(1)$, which is why the QQ-plot mentions the Chi-Squared distribution. We can see that the distribution has a shorter tail than expected, which means that the model might be missing some elements. However, the logarithm of the standardised residuals vs fitted and vs time do not demonstrate any obvious problems except for several potential outliers, which do not lie far away from the bounds and do not exceed 5% of the sample size in quantity. Those specific observations could be investigated to determine whether the model can be improved further. I leave this exercise to the reader.

Finally, if we want to produce forecasts from the scale model, we can use the same methods previously used for the location one (Figure 17.7).

```
forecast(adamMerged$scale,
         h=10, interval="prediction") |>
  plot()
```

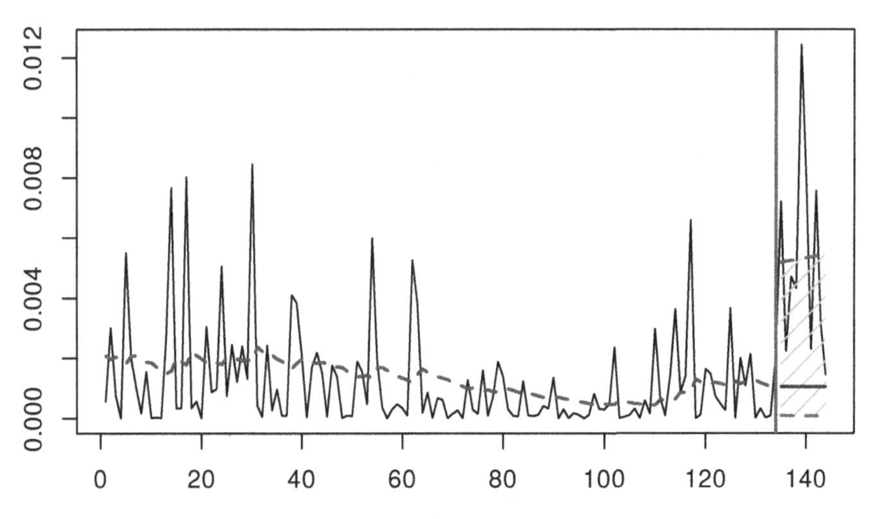

FIGURE 17.7 Forecast for the scale model.

The point forecast of the scale for the subsequent h observations can also be used to produce more appropriate prediction intervals for the merged model. This will be discussed in Chapter 18.

18

Forecasting with ADAM

Finally, we come to the technicalities of producing forecasts using ADAM. We have already discussed how conditional expectations can be generated from some of the models (e.g. Sections 5.3, 6.3, 9.2.1, 10.2, and 10.3.1), but we have not discussed this in necessary detail. Furthermore, as discussed in Section 1.1, forecasts should align with specific decisions, but we have not discussed how to do that.

In this chapter, we start with an explanation of how the simulation paths can be generated to obtain moments and quantiles in cases when they are not available analytically (Section 18.1). We then discuss the principles behind calculating the conditional moments from ADAM (including ETS, ARIMA, Regression, and their combinations, Section 18.2). After that we move to various methods for prediction interval construction, starting from the basic parametric and ending with empirical ones and those that take the uncertainty of parameters into account (building upon Chapter 16). Finally, we discuss prediction intervals for the intermittent state space model (Chapter 13), one-sided intervals, and cumulative forecasts over the forecast horizon, which is useful in practice, especially when inventory decisions need to be made. We also discuss the confidence interval for the conditional mean, which is not as important as the other topics mentioned above but is still useful in some contexts.

18.1 Creating simulation paths

As mentioned earlier in previous chapters, for some models, the conditional h steps ahead moments do not have analytical expressions. For example, as discussed in Section 6.3, pure multiplicative models typically do not have formulae for the conditional expectations and variance for longer horizons. The one exception is the ETS(M,N,N) model, where the point forecast corresponds to the conditional expectation for any horizon as long as the expectation of the error term $1 + \epsilon_t$ is equal to one. This problem is not only important for the moments, but it also arises when quantiles of distributions are needed, especially for models with multiplicative error. In all these cases, when the moments and/or quantiles are not available analytically, they need to be obtained via other means. One of those is simulating many trajectories and then calculating moments numerically.

18.1.1 Simulating trajectories

The general idea of this approach is based on the discussion in Section 16.1: we use the estimated parameters, the last obtained state vector (level, trend, seasonal, ARIMA components, etc.) and the estimate of the scale of distribution to generate the possible paths of the data for the next h observations. The simulation itself is done in several steps:

1. Generate h random variables for the error term, ϵ_{t+j} or $1+\epsilon_{t+j}$ – depending on the type of error, assumed distribution in the model (the latter was discussed in Sections 5.5 and 6.5), and estimated parameters of distribution (such as scale);
2. Insert the error terms in the state space model (7.1) from Section 7.1, both in the transition and observation parts, do that iteratively from $j=1$ to $j=h$:

$$\begin{aligned} y_{t+j} &= w(\mathbf{v}_{t+j-\boldsymbol{l}}) + r(\mathbf{v}_{t+j-\boldsymbol{l}}) \epsilon_{t+j} \\ \mathbf{v}_{t+j} &= f(\mathbf{v}_{t+j-\boldsymbol{l}}) + g(\mathbf{v}_{t+j-\boldsymbol{l}}) \epsilon_{t+j} \end{aligned} ;$$

3. Record actual values for $j=\{1,\dots,h\}$;
4. Repeat (1) – (3) n times;
5. Take desired moments or quantiles for each horizon from 1 to h.

Graphically, the result of this approach is shown in Figure 18.1, where each separate path is shown in grey colour, and the expectation is in black. See the R code for this in Subsection 18.1.2.

Remark. In the case of multiplicative trend or multiplicative seasonality, it makes sense to take trimmed mean instead of the basic arithmetic one on step (5). This is because the models with these components might exhibit explosive behaviour, and thus the expectation might become unrealistic. I suggest using 1% trimming, although this does not have any scientific merit and is only based on my personal expertise.

The simulation-based approach is universal, no matter what model is used, and can be applied to any ETS, ARIMA, Regression model, or combination (including dynamic ETSX, intermittent demand, and multiple frequency models). Furthermore, instead of extracting moments on step five, one can take geometric mean, median, or any other desired statistics.

The main issue with this approach is that the conditional expectation, or any other statistics calculated based on this, will differ from one simulation

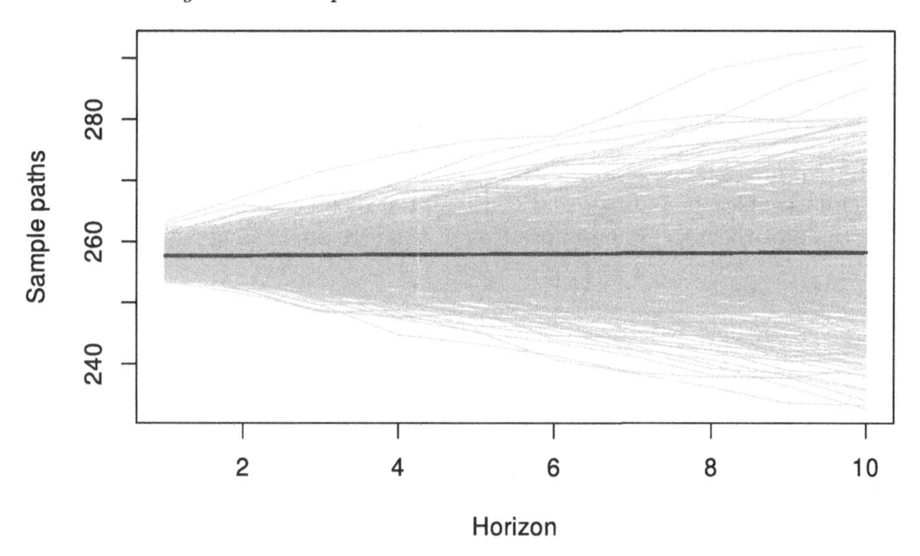

FIGURE 18.1 Sample paths (scenarios) generated from ADAM for the holdout sample.

run to another. If n is small, these values will be less stable (vary more with the new runs). But, with the increase of n, they will reach some asymptotic values, staying random nonetheless. However, this is a good thing because this randomness reflects the uncertain nature of these statistics in the sample: the true values are never known, and the estimates will inevitably change with the sample size change. Another limitation is the computational time and memory usage: the more iterations we want to produce, the more calculations will need to be done, and more memory will be consumed. Luckily, time complexity in this situation is linear: $O(h \times n)$.

18.1.2 Demonstration in R

In order to demonstrate the simulation approach, we consider an artificial case of an ETS(M,M,N) model with $l_t = 1000$, $b_t = 0.95$, $\alpha = 0.1$, $\beta = 0.01$, and Gamma distribution for error term with scale $s = 0.05$. We generate 1000 scenarios from this model for the horizon of $h = 10$ using the `sim.es()` function from the `smooth` package:

```
nsim <- 1000
h <- 10
s <- 0.1
initial <- c(1000,0.95)
persistence <- c(0.1,0.01)
y <- sim.es("MMN", obs=h, nsim=nsim, persistence=persistence,
```

```
            initial=initial, randomizer="rgamma",
            shape=1/s, scale=s)
```

After running the code above, we will obtain an object `y` that will contain several variables, including `y$data` with all the 1000 possible future trajectories. We can plot them to get an impression of what we are dealing with (see Figure 18.2):

```
plot(y$data[,1], ylab="Sales", ylim=range(y$data),
     col=rgb(0.8,0.8,0.8,0.4), xlab="Horizon")
# Plot all the generated lines
for(i in 2:nsim){
  lines(y$data[,i], col=rgb(0.8,0.8,0.8,0.4))
}
# Add conditional mean and quantiles
lines(apply(y$data,1,mean))
lines(apply(y$data,1,quantile,0.025),
      col="grey", lwd=2, lty=2)
lines(apply(y$data,1,quantile,0.975),
      col="grey", lwd=2, lty=2)
```

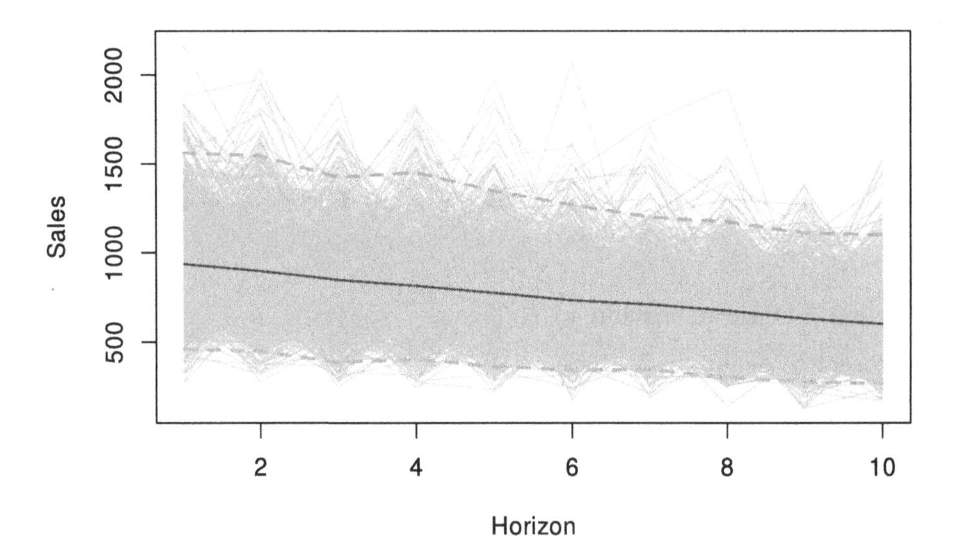

FIGURE 18.2 Data generated from 1000 ETS(M,M,N) models.

Based on the plot in Figure 18.2, we can see what the conditional h steps ahead expectation (black line) and what the 95% prediction interval will be for the data based on the ETS(M,M,N) model with the parameters mentioned above.

Similar paths are produced by the `forecast()` method for the `adam` class in the package `smooth`. If you need to extract them for further analysis, they are returned in the object `scenarios` if the parameters `interval="simulated"` and `scenarios=TRUE` are set.

18.2 Conditional moments and scale

We have already discussed how to obtain conditional expectation and variance in Sections 5.3 and 6.3. However, the topic is worth discussing in more detail, especially for non-Normal distributions.

18.2.1 Conditional expectation

The general rule that applies to ADAM in terms of generating conditional expectations is that if you deal with the pure additive model, then you can produce forecasts analytically. This not only applies to ETS but also to ARIMA (Subsection 9.2.1) and Regression (Section 10.2). If the model has multiplicative components (such as multiplicative error, or trend, or seasonality) or is formulated in logarithms (for example, ARIMA in logarithms), then simulations should be preferred (Section 18.1) – the point forecasts from these models would not necessarily correspond to the conditional expectations.

18.2.2 Explanatory variables

If the model contains explanatory variables, then the h steps ahead conditional expectations should use them in the calculation. The main challenge in this situation is that future values might not be known in some cases. This has been discussed in Section 10.2. Practically speaking, if the user provides the holdout sample values of explanatory variables, the `forecast.adam()` method will use them in forecasting. If they are not provided, the function will produce forecasts for each of the explanatory variables via the `adam()` function and use the conditional h steps ahead expectations in forecasting.

18.2.3 Conditional variance and scale

Similar to conditional expectations, as we have discussed in Sections 5.3 and 6.3, the conditional h steps ahead variance is in general available only for the pure additive models. While the conditional expectation might be required on its own to use as a point forecast, the conditional variance is typically needed to produce prediction intervals. However, it becomes useful only in cases of distributions that support convolution (addition of random variables), which limits its usefulness to pure additive models and to additive models applied to the data in logarithms. For example, if we deal with Inverse Gaussian distribution, then the h steps ahead values will not follow Inverse Gaussian distribution, and we would need to revert to simulations in order to obtain the proper statistics for it. Another situation would be a multiplicative error model that relies on Normal distribution – the product of Normal distributions is not a Normal distribution, so the statistics would need to be obtained using simulations again.

If we deal with a pure additive model with either Normal, Laplace, S, or Generalised Normal distributions, then the formulae derived in Section 5.3 can be used to produce h steps ahead conditional variance. Having obtained those values, we can then produce conditional h steps ahead scales for the distributions (which would be needed, for example, to generate quantiles), using the relations between the variance and scale in those distributions (discussed in Section 5.5):

1. Normal: scale is σ^2_h;
2. Laplace: $s_h = \sigma_h \sqrt{\frac{1}{2}}$;
3. Generalised Normal: $s_h = \sigma_h \sqrt{\frac{\Gamma(1/\beta)}{\Gamma(3/\beta)}}$;
4. S: $s_h = \sqrt{\sigma_h} \sqrt[4]{\frac{1}{120}}$.

If the variance is needed for the other combinations of model/distributions, simulations would need to be done to produce multiple trajectories, similar to how it was done in Section 18.1. An alternative to this would be the calculation of in-sample multistep forecast errors (similar to how it was discussed in Sections 11.3 and 14.7.3) and then calculating the variance based on them for each horizon $j = 1 \dots h$.

In the `smooth` package for R, there is a `multicov()` method that allows extracting the multiple steps ahead covariance matrix $\hat{\boldsymbol{\Sigma}}$ (see Subsection 11.3.5). The method can estimate the covariance matrix using analytical formulae (where available), or via empirical calculations (based on multiple steps ahead in-sample error), or via simulation. Here is an example for one of the models in R:

```
adam(BJsales) |>
  multicov(h=7) |>
  round(3)
```

```
##        h1    h2     h3     h4     h5     h6     h7
## h1 1.856 2.294  2.747  3.134  3.466  3.750  3.993
## h2 2.294 4.692  5.690  6.622  7.419  8.102  8.686
## h3 2.747 5.690  8.758 10.329 11.752 12.969 14.012
## h4 3.134 6.622 10.329 14.051 16.183 18.085 19.713
## h5 3.466 7.419 11.752 16.183 20.524 23.186 25.542
```

```
## h6 3.750 8.102 12.969 18.085 23.186 28.101 31.254
## h7 3.993 8.686 14.012 19.713 25.542 31.254 36.692
```

18.2.4 Scale model

In the case of the scale model (Chapter 17), the situation becomes more complicated because we no longer assume that the variance of the error term is constant (residuals are homoscedastic) – we now assume that it is a model on its own. In this case, we need to take a step back to the recursion (5.10) and when taking the conditional variance, introduce the time-varying variance σ^2_{t+h}.

Remark. Note the difference between σ^2_{t+h} and σ^2_h in our notations – the former is the variance of the error term for the specific step $t+h$, while the latter is the conditional variance h steps ahead, which is derived based on the assumption of homoscedasticity.

Making that substitution leads to the following analytical formula for the h steps ahead conditional variance in the case of the scale model:

$$\mathrm{V}(y_{t+h}|t) = \sum_{i=1}^{d} \left(\mathbf{w}'_{m_i} \sum_{j=1}^{\lceil \frac{h}{m_i} \rceil - 1} \mathbf{F}_{m_i}^{j-1} \mathbf{g}_{m_i} \mathbf{g}'_{m_i} (\mathbf{F}'_{m_i})^{j-1} \mathbf{w}_{m_i} \sigma^2_{t+h-j} \right) + \sigma^2_{t+h}. \tag{18.1}$$

This variance can then be used, for example, to produce quantiles from the assumed distribution.

As mentioned above, in the case of the not purely additive model or a model with other distributions than Normal, Laplace, S, or Generalised Normal, the conditional variance can be obtained using simulations. In the case of the scale model, the principles will be the same, just assuming that each error term ϵ_{t+h} has its own scale, obtained from the estimated scale model. The rest of the logic will be exactly the same as discussed in Section 18.1.

18.3 Prediction intervals

A prediction interval is needed to reflect the uncertainty about the data. In theory, the 95% prediction interval will cover the actual values in 95% of the cases if the model is correctly specified. The specific formula for the prediction interval will vary with the assumed distribution. For example, for the Normal distribution (assuming that $y_{t+j} \sim \mathcal{N}(\mu_{y,t+j}, \sigma^2_j)$) we will have the classical one:

$$y_{t+j} \in (\hat{y}_{t+j} + z_{\alpha/2}\hat{\sigma}_j, \hat{y}_{t+j} + z_{1-\alpha/2}\hat{\sigma}_j), \tag{18.2}$$

where $\hat{y}_{t+j}$ is the estimate of $\mu_{y,t+j}$, j steps ahead conditional expectation, $\hat{\sigma}_j^2$ is the estimate of σ_j^2, j steps ahead variance of the error term obtained using principles discussed in Subsection 18.2.3 (for example, calculated via the formula (5.12)), and z is z-statistics (quantile of Standard Normal distribution) for the selected significance level α.

Remark. Note that α has nothing to do with the smoothing parameters for the level of ETS model.

This type of prediction interval can be called **parametric**. It assumes a specific distribution and relies on the other assumptions about the constructed model (such as residuals are i.i.d., see Section 14). Most importantly, it assumes that the j steps ahead value follows a specific distribution related to the one for the error term. In the case of Normal distribution, the assumption $\epsilon_t \sim \mathcal{N}(0, \sigma^2)$ implies that $y_{t+1} \sim \mathcal{N}(\mu_{y,t+1}, \sigma_1^2)$ and due to the *convolution* of random variables (the sum of random variables follows the same distribution as individual variables, but with different parameters), that $y_{t+j} \sim \mathcal{N}(\mu_{y,t+j}, \sigma_j^2)$ for all j from 1 to h.

The interval produced via (18.2) corresponds to two quantiles from the Normal distribution and can be written in a more general form as:

$$y_{t+j} \in \left(q\left(\hat{y}_{t+j}, \hat{\sigma}_j^2, \frac{\alpha}{2}\right), q\left(\hat{y}_{t+j}, \hat{\sigma}_j^2, 1 - \frac{\alpha}{2}\right)\right), \tag{18.3}$$

where $q(\cdot)$ is a quantile function of an assumed distribution, $\hat{y}_{t+j}$ acts as a location, and $\hat{\sigma}^2$ acts as a scale of distribution. Using this general formula (18.3) for prediction intervals, we can construct them for other distributions as long as they support convolution. In the ADAM framework, this works for all pure additive models that have an error term that follows one of the following distributions:

1. Normal: $\epsilon_t \sim \mathcal{N}(0, \sigma^2)$, thus $y_{t+j} \sim \mathcal{N}(\mu_{y,t+j}, \sigma_j^2)$;
2. Laplace: $\epsilon_t \sim \mathcal{Laplace}(0, s)$ and $y_{t+j} \sim \mathcal{Laplace}(\mu_{y,t+j}, s_j)$;
3. Generalised Normal: $\epsilon_t \sim \mathcal{GN}(0, s, \beta)$, so that $y_{t+j} \sim \mathcal{GN}(\mu_{y,t+j}, s_j, \beta)$;
4. S: $\epsilon_t \sim \mathcal{S}(0, s)$, $y_{t+j} \sim \mathcal{S}(\mu_{y,t+j}, s_j)$.

If a model has multiplicative components or relies on a different distribution, then the several steps ahead actual value will not necessarily follow the assumed distribution, and the formula (18.3) will produce incorrect intervals. For example, if we work with a pure multiplicative ETS model, ETS(M,N,N), assuming that $\epsilon_t \sim \mathcal{N}(0, \sigma^2)$, the two steps ahead actual value can be expressed in terms of the values on observation t:

$$y_{t+2} = l_{t+1}(1 + \epsilon_{t+2}) = l_t(1 + \alpha\epsilon_{t+1})(1 + \epsilon_{t+2}), \tag{18.4}$$

which introduces the product of Normal distributions, and thus y_{t+2} does not follow Normal distribution anymore. In such cases, we might have several options of what to use to produce intervals. They are discussed in the subsections below.

18.3.1 Approximate intervals

Even if the actual multistep value does not follow the assumed distribution, we can use approximations in some cases: the produced prediction interval will not be too far from the correct one. The main idea behind the approximate intervals is to rely on the same distribution for y_{t+j} as for y_{t+1}, even though we know that the variable will not follow it. In the case of multiplicative error models, the limit (6.5) can be used to motivate the usage of that assumption:

$$\lim_{x \to 0} \log(1+x) = x.$$

For example, in the case of the ETS(M,N,N) model, we know that y_{t+2} will not follow the Normal distribution, but if the variance of the error term is low (e.g. $\sigma^2 < 0.05$) and the smoothing parameter α is close to zero, then the Normal distribution would be a satisfactory approximation of the real one. This becomes clear if we expand the brackets in (18.4):

$$y_{t+2} = l_t(1 + \alpha \epsilon_{t+1} + \epsilon_{t+2} + \alpha \epsilon_{t+1} \epsilon_{t+2}). \tag{18.5}$$

With the conditions discussed above (low α, low variance) the term $\alpha \epsilon_{t+1} \epsilon_{t+2}$ will be close to zero, thus making the sum of Normal distributions dominant in the formula (18.5). The advantage of this approach is in its speed: you only need to know the scale parameter of the error term and the conditional expectation. The disadvantage of the approach is that it becomes inaccurate with the increase of the parameters' values and scale of the model. The rule of thumb for when to use this approach: if the smoothing parameters are all below 0.1 (in the case of ARIMA, this is equivalent to MA terms being negative and AR terms being close to zero) or the scale of distribution is below 0.05, the differences between the proper interval and the approximate one should be negligible.

18.3.2 Simulated intervals

This approach relies on the idea discussed in Section 18.1. It is universal and supports any distribution because it only assumes that the error term follows the selected distribution (no need for the actual value to do that as well).

The simulated paths are then produced based on the generated values and the assumed model. After generating n paths, one can take the desired quantiles to get the bounds of the interval. The main issue of the approach is that it is time-consuming (slower than the approximate intervals) and might be highly inaccurate if the number of iterations n is low. This approach is used as a default in `adam()` for the non-additive models.

18.3.3 Semiparametric intervals

The three approaches above assume that the residuals of the applied model are i.i.d. (see discussion in Chapter 14). If this assumption is violated (for example, the residuals are autocorrelated), then the intervals might be miscalibrated (i.e. producing wrong values). In this case, we might need to use different approaches. One of these is the construction of **semiparametric** prediction intervals (see, for example, Lee & Scholtes, 2014). This approach relies on the in-sample multistep forecast errors discussed in Subsection 14.7.3. After producing $e_{t+j|t}$ for all in-sample values of t and for $j = 1, \dots, h$, we can use these errors to calculate the respective h steps ahead conditional variances σ_j^2 for $j = 1, \dots, h$. These values can then be inserted in the formula (18.3) to get the desired prediction interval. The approach works well in the case of pure additive models, as it relies on specific assumed distribution. However, it might have limitations similar to those discussed earlier for the mixed models and the models with positively defined distributions (such as Log-Normal, Gamma, and Inverse Gaussian). It can be considered a semiparametric alternative to the approximate method discussed above.

18.3.4 Nonparametric intervals

When some of the assumptions might be violated, and when we cannot rely on the parametric distributions, we can use the **nonparametric** approach proposed by Taylor and Bunn (1999). The authors proposed using the multistep forecast errors to construct the following quantile regression model:

$$\hat{e}_{t+j} = a_0 + a_1 j + a_2 j^2, \tag{18.6}$$

for each of the bounds of the interval. The motivation behind the polynomial in (18.6) is because, typically, the multistep conditional variance will involve the square of the forecast horizon. The main issue with this approach is that the polynomial function has an extremum, which might appear sometime in the future. For example, the upper bound of the interval would increase until that point and then start declining. To overcome this limitation, I propose using the power function instead:

$$\hat{e}_{t+j} = a_0 j^{a_1}. \tag{18.7}$$

This way, the bounds will always change monotonically, and the parameter a_1 will control the speed of expansion of the interval. The model (18.7) is estimated using quantile regression for the upper and the lower bounds separately, as Taylor and Bunn (1999) suggested. This approach does not require any assumptions about the model and works as long as there are enough observations in-sample (so that the matrix of forecast errors contains more rows than columns). The main limitation of this approach is that it relies on quantile regression and thus will have the same issues as, for example, pinball score has (see discussion in Section 2.2): the quantiles are not always uniquely defined. Another limitation is that we assume that the quantiles will follow the model (18.7), which might be violated in real life scenarios.

18.3.5 Empirical intervals

Another alternative to the parametric intervals uses the same matrix of multistep forecast errors, as discussed earlier. The **empirical** approach is more straightforward than the approaches discussed above and does not rely on any assumptions (it was discussed in Lee & Scholtes, 2014). The idea behind it is just to take quantiles of the forecast errors for each forecast horizon $j = 1, \dots, h$. These quantiles are then added to the point forecast if the error term is additive or are multiplied by it in the case of the multiplicative one. Trapero, Cardós, and Kourentzes (2019) show that the empirical prediction intervals perform on average better than the analytical ones. This is because of the potential violation of assumptions in real life. So, in general, I would recommend producing empirical intervals if it was not for the computational difficulties related to the multistep forecast errors. If you have an additive model and believe that the assumptions are satisfied, then the parametric interval will be as accurate but faster. Furthermore, the approach will be unreliable on small samples due to the same problem with the quantiles as discussed earlier.

18.3.6 Complete parametric intervals

So far, **all the intervals discussed above** relied on an unrealistic assumption that the parameters of the model are known. This is one of the reasons why the intervals produced for ARIMA and ETS are typically narrower than expected (see, for example, results of a tourism competition, Athanasopoulos et al., 2011). But as we discussed in Section 16, there are ways of capturing the uncertainty of estimated parameters of the model and propagating it to the future uncertainty (e.g. to the conditional h steps ahead variance). As discussed in Section 16.5, the more general approach is to create many in-sample model paths based on randomly generated parameters of the model. This way, we can obtain a variety of states for the final in-sample observation T and then use those values to construct final prediction intervals. The simplest and most general way of producing intervals, in this case, is using simulations (as discussed earlier in Subsection 18.3.2). The intervals produced via this approach will be wider than the conventional ones, and their width will be proportional to the uncertainty around the parameters. This also means that the intervals might become too wide if the uncertainty is not captured correctly (see discussion in Section 16.5). One of the main limitations of the approach is its computational time:

it will be proportional to the number of simulation paths for both refitted models and prediction intervals.

It is also theoretically possible to use other approaches for the intervals construction in the case of complete uncertainty (e.g. "empirical" one for each of the set of parameters of the model), but they would be even more computationally expensive than the approach described above and will have the limitations similar to the discussed above (i.e. non-uniqueness of quantiles, sample size requirements, etc).

18.3.7 Explanatory variables

In all the cases described above, when constructing prediction intervals for the model with explanatory variables, we assume that their values are known in the future. Even if they are not provided by the user and need to be forecasted, the produced conditional expectations of variables will be used for all the calculations. This is not an entirely correct approach, as was shown in Subsection 10.2.2, but it speeds up the calculation process and typically produces adequate results.

The more theoretically correct approach is to take the multistep variance of explanatory variables into account. This would work for pure additive models for explanatory and response variables but imply more complicated formulae for other models. This is one of the directions of future research.

18.3.8 Example in R

All the types of intervals discussed in this section are implemented for the `adam()` models in the `smooth` package. In order to demonstrate how they work and how they differ, we consider an example with an ETS model on `BJSales` data:

```
adamETSBJ <- adam(BJsales, h=10, holdout=TRUE)
modelType(adamETSBJ)
```

```
## [1] "AAdN"
```

The model selected above is ETS(A,Ad,N). In order to make sure that the parametric intervals are suitable, we can do model diagnostics (see Chapter 14), producing plots shown in Figure 18.3.

```
par(mfcol=c(2,3), mar=c(4,4,2,1))
plot(adamETSBJ, which=c(2,4,6,8,10:11))
```

The model's residuals do not exhibit any serious issues. Given that this is a

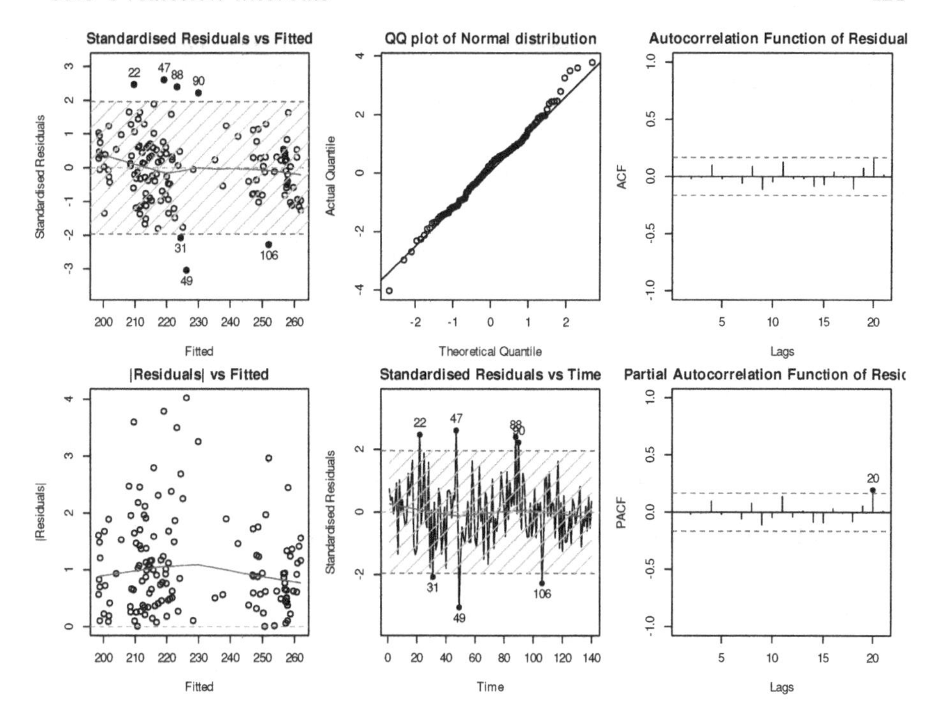

FIGURE 18.3 Diagnostics of the ADAM on BJSales data.

pure additive model, we can conclude that the parametric interval would be appropriate for this situation.

The only thing that this type of interval does not take into account is the uncertainty about the parameters, so we can construct the complete interval either via the `reforecast()` function or using the same `forecast()`, but with the option `interval="complete"`. Note that this is a computationally expensive operation (both in terms of time and memory), so the more iterations you set up, the longer it will take and the more memory it will consume. The two types of intervals are shown next to each other in Figure 18.4:

```
par(mfcol=c(1,2), mar=c(2,4,2,1))
forecast(adamETSBJ, h=10, interval="parametric") |>
    plot(main="Parametric prediction interval", ylim=c(200,280))
forecast(adamETSBJ, h=10, interval="complete", nsim=100) |>
    plot(main="Complete prediction interval", ylim=c(200,280))
```

The resulting complete parametric interval shown in Figure 18.4 is slightly wider than the conventional one. To understand what impacts the complete interval, we can analyse the summary of the model:

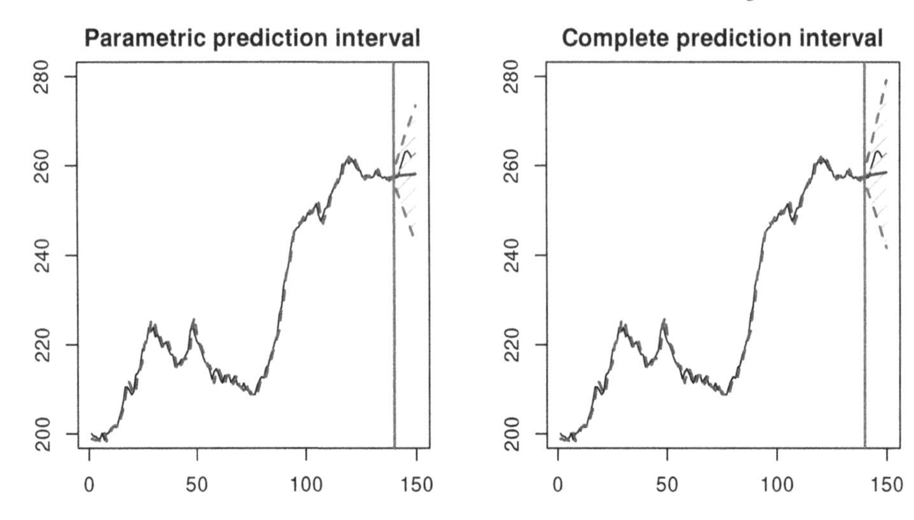

FIGURE 18.4 Prediction intervals for ADAM on BJSales data.

```
summary(adamETSBJ)
```

```
##
## Model estimated using adam() function: ETS(AAdN)
## Response variable: BJsales
## Distribution used in the estimation: Normal
## Loss function type: likelihood; Loss function value: 241.078
## Coefficients:
##       Estimate Std. Error Lower 2.5% Upper 97.5%
## alpha   0.9507     0.1144     0.7244      1.0000 *
## beta    0.2911     0.1395     0.0152      0.5667 *
## phi     0.8632     0.1257     0.6147      1.0000 *
## level 202.7529     4.7209   193.4158    212.0829 *
## trend  -2.3996     3.5642    -9.4491      4.6445
##
## Error standard deviation: 1.384
## Sample size: 140
## Number of estimated parameters: 6
## Number of degrees of freedom: 134
## Information criteria:
##      AIC     AICc      BIC     BICc
## 494.1560 494.7876 511.8058 513.3664
```

The smoothing parameters of the model are high, thus the model forgets the initial states fast, and the uncertainty of initial states does not propagate to the last observation as much as in the case of lower values of parameters. As a result, only the uncertainty of smoothing parameters will impact the width of the interval.

Figure 18.5 demonstrates what happens with the fitted values when we take the uncertainty into account.

```
reapply(adamETSBJ) |>
  plot()
```

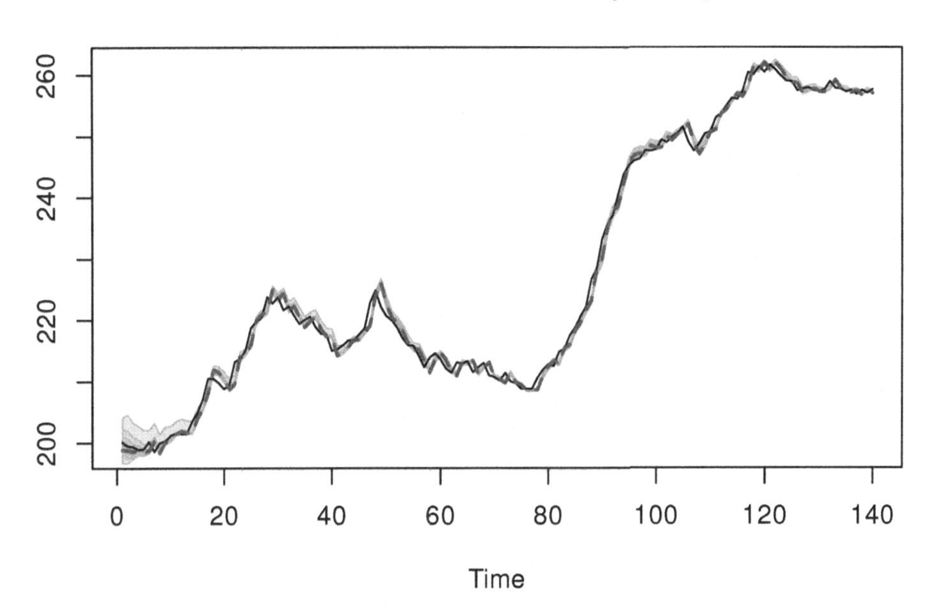

FIGURE 18.5 Refitted values for ADAM on BJSales data.

As we see from Figure 18.5, the uncertainty around the line is narrow at the end of the sample, so the impact of the initial uncertainty on the forecast deteriorates.

To make things more complicated and exciting, we introduce explanatory variable with lags and leads of the indicator `BJsales.lead`, automatically selecting the model and explanatory variables using information criteria (see discussion in Chapter 15).

```
# Form a matrix with response and the explanatory variables
BJsalesData <- cbind(as.data.frame(BJsales),
                     xregExpander(BJsales.lead,c(-3:3)))
colnames(BJsalesData)[1] <- "y"
```

```
# Seletct an ETSX model
adamETSXBJ <- adam(BJsalesData, "YYY",
                   h=10, holdout=TRUE,
                   regressors="select")
```

In the code above, I have asked the function specifically to do the selection between pure multiplicative models (see Section 15.1). We will then construct several types of prediction intervals and compare them:

```
intervalType <- c("approximate", "semiparametric",
                  "nonparametric", "simulated",
                  "empirical", "complete")

vector("list", length(intervalType)) |>
    setNames(intervalType) -> adamETSXBJPI

for(i in intervalType){
  adamETSXBJPI[[i]] <- forecast(adamETSXBJ, h=10,
                                interval=i)
}
```

These can be plotted in Figure 18.6.

```
par(mfcol=c(3,2), mar=c(2,2,2,1))
for(i in 1:6){
  plot(adamETSXBJPI[[i]],
       main=paste0(intervalType[i]," interval"))
}
```

The thing to notice is how the width and shape of intervals change depending on the approach. The *approximate* and *simulated* intervals look very similar because the selected model is ETSX(M,N,N) with a standard error of 0.005 (thus, the approximation works well). The *complete* interval is similar because the estimated smoothing parameter α equals to one (hence, the forgetting happens instantaneously). However, it has a slightly different shape because the number of iterations for the interval was low (`nsim=100` for `interval="complete"` by default). The *semiparametric* interval is the widest as it calculates the forecast errors directly but still uses the normal approximation. Both *nonparametric* and *empirical* are skewed because the in-sample forecast errors followed skewed distributions, which can be seen via the plot in Figure 18.7:

```
adamETSXBJForecastErrors <- rmultistep(adamETSXBJ,h=10)
boxplot(1+adamETSXBJForecastErrors)
```

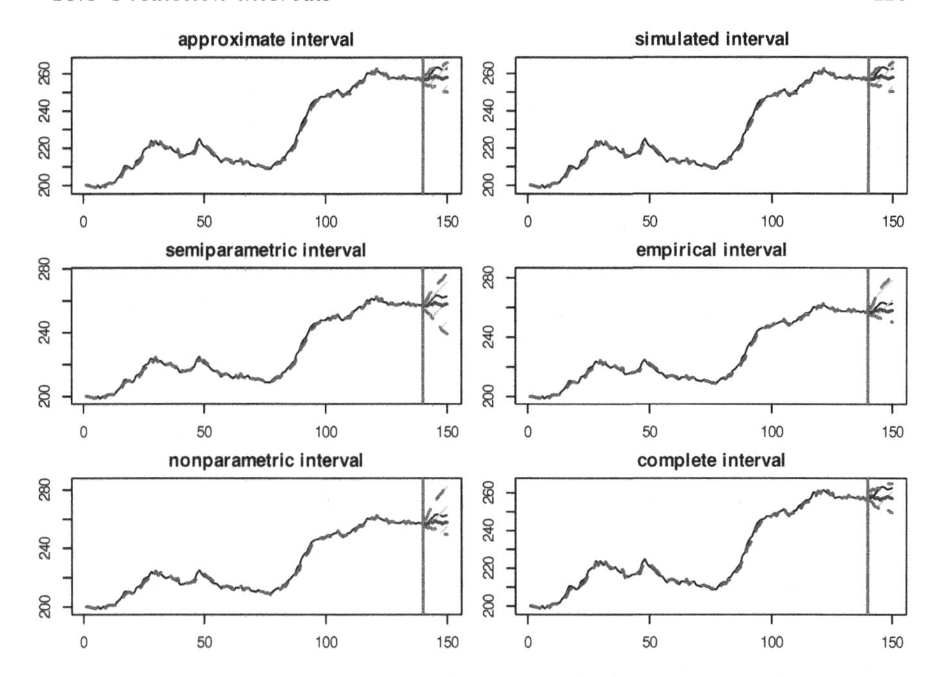

FIGURE 18.6 Different prediction intervals for ADAM ETS(M,N,N) on BJSales data.

```
abline(h=1,col="red")
points(apply(1+adamETSXBJForecastErrors,2,mean),
       col="red",pch=16)
```

Analysing the plots in Figure 18.6, it might be challenging to select the most appropriate type of prediction interval. But the model diagnostics (Section 14) might help in this situation:

1. If the residuals look i.i.d. and the model does not omit important variables, then choose between *parametric*, *approximate*, *simulated*, and *complete* interval types:
   a. "parametric" in case of the pure additive model,
   b. "approximate" in other cases, when the standard error is lower than 0.05 or smoothing parameters are close to zero,
   c. "simulated" if you deal with a non-additive model with high values of standard error and smoothing parameters,
   d. "complete parametric" when the smoothing parameters of the model are close to zero, and you want to take the uncertainty about the parameters into account;
2. If residuals seem to follow the assumed distribution but are not i.i.d.,

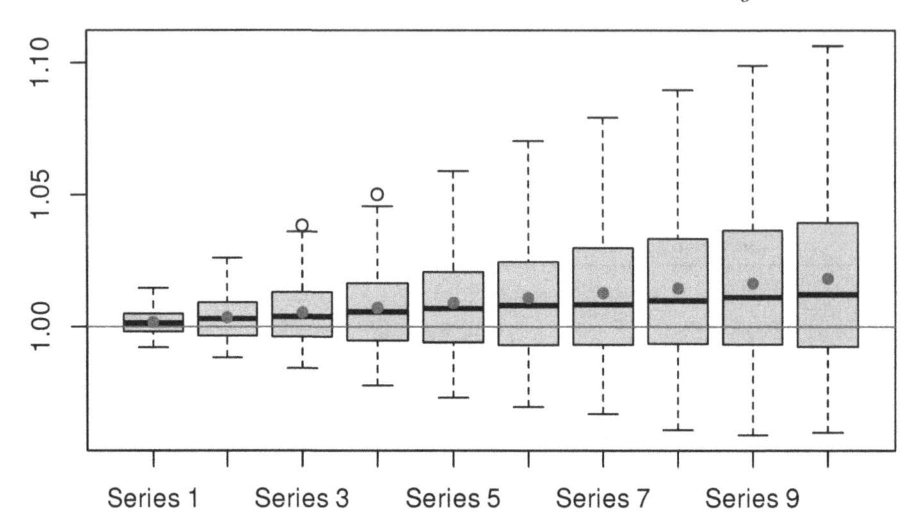

FIGURE 18.7 Distribution of in-sample multistep forecast errors from ADAM ETSX(M,N,N) model on BJSales data. Red point correspond to mean values.

then the *semiparametric* approach might help. Note that this only works on samples of $T >> h$;

3. If residuals do not follow the assumed distribution, but your sample is still larger than the forecast horizon, then use either *empirical* or *nonparametric* intervals.

Remark. `forecast.adam()` will automatically select between "parametric", "approximate", and "simulated" if you ask for `interval="prediction"`.

Finally, the discussion from this section also widely applies to ADAM ARIMA and/or Regression. The main difference is that ARIMA/Regression do not have mixed components (as ETS does), so the "parametric" prediction interval can be considered as a standard working option for the majority of cases. The only situation where simulations might be needed is when Log-ARIMA is constructed with Inverse Gaussian or Gamma distributions, because logarithms of these distributions do not support convolutions.

18.4 Other aspects of forecast uncertainty

There are other elements related to forecasting and taking uncertainty into account that we have not discussed in the previous sections. Here we discuss several special cases where forecasting approaches might differ from the conventional ones.

18.4.1 Prediction interval for intermittent demand model

When it comes to constructing a prediction interval for the intermittent state space model (from Chapter 13), then there is an important aspect that should be taken into account. Given that the model consists of two parts: demand sizes and demand occurrence, the prediction interval should take the uncertainty from both of them into account. In this case, we should first predict the probability of occurrence of demand for the h steps ahead and then decide what the width of the interval should be based on this probability. For example, if we estimate that the demand will occur with probability $\hat{p}_{t+h|t} = 0.8$, then this means that we expect that in 20% of the cases, we will observe zeroes. This should reduce the confidence level for the demand sizes. Formally speaking, this comes to the following equation:

$$F_{y_{t+h}}(y_{t+h} \leq q) = \hat{p}_{t+h|t} F_{z_{t+h}}(z_{t+h} \leq q) + (1 - \hat{p}_{t+h|t}), \tag{18.8}$$

where $F_{y_{t+h}}(\cdot)$ is the cumulative distribution function of demand, $F_{z_{t+h}}(\cdot)$ is the cumulative distribution function of the demand sizes, $\hat{p}_{t+h|t}$ is the h steps ahead expected probability of occurrence, and q is the quantile of distribution. In the formula (18.8), we know the expected probability and we know the confidence level $F_{y_{t+h}}(y_{t+h} \leq q)$. The unknown element is the $1 - \alpha = F_{z_{t+h}}(z_{t+h} \leq q)$. So after regrouping elements we get:

$$F_{z_{t+h}}(z_{t+h} \leq q) = \frac{F_{y_{t+h}}(y_{t+h} \leq q) - (1 - \hat{p}_{t+h|t})}{\hat{p}_{t+h|t}}, \tag{18.9}$$

which can be used for the calculation of the confidence level of a prediction interval. For example, if the confidence level is 0.95 and the expected probability of occurrence is 0.8, then $F_{z_{t+h}}(z_{t+h} \leq q) = \frac{0.95-0.2}{0.8} = 0.9375$. Assuming that demand sizes follow some distribution (e.g. Gamma), we can use formula (18.3) to construct a prediction interval of the width 93.75%, which will imply that 95% of demand is expected to be in the constructed bounds.

18.4.2 One-sided prediction interval

In some cases, we might not need both bounds of the interval. For example, when we deal with intermittent demand, we know that the lower bound will be equal to zero in many cases. Another example is the safety stock calculation: we only need the upper bound of the interval, and we need to make sure that the specific proportion of demand is satisfied (e.g. 95% of it). In these cases, we can just focus on the particular bound of the interval and drop the other one. Statistically speaking, this means that we cut only one tail of the assumed distribution.

Remark. In the case of an intermittent demand model, when the significance level is lower than the probability of inoccurrence $1 - p_{t+h|t}$, we will have the quantile equal to zero because the probability of having zeroes is higher than the significance level.

The one-sided interval has its implications and issues in several scenarios:

- When we are interested in the **upper bound** only and deal with **positive distribution** of demand (for example, Gamma, Log-Normal, or Inverse Gaussian), we know that the demand will always lie between zero and the constructed bound. In cases of low volume (or even intermittent) data, this makes sense because the original data might contain zeroes or have values close to it. The upper bound in this case will be lower than in the case of the two-sided prediction interval because we would not be splitting the probability into two parts (for the left and the right tails);
- The combination of the **lower bound** and **positive distribution** implies that the demand will be greater than the specified value in the pre-selected number of cases (defined by confidence level). There is no natural bound from above, so from a theoretical point of view, this implies that the demand can be infinite;
- The **upper** or **lower** bound with **real-valued distribution** (such as Normal, Laplace, S, or Generalised Normal) implies that the demand is either below or above the specified level, respectively, without any natural limit on the other side. If Normal distribution is used on positive low volume data, there is a natural lower bound, but the model itself will not be aware of it and will not restrict the space with the specific value, implying that the demand can be anything between the $-\infty$ and the selected value.

From the practical point of view, the case with the upper bound and a positively defined distribution makes more sense than the other two cases, because if we are interested in demand forecasting, having a non-negative demand makes more sense than having a real-valued one, while the upper bound aligns better with a safety stock calculation.

18.4.3 Cumulative over the horizon forecast

Another related thing to consider when producing forecasts in practice is that the point forecast is not needed in some contexts. Instead, the cumulative over the forecast horizon (or over the lead time) might be more suitable. The classic example is the safety stock calculation based on the lead time (time between the order of a product and its delivery). In this situation, we need to make sure that while the product is being delivered, we do not run out of stock, thus still satisfying the selected level of demand (e.g. 95%), but now over the whole period of time rather than on every separate observation.

In the case of **pure additive ADAM**, there are analytical formulae for the conditional expectations and conditional variance for this case that can be used in forecasting. These formulae come directly from the recursive relation (5.10) (for derivations for a simpler case, see for example, Hyndman et al. (2008) and Svetunkov and Petropoulos (2018)):

$$\begin{aligned} \mu_{Y,t,h} = \mathrm{E}(Y_{c,t,h}|t) = & \sum_{j=1}^{h} \sum_{i=1}^{d} \left(\mathbf{w}'_{m_i} \mathbf{F}_{m_i}^{\lceil \frac{j}{m_i} \rceil - 1} \right) \mathbf{v}_t \\ \sigma^2_{Y,h} = \mathrm{V}(Y_{c,t,h}|t) = & \left(1 + \sum_{k=1}^{h-1} \left(1 + (h-k) \sum_{i=1}^{d} \left(\mathbf{w}'_{m_i} \sum_{j=1}^{\lceil \frac{k}{m_i} \rceil - 1} \mathbf{F}_{m_i}^{j-1} \mathbf{g}_{m_i} \mathbf{g}'_{m_i} (\mathbf{F}'_{m_i})^{j-1} \mathbf{w}_{m_i} \right) \right) \right) \sigma^2 \end{aligned}, \tag{18.10}$$

where $Y_{c,t,h} = \sum_{j=1}^{h} y_{t+j}$ is the cumulative actual value and all the other variables have been defined in Section 5.2. Based on the expectation and variance above, we can construct a prediction interval as discussed in Section 18.3.

In cases of **multiplicative and mixed ADAM**, there are no closed forms for the conditional expectation and variance. As a result, simulations similar to the one discussed in Section 18.1 are needed to produce all possible paths for the next h steps ahead. The main difference would be that before taking the expectation or quantiles, the paths would need to be aggregated over the forecast horizon h. This approach, together with the idea of a one-sided prediction interval, can be directly used to calculate the safety stock over the lead time.

18.4.4 Example in R

For demonstration purposes, we consider an artificial intermittent demand example, similar to the one from Section 13.4:

```
set.seed(41)
y <- ts(c(rpois(20,0.25), rpois(20,0.5), rpois(20,1),
          rpois(20,2), rpois(20,3)))
```

For simplicity, we apply an iETS(M,Md,N) model with odds ratio occurrence:

```
adamiETSy <- adam(y, "MMdN", occurrence="odds-ratio",
                  h=7, holdout=TRUE)
plot(adamiETSy, 7, xlab="Time", ylab="Sales")
```

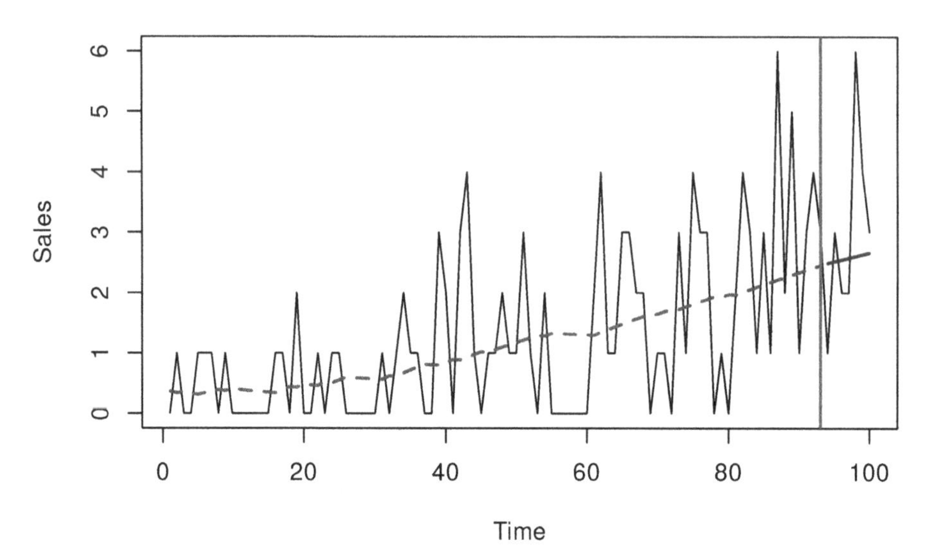

To make this setting closer to a possible real life situation, we assume that the lead time is seven days, and we need to satisfy the 99% of demand for the last seven observations based on our model. Thus we produce the upper bound for the cumulative values for the confidence level of 99%:

```
adamiETSyForecast <- forecast(adamiETSy, h=7,
                              cumulative=TRUE,
                              interval="prediction",
                              side="upper")
```

Given that we deal with cumulative values, the basic plot will not be helpful, and we should produce something different. One of the options is the following (see Figure 18.8):

```
# Point for the actual cumulative demand over the lead time
plot(sum(adamiETSy$holdout), ylab="Cumulative demand",
     xlab="", xaxt="n", pch=16,
     ylim=range(c(0, sum(adamiETSy$holdout),
                  adamiETSyForecast$upper)))
# Sum of expectations over the lead time
abline(h=adamiETSyForecast$mean, col="blue",
       lwd=2)
# Upper bound for the cumulative demand over the lead time
abline(h=adamiETSyForecast$upper, col="grey",
       lwd=2, lty=2)
```

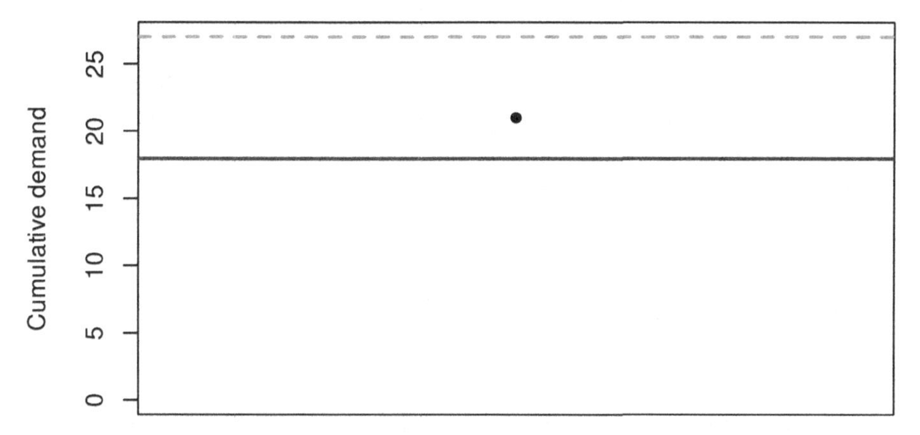

FIGURE 18.8 The actual cumulative demand (black dot), the expectation (the solid blue line), and the 95% quantile of the distribution of the cumulative demand (the dashed grey line) based on the iETS model.

What Figure 18.8 demonstrates is that for the holdout period of seven days, the cumulative demand was around 21 units, while the upper bound of the interval was approximately 27. Based on that upper bound, we could place an order (based on what we already have in stock) and have an appropriate safety stock.

This example is provided for demonstration purposes only. To see if the approach is suitable for a specific situation, we would need to apply it in either a rolling origin fashion (Section 2.4) or to a set of products to collect the distribution of related error measures.

18.4.5 Confidence interval

Finally, we can construct a confidence interval for some statistics. In general, it can be built for the mean, a parameter, fitted values, etc. In our context, we might be interested in the confidence interval for the conditional h steps ahead expectation. This implies that we are interested in the uncertainty of the line, not of the actual values, which can only be constructed for the model that takes the uncertainty of parameters into account (as discussed in Chapter 16). The construction of a confidence interval, in this case, relies on the Normal distribution (because of Central Limit Theorem), as long as the basic assumptions for the model and CLT are satisfied (see Section 6.2 and Chapter 15 of Svetunkov, 2022). Technically speaking, the construction of a confidence interval comes to capturing the model uncertainty discussed in Chapter 16.

18.4.5.1 Example in R

The only way that the confidence interval can be constructed for ADAM is via the `reforecast()` function. Consider the example with ADAM ETS(A,Ad,N) on `BJSales` data as in Section 18.3.8:

```
adamETSBJ <- adam(BJsales, h=10, holdout=TRUE)
```

The confidence interval for this model can be produced either directly via `reforecast()` or via `forecast()`, which will call it for you:

```
forecast(adamETSBJ, h=10,
         interval="confidence", nsim=1000) |>
    plot()
```

Remark. I have increased the number of iterations for the simulation to get a more accurate confidence interval around the conditional expectation. This will consume more memory, as the operation involves creating 1000 sample paths for the fitted values and another 1000 for the holdout sample forecasts.

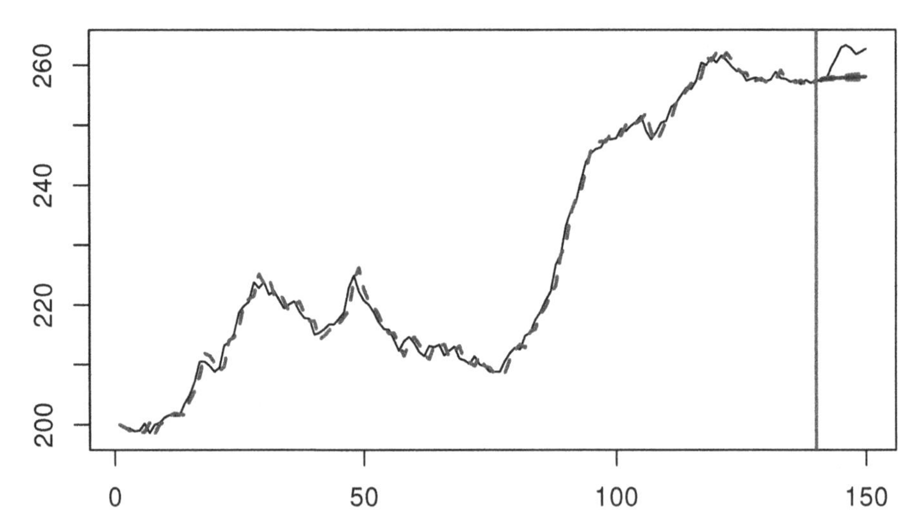

FIGURE 18.9 Confidence interval for the point forecast from an ADAM ETS(A,Ad,N) model.

Figure 18.9 shows the uncertainty around the point forecast based on the uncertainty of the parameters of the model. As can be seen, the interval is narrow, demonstrating that the conditional expectation would not change much if the model's parameters would vary slightly. The fact that the actual values are systematically above the forecast does not mean anything because the confidence interval does not consider the uncertainty of actual values.

19

Forecasting functions of the smooth package

While ADAM is the main focus of this monograph, there are several functions in the `smooth` package that implement some special cases of it or some related models. In fact, all the models implemented in the `smooth` package are formulated in one and the same SSOE framework, discussed in this monograph. In this chapter, I briefly explain how the specific functions of the package are related to ADAM and how they are formulated. I conclude the chapter with an `adam()` cheat sheet, listing the main commands that can be used to apply the model to a data in variety of scenarios.

Some of the functions discussed in this chapter are also explained in Svetunkov (2023c).

19.1 Exponential Smoothing

19.1.1 ETS, es()

The `es()` function from the `smooth` package implements the conventional ETS model (from Hyndman et al., 2008) with some modifications. The fundamental difference of the function from the `ets()` from the `forecast` package is that it is implemented in the SSOE framework discussed in Section 7.1, i.e. with lagged values of the state vector and a smaller transition matrix. In fact, starting from the `smooth` v3.2.0, it is just a wrapper of the `adam()` function, restricting it to:

- The single seasonal ETS (multiple seasonal cycles are supported by `adam()`, as discussed in Chapter 12);
- The Normal distribution for the ϵ_t (Sections 5.5 and 6.5);
- Continuous data only (no support for the occurrence model from Chapter 13);
- ETS and regression (no ARIMA elements as in Chapter 9);
- Location model only (no scale model from Chapter 17).

Still, the function supports the main features of ETS and introduces those that are not available in either `ets()` from the `forecast` package or `ETS()` from the `fable`:

1. Explanatory variables in the ETSX model from Chapter 10;
2. A variety of loss functions, including the custom one, as discussed in Chapter 11;
3. Different initialisation methods (see Section 11.4);
4. Diagnostic plots from Chapter 14;
5. Components and explanatory variables selection and combinations via approaches discussed in Chapter 15;
6. Covariance matrix of parameters and related methods for reapplication of an ETS model (Chapter 16);
7. Data simulation as in Section 16.1;
8. All types of prediction intervals and other related aspects discussed in Chapter 18.

Being a wrapper of the `adam()`, `es()` is agnostic of the input data, working with any object, including `ts`, `zoo`, `tibble`, etc as long as the provided object has only one column. The seasonal lag can be regulated in the function using the `lags` variable, the same way it is done in `adam()`. The explanatory variables in `es()` are provided in a form of a matrix in the variable `xreg` (similarly to how it is done in `arima()` from the `stats` package).

19.1.2 Complex Exponential Smoothing, ces()

Complex Exponential Smoothing (CES) is a model lying outside of the conventional ETS taxonomy. It does not have explicit level/trend components, but instead it has a non-linear trend, which changes its shape depending on the value of the complex smoothing parameter. I developed this model in my PhD, and it is explained in detail in the Svetunkov et al. (2022) paper. It is formulated as a set of the following equations:

$$\begin{aligned} y_t &= l_{t-1} + \epsilon_t \\ l_t &= l_{t-1} - (1 - \alpha_1) c_{t-1} + (\alpha_0 - \alpha_1) \epsilon_t \,, \\ c_t &= l_{t-1} + (1 - \alpha_0) c_{t-1} + (\alpha_0 + \alpha_1) \epsilon_t \end{aligned} \tag{19.1}$$

where ϵ_t is the white noise error term, l_t is the level component, c_t is the non-linear trend component, and α_0 and α_1 are the smoothing parameters. The model can also be formulated in the same SSOE framework (5.5) discussed in Section 5.1, which makes it easy to implement using the same code and to compare with ADAM ETS and/or ARIMA via information criteria. As for the term "complex", it arises because of how the model was originally formulated in its linear form, using complex variables:

$$\hat{y}_t + i\hat{e}_t = (\alpha_0 + i\alpha_1)(y_{t-1} + ie_{t-1}) + (1 - \alpha_0 + i - i\alpha_1)(\hat{y}_{t-1} + i\hat{e}_{t-1}), \tag{19.2}$$

where $i^2 = -1$ is the imaginary unit. The model itself allows distributing the weights over the actual observations in a harmonic or in an exponential way, capturing long-term non-linear patterns in the data. It works well in a median combination of models, together with ETS, ARIMA, and Theta, as it was shown in Petropoulos and Svetunkov (2020).

In addition to the non-seasonal CES, there is a modification that allows capturing either additive or multiplicative seasonality via lagged components, similarly to (19.1), which is formulated as:

$$\begin{aligned} y_t &= l_{0,t-1} + l_{1,t-m} + \epsilon_t \\ l_{0,t} &= l_{0,t-1} - (1 - \alpha_1) c_{0,t-1} + (\alpha_0 - \alpha_1) \epsilon_t \\ c_{0,t} &= l_{0,t-1} + (1 - \alpha_0) c_{0,t-1} + (\alpha_0 + \alpha_1) \epsilon_t \; . \\ l_{1,t} &= l_{1,t-m} - (1 - \beta_1) c_{1,t-m} + (\beta_0 - \beta_1) \epsilon_t \\ c_{1,t} &= l_{1,t-m} + (1 - \beta_0) c_{1,t-m} + (\beta_0 + \beta_1) \epsilon_t \end{aligned} \tag{19.3}$$

Similarly to the non-seasonal CES, (19.3) can be formulated in the SSOE framework discussed in Section 5.1. Based on that, it is possible to select between the seasonal and the non-seasonal CES using information criteria. The model is also extendible to include other components or explanatory variables. Finally, while it is possible to formulate the model with a variety of distributions, the implementation so far only supports the Normal distribution.

The function `ces()` in the `smooth` package supports many features discussed in this monograph:

1. Explanatory variables, similar to the ETSX model from Chapter 10, done via `xreg` variable;
2. A variety of loss functions, as discussed in Chapter 11, via `loss` variable;
3. Different initialisation methods (see Section 11.4);
4. Diagnostic plots (Chapter 14);
5. Explanatory variables selection (Section 15.3);
6. Data simulation (similar to the ADAM discussed in Section 16.1, done using the `sim.ces()` function or the `simulate()` method applied to an estimated `ces()` model);
7. All types of prediction intervals and other related aspects discussed in Chapter 18.

Finally, the `auto.ces()` function does selection between the seasonal and non-seasonal CES models using information criteria. By default, in addition to the models discussed above, it will also consider a "simple" seasonal CES with a classical additive seasonal component, similar to the one in the conventional ETS, and will select the model with the lowest AICc.

19.2 ARIMA

The smooth package also includes two distinct ARIMA functions, which are implemented in the SSOE framework, and have slightly different purposes.

19.2.1 Multiple Seasonal ARIMA, msarima()

The msarima() function implements the multiple seasonal ARIMA model as discussed in Chapter 9. In fact, starting from smooth v3.2.0, it is just a wrapper of adam(), similarly to how es() is its wrapper for the ETS counterpart. Being a wrapper, the function does not support all the features of ADAM, but still has the following functionality (similar to the es()):

1. Explanatory variables in the ARIMAX model (Chapter 10);
2. Different loss functions, including the custom one (Chapter 11);
3. Different initialisation methods (Section 11.4);
4. Diagnostic plots (Chapter 14);
5. Orders and explanatory variables selection (Chapter 15);
6. Covariance matrix of parameters and related methods for reapplication of the ARIMA model (Chapter 16);
7. Data generation from an estimated model (Section 16.1);
8. All types of prediction intervals and other related aspects (Chapter 18).

Being a part of ADAM, the function is tailored for multiple seasonality and works efficiently in this case, especially when compared with other ARIMA implementations in R. It only supports the Normal distribution and focuses on the location model for continuous data (non-intermittent). Furthermore, similar to the es() function, it is agnostic of the provided data and will work with any type of univariate object (vector, ts, zoo, etc). Finally, the explanatory variables can be provided via the xreg parameter, similar to how it is done in the arima() function from the stats package.

The selection of the ARIMA orders is implemented via the auto.msarima() function, which has a slightly different algorithm than the one explained in Section 15.2. It does selection based on an information criteria (IC), checking sequentially the orders of I, then MA and the AR for all the orders up until the ones provided by the user. So, for example, the following code:

```
auto.msarima(y, orders=list(ar=c(3,3),
                            i=c(2,1),
                            ma=c(3,3)),
             lags=c(1,12))
```

will test different orders of SARIMA up to SARIMA(3,2,3)$(3,1,3)_{12}$. If the parameter `fast` is `TRUE` then the function will skip orders that do not lead to the reduction of IC (e.g. if MA(3) does not improve the IC, it will not test MA(2) and MA(1)). As a final step, the function will compare the model with constant with the model without it and select the most appropriate one. Finally, the `auto.msarima()` also supports combination of forecasts from ARIMA models from the same pool as in the case of order selection described above.

19.2.2 State space ARIMA, ssarima()

The `ssarima()` function implements ARIMA in a different form, the one explained in Chapter 11 of Hyndman et al. (2008) and then discussed in Svetunkov and Boylan (2020). It can be considered a conventional SSOE state space ARIMA. The main difference with MSARIMA is in its architecture, i.e. the measurement vector $\mathbf{w}$ and the transition matrix $\mathbf{F}$ differ between the two implementations. The measurement vector is always of the style:

$$\mathbf{w} = \begin{pmatrix} 1 \\ 0 \\ \vdots \\ 0 \end{pmatrix}, \tag{19.4}$$

while the size of the transition matrix is K^2, where K is the maximum order of the ARI or MA polynomial. This contrasts with the ADAM ARIMA (and consequently `msarima()`), where the size of the matrix equals to the number of elements in the polynomial equation. So, for example, SARIMA(1,1,2)(0,1,0)$_1$2 can be expanded into the following equation (based on a similar model discussed in Section 9.1):

$$y_t = (1+\phi_1)y_{t-1} - \phi_1 y_{t-2} + y_{t-12} - (1+\phi_1)y_{t-13} + \phi_1 y_{t-14} + \theta_1 \epsilon_{t-1} + \theta_2 \epsilon_{t-2} + \epsilon_t.$$

Based on this, in the case of SSARIMA, the size of the transition matrix will be 14×14, while in the case of MSARIMA, the matrix will be 5×5. This difference in complexity pays of for MSARIMA when a multiple frequency model is needed, because the size of the transition matrix will impact the computational speed. Summarising all of this, `ssarima()` is not suitable for multiple frequencies, because it relies on the conventional state space architecture. Still, it has an advantage: it is in general easier to initialise than the ADAM ARIMA, because the vector of initials equals to K, while in case of the latter it is up to $\frac{K(K+1)}{2}$.

Finally, the `auto.ssarima()` function implements the order selection for state space ARIMA based on the algorithm explained in Svetunkov and Boylan (2020).

19.3 Cheat sheet for adam() function

This section summarises the main ways of using `adam()` in R, with references to the chapters and sections where specific elements were discussed.

Estimate the best ETS model for the provided data (Section 15.1):

```
ourModel <- adam(y)
```

The best ETS, taking the last 10 observations of time series as a holdout to test the performance of model:

```
ourModel <- adam(y, h=10, holdout=TRUE)
```

Build a pure multiplicative seasonal ETS with an arbitrary seasonal lag of 7 (can be applied to an object `y` of any class, including `ts`, `vector`, `matrix`, `zoo`, `tibble`, etc):

```
ourModel <- adam(y, model="YYM", lags=7)
```

Estimate the best ARIMA model for the provided data (assuming seasonal lag of 7, Section 15.2):

```
ourModel <- adam(y, model="NNN", lags=7,
                 orders=list(ar=c(3,2),
                             i=c(2,1),
                             ma=c(3,2),
                             select=TRUE))
```

Build ARIMA(0,1,1) with drift (Chapter 8 for the general ARIMA and Section 8.1.4 for the one with constant):

```
ourModel <- adam(y, model="NNN",
                 orders=c(0,1,1), constant=TRUE)
```

Estimate ETS(A,N,N)+ARIMA(1,0,0) model (Sections 8.4 and 9.4):

```
ourModel <- adam(y, model="ANN", orders=c(1,0,0))
```

Use Generalised Normal distribution for the residuals of ADAM ETS(A,A,N) (Sections 5.5, 6.5, and 11.1):

```
ourModel <- adam(y, model="AAN", distribution="dgnorm")
```

Select the best distribution for the specific ADAM ETS(A,A,N) (Chapter 15):

```
ourModel <- auto.adam(y, model="AAN")
```

Select the most appropriate ETSX model for the provided `data` (which can be any two-dimensional object, such as `matrix`, `data.frame`, or `tibble`, see Chapter 10):

```
ourModel <- adam(data)
```

Specify, which explanatory variables to include and in what form (Section 10.1):

```
ourModel <- adam(data, formula=y~x1+x2+I(x2^2))
```

Select the set of explanatory variables for ETSX(M,N,N) based on AIC (Section 15.3):

```
ourModel <- adam(data, model="MNN",
                 regressors="select", ic="AIC")
```

Estimate ETS(A,Ad,N) model using a multistep loss function, GTMSE (Section 11.3):

```
ourModel <- adam(y, model="AAdN",
                 h=10, loss="GTMSE")
```

Estimate ARIMA(1,1,2) using a multistep loss function (Section 11.3) with backcasting of initials (Section 11.4):

```
ourModel <- adam(y, model="NNN", orders=c(1,1,2),
                 h=10, loss="GTMSE", initial="backcasting")
```

Select and estimate the most appropriate ETS model on the data with multiple frequencies (Chapter 12):

```
ourModel <- adam(y, model="ZXZ", lags=c(24,24*7,24*365))
```

Select and estimate the triple seasonal ARIMA on the data with multiple frequencies (Chapter 12):

```
ourModel <- adam(y, model="NNN", lags=c(1,24,24*7,24*365),
                 orders=list(ar=c(3,2,2,2),
                             i=c(2,1,1,1),
                             ma=c(3,2,2,2),
                             select=TRUE),
                 initial="backcasting")
```

Apply an automatically selected occurrence part of the model to intermittent data (Section 13.1):

```
oesModel <- oes(y, model="YYY", occurrence="auto")
```

Use the estimated occurrence model in `adam()` to model intermittent data (Section 13.2):

```
ourModel <- adam(y, model="YYY", occurrence=oesModel)
```

Or alternatively just use the same model for occurrence and demand sizes part (Chapter 13):

```
ourModel <- adam(y, model="YYY", occurrence="auto")
```

Estimate the scale model for previously estimated ADAM (Chapter 17):

```
scaleModel <- sm(ourModel, model="YYY")
```

Implant the scale model into the ADAM for future use (e.g. for forecasting):

```
mergedModel <- implant(ourModel, scaleModel)
```

Produce diagnostic plots to see if the ADAM can be improved any further (Chapter 14):

```
par(mfcol=c(2,2))
plot(ourModel, which=c(1,2,4,6))
```

Extract conventional, standardised, and studentised residuals (Chapter 14):

```
residuals(ourModel)
rstandard(ourModel)
rstudent(ourModel)
```

Plot time series decomposition according to ADAM ETS (Section 4.1):

```
plot(ourModel, which=12)
```

Produce point forecast and prediction interval from ADAM for 10 steps ahead (Chapter 18):

```
forecast(ourModel, h=10, interval="prediction")
```

Produce point forecast and prediction interval for ADAM, cumulative over the lead time of 10 (Subsection 18.4.3):

```
forecast(ourModel, h=10, interval="prediction",
         cumulative=TRUE)
```

Produce point forecast and empirical prediction interval for upper bound (upper quantile of distribution, Sections 18.3.5 and 18.4.2):

```
forecast(ourModel, h=10, interval="empirical",
         side="upper")
```

Produce summary of ADAM (Chapter 16):

```
summary(ourModel)
```

Reapply ADAM with randomly selected initials and parameters (to capture the uncertainty of parameters) and produce forecasts from each of these models (Section 16.5):

```
reapply(ourModel)
reforecast(ourModel, h=10, interval="prediction")
```

Extract multistep forecast errors from ADAM (Subsection 14.7.3):

```
rmultistep(ourModel, h=10)
```

Extract covariance matrix of multistep forecast errors from ADAM (Section 11.3):

```
multicov(ourModel, h=10)
```

Extract actual and fitted values from ADAM:

```
actuals(ourModel)
fitted(ourModel)
```

20

What's next?

Now that we reach the final pages of this monograph, I want to pause to look back at what we have discussed and what is left untold in the story of the Augmented Dynamic Adaptive Model.

The reason why I did not call this monograph "Forecasting with ETS" or "Forecasting with State Space Models" is because the framework proposed here is not the same as for ETS, and it does not rely on the standard state space model. The combination of ETS, ARIMA, and Regression in one unified model has not been discussed in the literature before this book. But I did not stop at that, I extended it by introducing a variety of distributions: typically, dynamic models rely on Normality, which is not realistic in real life, but ADAM supports several real-valued and several positive distributions. Furthermore, the model that can be applied to both regular and intermittent demand has been developed only by Svetunkov and Boylan (2023) in the ADAM framework. In addition, ADAM can be extended with multiple seasonal components, making it applicable to high-frequency data. Moreover, ADAM supports not only the location, but the scale model as well, allowing us to model and predict the scale of distribution explicitly (giving it a connection with GARCH models). All of the aspects mentioned above are united in one approach, giving immense flexibility to an analyst.

But what's next? While we have discussed the important aspects of ADAM, there are still several things left that I did not have time to make work yet.

The first one is the ADAM with Asymmetric Laplace and similar distributions with the non-zero mean (related to this is the estimation of ADAM with non-standard loss functions, such as pinball). While it is possible to use such distributions, in theory, they do not work as intended in dynamic models, because the latter rely on the assumption that the mean of the error term is zero. They work perfectly in the case of the regression model (e.g. see how the `alm()` from the `greybox` works with the Asymmetric Laplace) but fail when a model has MA-related terms. This is because the model becomes more adaptive to the changes, pulls to the centre, and cannot maintain the desired quantile. An introduction of such distributions would imply changing the architecture of the state space model (this was briefly discussed in Section 14.7).

Second, model combination and selection literature has seen several bright additions to the field, such as a stellar paper by Kourentzes, Barrow, and

Petropoulos (2019) on pooling. This is neither implemented in the `adam()` function nor discussed in the monograph. Yes, one can use ADAM to do pooling, but it would make sense to introduce it as a part of the ADAM framework.

Third, related to the previous point, is the selection and combinations based on cross-validation techniques (and specifically using rolling origin discussed in Section 2.4). The selected and combined models in this case would differ from the AIC-based ones, hopefully doing better in terms of long-term forecasts.

Fourth, we have not discussed multiple frequency models in the detail that they deserve. For example, we have not mentioned how to diagnose such models when the sample includes thousands of observations. The classical statistical approaches discussed in Section 14 typically fail in this situation, and other tools should be used in this context.

Fifth, `adam()` has a built-in missing values approach that relies on interpolation and the intermittent state space model (from Section 13). While this already works in practice, there are some aspects of this that are worth discussing that have been left outside this monograph. Most importantly, it is not very clear how to interpolate the components of ADAM in these cases.

Sixth, when discussing pure multiplicative ETS models (Chapter 6), we mentioned that it is possible to apply pure additive ETS to the data in logarithms to achieve similar results to the conventional pure multiplicative model. Akram et al. (2009) have done investigations in this direction, but these models are not yet supported by ADAM. They can be done manually, but their implementation in R functions and detailed explanation of their work can be considered another potential direction for future work.

Finally, while I tried to introduce examples of the application of ADAM, case studies for several contexts would be helpful. This would show how ADAM can be used for decisions in inventory management (we have touched on the topic in Subsection 18.4.4), scheduling, staff allocation, etc.

All of this will hopefully come in the next editions of this monograph.

References

Akram, M., Hyndman, R. J., & Ord, J. K. (2009). Exponential Smoothing and Non-negative Data. *Australian & New Zealand Journal of Statistics*, *51*(4), 415 – 432. https://doi.org/10.1111/j.1467-842X.2009.00555.x

Anderson, T. W., & Darling, D. A. (1952). Asymptotic Theory of Certain "Goodness of Fit" Criteria Based on Stochastic Processes. *The Annals of Mathematical Statistics*, *23*(2), 193 – 212. https://doi.org/10.1214/aoms/1177729437

Assimakopoulos, V., & Nikolopoulos, K. (2000). The Theta Model: A Decomposition Approach to Forecasting. *International Journal of Forecasting*, *16*, 521 – 530. https://doi.org/10.1016/S0169-2070(00)00066-2

Athanasopoulos, G., Hyndman, R. J., Song, H., & Wu, D. C. (2011). The Tourism Forecasting Competition. *International Journal of Forecasting*, *27*(3), 822 – 844. https://doi.org/10.1016/j.ijforecast.2010.04.009

Barrow, D., Kourentzes, N., Sandberg, R., & Niklewski, J. (2020). Automatic Robust Estimation for Exponential Smoothing: Perspectives from Statistics and Machine Learning. *Expert Systems with Applications*, *160*, 113637. https://doi.org/10.1016/j.eswa.2020.113637

Bartlett, M. S. (1937). Properties of Sufficiency and Statistical Tests. *Proceedings of the Royal Society of London. Series A, Mathematical and Physical Sciences*, *160*(901), 268 – 282. https://doi.org/10.1098/rspa.1937.0109

Bergmeir, C., Hyndman, R. J., & Benítez, J. M. (2016). Bagging Exponential Smoothing Methods Using STL Decomposition and Box-Cox Transformation. *International Journal of Forecasting*, *32*(2), 303 – 312. https://doi.org/10.1016/j.ijforecast.2015.07.002

Bollerslev, T. (1986, apr). Generalized Autoregressive Conditional Heteroskedasticity. *Journal of Econometrics*, *31*(3), 307 – 327. https://doi.org/10.1016/0304-4076(86)90063-1

Borchers, H. W. (2022). pracma: Practical numerical math functions [Computer software manual]. Retrieved from https://CRAN.R-project.org/package=pracma (R package version 2.4.2)

Box, G., & Jenkins, G. (1976). *Time Series Analysis: Forecasting and Control.* Holden-day, Oakland, California.

Box, G. E. P., & Pierce, D. A. (1970). Distribution of Residual Autocorrelations in Autoregressive-Integrated Moving Average Time Series Models. *Journal of the American Statistical Association*, *65*(332), 1509-1526. https://doi.org/10.1080/01621459.1970.10481180

Boylan, J. E., & Syntetos, A. A. (2021). *Intermittent Demand Forecasting. Context, Methods and Applications.* 111 River St., Hoboken, NJ 07030, USA: John Wiley & Sons Ltd.

Brenner, J. L., D'Esopo, D. A., & Fowler, A. G. (1968). Difference Equations in Forecasting Formulas. *Management Science, 15*(3), 141 – 159. https://doi.org/10.1287/mnsc.15.3.141

Breusch, T. S. (1978). Testing for Autocorrelation in Dynamic Linear Models. *Australian Economic Papers, 17*(31), 334-355. https://doi.org/10.1111/j.1467-8454.1978.tb00635.x

Breusch, T. S., & Pagan, A. R. (1979). A Simple Test for Heteroscedasticity and Random Coefficient Variation. *Econometrica, 47*(5), 1287 – 1294. https://doi.org/10.2307/1911963

Brown, R. G. (1956). *Exponential Smoothing for Predicting Demand.* Cambridge 42, Massachusetts: Arthur D. Little, Inc.

Burnham, K. P., & Anderson, D. R. (2004). *Model Selection and Multimodel Inference.* Springer New York. https://doi.org/10.1007/b97636

Chakraborty, S. (2015). Generating Discrete Analogues of Continuous Probability Distributions: A Survey of Methods and Constructions. *Journal of Statistical Distributions and Applications, 2*(1), 6. https://doi.org/10.1186/s40488-015-0028-6

Chatfield, C. (1977). Some Recent Developments in Time-Series Analysis. *Journal of the Royal Statistical Society. Series A (General), 140*(4), 492. https://doi.org/10.2307/2345281

Chatfield, C. (1996). Model Uncertainty and Forecast Accuracy. *Journal of Forecasting, 15*(7), 495 – 508. https://doi.org/10.1002/(SICI)1099-131X(199612)15:7<495::AID-FOR640>3.3.CO;2-F

Chatfield, C., Koehler, A. B., Ord, J. K., & Snyder, R. D. (2001). A New Look at Models for Exponential Smoothing. *Journal of the Royal Statistical Society, Series D (The Statistician), 50*(2), 147 – 159. Retrieved from https://www.jstor.org/stable/2681090

Claeskens, G., Magnus, J. R., Vasnev, A. L., & Wang, W. (2016, jul). The Forecast Combination Puzzle: A Simple Theoretical Explanation. *International Journal of Forecasting, 32*(3), 754 – 762. https://doi.org/10.1016/j.ijforecast.2015.12.005

Clements, M., & Hendry, D. (1998). *Forecasting Economic Time Series.* Cambridge University Press. https://doi.org/10.1017/CBO9780511599286

Cleveland, R. B., Cleveland, W. S., McRae, J. E., & Terpenning, I. (1990). STL: A Seasonal-trend Decomposition Procedure Based on LOESS. *Journal of Official Statistics, 6*(1), 3 – 73.

Cleveland, W. S. (1979). Robust Locally Weighted Regression and Smoothing Scatterplots. *Journal of the American Statistical Association, 74*(368), 829 – 836. https://doi.org/10.2307/2286407

Croston, J. D. (1972). Forecasting and Stock Control for Intermittent Demands. *Operational Research Quarterly (1970-1977), 23*(3), 289. https://doi.org/10.2307/3007885

Davydenko, A., & Fildes, R. (2013). Measuring Forecasting Accuracy: The Case Of Judgmental Adjustments To SKU-Level Demand Forecasts. *International Journal of Forecasting*, *29*(3), 510 – 522. https://doi.org/10.1016/j.ijforecast.2012.09.002

De Livera, A. M. (2010). Exponentially Weighted Methods for Multiple Seasonal Time Series. *International Journal of Forecasting*, *26*(4), 655 – 657. https://doi.org/10.1016/j.ijforecast.2010.05.010

De Livera, A. M., Hyndman, R. J., & Snyder, R. D. (2011). Forecasting Time Series With Complex Seasonal Patterns Using Exponential Smoothing. *Journal of the American Statistical Association*, *106*(496), 1513 – 1527. https://doi.org/10.1198/jasa.2011.tm09771

Demšar, J. (2006). Statistical Comparisons of Classifiers Over Multiple Data Sets. *Journal of Machine Learning Research*, *7*, 1 – 30. Retrieved from https://www.jmlr.org/papers/volume7/demsar06a/demsar06a.pdf

Dhar, P. (1999). The Carbon Impact of Artificial Intelligence. *Nature Machine Intelligence*, *2*, 423-425. https://doi.org/10.1038/s42256-020-0219-9

Dickey, D. A., & Fuller, W. A. (1979). Distribution of the Estimators for Autoregressive Time Series with a Unit Root. *Journal of the American Statistical Association*, *74*(366a), 427 – 431. https://doi.org/10.1080/01621459.1979.10482531

Dictionary. (2021). *Method.* Cambridge Dictionary. Retrieved from https://dictionary.cambridge.org/dictionary/english/method (version: 2021-09-02)

Durbin, J., & Watson, G. S. (1950, 12). Testing for Serial Correlation in Least Squares Regression. I. *Biometrika*, *37*(3-4), 409-428. https://doi.org/10.1093/biomet/37.3-4.409

Engle, R. F. (1982, jul). Autoregressive Conditional Heteroscedasticity with Estimates of the Variance of United Kingdom Inflation. *Econometrica*, *50*(4), 987. https://doi.org/10.2307/1912773

Fildes, R., Hibon, M., Makridakis, S., & Meade, N. (1998). Generalising about Univariate Forecasting Methods: Further Empirical Evidence. *International Journal of Forecasting*, *14*(3), 339 – 358. https://doi.org/10.1016/S0169-2070(98)00009-0

Friedman, M. (1937). The Use of Ranks to Avoid the Assumption of Normality Implicit in the Analysis of Variance. *Journal of the American Statistical Association*, *32*(200), 675-701. https://doi.org/10.1080/01621459.1937.10503522

Gardner, E. S. (1985). Exponential Smoothing: The State of the Art. *Journal of Forecasting*, *4*(1), 1 – 28. https://doi.org/10.1002/for.3980040103

Gardner, E. S. (2006). Exponential Smoothing: The State of the Art-Part II. *International Journal of Forecasting*, *22*(4), 637 – 666. https://doi.org/10.1016/j.ijforecast.2006.03.005

Gardner, E. S., & Diaz-Saiz, J. (2008). Exponential Smoothing in the Telecommunications Data. *International Journal of Forecasting*, *24*(1), 170 – 174. https://doi.org/10.1016/j.ijforecast.2007.05.002

Gardner, E. S., & McKenzie, E. (1985). Forecasting Trends in Time Series. *Management Science*, *31*(10), 1237 – 1246. https://doi.org/10.1016/0169-2070(86)90056-7

Gardner, E. S., & McKenzie, E. (1989). Seasonal Exponential Smoothing with Damped Trends. *Management Science*, *35*(3), 372 – 376. https://doi.org/10.1287/mnsc.35.3.372

Geweke, J. (1986). Comment. *Econometric Reviews*, *5*(1), 57 – 61. https://doi.org/10.1080/07474938608800097

Gneiting, T., & Raftery, A. E. (2007). Strictly Proper Scoring Rules, Prediction, and Estimation. *Journal of the American Statistical Association*, *102*(477), 359 – 378. https://doi.org/10.1198/016214506000001437

Godfrey, L. G. (1978). Testing Against General Autoregressive and Moving Average Error Models when the Regressors Include Lagged Dependent Variables. *Econometrica*, *46*(6), 1293 – 1301. https://doi.org/10.2307/1913829

Goodwin, P., & Lawton, R. (1999). On the Asymmetry of the Symmetric MAPE. *International Journal of Forecasting*, *15*(4), 405 – 408. https://doi.org/10.1016/S0169-2070(99)00007-2

Gould, P. G., Koehler, A. B., Ord, J. K., Snyder, R. D., Hyndman, R. J., & Vahid-Araghi, F. (2008). Forecasting Time Series with Multiple Seasonal Patterns. *European Journal of Operational Research*, *191*(1), 205 – 220. https://doi.org/10.1016/j.ejor.2007.08.024

Hanck, C., Arnold, M., Gerber, A., & Schmelzer, M. (2020). *Introduction to Econometrics with R*. Bookdown. Retrieved from https://www.econometrics-with-r.org/index.html (version: 2020-08-12)

Holt, C. C. (2004). Forecasting Seasonals and Trends by Exponentially Weighted Moving Averages. *International Journal of Forecasting*, *20*(1), 5 – 10. https://doi.org/10.1016/j.ijforecast.2003.09.015

Huber, P. J. (1992). Robust Estimation of a Location Parameter. In S. Kotz & N. L. Johnson (Eds.), *Breakthroughs in statistics: Methodology and distribution* (pp. 492 – 518). New York, NY: Springer New York. https://doi.org/10.1007/978-1-4612-4380-9_35

Hyndman, R. J., & Khandakar, Y. (2008). Automatic Time Series Forecasting: the forecast Package for R. *Journal of Statistical Software*, *26*(3), 1 – 22. Retrieved from https://www.jstatsoft.org/article/view/v027i03

Hyndman, R. J., & Koehler, A. B. (2006). Another Look at Measures of Forecast Accuracy. *International Journal of Forecasting*, *22*(4), 679 – 688. https://doi.org/10.1016/j.ijforecast.2006.03.001

Hyndman, R. J., Koehler, A. B., Ord, J. K., & Snyder, R. D. (2008). *Forecasting with Exponential Smoothing*. Springer Berlin Heidelberg.

Hyndman, R. J., Koehler, A. B., Snyder, R. D., & Grose, S. (2002). A State Space Framework for Automatic Forecasting Using Exponential Smoothing Methods. *International Journal of Forecasting*, *18*(3), 439 – 454. https://doi.org/10.1016/S0169-2070(01)00110-8

James, G., Witen, D., Hastie, T., & Tibshirani, R. (2017). *An Introduction to Statistical Learning with Applications in R* (Vol. 64). Springer New York. https://doi.org/10.1016/j.peva.2007.06.006

Johnson, S. G. (2021). *The NLopt Nonlinear Optimization Package*. NLopt Documentation. Retrieved from https://nlopt.readthedocs.io/ (version: 2021-11-01)

Jose, V. R. R., & Winkler, R. L. (2008). Simple Robust Averages of Forecasts: Some Empirical Results. *International Journal of Forecasting*, *24*(1), 163 – 169. https://doi.org/10.1016/j.ijforecast.2007.06.001

Koehler, A. B., Snyder, R. D., Ord, J. K., & Beaumont, A. (2012). A Study of Outliers in the Exponential Smoothing Approach to Forecasting. *International Journal of Forecasting*, *28*(2), 477 – 484. https://doi.org/10.1016/j.ijforecast.2011.05.001

Koenker, R., & Bassett, G. (1978). Regression Quantiles. *Econometrica*, *46*(1), 33. https://doi.org/10.2307/1913643

Kolassa, S. (2011). Combining Exponential Smoothing Forecasts Using Akaike Weights. *International Journal of Forecasting*, *27*(2), 238 – 251. https://doi.org/10.1016/j.ijforecast.2010.04.006

Kolassa, S. (2016). Evaluating Predictive Count Data Distributions in Retail Sales Forecasting. *International Journal of Forecasting*, *32*(3), 788 – 803. https://doi.org/10.1016/j.ijforecast.2015.12.004

Koning, A. J., Franses, P. H., Hibon, M., & Stekler, H. O. (2005). The M3 Competition: Statistical Tests of the Results. *International Journal of Forecasting*, *21*(3), 397 – 409. https://doi.org/10.1016/j.ijforecast.2004.10.003

Kourentzes, N. (2012). *Statistical Significance of Forecasting Methods – an Empirical Evaluation of the Robustness and Interpretability of the MCB, ANOM and Friedman-Nemenyi Test.* Website. Retrieved from https://kourentzes.com/forecasting/2012/04/19/statistical-significance-of-forecasting-methods-an-empirical-evaluation-of-the-robustness-and-interpretability-of-the-mcb-anom-and-friedman-nemenyi-test/ (version: 2021-08-12)

Kourentzes, N. (2014, oct). On Intermittent Demand Model Optimisation and Selection. *International Journal of Production Economics*, *156*, 180 – 190. https://doi.org/10.1016/j.ijpe.2014.06.007

Kourentzes, N., Barrow, D., & Petropoulos, F. (2019). Another Look at Forecast Selection and Combination: Evidence from Forecast Pooling. *International Journal of Production Economics*, *209*(September 2016), 226 – 235. https://doi.org/10.1016/j.ijpe.2018.05.019

Kourentzes, N., Li, D., & Strauss, A. K. (2019). Unconstraining methods for revenue management systems under small demand. *Journal of Revenue and Pricing Management*, *18*(1), 27 – 41.

Kourentzes, N., & Petropoulos, F. (2016). Forecasting with Multivariate Temporal Aggregation: The Case of Promotional Modelling. *International Journal of Production Economics*, *181*, 145 – 153. https://doi.org/10.1016/j.ijpe.2015.09.011

Kourentzes, N., & Trapero, J. R. (2018). On the use of multi-step cost functions for generating forecasts. *Department of Management Science Working Paper Series*(1).

Kourentzes, N., Trapero, J. R., & Barrow, D. K. (2019). Optimising Forecasting Models for Inventory Planning. *International Journal of Production Economics*(November 2019), 107597. https://doi.org/10.1016/j.ijpe.2019.107597

Kwiatkowski, D., Phillips, P. C. B., Schmidt, P., & Shin, Y. (1992). Testing the Null Hypothesis of Stationarity Against the Alternative of a Unit Root : How Sure Are We That Economic Time Series Are Nonstationary? *Journal of Econometrics*, *54*, 159 – 178. https://doi.org/10.1016/0304-4076(92)90104-Y

Lazo, A., & Rathie, P. (1978, jan). On the Entropy of Continuous Probability Distributions (Corresp.). *IEEE Transactions on Information Theory*, *24*(1), 120 – 122. https://doi.org/10.1109/TIT.1978.1055832

Lee, Y. S., & Scholtes, S. (2014). Empirical Prediction Intervals Revisited. *International Journal of Forecasting*, *30*, 217-234. https://doi.org/10.1016/j.ijforecast.2013.07.018

Lichtendahl, K. C., Grushka-Cockayne, Y., & Winkler, R. L. (2013). Is It Better to Average Probabilities or Quantiles? *Management Science*, *59*(7), 1594 – 1611. https://doi.org/10.1287/mnsc.1120.1667

Ljung, G. M., & Box, G. E. P. (1978, 08). On a Measure of Lack of Fit in Time Series Models. *Biometrika*, *65*(2), 297-303. https://doi.org/10.1093/biomet/65.2.297

Makridakis, S. (1993). Accuracy Concerns Measures: Theoretical and Practical Concerns. *International Journal of Forecasting*, *9*, 527 – 529. https://doi.org/10.1016/0169-2070(93)90079-3

Makridakis, S., Andersen, A. P., Carbone, R., Fildes, R., Hibon, M., Lewandowski, R., ... Winkler, R. L. (1982). The Accuracy of Extrapolation (Time Series) Methods: Results of a Forecasting Competition. *Journal of Forecasting*, *1*(2), 111 – 153. https://doi.org/10.1002/for.3980010202

Makridakis, S., & Hibon, M. (1997). ARMA Models and the Box–Jenkins Methodology. *Journal of Forecasting*, *16*, 147 – 163. https://doi.org/10.1002/(SICI)1099-131X(199705)16:3<147::AID-FOR652>3.0.CO;2-X

Makridakis, S., & Hibon, M. (2000). The M3-Competition: Results, Conclusions and Implications. *International Journal of Forecasting*, *16*, 451 – 476. https://doi.org/10.1016/S0169-2070(00)00057-1

Makridakis, S., Spiliotis, E., & Assimakopoulos, V. (2022, oct). M5 Accuracy Competition: Results, Findings, and Conclusions. *International Journal of Forecasting*, *38*(4), 1346 – 1364. https://doi.org/10.1016/j.ijforecast.2021.11.013

McKenzie, E. (1976, mar). A Comparison of Some Standard Seasonal Forecasting Systems. *The Statistician*, *25*(1), 3. https://doi.org/10.2307/2988127

Mudholkar, G. S., & Tian, L. (2002). An Entropy Characterization of the Inverse Gaussian Distribution and Related Goodness-of-fit Test. *Journal of Statistical Planning and Inference, 102*(2), 211 – 221. https://doi.org/10.1016/S0378-3758(01)00099-4

Muth, J. F. (1960). Optimal Properties of Exponentially Weighted Forecasts. *Journal of the American Statistical Association, 55*(1), 299 – 306. https://doi.org/10.2307/2281742

Nerlove, M., & Wage, S. (1964). On the Optimality of Adaptive Forecasting. *Management Science, 10*(2), 207 – 224. https://doi.org/10.1287/mnsc.10.2.207

Newbold, P., Carlson, W., & Thorne, B. (2020). *Statistics for Business and Economics, 9th Global Edition.* Pearson.

Ord, J. K., Koehler, A. B., & Snyder, R. D. (1997). Estimation and Prediction for a Class of Dynamic Nonlinear Statistical Models. *Journal of the American Statistical Association, 92*(440), 1621 – 1629. https://doi.org/10.1080/01621459.1997.10473684

Ord, K., Fildes, R., & Kourentzes, N. (2017). *Principles of Business Forecasting* (2nd ed.). New York, New York, USA: Wessex Press, Inc.

Osman, A. F., & King, M. L. (2015). A New Approach to Forecasting Based on Exponential Smoothing with Independent Regressors. *Department of Econometrics and Business Statistics*(2). Retrieved from http://econpapers.repec.org/paper/mshebswps/2015-2.htm

Pantula, S. G. (1986, jan). Comment. *Econometric Reviews, 5*(1), 71 – 74. https://doi.org/10.1080/07474938608800099

Pegels, C. C. (1969). Exponential Forecasting: Some New Variations. *Management Science, 15*(5), 311 – 315. Retrieved from https://www.jstor.org/stable/2628137

Petropoulos, F., Hyndman, R. J., & Bergmeir, C. (2018). Exploring the Sources of Uncertainty: Why Does Bagging for Time Series Forecasting Work? *European Journal of Operational Research, 268*(2), 545 – 554. https://doi.org/10.1016/j.ejor.2018.01.045

Petropoulos, F., & Kourentzes, N. (2015). Forecast Combinations for Intermittent Demand. *Journal of the Operational Research Society, 66*(6), 914 – 924. https://doi.org/10.1057/jors.2014.62

Petropoulos, F., Kourentzes, N., Nikolopoulos, K., & Siemsen, E. (2018). Judgmental Selection of Forecasting Models. *Journal of Operations Management, 60*(May), 34 – 46. https://doi.org/10.1016/j.jom.2018.05.005

Petropoulos, F., & Svetunkov, I. (2020). A Simple Combination of Univariate Models. *International Journal of Forecasting, 36*(1), 110 – 115. https://doi.org/10.1016/j.ijforecast.2019.01.006

Pritularga, K., Svetunkov, I., & Kourentzes, N. (2023). Shrinkage Estimator for Exponential Smoothing Models. *International Journal of Forecasting, 39*(3), NA. https://doi.org/10.1016/j.ijforecast.2022.07.005

Rao, R. C. (1945). Information and Accuracy Attainable in the Estimation of Statistical Parameters. *Bulletin of the Calcutta Mathematical Society, 37*(3), 81 – 91.

Roberts, S. A. (1982). A General Class of Holt-Winters Type Forecasting Models. *Management Science, 28*(7), 808 – 820. https://doi.org/10.1287/mnsc.28.7.808

Sagaert, Y., & Svetunkov, I. (2022). Trace Forward Stepwise: Automatic Selection of Variables in No Time. *Department of Management Science Working Paper Series*(1), 1 – 25. https://doi.org/10.13140/RG.2.2.34995.35369

Sangal, B., & Biswas, A. K. (1970). The 3-Parameter Lognormal Distribution Applications in Hydrology. *Water Resources Research, 6*(2), 505 – 515. https://doi.org/10.1029/WR006i002p00505

Schwertman, N. C., Gilks, A. J., & Cameron, J. (1990). A Simple Noncalculus Proof That the Median Minimizes the Sum of the Absolute Deviations. *The American Statistician, 44*(1), 38 – 39. https://doi.org/10.1080/00031305.1990.10475690

Shapiro, S. S., & Wilk, M. B. (1965, 12). An Analysis of Variance Test for Normality (Complete Samples). *Biometrika, 52*(3-4), 591-611. https://doi.org/10.1093/biomet/52.3-4.591

Sichel, H., Dohm, C., & Kleingeld, W. (1997). The Logarithmic Generalized Inverse Gaussian Distribution (LNGIG). *South African Statistical Journal, 31*(1), 125 – 149. Retrieved from https://hdl.handle.net/10520/AJA0038271X_560

Silver, E. A., Pyke, D. F., & Thomas, D. J. (2016). *Inventory and Production Management in Supply Chains. 4th Edition.* Routledge and CRC Press.

Snyder, R. D. (1985). Recursive Estimation of Dynamic Linear Models. *Journal of the Royal Statistical Society, Series B (Methodological), 47*(2), 272 – 276. https://doi.org/10.1111/j.2517-6161.1985.tb01355.x

Snyder, R. D., Ord, J. K., Koehler, A. B., McLaren, K. R., & Beaumont, A. N. (2017). Forecasting Compositional Time Series: A State Space Approach. *International Journal of Forecasting, 33*(2), 502 – 512. https://doi.org/10.1016/j.ijforecast.2016.11.008

Socci, N., Lee, D., & Seung, H. S. (1997). The Rectified Gaussian Distribution. In M. Jordan, M. Kearns, & S. Solla (Eds.), *Advances in Neural Information Processing Systems* (Vol. 10). MIT Press. Retrieved from https://proceedings.neurips.cc/paper/1997/file/28fc2782ea7ef51c1104ccf7b9bea13d-Paper.pdf

Spavound, S., & Kourentzes, N. (2022). Making Forecasts More Trustworthy. *Foresight: The International Journal of Applied Forecasting, 66*(3), 20 – 23.

Stock, J. H., & Watson, M. W. (2004). Combination Forecasts of Output Growth in a Seven-country Data Set. *Journal of Forecasting, 23*(6), 405 – 430. https://doi.org/10.1002/for.928

Svetunkov, I. (2017). *Naughty APEs and the Quest for the Holy Grail.* Open Forecasting. Retrieved from `https://forecasting.svetunkov.ru/en/2017/07/29/naughty-apes-and-the-quest-for-the-holy-grail/` (version: 2017-07-29)

Svetunkov, I. (2019). *Are You Sure You're Precise? Measuring Accuracy of Point Forecasts.* Open Forecasting. Retrieved from `https://forecasting.svetunkov.ru/en/2019/08/25/are-you-sure-youre-precise-measuring-accuracy-of-point-forecasts/` (version: 2019-08-25)

Svetunkov, I. (2022). *Statistics for business analytics.* Lecture notes. Open-Forecast. Retrieved from `https://openforecast.org/sba/` (version: 31.10.2022)

Svetunkov, I. (2023a). greybox: Toolbox for model building and forecasting [Computer software manual]. Retrieved from `https://github.com/config-i1/greybox` (R package version 1.0.8)

Svetunkov, I. (2023b). smooth: Forecasting using state space models [Computer software manual]. Retrieved from `https://github.com/config-i1/smooth` (R package version 3.2.2)

Svetunkov, I. (2023c). Smooth Forecasting with the Smooth Package in R. *arXiv.* `https://doi.org/10.48550/arXiv.2301.01790`

Svetunkov, I., & Boylan, J. E. (2020). State-space ARIMA for Supply-chain Forecasting. *International Journal of Production Research*, *58*(3), 818 – 827. `https://doi.org/10.1080/00207543.2019.1600764`

Svetunkov, I., & Boylan, J. E. (2022). Dealing with Positive Data Using Pure Multiplicative ETS Models. *Department of Management Science Working Paper Series*(3), 1 – 29.

Svetunkov, I., & Boylan, J. E. (2023). iETS: State Space Model for Intermittent Demand Forecastings. *International Journal of Production Economics*(109013), 1 – 43. `https://doi.org/10.1016/j.ijpe.2023.109013`

Svetunkov, I., Kourentzes, N., & Killick, R. (2023, jun). Multi-step Estimators and Shrinkage Effect in Time Series Models. *Computational Statistics*(0123456789). `https://doi.org/10.1007/s00180-023-01377-x`

Svetunkov, I., Kourentzes, N., & Ord, J. K. (2022, 8). Complex Exponential Smoothing. *Naval Research Logistics (NRL)*, 31. `https://doi.org/10.1002/nav.22074`

Svetunkov, I., Kourentzes, N., & Svetunkov, S. (2023). Half Central Moment for Data Analysis. *Department of Management Science Working Paper Series*(3), 1 – 20.

Svetunkov, I., & Petropoulos, F. (2018). Old Dog, New Tricks: a Modelling View of Simple Moving Averages. *International Journal of Production Research*, *56*(18), 6034 – 6047. `https://doi.org/10.1080/00207543.2017.1380326`

Svetunkov, I., & Pritularga, K. (2023a). Incorporating Parameters Uncertainty in ETS. *Department of Management Science Working Paper Series*(2), 1 – 19.

Svetunkov, I., & Pritularga, K. F. (2023b). legion: Forecasting using multivariate models [Computer software manual]. Retrieved from `https://github.com/config-i1/legion` (R package version 0.1.2)

Svetunkov, I., & Svetunkov, S. (2014). *Forecasting Methods. Textbook for Universities*. Moscow: Urait.

Svetunkov, S. (1985). *Adaptive Methods in the Process of Optimisation of Regimes of Electricity Consumption*.

Tashman, L. J. (2000). Out-of-sample Tests of Forecasting Accuracy: An Analysis and Review. *International Journal of Forecasting*, *16*(4), 437 – 450. https://doi.org/10.1016/S0169-2070(00)00065-0

Taylor, J. W. (2003a). Exponential Smoothing with a Damped Multiplicative Trend. *International Journal of Forecasting*, *19*(4), 715 – 725. https://doi.org/10.1016/S0169-2070(03)00003-7

Taylor, J. W. (2003b). Short-term Electricity Demand Forecasting Using Double Seasonal Exponential Smoothing. *Journal of the Operational Research Society*, *54*(8), 799 – 805. https://doi.org/10.1057/palgrave.jors.2601589

Taylor, J. W. (2008). An Evaluation of Methods for Very Short-term Load Forecasting Using Minute-by-minute British Data. *International Journal of Forecasting*, *24*(4), 645 – 658. https://doi.org/10.1016/j.ijforecast.2008.07.007

Taylor, J. W. (2010). Triple Seasonal Methods for Short-term Electricity Demand Forecasting. *European Journal of Operational Research*, *204*(1), 139 – 152. https://doi.org/10.1016/j.ejor.2009.10.003

Taylor, J. W. (2020). Evaluating Quantile-bounded and Expectile-bounded Interval Forecasts. *International Journal of Forecasting*, *37*. https://doi.org/10.1016/j.ijforecast.2020.09.007

Taylor, J. W., & Bunn, D. W. (1999). A Quantile Regression Approach to Generating Prediction Intervals. *Management Science*, *45*(2), 225 – 237. https://doi.org/10.1287/mnsc.45.2.225

Teunter, R. H., Syntetos, A. A., & Babai, M. Z. (2011). Intermittent Demand: Linking Forecasting to Inventory Obsolescence. *European Journal of Operational Research*, *214*(3), 606 – 615. https://doi.org/10.1016/j.ejor.2011.05.018

Tibshirani, R. (1996). Regression Shrinkage and Selection via the LASSO. *Journal of the Royal Statistical Society: Series B (Methodological)*, *58*(1), 267 – 288. https://doi.org/10.1111/j.2517-6161.1996.tb02080.x

Tofallis, C. (2015). A better measure of relative prediction accuracy for model selection and model estimation. *The Journal of the Operational Research Society*, *66*(8), 1352 – 1362. https://doi.org/10.1057/jors.2014.103

Trapero, J. R., Cardós, M., & Kourentzes, N. (2019). Empirical Safety Stock Estimation Based on Kernel and GARCH Models. *Omega (United Kingdom)*, *84*, 199 – 211. https://doi.org/10.1016/j.omega.2018.05.004

Wallström, P., & Segerstedt, A. (2010). Evaluation of Forecasting Error Measurements and Techniques for Intermittent Demand. *International Journal of Production Economics*, *128*(2), 625 – 636. https://doi.org/10.1016/j.ijpe.2010.07.013

Warren M. Persons. (1919). General Considerations and Assumptions. *The Review of Economics and Statistics*, *1*(1), 5 – 107. https://doi.org/10.2307/1928754

Wasserstein, R. L., & Lazar, N. A. (2016). The ASA's Statement on p-Values: Context, Process, and Purpose. *American Statistician*, *70*(2), 129 – 133. https://doi.org/10.1080/00031305.2016.1154108

Weller, M., & Crone, S. F. (2012, November). *Supply Chain Forecasting: Best Practices & Benchmarking Study* (Tech. Rep.). Lancaster Centre for Forecasting.

White, H. (1980). A Heteroskedasticity-Consistent Covariance Matrix Estimator and a Direct Test for Heteroskedasticity. *Econometrica*, *48*(4), 817 – 838. https://doi.org/10.2307/1912934

Wilcoxon, F. (1945). Individual Comparisons by Ranking Methods. *Biometrics Bulletin*, *1*(6), 80 – 83. https://doi.org/10.2307/3001968

Winters, P. R. (1960). Forecasting Sales by Exponentially Weighted Moving Averages. *Management Science*, *6*(3), 324 – 342. https://doi.org/10.1287/mnsc.6.3.324

Index

Printed in the United States
by Baker & Taylor Publisher Services